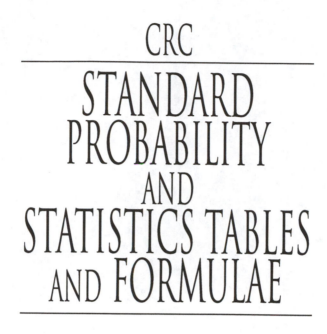

CRC

STANDARD PROBABILITY AND STATISTICS TABLES AND FORMULAE

STANDARD PROBABILITY AND STATISTICS TABLES AND FORMULAE

DANIEL ZWILLINGER

Rensselaer Polytechnic Institute
Troy, New York

STEPHEN KOKOSKA

Bloomsburg University
Bloomsburg, Pennsylvania

CHAPMAN & HALL/CRC

Boca Raton   London   New York   Washington, D.C.

Library of Congress Cataloging-in-Publication Data

Zwillinger, Daniel, 1957-
   CRC standard probability and statistics tables and formulae / Daniel Zwillinger, Stephen Kokoska.
      p. cm.
   Includes bibliographical references and index.
   ISBN 1-58488-059-7 (alk. paper)
      1. Probabilities—Tables. 2. Mathematical statistics—Tables. I. Kokoska, Stephen.
   II. Title.
   QA273.3 .Z95 1999
   519.2′02′1—dc21   99-045786

Visit the CRC Press Web site at www.crcpress.com

© 2000 by Chapman & Hall/CRC

No claim to original U.S. Government works
International Standard Book Number 1-58488-059-7
Library of Congress Card Number 99-045786
Printed in the United States of America   2 3 4 5 6 7 8 9 0
Printed on acid-free paper

Preface

It has long been the established policy of CRC Press to publish, in handbook form, the most up-to-date, authoritative, logically arranged, and readily usable reference material available. This book fills the need in probability and statistics.

Prior to the preparation of this book the contents of similar books were considered. It is easy to fill a statistics reference book with many hundred pages of tables—indeed, some large books contain statistical tables for only a single test. The authors of this book focused on the basic principles of statistics. We have tried to ensure that each topic had an understandable textual introduction as well as easily understood examples. There are more than 80 examples; they usually follow the same format: start with a word problem, interpret the words as a statistical problem, find the solution, interpret the solution in words.

We have organized this reference in an efficient and useful format. We believe both students and researchers will find this reference easy to read and understand. Material is presented in a multi-sectional format, with each section containing a valuable collection of fundamental reference material—tabular and expository. This Handbook serves as a guide for determining appropriate statistical procedures and interpretation of results. We have assembled the most important concepts in probability and statistics, as experienced through our own teaching, research, and work in industry.

For most topics, concise yet useful tables were created. In most cases, the tables were re-generated and verified against existing tables. Even very modest statistical software can generate many of the tables in the book—often to more decimal places and for more values of the parameters. The values in this book are designed to illustrate the range of possible values and act as a handy reference for the most commonly needed values.

This book also contains many useful topics from more advanced areas of statistics, but these topics have fewer examples. Also included are a large collection of short topics containing many classical results and puzzles. Finally, a section on notation used in the book and a comprehensive index are also included.

In line with the established policy of CRC Press, this Handbook will be kept as current and timely as is possible. Revisions and anticipated uses of newer materials and tables will be introduced as the need arises. Suggestions for the inclusion of new material in subsequent editions and comments concerning the accuracy of stated information are welcomed.

If any errata are discovered for this book, they will be posted to http://vesta.bloomu.edu/~skokoska/prast/errata.

Many people have helped in the preparation of this manuscript. The authors are especially grateful to our families who have remained lighthearted and cheerful throughout the process. A special thanks to Janet and Kent, and to Joan, Mark, and Jen.

Daniel Zwillinger
zwillinger@alum.mit.edu

Stephen Kokoska
skokoska@planetx.bloomu.edu

ACKNOWLEDGMENTS

Plans 6.1–6.6, 6A.1–6A.6, and 13.1–13.5 (appearing on pages 331–337) originally appeared on pages 234–237, 276–279, and 522–523 of W. G. Cochran and G. M. Cox, *Experimental Designs*, Second Edition, John Wiley & Sons, Inc, New York, 1957. Reprinted by permission of John Wiley & Sons, Inc.

The tables of Bartlett's critical values (in section 10.6.2) are from D. D. Dyer and J. P. Keating, "On the Determination of Critical Values for Bartlett's Test", *JASA*, Volume 75, 1980, pages 313–319. Reprinted with permission from the *Journal of American Statistical Association*. Copyright 1980 by the American Statistical Association. All rights reserved.

The tables of Cochran's critical values (in section 10.7.1) are from C. Eisenhart, M. W. Hastay, and W. A. Wallis, *Techniques of Statistical Analysis*, McGraw-Hill Book Company, 1947, Tables 15.1 and 15.2 (pages 390-391). Reprinted courtesy of *The McGraw-Hill Companies*.

The tables of Dunnett's critical values (in section 12.1.4.5) are from C. W. Dunnett, "A Multiple Comparison Procedure for Comparing Several Treatments with a Control", *JASA*, Volume 50, 1955, pages 1096–1121. Reprinted with permission from the *Journal of American Statistical Association*. Copyright 1980 by the American Statistical Association. All rights reserved.

The tables of Duncan's critical values (in section 12.1.4.3) are from L. Hunter, "Critical Values for Duncan's New Multiple Range Test", *Biometrics*, 1960, Volume 16, pages 671–685. Reprinted with permission from the *Journal of American Statistical Association*. Copyright 1960 by the American Statistical Association. All rights reserved.

Table 15.1 is reproduced, by permission, from *ASTM Manual on Quality Control of Materials*, American Society for Testing and Materials, Philadelphia, PA, 1951.

The table in section 15.1.2 and much of Chapter 18 originally appeared in D. Zwillinger, *Standard Mathematical Tables and Formulae*, 30th edition, CRC Press, Boca Raton, FL, 1995. Reprinted courtesy of CRC Press, LLC.

Much of section 17.17 is taken from the URL http://members.aol.com/johnp71/javastat.html Permission courtesy of John C. Pezzullo.

Contents

CHAPTER 1

Introduction

Contents

1.1 BACKGROUND

The purpose of this book is to provide a modern set of tables and a comprehensive list of definitions, concepts, theorems, and formulae in probability and statistics. While the numbers in these tables have not changed since they were first computed (in some cases, several hundred years ago), the presentation format here is modernized. In addition, nearly all table values have been re-computed to ensure accuracy.

Almost every table is presented along with a textual description and at least one example using a value from the table. Most concepts are illustrated with examples and step-by-step solutions. Several data sets are described in this chapter; they are used in this book in order for users to be able to check algorithms.

The emphasis of this book is on what is often called *basic statistics*. Most real-world statistics users will be able to refer to this book in order to quickly verify a formula, definition, or theorem. In addition, the set of tables here should make this a complete statistics reference tool. Some more advanced useful and current topics, such as Brownian motion and decision theory are also included.

1.2 DATA SETS

We have established a few data sets which we have used in examples throughout this book. With these, a user can check a local statistics program by verifying that it returns the same values as given in this book. For example, the correlation coefficient between the first 100 elements of the sequence of integers $\{1, 2, 3 \ldots\}$ and the first 100 elements of the sequence of squares $\{1, 4, 9 \ldots\}$ is 0.96885. Using this value is an easy way to check for correct computation of a computer program. These data sets may be obtained from `http://vesta.bloomu.edu/~skokoska/prast/data`.

1

Ticket data: Forty random speeding tickets were selected from the courthouse records in Columbia County. The speed indicated on each ticket is given in the table below.

58	72	64	65	67	92	55	51	69	73
64	59	65	55	75	56	89	60	84	68
74	67	55	68	74	43	67	71	72	66
62	63	83	64	51	63	49	78	65	75

Swimming pool data: Water samples from 35 randomly selected pools in Beverly Hills were tested for acidity. The following table lists the PH for each sample.

6.4	6.6	6.2	7.2	6.2	8.1	7.0
7.0	5.9	5.7	7.0	7.4	6.5	6.8
7.0	7.0	6.0	6.3	5.6	6.3	5.8
5.9	7.2	7.3	7.7	6.8	5.2	5.2
6.4	6.3	6.2	7.5	6.7	6.4	7.8

Soda pop data: A new soda machine placed in the Mathematics Building on campus recorded the following sales data for one week in April.

Soda	Number of cans
Pepsi	72
Wild Cherry Pepsi	60
Diet Pepsi	85
Seven Up	54
Mountain Dew	32
Lipton Ice Tea	64

1.3 REFERENCES

Gathered here are some of the books referenced in later sections; each has a broad coverage of the topics it addresses.

1. W. G. Cochran and G. M. Cox, *Experimental Designs*, Second Edition, John Wiley & Sons, Inc., New York, 1957.
2. C. J. Colbourn and J. H. Dinitz, *CRC Handbook of Combinatorial Designs*, CRC Press, Boca Raton, FL, 1996.
3. L. Devroye, *Non-Uniform Random Variate Generation*, Springer–Verlag, New York, 1986.
4. W. Feller, *An Introduction to Probability Theory and Its Applications*, Volumes 1 and 2, John Wiley & Sons, New York, 1968.
5. C. W. Gardiner, *Handbook of Stochastic Methods*, Second edition, Springer–Verlag, New York, 1985.
6. D. J. Sheskin, *Handbook of Parametric and Nonparametric Statistical Procedures*, CRC Press LLC, Boca Raton, FL, 1997.

CHAPTER 2

Summarizing Data

Contents

Numerical descriptive statistics and graphical techniques may be used to summarize information about central tendency and/or variability.

2.1 TABULAR AND GRAPHICAL PROCEDURES

2.1.1 Stem-and-leaf plot

A stem-and-leaf plot is a a graphical summary used to describe a set of observations (as symmetric, skewed, etc.). Each observation is displayed on the graph and should have at least two digits. Split each observation (at the same point) into a stem (one or more of the leading digit(s)) and a leaf (remaining digits). Select the split point so that there are 5–20 total stems. List the stems in a column to the left, and write each leaf in the corresponding stem row.

Example 2.1: Construct a stem-and-leaf plot for the Ticket Data (page 2).

Solution:

Stem	Leaf
4	3 9
5	1 1 5 5 5 6 8 9
6	0 2 3 3 4 4 4 5 5 5 6 7 7 7 8 8 9
7	1 2 2 3 4 4 5 5 8
8	3 4 9
9	2

Stem = 10, Leaf = 1

Figure 2.1: Stem–and–leaf plot for Ticket Data.

2.1.2 Frequency distribution

A frequency distribution is a tabular method for summarizing continuous or discrete numerical data or categorical data.

(1) Partition the measurement axis into 5–20 (usually equal) reasonable subintervals called classes, or class intervals. Thus, each observation falls into exactly one class.

(2) Record, or tally, the number of observations in each class, called the frequency of each class.

(3) Compute the proportion of observations in each class, called the relative frequency.

(4) Compute the proportion of observations in each class *and* all preceding classes, called the cumulative relative frequency.

Example 2.2:   Construct a frequency distribution for the Ticket Data (page 2).

Solution:

(S1) Determine the classes. It seems reasonable to use 40 to less than 50, 50 to less than 60, ..., 90 to less than 100.

Note: For continuous data, one end of each class must be open. This ensures that each observation will fall into only one class. The open end of each class may be either the left or right, but should be consistent.

(S2) Record the number of observations in each class.

(S3) Compute the relative frequency and cumulative relative frequency for each class.

(S4) The resulting frequency distribution is in Figure 2.2.

Class	Frequency	Relative frequency	Cumulative relative frequency
$[40, 50)$	2	0.050	0.050
$[50, 60)$	8	0.200	0.250
$[60, 70)$	17	0.425	0.625
$[70, 80)$	9	0.225	0.900
$[80, 90)$	3	0.075	0.975
$[90, 100)$	1	0.025	1.000

Figure 2.2: Frequency distribution for Ticket Data.

2.1.3   Histogram

A histogram is a graphical representation of a frequency distribution. A (relative) frequency histogram is a plot of (relative) frequency versus class interval. Rectangles are constructed over each class with height proportional (usually equal) to the class (relative) frequency. A frequency and relative frequency histogram have the same shape, but different scales on the vertical axis.

Example 2.3:   Construct a frequency histogram for the Ticket Data (page 2).

Solution:

(S1) Using the frequency distribution in Figure 2.2, construct rectangles above each class, with height equal to class frequency.

(S2) The resulting histogram is in Figure 2.3.

Note: A probability histogram is constructed so that the area of each rectangle equals the relative frequency. If the class widths are unequal, this histogram presents a more accurate description of the distribution.

2.1.4   Frequency polygons

A frequency polygon is a line plot of points with x coordinate being class midpoint and y coordinate being class frequency. Often the graph extends to

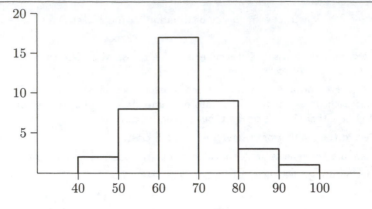

Figure 2.3: Frequency histogram for Ticket Data.

an additional empty class on both ends. The relative frequency may be used in place of frequency.

Example 2.4:   Construct a frequency polygon for the Ticket Data (page 2).

Solution:

(S1) Using the frequency distribution in Figure 2.2, plot each point and connect the graph.

(S2) The resulting frequency polygon is in Figure 2.4.

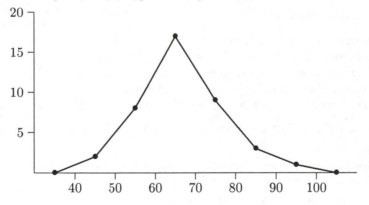

Figure 2.4: Frequency polygon for Ticket Data.

An **ogive**, or **cumulative frequency polygon**, is a plot of cumulative frequency versus the upper class limit. Figure 2.5 is an ogive for the Ticket Data (page 2).

Another type of frequency polygon is a *more-than* cumulative frequency polygon. For each class this plots the number of observations in that class and every class above versus the lower class limit.

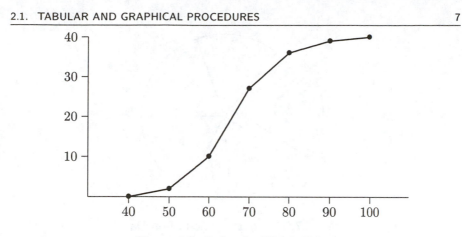

Figure 2.5: Ogive for Ticket Data.

A **bar chart** is often used to graphically summarize discrete or categorical data. A rectangle is drawn over each bin with height proportional to frequency. The chart may be drawn with horizontal rectangles, in three dimensions, and may be used to compare two or more sets of observations. Figure 2.6 is a bar chart for the Soda Pop Data (page 2).

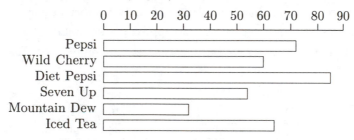

Figure 2.6: Bar chart for Soda Pop Data.

A **pie chart** is used to illustrate parts of the total. A circle is divided into slices proportional to the bin frequency. Figure 2.7 is a pie chart for the Soda Pop Data (page 2).

2.1.5 Chernoff faces

Chernoff faces are used to illustrate trends in multidimensional data. They are effective because people are used to differentiating between facial features. Chernoff faces have been used for cluster, discriminant, and time-series analyses. Facial features that might be controllable by the data include:

(a) ear: level, radius

(b) eyebrow: height, slope, length

(c) eyes: height, size, separation, eccentricity, pupil position or size

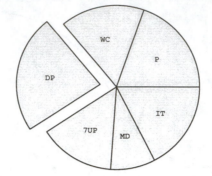

Figure 2.7: Pie chart for Soda Pop Data.

(d) face: width, half-face height, lower or upper eccentricity

(e) mouth: position of center, curvature, length, openness

(f) nose: width, length

The Chernoff faces in Figure 2.8 come from data about this book. For the even chapters:

(a) eye size is proportional to the approximate number of pages

(b) mouth size is proportional to the approximate number of words

(c) face shape is proportional to the approximate number of occurrences of the word "the"

The data are as follows:

Chapter	2	4	6	8	10	12	14	16	18
Number of pages	18	30	56	8	36	40	40	26	23
Number of words	4514	5426	12234	2392	9948	18418	8179	11739	5186
Occurrences of "the"	159	147	159	47	153	118	264	223	82

An interactive program for creating Chernoff faces is available at http:// www.hesketh.com/schampeo/projects/Faces/interactive.shtml. See H. Chernoff, "The use of faces to represent points in a K-dimensional space graphically," *Journal of the American Statistical Association*, Vol. 68, No. 342, 1973, pages 361–368.

2.2 NUMERICAL SUMMARY MEASURES

The following conventions will be used in the definitions and formulas in this section.

(C1) *Ungrouped data*: Let $x_1, x_2, x_3, \ldots, x_n$ be a set of observations.

(C2) *Grouped data*: Let $x_1, x_2, x_3, \ldots, x_k$ be a set of class marks from a frequency distribution, or a representative set of observations, with corre-

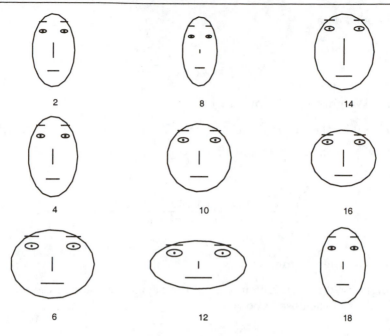

Figure 2.8: Chernoff faces for chapter data.

sponding frequencies $f_1, f_2, f_3, \ldots, f_k$. The total number of observations is $n = \sum_{i=1}^{k} f_i$. Let c denote the (constant) width of each bin and x_o one of the class marks selected to be the *computing origin*. Each class mark, x_i, may be coded by $u_i = (x_i - x_o)/c$. Each u_i will be an integer and the bin mark taken as the computing origin will be coded as a 0.

2.2.1 (Arithmetic) mean

The (arithmetic) mean of a set of observations is the sum of the observations divided by the total number of observations.

(1) Ungrouped data:

$$\overline{x} = \frac{1}{n} \sum_{i=1}^{n} x_i = \frac{x_1 + x_2 + x_3 + \cdots + x_n}{n} \qquad (2.1)$$

(2) Grouped data:

$$\overline{x} = \frac{1}{n} \sum_{i=1}^{k} f_i x_i = \frac{f_1 x_1 + f_2 x_2 + f_3 x_3 + \cdots + f_n x_n}{n} \qquad (2.2)$$

(3) Coded data:

$$\bar{x} = x_o + c \cdot \frac{\sum\limits_{i=1}^{k} f_i u_i}{n} \qquad (2.3)$$

2.2.2   Weighted (arithmetic) mean

Let $w_i \geq 0$ be the weight associated with observation x_i. The *total weight* is given by $\sum\limits_{i=1}^{n} w_i$ and the weighted mean is

$$\bar{x}_w = \frac{\sum\limits_{i=1}^{n} w_i x_i}{\sum\limits_{i=1}^{n} w_i} = \frac{w_1 x_1 + w_2 x_2 + w_3 x_3 + \cdots + w_n x_n}{w_1 + w_2 + w_3 + \cdots + w_n}. \qquad (2.4)$$

2.2.3   Geometric mean

For ungrouped data such that $x_i > 0$, the geometric mean is the nth root of the product of the observations:

$$\text{GM} = \sqrt[n]{x_1 \cdot x_2 \cdot x_3 \cdots x_n}. \qquad (2.5)$$

In logarithmic form:

$$\log(\text{GM}) = \frac{1}{n} \sum_{i=1}^{n} \log x_i = \frac{\log x_1 + \log x_2 + \log x_3 + \cdots + \log x_n}{n}. \qquad (2.6)$$

For grouped data with each class mark $x_i > 0$:

$$\text{GM} = \sqrt[n]{x_1^{f_1} \cdot x_2^{f_2} \cdot x_3^{f_3} \cdots x_k^{f_k}}. \qquad (2.7)$$

In logarithmic form:

$$\log(\text{GM}) = \frac{1}{n} \sum_{i=1}^{k} f_i \log(x_i) \qquad (2.8)$$

$$= \frac{f_1 \log(x_1) + f_2 \log(x_2) + f_3 \log(x_3) + \cdots + f_k \log(x_k)}{n}.$$

2.2.4   Harmonic mean

For ungrouped data the harmonic mean is given by

$$\text{HM} = \frac{n}{\sum\limits_{i=1}^{n} \frac{1}{x_i}} = \frac{n}{\frac{1}{x_1} + \frac{1}{x_2} + \frac{1}{x_3} + \cdots + \frac{1}{x_n}}. \qquad (2.9)$$

For grouped data:

$$\text{HM} = \frac{n}{\displaystyle\sum_{i=1}^{k} \frac{f_i}{x_i}} = \frac{n}{\dfrac{f_1}{x_1} + \dfrac{f_2}{x_2} + \dfrac{f_3}{x_3} + \cdots + \dfrac{f_k}{x_k}} . \tag{2.10}$$

Note: The equation involving the arithmetic, geometric, and harmonic mean is

$$\text{HM} \leq \text{GM} \leq \bar{x}. \tag{2.11}$$

Equality holds if all n observations are equal.

2.2.5 Mode

For ungrouped data, the mode, M_o, is the value that occurs most often, or with the greatest frequency. A mode may not exist, for example, if all observations occur with the same frequency. If the mode does exist, it may not be unique, for example, if two observations occur with the greatest frequency.

For grouped data, select the class containing the largest frequency, called the modal class. Let L be the lower boundary of the modal class, d_L the difference in frequencies between the modal class and the class immediately below, and d_H the difference in frequencies between the modal class and the class immediately above. The mode may be approximated by

$$M_o \approx L + c \cdot \frac{d_L}{d_L + d_H} . \tag{2.12}$$

2.2.6 Median

The median, $\tilde{x}$, is another measure of central tendency, resistant to outliers. For ungrouped data, arrange the observations in order from smallest to largest. If n is odd, the median is the middle value. If n is even, the median is the mean of the two middle values.

For grouped data, select the class containing the median (median class). Let L be the lower boundary of the median class, f_m the frequency of the median class, and CF the sum of frequencies for all classes below the median class (a cumulative frequency). The median may be approximated by

$$\tilde{x} \approx L + c \cdot \frac{\dfrac{n}{2} - \text{CF}}{f_m} . \tag{2.13}$$

Note: If $\bar{x} > \tilde{x}$ the distribution is positively skewed. If $\bar{x} < \tilde{x}$ the distribution is negatively skewed. If $\bar{x} \approx \tilde{x}$ the distribution is approximately symmetric.

2.2.7 $p\%$ trimmed mean

A trimmed mean is a measure of central tendency and a compromise between a mean and a median. The mean is more sensitive to outliers, and the median is less sensitive to outliers. Order the observations from smallest to largest.

Delete the smallest $p\%$ *and* the largest $p\%$ of the observations. The $p\%$ trimmed mean, $\bar{x}_{\mathrm{tr}(p)}$, is the arithmetic mean of the remaining observations.

Note: If $p\%$ of n (observations) is not an integer, several (computer) algorithms exist for interpolating at each end of the distribution and for determining $\bar{x}_{\mathrm{tr}(p)}$.

Example 2.5:   Using the Swimming Pool data (page 2) find the mean, median, and mode. Compute the geometric mean and the harmonic mean, and verify the relationship between these three measures.

Solution:

(S1) $\bar{x} = \dfrac{1}{35}(6.4 + 6.6 + 6.2 + \cdots + 7.8) = 6.5886$

(S2) $\tilde{x} = 6.5$, the middle values when the observations are arranged in order from smallest to largest.

(S3) $M_o = 7.0$, the observation that occurs most often.

(S4) GM $= \sqrt[35]{(6.4)(6.6)(6.2) \cdots (7.8)} = 6.5513$

(S5) HM $= \dfrac{35}{(1/6.4) + (1/6.6) + (1/6.2) + \cdots + (1/7.8)} = 6.5137$

(S6) To verify the inequality: $\underbrace{6.5137}_{\text{HM}} \leq \underbrace{6.5513}_{\text{GM}} \leq \underbrace{6.5886}_{\bar{x}}$

2.2.8   Quartiles

Quartiles split the data into four parts. For ungrouped data, arrange the observations in order from smallest to largest.

(1) The second quartile is the median: $Q_2 = \tilde{x}$.

(2) If n is even:

   The first quartile, Q_1, is the median of the smallest $n/2$ observations; and the third quartile, Q_3, is the median of the largest $n/2$ observations.

(3) If n is odd:

   The first quartile, Q_1, is the median of the smallest $(n + 1)/2$ observations; and the third quartile, Q_3, is the median of the largest $(n + 1)/2$ observations.

For grouped data, the quartiles are computed by applying equation (2.13) for the median. Compute the following:

$L_1 = $ the lower boundary of the class containing Q_1.

$L_3 = $ the lower boundary of the class containing Q_3.

$f_1 = $ the frequency of the class containing the first quartile.

$f_3 = $ the frequency of the class containing the third quartile.

$\mathrm{CF}_1 = $ cumulative frequency for classes below the one containing Q_1.

$\mathrm{CF}_3 = $ cumulative frequency for classes below the one containing Q_3.

The (approximate) quartiles are given by

$$Q_1 = L_1 + c \cdot \frac{\frac{n}{4} - CF_1}{f_1} \qquad Q_3 = L_3 + c \cdot \frac{\frac{3n}{4} - CF_3}{f_3}. \qquad (2.14)$$

2.2.9 Deciles

Deciles split the data into 10 parts.

(1) For ungrouped data, arrange the observations in order from smallest to largest. The ith decile, D_i (for $i = 1, 2, \ldots, 9$), is the $i(n+1)/10^{th}$ observation. It may be necessary to interpolate between successive values.

(2) For grouped data, apply equation (2.13) (as in equation (2.14)) for the median to find the approximate deciles. D_i is in the class containing the $in/10^{th}$ largest observation.

2.2.10 Percentiles

Percentiles split the data into 100 parts.

(1) For ungrouped data, arrange the observations in order from smallest to largest. The ith percentile, P_i (for $i = 1, 2, \ldots, 99$), is the $i(n+1)/100^{th}$ observation. It may be necessary to interpolate between successive values.

(2) For grouped data, apply equation (2.13) (as in equation (2.14)) for the median to find the approximate percentiles. P_i is in the class containing the $in/100^{th}$ largest observation.

2.2.11 Mean deviation

The mean deviation is a measure of variability based on the absolute value of the deviations about the mean or median.

(1) For ungrouped data:

$$MD = \frac{1}{n} \sum_{i=1}^{n} |x_i - \bar{x}| \quad \text{or} \quad MD = \frac{1}{n} \sum_{i=1}^{n} |x_i - \tilde{x}|. \qquad (2.15)$$

(2) For grouped data:

$$MD = \frac{1}{n} \sum_{i=1}^{k} f_i |x_i - \bar{x}| \quad \text{or} \quad MD = \frac{1}{n} \sum_{i=1}^{k} f_i |x_i - \tilde{x}|. \qquad (2.16)$$

2.2.12 Variance

The variance is a measure of variability based on the squared deviations about the mean.

(1) For ungrouped data:

$$s^2 = \frac{1}{n-1} \sum_{i=1}^{n} (x_i - \overline{x})^2. \tag{2.17}$$

The computational formula for s^2:

$$s^2 = \frac{1}{n-1} \left[\sum_{i=1}^{n} x_i^2 - \frac{1}{n} \left(\sum_{i=1}^{n} x_i \right)^2 \right] = \frac{1}{n-1} \left(\sum_{i=1}^{n} x_i^2 - n\overline{x}^2 \right). \tag{2.18}$$

(2) For grouped data:

$$s^2 = \frac{1}{n-1} \sum_{i=1}^{k} f_i (x_i - \overline{x})^2. \tag{2.19}$$

The computational formula for s^2:

$$s^2 = \frac{1}{n-1} \left[\sum_{i=1}^{k} f_i x_i^2 - \frac{1}{n} \left(\sum_{i=1}^{k} f_i x_i \right)^2 \right]$$

$$= \frac{1}{n-1} \left(\sum_{i=1}^{k} f_i x_i^2 - n\overline{x}^2 \right). \tag{2.20}$$

(3) For coded data:

$$s^2 = \frac{c}{n-1} \left[\sum_{i=1}^{k} f_i u_i^2 - \frac{1}{n} \left(\sum_{i=1}^{k} f_i u_i \right)^2 \right]. \tag{2.21}$$

2.2.13   Standard deviation

The *standard deviation* is the positive square root of the variance: $s = \sqrt{s^2}$. The *probable error* is 0.6745 times the standard deviation.

2.2.14   Standard errors

The standard error of a statistic is the standard deviation of the sampling distribution of that statistic. The standard error of a statistic is often designated by σ with a subscript indicating the statistic.

2.2.14.1   Standard error of the mean

The standard error of the mean is used in hypothesis testing and is an indication of the accuracy of the estimate $\overline{x}$.

$$\text{SEM} = s/\sqrt{n}. \tag{2.22}$$

2.2.15 Root mean square

(1) For ungrouped data:

$$\text{RMS} = \left(\frac{1}{n} \sum_{i=1}^{n} x_i^2 \right)^{1/2} . \qquad (2.23)$$

(2) For grouped data:

$$\text{RMS} = \left(\frac{1}{n} \sum_{i=1}^{k} f_i x_i^2 \right)^{1/2} . \qquad (2.24)$$

2.2.16 Range

The range is the difference between the largest and smallest values.

$$\text{R} = \max\{x_1, x_2, \ldots, x_n\} - \min\{x_1, x_2, \ldots, x_n\} = x_{(n)} - x_{(1)} . \qquad (2.25)$$

2.2.17 Interquartile range

The interquartile range, or fourth spread, is the difference between the third and first quartile.

$$\text{IQR} = Q_3 - Q_1 . \qquad (2.26)$$

2.2.18 Quartile deviation

The quartile deviation, or semi-interquartile range, is half the interquartile range.

$$\text{QD} = \frac{Q_3 - Q_1}{2} . \qquad (2.27)$$

2.2.19 Box plots

Box plots, also known as quantile plots, are graphics which display the center portions of the data and some information about the range of the data. There are a number of variations and a box plot may be drawn with either a horizontal or vertical scale. The *inner* and *outer fences* are used in constructing a box plot and are markers used in identifying mild and extreme outliers.

$$\begin{aligned} &\text{Inner Fences: } Q_1 - 1.5 \cdot \text{IQR}, \; Q_1 + 1.5 \cdot \text{IQR} \\ &\text{Outer Fences: } Q_3 - 3 \cdot \text{IQR}, \quad Q_3 + 3 \cdot \text{IQR} \end{aligned} \qquad (2.28)$$

A general description:

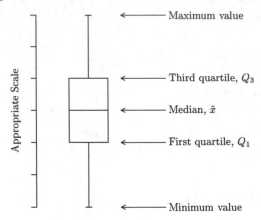

Multiple box plots on the same measurement axis may be used to compare the center and spread of distributions. Figure 2.9 presents box plots for randomly selected August residential electricity bills for three different parts of the country.

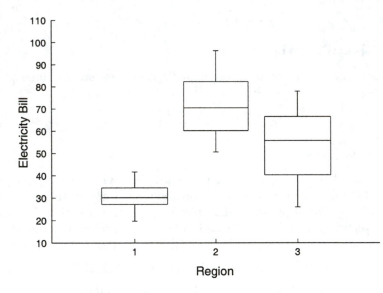

Figure 2.9: Example of multiple box plots.

2.2.20   Coefficient of variation

The coefficient of variation is a measure of relative variability. Reported as percentage it is defined as:

$$CV = 100 \, \frac{s}{\overline{x}} \, . \tag{2.29}$$

2.2.21   Coefficient of quartile variation

The coefficient of quartile variation is a measure of variability.

$$CQV = 100 \, \frac{Q_3 - Q_1}{Q_3 + Q_1} \, . \tag{2.30}$$

2.2.22 Z score

The z score, or standard score, associated with an observation is a measure of relative standing.

$$z = \frac{x_i - \overline{x}}{s} \tag{2.31}$$

2.2.23   Moments

Moments are used to characterize a set of observations.

(1) For ungrouped data:
The rth moment about the origin:

$$m'_r = \frac{1}{n} \sum_{i=1}^{n} x_i^r \, . \tag{2.32}$$

The rth moment about the mean $\overline{x}$:

$$m_r = \frac{1}{n} \sum_{i=1}^{n} (x_i - \overline{x})^r = \sum_{j=0}^{r} \binom{r}{j} (-1)^j m'_{r-j} \overline{x}^j \, . \tag{2.33}$$

(2) For grouped data:
The rth moment about the origin:

$$m'_r = \frac{1}{n} \sum_{i=1}^{k} f_i x_i^r \, . \tag{2.34}$$

The rth moment about the mean $\overline{x}$:

$$m_r = \frac{1}{n} \sum_{i=1}^{k} f_i (x_i - \overline{x})^r = \sum_{j=0}^{r} \binom{r}{j} (-1)^j m'_{r-j} \overline{x}^j . \tag{2.35}$$

(3) For coded data:

$$m'_r = \frac{c^r}{n} \sum_{i=1}^{n} f_i u_i^r \, . \tag{2.36}$$

2.2.24   Measures of skewness

The following descriptive statistics measure the lack of symmetry. Larger values (in magnitude) indicate more skewness in the distribution of observations.

2.2.24.1   Coefficient of skewness

$$g_1 = \frac{m_3}{m_2^{3/2}} \tag{2.37}$$

2.2.24.2   Coefficient of momental skewness

$$\frac{g_1}{2} = \frac{m_3}{2m_2^{3/2}} \tag{2.38}$$

2.2.24.3   Pearson's first coefficient of skewness

$$S_{k_1} = \frac{3(\bar{x} - M_o)}{s} \tag{2.39}$$

2.2.24.4   Pearson's second moment of skewness

$$S_{k_2} = \frac{3(\bar{x} - \tilde{x})}{s} \tag{2.40}$$

2.2.24.5   Quartile coefficient of skewness

$$S_{k_q} = \frac{Q_3 - 2\tilde{x} + Q_1}{Q_3 - Q_1} \tag{2.41}$$

Example 2.6:   Using the Swimming Pool data (page 2) find the coefficient of skewness, coefficient of momental skewness, Pearson's first coefficient of skewness, Pearson's second moment of skewness, and the quartile coefficient of skewness.

Solution:

(S1) $\bar{x} = 6.589$, $\tilde{x} = 6.5$, $s = 0.708$, $Q_1 = 6.2$, $Q_3 = 7.0$, $M_o = 7.0$

(S2) $m_2 = \dfrac{1}{n} \displaystyle\sum_{i=1}^{35}(x_i - \bar{x})^2 = 0.4867$ $m_3 = \dfrac{1}{n} \displaystyle\sum_{i=1}^{35}(x_i - \bar{x})^3 = 0.0126$

(S3) $g_1 = 0.0126/(0.4867)^{3/2} = 0.0371$, $g_1/2 = 0.0372/2 = 0.0186$

(S4) $S_{k_1} = \dfrac{3(6.589 - 7)}{0.708} = -1.7415$, $S_{k_2} = \dfrac{3(6.589 - 6.5)}{0.708} = 0.3771$

(S5) $S_{k_q} = \dfrac{7.0 - 2(6.5) + 6.2}{7.0 - 6.2} = 0.25$

2.2.25   Measures of kurtosis

The following statistics describe the extent of the peak in a distribution. Smaller values (in magnitude) indicate a flatter, more uniform distribution.

2.2.25.1 Coefficient of kurtosis

$$g_2 = \frac{m_4}{m_2^2} \tag{2.42}$$

2.2.25.2 Coefficient of excess kurtosis

$$g_2 - 3 = \frac{m_4}{m_2^2} - 3 \tag{2.43}$$

2.2.26 Data transformations

Suppose $y_i = ax_i + b$ for $i = 1, 2, \ldots, n$. The following summary statistics for the distribution of y's are related to summary statistics for the distribution of x's.

$$\bar{y} = a\bar{x} + b, \qquad s_y^2 = a^2 s_x^2, \qquad s_y = |a| s_x \tag{2.44}$$

2.2.27 Sheppard's corrections for grouping

For grouped data, suppose every class interval has width c. If both tails of the distribution are very flat and close to the measurement axis, the grouped data approximation to the sample variance may be improved by using Sheppard's correction, $-c^2/12$:

$$\text{corrected variance} = \text{grouped data variance} - \frac{c^2}{12} \tag{2.45}$$

There are similar corrected sample moments, denoted m'_{r_c} and m_{r_c}:

$$
\begin{aligned}
m'_{1_c} &= m'_1 & m_{1_c} &= m_1 \\
m'_{2_c} &= m'_2 - \frac{c^2}{12} & m_{2_c} &= m_2 - \frac{c^2}{12} \\
m'_{3_c} &= m'_3 - \frac{c^2}{4} m'_1 & m_{3_c} &= m_3 \\
m'_{4_c} &= m'_4 - \frac{c^2}{2} m'_1 + \frac{7c^2}{240} & m_{4_c} &= m_4 - \frac{c^2}{2} m_2 + \frac{7c^2}{240}
\end{aligned}
\tag{2.46}
$$

Example 2.7:   Consider the *grouped* Ticket Data (page 2) as presented in the frequency distribution in Example 2.2 (on page 5). Find the corrected sample variance and corrected sample moments.

Solution:

(S1) $\bar{x} = 66.5$, $\quad s^2 = 115.64$ (for grouped data), $\quad c = 10$

(S2) corrected variance $= 115.64 - (10^2/12) = 107.31$

(S3)

r	m'_r	m'_{r_c}	m_r	m_{r_c}
1	66.5	66.5	0.0	0.0
2	4535.0	4526.7	112.8	104.4
3	316962.5	315300.0	389.3	389.3
4	22692125.0	22688802.9	40637.3	35002.7

CHAPTER 3

Probability

Contents

3.1   ALGEBRA OF SETS

Properties of and operations on sets are important since events may be thought of as sets. Some set facts:

(1) A set A is a collection of objects called the elements of the set.

    $a \in A$ means a is an element of the set A.

    $a \notin A$ means a is *not* an element of the set A.

    $A = \{a, b, c\}$ is used to denote the elements of the set A.

(2) The *null set*, denoted by ϕ or $\{\ \}$, is the empty set; the set that contains no elements.

(3) Two sets A and B are *equal*, written $A = B$, if

    1) every element of A is an element of B, and

    2) every element of B is an element of A.

(4) The set A is a *subset* of the set B if every element of A is also in B; written $A \subset B$ (or $B \supset A$). For every set A, $\phi \subset A$.

(5) If $A \subset B$ and $B \subset A$ then $A = B$ and A is an *improper* subset of B. If $A \subset B$ and there is at least one element of B not in A then A is a *proper* subset of B. The subset symbol $\subset$ is often used to denote a *proper* subset while the symbol $\subseteq$ indicates an *improper* subset.

(6) Let S be the universal set, the set consisting of all elements of interest. For any set A, $A \subset S$.

(7) The *complement* of the set A, denoted A', is the set consisting of all elements in S but not in A (Figure 3.1).

(8) For any two sets A and B:

    The *union* of A and B, denoted $A \cup B$, is the set consisting of all elements in A, or B, or both (Figure 3.2).

    The *intersection* of A and B, denoted $A \cap B$, is the set consisting of all elements in both A and B (Figure 3.3).

(9) A and B are *disjoint* or *mutually exclusive* if $A \cap B = \phi$ (Figure 3.4).

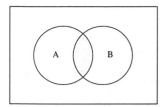

Figure 3.1: Shaded region = A'.

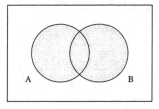

Figure 3.2: Shaded region = $A \cup B$.

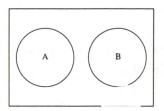

Figure 3.3: Shaded region = $A \cap B$.

Figure 3.4: Mutually exclusive sets.

For the following properties, suppose A, B, and C are sets. It is necessary to assume these sets lie in a universal set S only in those properties that explicitly involve S.

(1) Closure

    (a) There is a unique set $A \cup B$.

    (b) There is a unique set $A \cap B$.

(2) Commutative laws

    (a) $A \cup B = B \cup A$

    (b) $A \cap B = B \cap A$

(3) Associative laws

    (a) $(A \cup B) \cup C = A \cup (B \cup C)$

    (b) $(A \cap B) \cap C = A \cap (B \cap C)$

(4) Distributive laws

    (a) $A \cup (B \cap C) = (A \cup B) \cap (A \cup C)$

    (b) $A \cap (B \cup C) = (A \cap B) \cup (A \cap C)$

(5) Idempotent laws

    (a) $A \cup A = A$

    (b) $A \cap A = A$

(6) Properties of S and ϕ

    (a) $A \cap S = A$

    (b) $A \cup \phi = A$

    (c) $A \cap \phi = \phi$

    (d) $A \cup S = S$

(7) Properties of $\subset$

    (a) $A \subset (A \cup B)$

    (b) $(A \cap B) \subset A$

    (c) $A \subset S$

    (d) $\phi \subset A$

    (e) If $A \subset B$, then $A \cup B = B$ and $A \cap B = A$.

(8) Properties of ' (set complement)

    (a) For every set A, there is a unique set A'.

    (b) $A \cup A' = S$

    (c) $A \cap A' = \phi$

    (d) $\left. \begin{aligned} (A \cup B)' &= A' \cap B' \\ (A \cap B)' &= A' \cup B' \end{aligned} \right\}$ DeMorgan's laws

(9) Some generalizations

    Suppose $A_1, A_2, A_3, \ldots, A_n$ is a collection of sets.

    (a) The *generalized union*, $A_1 \cup A_2 \cup \cdots \cup A_n$, is the set consisting of all elements in at least one A_i.

    (b) The *generalized intersection*, $A_1 \cap A_2 \cap \cdots \cap A_n$, is the set consisting of all elements in every A_i.

    (c) $\left(\bigcup_{i=1}^{n} A_i \right)' = (A_1 \cup \cdots \cup A_n)' = A_1' \cap \cdots \cap A_n' = \bigcap_{i=1}^{n} A_i'$

    (d) $\left(\bigcap_{i=1}^{n} A_i \right)' = (A_1 \cap \cdots \cap A_n)' = A_1' \cup \cdots \cup A_n' = \bigcup_{i=1}^{n} A_i'$

3.2   COMBINATORIAL METHODS

In an equally likely outcome experiment, computing the probability of an event involves counting. The following techniques are useful for determining the number of outcomes in an event and/or the sample space.

3.2.1   The product rule for ordered pairs

If the first element of an ordered pair can be selected in n_1 ways, and for each of these n_1 ways the second element of the pair can be selected in n_2 ways, then the number of possible pairs is $n_1 n_2$.

3.2.2 The generalized product rule for k-tuples

Suppose a sample space, or set, consists of ordered collections of k-tuples. If there are n_1 choices for the first element, and for each choice of the first element there are n_2 choices for the second element, ..., and for each of the first $k-1$ elements there are n_k choices for the kth element, then there are $n_1 n_2 \cdots n_k$ possible k-tuples.

3.2.3 Permutations

The number of permutations of n distinct objects taken k at a time is

$$P(n,k) = \frac{n!}{(n-k)!}. \tag{3.1}$$

A table of values is on page 500.

3.2.4 Circular permutations

The number of permutations of n distinct objects arranged in a circle is $(n-1)!$.

3.2.5 Combinations (binomial coefficients)

The binomial coefficient $\binom{n}{k}$ is the number of combinations of n distinct objects taken k at a time without regard to order:

$$C(n,k) = \binom{n}{k} = \frac{n!}{k!(n-k)!} = \frac{P(n,k)}{k!}. \tag{3.2}$$

A table of values is on page 500. Other formulas involving binomial coefficients:

(a) $\displaystyle \binom{n}{k} = \frac{n(n-1)\cdots(n-k+1)}{k!} = \binom{n}{n-k}$

(b) $\displaystyle \binom{n}{0} = \binom{n}{n} = 1$ and $\displaystyle \binom{n}{1} = n$

(c) $\displaystyle \binom{n}{k} = \binom{n-1}{k} + \binom{n-1}{k-1}$

(d) $\displaystyle \binom{n}{0} + \binom{n}{1} + \cdots + \binom{n}{n} = 2^n$

(e) $\displaystyle \binom{n}{0} - \binom{n}{1} + \cdots + (-1)^n \binom{n}{n} = 0$

(f) $\displaystyle \binom{2n}{n} = \frac{2^n (2n-1)(2n-3)\cdots 3 \cdot 1}{n!}$

(g) $\displaystyle \binom{n}{n} + \binom{n+1}{n} + \binom{n+2}{n} + \cdots + \binom{n+m}{n} = \binom{n+m+1}{n+1}$

(h) $\displaystyle \binom{n}{0} + \binom{n}{2} + \binom{n}{4} + \cdots + = 2^{n-1}$ (last term in sum is $\binom{n}{n-1}$ or $\binom{n}{n}$)

(i) $\displaystyle \binom{n}{1} + \binom{n}{3} + \binom{n}{5} + \cdots + = 2^{n-1}$ (last term in sum is $\binom{n}{n-1}$ or $\binom{n}{n}$)

(j) $\binom{n}{0}^2 + \binom{n}{1}^2 + \cdots + \binom{n}{n}^2 = \binom{2n}{n}$

(k) $\binom{m}{0}\binom{n}{p} + \binom{m}{1}\binom{n}{p-1} + \cdots + \binom{m}{p}\binom{n}{0} = \binom{m+n}{p}$

(l) $1\binom{n}{1} + 2\binom{n}{2} + \cdots + n\binom{n}{n} = n2^{n-1}$

(m) $1\binom{n}{1} - 2\binom{n}{2} + \cdots + (-1)^{n+1}n\binom{n}{n} = 0$

Example 3.8: For the 5 element set $\{a, b, c, d, e\}$ find the number of subsets containing exactly 3 elements.

Solution:

(S1) There are $\binom{5}{3} = \dfrac{5!}{3!2!} = 10$ subsets containing exactly 3 elements.

(S2) The subsets are

$$
\begin{array}{ccccc}
(a,b,c) & (a,b,d) & (a,b,e) & (a,c,d) & (a,c,e) \\
(a,d,e) & (b,c,d) & (b,c,e) & (b,d,e) & (c,d,e)
\end{array}
$$

3.2.6 Sample selection

There are 4 ways in which a sample of k elements can be obtained from a set of n distinguishable objects.

Order counts?	Repetitions allowed?	The sample is called a	Number of ways to choose the sample
No	No	k-combination	$C(n, k)$
Yes	No	k-permutation	$P(n, k)$
No	Yes	k-combination with replacement	$C^R(n, k)$
Yes	Yes	k-permutation with replacement	$P^R(n, k)$

where

$$
\begin{aligned}
C(n, k) &= \binom{n}{k} = \frac{n!}{k!\,(n-k)!} \\
P(n, k) &= (n)_k = n^{\underline{k}} = \frac{n!}{(n-k)!} \\
C^R(n, k) &= C(n+k-1, k) = \frac{(n+k-1)!}{k!(n-1)!} \\
P^R(n, k) &= n^k
\end{aligned}
\tag{3.3}
$$

Example 3.9: There are 4 ways in which to choose a 2 element sample from the set $\{a, b\}$:

combination	$C(2, 2) = 1$	ab
permutation	$P(2, 2) = 2$	ab and ba
combination with replacement	$C^R(2, 2) = 3$	aa, ab, and bb
permutation with replacement	$P^R(2, 2) = 4$	aa, ab, ba, and bb

3.2.7   Balls into cells

There are 8 different ways in which n balls can be placed into k cells.

Distinguish balls?	Distinguish cells?	Can cells be empty?	Number of ways to place n balls into k cells
Yes	Yes	Yes	k^n
Yes	Yes	No	$k! \left\{ {n \atop k} \right\}$
No	Yes	Yes	$C(k+n-1, n) = \binom{k+n-1}{n}$
No	Yes	No	$C(n-1, k-1) = \binom{n-1}{k-1}$
Yes	No	Yes	$\left\{ {n \atop 1} \right\} + \left\{ {n \atop 2} \right\} + \cdots + \left\{ {n \atop k} \right\}$
Yes	No	No	$\left\{ {n \atop k} \right\}$
No	No	Yes	$p_1(n) + p_2(n) + \cdots + p_k(n)$
No	No	No	$p_k(n)$

where $\left\{ {n \atop k} \right\}$ is the Stirling cycle number (see page 525) and $p_k(n)$ is the number of partitions of the number n into exactly k integer pieces (see page 523).

Given n distinguishable balls and k distinguishable cells, the number of ways in which we can place n_1 balls into cell 1, n_2 balls into cell 2, ..., n_k balls into cell k, is given by the multinomial coefficient $\binom{n}{n_1, n_2, \ldots, n_k}$.

3.2.8   Multinomial coefficients

The multinomial coefficient, $\binom{n}{n_1, n_2, \ldots, n_k} = C(n; n_1, n_2, \ldots, n_k)$, is the number of ways of choosing n_1 objects, then n_2 objects, ..., then n_k objects from a collection of n distinct objects without regard to order. This requires that $\sum_{j=1}^{k} n_j = n$.

Other ways to interpret the multinomial coefficient:

(1) Permutations (all objects not distinct): Given n_1 objects of one kind, n_2 objects of a second kind, ..., n_k objects of a kth kind, and $n_1 + n_2 + \cdots + n_k = n$. The number of permutations of the n objects is $\binom{n}{n_1, n_2, \ldots, n_k}$.

(2) Partitions: The number of ways of partitioning a set of n distinct objects into k subsets with n_1 objects in the first subset, n_2 objects in the second subset, ..., and n_k objects in the kth subset is $\binom{n}{n_1, n_2, \ldots, n_k}$.

The multinomial symbol is numerically evaluated as

$$\binom{n}{n_1, n_2, \ldots, n_k} = \frac{n!}{n_1! \, n_2! \cdots n_k!} \tag{3.4}$$

Example 3.10:   The number of ways to choose 2 objects, then 1 object, then 1 object from the set $\{a, b, c, d\}$ is $\binom{4}{2,1,1} = 12$; they are as follows (commas separate the ordered selections):

$$\{ab, c, d\} \qquad \{ab, d, c\} \qquad \{ac, b, d\} \qquad \{ac, d, b\}$$
$$\{ad, b, c\} \qquad \{ad, c, b\} \qquad \{bc, a, d\} \qquad \{bc, d, a\}$$
$$\{bd, a, c\} \qquad \{bd, c, a\} \qquad \{cd, a, b\} \qquad \{cd, b, a\}$$

3.2.9   Arrangements and derangements

(a) The number of ways to arrange n distinct objects in a row is $n!$; this is the number of permutations of n objects.

Example 3.11:   For the three objects $\{a, b, c\}$ the number of arrangements is $3! = 6$. These permutations are $\{abc, bac, cab, acb, bca, cba\}$.

(b) The number of ways to arrange n non-distinct objects (assuming that there are k types of objects, and n_i copies of each object of type i) is the multinomial coefficient $\binom{n}{n_1, n_2, \ldots, n_k}$.

Example 3.12:   For the set $\{a, a, b, c\}$ the parameters are $n = 4$, $k = 3$, $n_1 = 2$, $n_2 = 1$, and $n_3 = 1$. Hence, there are $\binom{4}{2,1,1} = \frac{4!}{2!\,1!\,1!} = 12$ arrangements, they are:

$$\begin{array}{cccccc} aabc & aacb & abac & abca & acab & acba \\ baac & baca & bcaa & caab & caba & cbaa \end{array}$$

(c) A *derangement* is a permutation of objects, in which object i is not in the ith location.

Example 3.13:   All the derangements of $\{1, 2, 3, 4\}$ are:

$$\begin{array}{ccc} 2143 & 2341 & 2413 \\ 3142 & 3412 & 3421 \\ 4123 & 4312 & 4321 \end{array}$$

The number of derangements of n elements, D_n, satisfies the recursion relation: $D_n = (n - 1)\,(D_{n-1} + D_{n-2})$, with the initial values $D_1 = 0$ and $D_2 = 1$. Hence,

$$D_n = n! \left(1 - \frac{1}{1!} + \frac{1}{2!} - \frac{1}{3!} + \cdots + (-1)^n \frac{1}{n!} \right)$$

The numbers D_n are also called *subfactorials* and *rencontres numbers*. For large values of n, $D_n/n! \sim e^{-1} \approx 0.37$. Hence, more than one of every three permutations is a derangement.

n	1	2	3	4	5	6	7	8	9	10
D_n	0	1	2	9	44	265	1854	14833	133496	1334961

3.3   PROBABILITY

The *sample space* of an experiment, denoted S, is the set of all possible outcomes. Each outcome of the sample space is also called an element of the sample space or a sample point. An *event* is any collection of outcomes contained in the sample space. A *simple event* consists of exactly one outcome and a *compound event* consists of more than one outcome.

3.3.1   Relative frequency concept of probability

Suppose an experiment is conducted n identical and independent times and $n(A)$ is the number of times the event A occurs. The quotient $n(A)/n$ is the relative frequency of occurrence of the event A. As n increases, the relative frequency converges to the limiting relative frequency of the event A. The probability of the event A, Prob $[A]$, is this limiting relative frequency.

3.3.2   Axioms of probability (discrete sample space)

(1) For any event A, Prob $[A] \geq 0$.

(2) Prob $[S] = 1$.

(3) If $A_1, A_2, A_3, \ldots$, is a finite or infinite collection of pairwise mutually exclusive events of S, then

$$\text{Prob}[A_1 \cup A_2 \cup A_3 \cup \cdots] = \text{Prob}[A_1] + \text{Prob}[A_2] + \text{Prob}[A_3] + \cdots$$
$$(3.5)$$

3.3.3   The probability of an event

The probability of an event A is the sum of Prob $[a_i]$ for all sample points a_i in the event A:

$$\text{Prob}[A] = \sum_{a_i \in A} \text{Prob}[a_i]. \qquad (3.6)$$

If all of the outcomes in S are *equally likely*:

$$\text{Prob}[A] = \frac{n(A)}{n(S)} = \frac{\text{number of outcomes in } A}{\text{number of outcomes in } S}. \qquad (3.7)$$

3.3.4   Probability theorems

(1) Prob $[\phi] = 0$ for any sample space S.

(2) If A and A' are complementary events, Prob $[A] +$ Prob $[A'] = 1$.

(3) For any events A and B, if $A \subset B$ then Prob $[A] \leq$ Prob $[B]$.

(4) For any events A and B,

$$\text{Prob}[A \cup B] = \text{Prob}[A] + \text{Prob}[B] - \text{Prob}[A \cap B]. \qquad (3.8)$$

If A and B are mutually exclusive events, Prob $[A \cap B] = 0$ and

$$\text{Prob}[A \cup B] = \text{Prob}[A] + \text{Prob}[B]. \qquad (3.9)$$

(5) For any events A and B,

$$\text{Prob}[A] = \text{Prob}[A \cap B] + \text{Prob}[A \cap B']. \qquad (3.10)$$

(6) For any events A, B, and C,

$$\begin{aligned} \text{Prob}\,[A \cup B \cup C] =& \text{Prob}\,[A] + \text{Prob}\,[B] + \text{Prob}\,[C] \\ & - \text{Prob}\,[A \cap B] - \text{Prob}\,[A \cap C] - \text{Prob}\,[B \cap C] \\ & + \text{Prob}\,[A \cap B \cap C]. \end{aligned} \qquad (3.11)$$

(7) For any events $A_1, A_2, \ldots, A_n$,

$$\text{Prob}\left[\bigcup_{i=1}^{n} A_i\right] \le \sum_{i=1}^{n} \text{Prob}\,[A_i]. \qquad (3.12)$$

Equality holds if the events are pairwise mutually exclusive.

3.3.5 Probability and odds

If the probability of an event A is $\text{Prob}\,[A]$ then

$$\begin{aligned} \text{odds for } A &= \text{Prob}\,[A]/\text{Prob}\,[A'], \quad \text{Prob}\,[A'] \neq 0 \\ \text{odds against } A &= \text{Prob}\,[A']/\text{Prob}\,[A], \quad \text{Prob}\,[A] \neq 0. \end{aligned} \qquad (3.13)$$

If the odds for the event A are $a{:}b$, then $\text{Prob}\,[A] = a/(a+b)$.

Example 3.14: The odds of a fair coin coming up heads are 1:1; that it, is has a probability of $1/2$.

The odds of a die showing a "1" are 5:1 against; that it, there is a probability of $5/6$ that a "1" does not appear.

3.3.6 Conditional probability

The conditional probability of A given the event B has occurred is

$$\text{Prob}\,[A\,|\,B] = \frac{\text{Prob}\,[A \cap B]}{\text{Prob}\,[B]}, \quad \text{Prob}\,[B] > 0. \qquad (3.14)$$

(1) If $\text{Prob}\,[A_1 \cap A_2 \cap \cdots \cap A_{n-1}] > 0$ then

$$\begin{aligned} \text{Prob}\,[A_1 \cap A_2 \cap \cdots \cap A_n] =& \text{Prob}\,[A_1] \cdot \text{Prob}\,[A_2\,|\,A_1] \\ & \cdot \text{Prob}\,[A_3\,|\,A_1 \cap A_2] \\ & \cdots \text{Prob}\,[A_n\,|\,A_1 \cap A_2 \cap \cdots \cap A_{n-1}]. \end{aligned} \qquad (3.15)$$

(2) If $A \subset B$, then $\text{Prob}\,[A\,|\,B] = \text{Prob}\,[A]/\text{Prob}\,[B]$ and $\text{Prob}\,[B\,|\,A] = 1$.

(3) $\text{Prob}\,[A'\,|\,B] = 1 - \text{Prob}\,[A\,|\,B]$.

Example 3.15: A local bank offers loans for three purposes: home (H), automobile (A), and personal (P), and two different types: fixed rate (FR) and adjustable rate (ADJ). The joint probability table given below presents the proportions for the various categories of loan and type:

		Loan Purpose		
		H	A	P
Type	FR	.27	.19	.14
	ADJ	.13	.09	.18

Suppose a person who took out a loan at this bank is selected at random.

(a) What is the probability the person has an automobile loan and it is fixed rate?

(b) Given the person has an adjustable rate loan, what is the probability it is for a home?

(c) Given the person does not have a personal loan, what is the probability it is adjustable rate?

Solution:

(S1) $\text{Prob}[A \cap FR] = .19$

(S2) $\text{Prob}[H \mid ADJ] = \text{Prob}[H \cap ADJ]/\text{Prob}[ADJ] = .13/.4 = .325$

(S3) $\text{Prob}[ADJ \mid P'] = \text{Prob}[ADJ \cap P']/\text{Prob}[P'] = .22/.68 = .3235$

3.3.7 The multiplication rule

$$\begin{aligned}
\text{Prob}[A \cap B] &= \text{Prob}[A \mid B] \cdot \text{Prob}[B], & \text{Prob}[B] \neq 0 \\
&= \text{Prob}[B \mid A] \cdot \text{Prob}[A], & \text{Prob}[A] \neq 0
\end{aligned} \tag{3.16}$$

3.3.8 The law of total probability

Suppose $A_1, A_2, \ldots, A_n$ is a collection of mutually exclusive, exhaustive events, $\text{Prob}[A_i] \neq 0$, $i = 1, 2, \ldots, n$. For any event B:

$$\text{Prob}[B] = \sum_{i=1}^{n} \text{Prob}[B \mid A_i] \cdot \text{Prob}[A_i]. \tag{3.17}$$

Example 3.16: *A ball drawing strategy.* There are two urns. A marked ball may be in urn 1 (with probability p) or urn 2 (with probability $1 - p$). The probability of drawing the marked ball from the urn it is in is r (with $r < 1$). After a ball is drawn from an urn, it is replaced. What is the best way to use n draws of balls from any urn so that the probability of drawing the marked ball is largest?

Solution:

(S1) Let the event of selecting the marked ball be A.

(S2) Let H_i be the hypothesis that the marked ball is in urn i.

(S3) By assumption, $\text{Prob}[H_1] = p$ and $\text{Prob}[H_2] = 1 - p$.

(S4) Choose m balls from urn 1, and $n - m$ balls from urn 2. The conditional probabilities are then:

$$\text{Prob}[A \mid H_1] = 1 - (1 - r)^m, \qquad \text{Prob}[A \mid H_2] = 1 - (1 - r)^{n-m} \tag{3.18}$$

so that (using the law of total probability)

$$\text{Prob}[A] = \text{Prob}[H_1] \cdot \text{Prob}[A \mid H_1] + \text{Prob}[H_2] \cdot \text{Prob}[A \mid H_2]$$
$$= p[1 - (1 - r)^m] + (1 - p)[1 - (1 - r)^{n-m}].$$

(3.19)

(S5) Differentiating this with respect to m, and setting $\frac{d\text{Prob}[A]}{dm} = 0$ results in $(1 - r)^{2m-n} = (1 - p)/p$ or

$$m = \frac{n}{2} + \frac{\ln\left(\frac{1-p}{p}\right)}{2\ln(1 - r)}.$$

(3.20)

3.3.9 Bayes' theorem

Suppose $A_1, A_2, \ldots, A_n$ is a collection of mutually exclusive, exhaustive events, $\text{Prob}[A_i] \neq 0$, $i = 1, 2, \ldots, n$. For any event B such that $\text{Prob}[B] \neq 0$:

$$\text{Prob}[A_k \mid B] = \frac{\text{Prob}[A_k \cap B]}{\text{Prob}[B]} = \frac{\text{Prob}[B \mid A_k] \cdot \text{Prob}[A_k]}{\sum_{i=1}^{n} \text{Prob}[B \mid A_i] \cdot \text{Prob}[A_i]},$$

(3.21)

for $k = 1, 2, \ldots, n$.

Example 3.17: A large manufacturer uses three different trucking companies (A, B, and C) to deliver products. The probability a randomly selected shipment is delivered by each company is

$$\text{Prob}[A] = .60, \quad \text{Prob}[B] = .25, \quad \text{Prob}[C] = .15$$

Occasionally, a shipment is damaged (D) in transit.

$$\text{Prob}[D \mid A] = .01, \quad \text{Prob}[D \mid B] = .005, \quad \text{Prob}[D \mid C] = .015$$

Suppose a shipment is selected at random.
   (a) Find the probability the shipment is sent by trucking company B and is damaged.
   (b) Find the probability the shipment is damaged.
   (c) Suppose a shipment arrives damaged. What is the probability it was shipped by company B?

Solution:
(S1) $\text{Prob}[B \cap D] = \text{Prob}[B] \cdot \text{Prob}[D \mid B] = (.25)(.005) = .00125$
(S2) $\text{Prob}[D] = \text{Prob}[A \cap D] + \text{Prob}[B \cap D] + \text{Prob}[C \cap D]$
$$= \text{Prob}[A] \cdot \text{Prob}[D \mid A] + \text{Prob}[B] \cdot \text{Prob}[D \mid B]$$
$$+ \text{Prob}[B] \cdot \text{Prob}[D \mid B]$$
$$= (.60)(.01) + (.25)(.005) + (.15)(.015) = .0095$$
(S3) $\text{Prob}[B \mid D] = \text{Prob}[B \cap D]/\text{Prob}[D] = .00125/.0095 = .1316$

3.3.10 Independence

(1) A and B are independent events if $\text{Prob}[A|B] = \text{Prob}[A]$ or, equivalently, if $\text{Prob}[B|A] = \text{Prob}[B]$.

(2) A and B are independent events if and only if $\text{Prob}[A \cap B] = \text{Prob}[A] \cdot \text{Prob}[B]$.

(3) $A_1, A_2, \ldots, A_n$ are pairwise independent events if

$$\text{Prob}[A_i \cap A_j] = \text{Prob}[A_i] \cdot \text{Prob}[A_j] \quad \text{for every pair } i, j \text{ with } i \neq j. \tag{3.22}$$

(4) $A_1, A_2, \ldots, A_n$ are mutually independent events if for every k, $k = 2, 3, \ldots, n$, and every subset of indices $i_1, i_2, \ldots, i_k$,

$$\text{Prob}[A_{i_1} \cap A_{i_2} \cap \cdots \cap A_{i_k}] = \text{Prob}[A_{i_1}] \cdot \text{Prob}[A_{i_2}] \cdots \text{Prob}[A_{i_k}]. \tag{3.23}$$

3.4   RANDOM VARIABLES

Given a sample space S, a random variable is a function with domain S and range some subset of the real numbers. A random variable is *discrete* if it can assume only a finite or countably infinite number of values. A random variable is *continuous* if its set of possible values is an entire interval of numbers. Random variables are denoted by upper-case letters, for example X.

3.4.1   Discrete random variables

3.4.1.1   Probability mass function

The probability distribution or probability mass function (pmf), $p(x)$, of a discrete random variable is a rule defined for every number x by $p(x) = \text{Prob}[X = x]$ such that

(1) $p(x) \geq 0$; and

(2) $\sum_x p(x) = 1$

3.4.1.2   Cumulative distribution function

The cumulative distribution function (cdf), $F(x)$, for a discrete random variable X with pmf $p(x)$ is defined for every number x:

$$F(x) = \text{Prob}[X \leq x] = \sum_{y|y \leq x} p(y). \tag{3.24}$$

(1) $\lim_{x \to -\infty} F(x) = 0$

(2) $\lim_{x \to \infty} F(x) = 1$

(3) If a and b are real numbers such that $a < b$, then $F(a) \leq F(b)$.

(4) $\text{Prob}[a \leq X \leq b] = \text{Prob}[X \leq b] - \text{Prob}[X < a] = F(b) - F(a^-)$ where a^- is the first value X assumes less than a. Valid for $a, b, \in \mathcal{R}$ and $a < b$.

3.4.2   Continuous random variables

3.4.2.1   Probability density function

The probability distribution or probability density function (pdf) of a continuous random variable X is a real-valued function $f(x)$ such that

$$\text{Prob}\,[a \le X \le b] = \int_a^b f(x)\,dx, \quad a, b \in \mathcal{R}, \quad a \le b. \tag{3.25}$$

(1) $f(x) \ge 0$ for $-\infty < x < \infty$

(2) $\displaystyle\int_{-\infty}^{\infty} f(x)\,dx = 1$

(3) $\text{Prob}\,[X = c] = 0$ for $c \in \mathcal{R}$.

3.4.2.2   Cumulative distribution function

The cumulative distribution function (cdf), $F(x)$, for a continuous random
variable X is defined by

$$F(x) = \text{Prob}\,[X \le x] = \int_{-\infty}^{x} f(y)\,dy \quad -\infty < x < \infty. \tag{3.26}$$

(1) $\displaystyle\lim_{x \to -\infty} F(x) = 0$

(2) $\displaystyle\lim_{x \to \infty} F(x) = 1$

(3) If a and b are real numbers such that $a < b$, then $F(a) \le F(b)$.

(4) $\text{Prob}\,[a \le X \le b] = \text{Prob}\,[X \le b] - \text{Prob}\,[X < a] = F(b) - F(a)$,
    $a, b, \in \mathcal{R}$ and $a < b$.

(5) The pdf $f(x)$ may be found from the cdf:

$$f(x) = \frac{dF(x)}{dx} \quad \text{whenever the derivative exists.} \tag{3.27}$$

3.4.3   Random functions

A random function of a real variable t is a function, denoted $X(t)$, that is a
random variable for each value of t. If the variable t can assume any value in
an interval, then $X(t)$ is called a *stochastic process*; if the variable t can only
assume discrete values then $X(t)$ is called a *random sequence*.

3.5   MATHEMATICAL EXPECTATION

3.5.1   Expected value

(1) If X is a discrete random variable with pmf $p(x)$:

    (a) The expected value of X is

$$\text{E}\,[X] = \mu = \sum_x xp(x), \tag{3.28}$$

(b) The expected value of a function $g(X)$ is

$$E\left[g(X)\right] = \mu_{g(X)} = \sum_x g(x)p(x)\,. \tag{3.29}$$

(2) If X is a continuous random variable with pdf $f(x)$:

(a) The expected value of X is

$$E\left[X\right] = \mu = \int_{-\infty}^{\infty} xf(x)\,dx\,, \tag{3.30}$$

(b) The expected value of a function $g(X)$ is

$$E\left[g(X)\right] = \mu_{g(X)} = \int_{-\infty}^{\infty} g(x)f(x)\,dx\,. \tag{3.31}$$

(3) Jensen's inequality

Let $h(x)$ be a function such that $\dfrac{d^2}{dx^2}\left[h(x)\right] \geq 0$, then
$E\left[h(X)\right] \geq h(E\left[X\right])$.

(4) Theorems:

(a) $E\left[aX + bY\right] = aE\left[X\right] + bE\left[Y\right]$

(b) $E\left[X \cdot Y\right] = E\left[X\right] \cdot E\left[Y\right]$ if X and Y are independent.

3.5.2 Variance

The variance of a random variable X is

$$\sigma^2 = E\left[(X - \mu)^2\right] = \begin{cases} \displaystyle\sum_x (x - \mu)^2 p(x) & \text{if } X \text{ is discrete} \\[2ex] \displaystyle\int_{-\infty}^{\infty} (x - \mu)^2 f(x)\,dx & \text{if } X \text{ is continuous} \end{cases} \tag{3.32}$$

The standard deviation of X is $\sigma = \sqrt{\sigma^2}$.

3.5.2.1 Theorems

Suppose X is a random variable, and a, b are constants.

(1) $\sigma_X^2 = E\left[X^2\right] - (E\left[X\right])^2$.

(2) $\sigma_{aX}^2 = a^2 \cdot \sigma_X^2$, $\qquad \sigma_{aX} = |a| \cdot \sigma_X$.

(3) $\sigma_{X+b}^2 = \sigma_X^2$.

(4) $\sigma_{aX+b}^2 = a^2 \cdot \sigma_X^2$, $\qquad \sigma_{aX+b} = |a| \cdot \sigma_X$.

3.5.3 Moments

3.5.3.1 Moments about the origin

The moments about the origin completely characterize a probability distribution. The rth moment about the origin, $r = 0, 1, 2, \ldots$, of a random variable

X is

$$\mu'_r = \mathrm{E}\left[X^r\right] = \begin{cases} \displaystyle\sum_x x^r p(x) & \text{if } X \text{ is discrete} \\[2ex] \displaystyle\int_{-\infty}^{\infty} x^r f(x)\, dx & \text{if } X \text{ is continuous} \end{cases} \tag{3.33}$$

The first moment about the origin is the mean of the random variable: $\mu'_1 = \mathrm{E}\left[X\right] = \mu$.

3.5.3.2   Moments about the mean

The rth moment about the mean, $r = 0, 1, 2, \ldots$, of a random variable X is

$$\mu_r = \mathrm{E}\left[(X - \mu)^r\right] = \begin{cases} \displaystyle\sum_x (x - \mu)^r p(x) & \text{if } X \text{ is discrete} \\[2ex] \displaystyle\int_{-\infty}^{\infty} (x - \mu)^r f(x)\, dx & \text{if } X \text{ is continuous} \end{cases} \tag{3.34}$$

The second moment about the mean is the variance of the random variable: $\mu_2 = \mathrm{E}\left[(X - \mu)^r\right] = \sigma^2 = \mu'_2 - \mu^2$.

3.5.3.3   Factorial moments

The rth factorial moment, $r = 0, 1, 2, \ldots$, of a random variable is

$$\mu_{[r]} = \mathrm{E}\left[X^{[r]}\right] = \begin{cases} \displaystyle\sum_x x^{[r]} p(x) & \text{if } X \text{ is discrete} \\[2ex] \displaystyle\int_{-\infty}^{\infty} x^{[r]} f(x)\, dx & \text{if } X \text{ is continuous} \end{cases} \tag{3.35}$$

where $x^{[r]}$ is the factorial expression

$$x^{[r]} = x(x - 1)(x - 2)\cdots(x - r + 1). \tag{3.36}$$

3.5.4   Generating functions

3.5.4.1   Moment generating function

The moment generating function (mgf) of a random variable X, where it exists, is

$$m_X(t) = \mathrm{E}\left[e^{tX}\right] = \begin{cases} \displaystyle\sum_x e^{tx} p(x) & \text{if } X \text{ is discrete} \\[2ex] \displaystyle\int_{-\infty}^{\infty} e^{tx} f(x)\, dx & \text{if } X \text{ is continuous} \end{cases} \tag{3.37}$$

The moment generating function $m_X(t)$ is the expected value of e^{tX} and may be written as

$$m_X(t) = \mathrm{E}\left[e^{tX}\right]$$

$$= \mathrm{E}\left[1 + Xt + \frac{(Xt)^2}{2!} + \frac{(Xt)^3}{3!} + \cdots\right] \tag{3.38}$$

$$= 1 + \mu'_1 t + \mu'_2 \frac{t^2}{2!} + \mu'_3 \frac{t^3}{3!} + \cdots$$

The moments μ'_r are the coefficients of $t^r/r!$ in equation (3.38). Therefore, $m_X(t)$ *generates* the moments since the rth derivative of $m_X(t)$ evaluated at $t = 0$ yields μ'_r:

$$\mu'_r = m_x^{(r)}(0) = \left.\frac{d^r m_X(t)}{dt^r}\right|_{t=0} \tag{3.39}$$

Theorems: Suppose $m_X(t)$ is the moment generating function for the random variable X and a, b are constants.

(1) $m_{aX}(t) = m_X(at)$

(2) $m_{X+b}(t) = e^{bt} \cdot m_X(t)$

(3) $m_{(X+b)/a}(t) = e^{(b/a)t} \cdot m_X(t/a)$

(4) If $X_1, X_2, \ldots, X_n$ are independent random variables and $Y = X_1 + X_2 + \cdots + X_n$, then $m_Y(t) = [m_X(t)]^n$.

The moment generating function for $X - \mu$ is

$$m_{X-\mu}(t) = e^{-\mu t} \cdot m_X(t). \tag{3.40}$$

Equation (3.40) may be used to generate the moments about the mean for the random variable X:

$$\mu_r = m_{X-\mu}^{(r)}(0) = \left.\frac{d^r \left(e^{-\mu t} \cdot m_X(t)\right)}{dt^r}\right|_{t=0} \tag{3.41}$$

3.5.4.2 *Factorial moment generating functions*

The factorial moment generating function of a random variable X is

$$P(t) = \mathrm{E}\left[t^X\right] = \begin{cases} \displaystyle\sum_x t^x p(x) & \text{if } X \text{ is discrete} \\ \displaystyle\int_{-\infty}^{\infty} t^x f(x)\, dx & \text{if } X \text{ is continuous} \end{cases} \tag{3.42}$$

The rth derivative of the function P (in equation (3.42)) with respect to t, evaluated at $t = 1$ is the rth factorial moment. Therefore, the function P

generates the factorial moments:

$$\mu_{[r]} = P^{(r)}(1) = \left.\frac{d^r P(t)}{dt^r}\right|_{t=1}.\tag{3.43}$$

In particular:

$$1 = P(1) \quad \text{"conservation of probability"}$$
$$\mu = P'(1)\tag{3.44}$$
$$\sigma^2 = P''(1) + P'(1) - [P'(1)]^2$$

3.5.4.3 Factorial moment generating function theorems

Theorems: Suppose $P_X(t)$ is the factorial moment generating function for the random variable X and a, b are constants.

(1) $P_{aX}(t) = P_X(t^a)$

(2) $P_{X+b}(t) = t^b \cdot P_X(t)$

(3) $P_{(X+b)/a}(t) = t^{b/a} \cdot P_X(t^{1/a})$

(4) $P_X(t) = m_X(\ln t)$, where $m_x(t)$ is the moment generating function for X.

(5) If $X_1, X_2, \ldots, X_n$ are *independent* random variables with factorial moment generating function $P_X(t)$ and $Y = X_1 + X_2 + \cdots + X_n$, then $P_Y(t) = [P_X(t)]^n$.

3.5.4.4 Cumulant generating function

Let $m_X(t)$ be a moment generating function. If $\ln m_X(t)$ can be expanded in the form

$$c(t) = \ln m_X(t) = \kappa_1 t + \kappa_2 \frac{t^2}{2!} + \kappa_3 \frac{t^3}{3!} + \cdots + \kappa_r \frac{t^r}{r!} + \cdots,\tag{3.45}$$

then $c(t)$ is the cumulant generating function (or semi–invariant generating function). The constants κ_r are the cumulants (or semi–invariants) of the distribution. The rth derivative of c with respect to t, evaluated at 0 is the rth cumulant. The function c *generates* the cumulants:

$$\kappa_r = c^{(r)}(0) = \left.\frac{d^r c(t)}{dt^r}\right|_{t=0}.\tag{3.46}$$

Marcienkiewicz's theorem states that either all but the first two cumulants vanish (i.e., it is a normal distribution) or there are an infinite number of non-vanishing cumulants.

3.5.4.5 Characteristic function

The characteristic function exists for every random variable X and is defined by

$$\phi(t) = \mathrm{E}\left[e^{itX}\right] = \begin{cases} \displaystyle\sum_x e^{itx} p(x) & \text{if } X \text{ is discrete} \\[2mm] \displaystyle\int_{-\infty}^{\infty} e^{itx} f(x)\,dx & \text{if } X \text{ is continuous} \end{cases} \tag{3.47}$$

where t is a real number and $i^2 = -1$. The rth derivative of ϕ with respect to t, evaluated at $t = 0$ is $i^r \mu_r'$. Therefore, the characteristic function also generates the moments:

$$i^r \mu_r' = \phi^{(r)}(0) = \left. \frac{d^r \phi(t)}{dt^r} \right|_{t=0}. \tag{3.48}$$

3.6 MULTIVARIATE DISTRIBUTIONS

Note that the specialization to bivariate distributions is on page 45.

3.6.1 Discrete case

A n-dimensional random variable $(X_1, X_2, \ldots, X_n)$ is n-dimensional discrete if it can assume only a finite or countably infinite number of values. The joint probability distribution, joint probability mass function, or joint density, for $(X_1, X_2, \ldots, X_n)$ is

$$p(x_1, x_2, \ldots, x_n) = \mathrm{Prob}\left[X_1 = x_1, X_2 = x_2, \ldots, X_n = x_n\right]$$
$$\forall (x_1, x_2, \ldots, x_n). \tag{3.49}$$

Suppose E is a subset of values the random variable may assume. The probability the event E occurs is

$$\mathrm{Prob}\,[E] = \mathrm{Prob}\left[(X_1, X_2, \ldots, X_n) \in E\right]$$
$$= \sum\sum \cdots \sum_{(x_1, x_2, \ldots, x_n) \in E} p(x_1, x_2, \ldots, x_n). \tag{3.50}$$

The cumulative distribution function for $(X_1, X_2, \ldots, X_n)$ is

$$F(x_1, x_2, \ldots, x_n) = \sum_{t_1 \mid t_1 \leq x_1} \sum_{t_2 \mid t_2 \leq x_2} \cdots \sum_{t_n \mid t_n \leq x_n} p(x_1, x_2, \ldots, x_n). \tag{3.51}$$

3.6.2 Continuous case

The continuous random variables $X_1, X_2, \ldots, X_n$ are jointly distributed if there exists a function f such that $f(x_1, x_2, \ldots, x_n) \geq 0$ for $-\infty < x_i < \infty$,

$i = 1, 2, \ldots, n$, and for any event E

$$\text{Prob}\,[E] = \text{Prob}\,[(X_1, X_2, \ldots, X_n) \in E]$$
$$= \int\!\!\int \cdots \int_E f(x_1, x_2, \ldots, x_n)\, dx_n \cdots dx_1 \tag{3.52}$$

where f is the joint distribution function or joint probability density function for the random variables $X_1, X_2, \ldots, X_n$. The cumulative distribution function for $X_1, X_2, \ldots, X_n$ is

$$F(x_1, x_2, \ldots, x_n) = \int_{-\infty}^{x_1} \int_{-\infty}^{x_2} \cdots \int_{-\infty}^{x_n} f(x_1, x_2, \ldots, x_n)\, dx_n \cdots dx_1 . \tag{3.53}$$

Given the cumulative distribution function, F, the probability density function may be found by

$$f(x_1, x_2, \ldots, x_n) = \frac{\partial^n}{\partial x_1\, \partial x_2 \cdots \partial x_n} F(x_1, x_2, \ldots, x_n) \tag{3.54}$$

wherever the partials exist.

3.6.3   Expectation

Let $g(X_1, X_2, \ldots, X_n)$ be a function of the random variables $X_1, \ldots, X_n$. The expected value of $g(X_1, X_2, \ldots, X_n)$ is

$$E\,[g(X_1, X_2, \ldots, X_n)] = \sum_{x_1} \sum_{x_2} \cdots \sum_{x_n} g(x_1, x_2, \ldots, x_n) p(x_1, x_2, \ldots, x_n) \tag{3.55}$$

if $X_1, X_2, \ldots, X_n$ are discrete, and

$$E\,[g(X_1, X_2, \ldots, X_n)] =$$
$$\int_{-\infty}^{\infty} \int_{-\infty}^{\infty} \cdots \int_{-\infty}^{\infty} g(x_1, \ldots, x_n) f(x_1, \ldots, x_n)\, dx_n \cdots dx_1 \tag{3.56}$$

if $X_1, X_2, \ldots, X_n$ are continuous.

If $c_1, c_2, \ldots, c_n$ are constants, then

$$E\left[\sum_{i=1}^{n} c_i g_i(X_1, X_2, \ldots, X_n)\right] = \sum_{i=1}^{n} c_i E\,[g_i(X_1, X_2, \ldots, X_n)] . \tag{3.57}$$

3.6.4   Moments

If $X_1, X_2, \ldots, X_n$ are jointly distributed, the rth moment of X_i is

$$E\,[X_i^r] = \sum_{x_1} \sum_{x_2} \cdots \sum_{x_n} x_i^r p(x_1, x_2, \ldots, x_n) \tag{3.58}$$

if $X_1, X_2, \ldots, X_n$ are discrete, and

$$E\,[X_i^r] = \int_{-\infty}^{\infty} \int_{-\infty}^{\infty} \cdots \int_{-\infty}^{\infty} x_i^r f(x_1, x_2, \ldots, x_n)\, dx_n \cdots dx_1 \tag{3.59}$$

if $X_1, X_2, \ldots, X_n$ are continuous.

The joint (product) moments about the origin are

$$E[X_1^{r_1} X_2^{r_2} \cdots X_n^{r_n}] = \sum_{x_1} \sum_{x_2} \cdots \sum_{x_n} x_1^{r_1} x_2^{r_2} \cdots x_n^{r_n} p(x_1, x_2, \ldots, x_n) \quad (3.60)$$

if $X_1, X_2, \ldots, X_n$ are discrete, and

$$E[X_1^{r_1} X_2^{r_2} \cdots X_n^{r_n}]$$
$$= \int_{-\infty}^{\infty} \int_{-\infty}^{\infty} \cdots \int_{-\infty}^{\infty} x_1^{r_1} x_2^{r_2} \cdots x_n^{r_n} f(x_1, x_2, \ldots, x_n) \, dx_n \cdots dx_1 \quad (3.61)$$

if $X_1, X_2, \ldots, X_n$ are continuous. The value $r = r_1 + r_2 + \cdots + r_n$ is the *order* of the moment.

If $E[X_i] = \mu_i$, then the joint moments about the mean are

$$E[(X_1 - \mu_1)^{r_1} (X_2 - \mu_2)^{r_2} \cdots (X_n - \mu_n)^{r_n}] =$$
$$\sum_{x_1} \sum_{x_2} \cdots \sum_{x_n} (x_1 - \mu_1)^{r_1} (x_2 - \mu_2)^{r_1} \cdots (x_n - \mu_n)^{r_n} p(x_1, x_2, \ldots, x_n) \quad (3.62)$$

if the $X_1, X_2, \ldots, X_n$ are discrete, and

$$E[(X_1 - \mu_1)^{r_1} (X_2 - \mu_2)^{r_2} \cdots (X_n - \mu_n)^{r_n}] =$$
$$\int_{-\infty}^{\infty} \int_{-\infty}^{\infty} \cdots \int_{-\infty}^{\infty} (x_1 - \mu_1)^{r_1} \cdots (x_n - \mu_n)^{r_n} f(x_1, \ldots, x_n) \, dx_n \cdots dx_1 , \quad (3.63)$$

if the $X_1, X_2, \ldots, X_n$ are continuous.

3.6.5   Marginal distributions

Let $X_1, X_2, \ldots, X_n$ be a collection of random variables. The marginal distribution of a subset of the random variables $X_1, X_2, \ldots, X_k$ (with $(k < n)$) is

$$g(x_1, x_2, \ldots, x_k) = \sum_{x_{k+1}} \sum_{x_{k+2}} \cdots \sum_{x_n} p(x_1, x_2, \ldots, x_n) \quad (3.64)$$

if $X_1, X_2, \ldots, X_n$ are discrete, and

$$g(x_1, x_2, \ldots, x_k) = \int_{-\infty}^{\infty} \int_{-\infty}^{\infty} \cdots \int_{-\infty}^{\infty} f(x_1, x_2, \ldots, x_n) \, dx_{k+1} dx_{k+2} \cdots dx_n \quad (3.65)$$

if $X_1, X_2, \ldots, X_n$ are continuous.

Example 3.18:   The joint density functions $g(x, y) = x+y$ and $h(x, y) = (x+\frac{1}{2})(y+\frac{1}{2})$ when $0 \leq x \leq 1$ and $0 \leq y \leq 1$ have the same marginal distributions. Using equation

(3.65):

$$g_x(x) = \int_0^1 g(x,y)\,dy = \left(xy + \frac{y^2}{2}\right)\Bigg|_{y=0}^{y=1} = x + \frac{1}{2}$$

$$h_x(x) = \int_0^1 h(x,y)\,dy = \left(x + \frac{1}{2}\right)\left(\frac{y^2}{2} + \frac{y}{2}\right)\Bigg|_{y=0}^{y=1} = x + \frac{1}{2} \tag{3.66}$$

and, by symmetry, $g_y(y)$ has the same form as $g_x(x)$ (likewise for $h_y(y)$ and $h_x(x)$).

3.6.6   Independent random variables

Let $X_1, X_2, \ldots, X_n$ be a collection of discrete random variables with joint probability distribution function $p(x_1, x_2, \ldots, x_n)$. Let $g_{X_i}(x_i)$ be the marginal distribution for X_i. The random variables $X_1, X_2, \ldots, X_n$ are independent if and only if

$$p(x_1, x_2, \ldots, x_n) = g_{X_1}(x_1) \cdot g_{X_2}(x_2) \cdots g_{X_n}(x_n). \tag{3.67}$$

Let $X_1, X_2, \ldots, X_n$ be a collection of continuous random variables with joint probability distribution function $f(x_1, x_2, \ldots, x_n)$. Let $g_{X_i}(x_i)$ be the marginal distribution for X_i. The random variables $X_1, X_2, \ldots, X_n$ are independent if and only if

$$f(x_1, x_2, \ldots, x_n) = g_{X_1}(x_1) \cdot g_{X_2}(x_2) \cdots g_{X_n}(x_n). \tag{3.68}$$

Example 3.19:   Suppose X_1, X_2, and X_3 are independent random variables with probability density functions given by

$$g_{X_1}(x_1) = \begin{cases} e^{-x_1} & x_1 > 0 \\ 0 & \text{elsewhere} \end{cases}$$

$$g_{X_2}(x_2) = \begin{cases} 3e^{-3x_2} & x_2 > 0 \\ 0 & \text{elsewhere} \end{cases} \qquad g_{X_3}(x_3) = \begin{cases} 7e^{-7x_3} & x_3 > 0 \\ 0 & \text{elsewhere} \end{cases}$$

Using equation (3.68), the joint probability distribution for X_1, X_2, X_3 is

$$\begin{aligned} f(x_1, x_2, x_3) &= g_{X_1}(x_1) \cdot g_{X_2}(x_2) \cdot g_{X_3}(x_3) \\ &= (e^{-x_1}) \cdot (3e^{-3x_2}) \cdot (7e^{-7x_3}) \\ &= 21e^{-x_1 - 3x_2 - 7x_3} \qquad [x_1 > 0, x_2 > 0, x_3 > 0]. \end{aligned}$$

3.6.7   Conditional distributions

Let $X_1, X_2, \ldots, X_n$ be a collection of random variables. The conditional distribution of any subset of the random variables $X_1, X_2, \ldots, X_k$ given $X_{k+1} = x_{k+1}, X_{k+2} = x_{k+2}, \ldots, X_n = x_n$ is

$$p(x_1, x_2, \ldots, x_k \,|\, x_{k+1}, x_{k+2} \ldots, x_n) = \frac{p(x_1, x_2, \ldots, x_n)}{g(x_{k+1}, x_{k+2}, \ldots, x_n)} \tag{3.69}$$

if $X_1, X_2, \ldots, X_n$ are discrete with joint distribution function $p(x_1, x_2, \ldots, x_n)$ and $X_{k+1}, X_{k+2}, \ldots, X_n$ have marginal distribution $g(x_{k+1}, x_{k+2}, \ldots, x_n) \neq 0$, and

$$f(x_1, x_2, \ldots, x_k \mid x_{k+1}, x_{k+2}, \ldots, x_n) = \frac{f(x_1, x_2, \ldots, x_n)}{g(x_{k+1}, x_{k+2}, \ldots, x_n)} \tag{3.70}$$

if $X_1, X_2, \ldots, X_n$ are continuous with joint distribution function $f(x_1, x_2, \ldots, x_n)$ and $X_{k+1}, X_{k+2}, \ldots, X_n$ have marginal distribution $g(x_{k+1}, x_{k+2}, \ldots, x_n) \neq 0$.

Example 3.20: Suppose X_1, X_2, X_3 have a joint distribution function given by

$$f(x_1, x_2, x_3) = \begin{cases} (x_1 + x_2)e^{-x_3} & \text{when } 0 < x_1 < 1, \ 0 < x_2 < 1, \ x_3 > 0 \\ 0 & \text{elsewhere} \end{cases}$$

The marginal distribution of X_2 is

$$g(x_2) = \int_0^1 \int_0^\infty (x_1 + x_2)e^{-x_3} \, dx_3 \, dx_1$$

$$= \int_0^1 (x_1 + x_2) \, dx_1 = \frac{1}{2} + x_2, \quad 0 < x_2 < 1.$$

The conditional distribution of X_1, X_3 given $X_2 = x_2$ is

$$f(x_1, x_3 \mid x_2) = \frac{f(x_1, x_2, x_3)}{g(x_2)} = \frac{(x_1 + x_2)e^{-x_3}}{\frac{1}{2} + x_2}.$$

If $X_2 = 3/4$, then

$$f(x_1, x_3 \mid 3/4) = \frac{\left(x_1 + \frac{3}{4}\right)e^{-x_3}}{\frac{1}{2} + \frac{3}{4}} = \frac{4}{5}\left(x_1 + \frac{3}{4}\right)e^{-x_3}, \quad 0 < x_1 < 1, \ x_3 > 0.$$

3.6.8 Variance and covariance

Let $X_1, X_2, \ldots, X_n$ be a collection of random variables. The variance, σ_{ii}, of X_i is

$$\sigma_{ii} = \sigma_i^2 = \mathrm{E}\left[(X_i - \mu_i)^2\right] \tag{3.71}$$

and the covariance, σ_{ij}, of X_i and X_j is

$$\sigma_{ij} = \rho_{ij}\sigma_i\sigma_j = \mathrm{E}[(X_i - \mu_i)(X_j - \mu_j)] \tag{3.72}$$

where ρ_{ij} is the correlation coefficient and σ_i and σ_j are the standard deviations of X_i and X_j, respectively.

Theorems:

(1) If $X_1, X_2, \ldots, X_n$ are independent, then

$$\mathrm{E}[X_1 X_2 \cdots X_n] = \mathrm{E}[X_1]\mathrm{E}[X_2] \cdots \mathrm{E}[X_n]. \tag{3.73}$$

(2) For two random variables X_i and X_j:

$$\sigma_{ij} = \mathrm{E}[X_i X_j] - \mathrm{E}[X_i]\mathrm{E}[X_j]. \tag{3.74}$$

(3) If X_i and X_j are independent random variables, then $\sigma_{ij} = 0$.

(4) Two variables may be dependent and have zero covariance. For example, let X take the four values $\{-2, -1, 1, 2\}$ with equal probability. If $Y = X^2$ then the covariance of X and Y is zero.

The *correlation function* of a random function (see page 34) is

$$K_X(t_1, t_2) = \mathrm{E}\left[[X^*(t_1) - \mu_{X^*}(t_1)][X(t_2) - \mu_X(t_2)]\right] \tag{3.75}$$

where * denotes the complex conjugate. If $X(t)$ is *stationary* then

$$K_X(t_1, t_2) = K_X(t_1 - t_2) \quad \text{and} \quad \mu_X(t) = \text{constant}. \tag{3.76}$$

3.6.9   Correlation coefficient

The correlation coefficient, defined by (see equation (3.72))

$$\rho_{ij} = \frac{\sigma_{ij}}{\sigma_i \sigma_j} \tag{3.77}$$

is no greater than one in magnitude: $|\rho_{ij}| \leq 1$. Figure 3.5 contains 4 data sets of 100 points each; the correlation coefficients vary from -0.7 to 0.99.

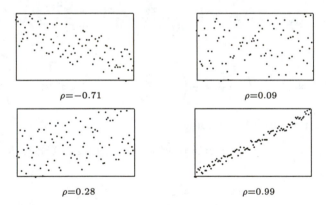

$\rho = -0.71$ $\rho = 0.09$

$\rho = 0.28$ $\rho = 0.99$

Figure 3.5: Data sets illustrating different correlation coefficients.

Example 3.21:   The correlation coefficient of the first 100 integers $\{1, 2, 3 \dots\}$ and the first 100 squares $\{1, 4, 9 \dots\}$ is 0.96885.

3.6.10   Moment generating function

Let $X_1, X_2, \dots, X_n$ be a collection of random variables. The joint moment generating function is

$$m(t_1, t_2, \dots, t_n) = m(\mathbf{t}) = \mathrm{E}\left[e^{t_1 X_1 + t_2 X_2 + \cdots + t_n X_n}\right] = \mathrm{E}\left[e^{\mathbf{t} \cdot \mathbf{X}}\right] \tag{3.78}$$

if it exists for all values of t_i such that $|t_i| < h^2$ (for some value h).

The rth moment of X_i may be obtained (generated) by differentiating $m(t_1, t_2, \ldots, t_n)$ r times with respect to t_i, and then evaluating the result with all t's equal to zero:

$$\mathrm{E}\left[X_i^r\right] = \left.\frac{\partial^r m(t_1, t_2, \ldots, t_n)}{\partial t_i^r}\right|_{(t_1, t_2, \ldots, t_n) = (0, 0, \ldots, 0)} \tag{3.79}$$

The rth joint moment, $r = r_1 + r_2 + \cdots + r_n$, may be obtained by differentiating $m(t_1, t_2, \ldots, t_n)$ r_1 times with respect to t_1, r_2 times with respect to t_2, $\ldots$, and r_n times with respect to t_n, and then evaluating the result with all t's equal to zero:

$$\mathrm{E}\left[X_1^{r_1} X_2^{r_2} \cdots X_n^{r_n}\right] = \left.\frac{\partial^r m(t_1, t_2, \ldots, t_n)}{\partial t_1^{r_1} \partial t_2^{r_2} \cdots \partial t_n^{r_n}}\right|_{(t_1, t_2, \ldots, t_n) = (0, 0, \ldots, 0)} \tag{3.80}$$

3.6.11 Linear combination of random variables

Let $X_1, X_2, \ldots, X_m$ and $Y_1, Y_2, \ldots, Y_n$ be random variables, let $a_1, a_2, \ldots, a_m$ and $b_1, b_2, \ldots, b_n$ be constants, and let U and V be the linear combinations

$$U = \sum_{i=1}^{m} a_i X_i, \qquad V = \sum_{j=1}^{n} b_j Y_j. \tag{3.81}$$

Theorems:

(1) $\mathrm{E}\left[U\right] = \displaystyle\sum_{i=1}^{m} a_i \mathrm{E}\left[X_i\right]$.

(2) $\sigma_i^2 = \displaystyle\sum_{i=1}^{m} a_i^2 \sigma_i^2 + 2 \sum_{i<j} \sum a_i a_j \sigma_{ij}$,

where the double sum extends over all pairs (i, j) with $i < j$.

(3) If the random variables $X_1, X_2, \ldots, X_m$ are independent, then

$$\sigma_U^2 = \sum_{i=1}^{m} a_i^2 \sigma_i^2.$$

(4) $\sigma_{UV} = \displaystyle\sum_{i=1}^{m} \sum_{j=1}^{n} a_i b_j \sigma_{ij}$.

3.6.12 Bivariate distribution

3.6.12.1 Joint probability distribution

(a) Discrete case

Let X and Y be discrete random variables. The joint probability distribution for X and Y is

$$p(x, y) = \mathrm{Prob}\left[X = x, Y = y\right] \qquad \forall (x, y). \tag{3.82}$$

(1) For any subset E consisting of pairs (x, y),

$$\text{Prob}\,[(X, Y) \in E] = \sum_{(x,y) \in E} \sum p(x, y). \qquad (3.83)$$

(2) $p(x, y) \geq 0 \qquad \forall (x, y).$

(3) $\displaystyle\sum_x \sum_y p(x, y) = 1.$

(b) Continuous case

Let X and Y be continuous random variables. The joint probability distribution for X and Y is a function $f(x, y)$ such that for any two-dimensional set E

$$\text{Prob}\,[(X, Y) \in E] = \iint_E f(x, y)\, dx\, dy \,. \qquad (3.84)$$

(1) If E is a rectangle $\{(x, y) \mid a \leq x \leq b, c \leq y \leq d\}$, then

$$\text{Prob}\,[(X, Y) \in E] = \text{Prob}\,[a \leq X \leq b, c \leq Y \leq d]$$
$$= \int_a^b \int_c^d f(x, y)\, dy\, dx \,. \qquad (3.85)$$

(2) $f(x, y) \geq 0 \qquad \forall (x, y).$

(3) $\displaystyle\int_{-\infty}^{\infty} \int_{-\infty}^{\infty} f(x, y)\, dx\, dy = 1.$

3.6.12.2 Cumulative distribution function

For any two random variables X and Y the cumulative distribution function is $F(x, y) = \text{Prob}\,[X \leq x, Y \leq y]$:

$$F(a, b) = \begin{cases} \displaystyle\sum_{x \mid x \leq a} \sum_{y \mid y \leq b} p(x, y) & \text{if } X \text{ and } Y \text{ are discrete} \\[2ex] \displaystyle\int_{-\infty}^{a} \int_{-\infty}^{b} f(x, y)\, dy\, dx & \text{if } X \text{ and } Y \text{ are continuous} \end{cases} \qquad (3.86)$$

Properties:

(1) $\displaystyle\lim_{(x,y) \to (-\infty, -\infty)} F(x, y) = \lim_{x \to -\infty} F(x, y) = \lim_{y \to -\infty} F(x, y) = 0.$

(2) $\displaystyle\lim_{(x,y) \to (\infty, \infty)} F(x, y) = 1.$

(3) If $a \leq b$ and $c \leq d$, then

$$\text{Prob}\,[a < X \leq b, c < Y \leq d] = F(b, d) - F(b, c) - F(a, d) + F(a, c)$$
$$\geq 0. \qquad (3.87)$$

(4) Given the cumulative distribution function, F, for the continuous random variables X and Y, the probability density function may be found

by

$$f(x,y) = \frac{\partial^2}{\partial x \partial y} F(x,y) \tag{3.88}$$

wherever the partials exist.

3.6.12.3 Marginal distributions

Let X and Y be discrete random variables with joint distribution $p(x,y)$. The marginal distributions for X and Y are

$$p_X(x) = \sum_y p(x,y), \qquad p_Y(y) = \sum_x p(x,y). \tag{3.89}$$

Let X and Y be continuous random variables with joint distribution $f(x,y)$. The marginal distributions for X and Y are

$$f_X(x) = \int_{-\infty}^{\infty} f(x,y)\,dy, \quad -\infty < x < \infty$$

$$f_Y(y) = \int_{-\infty}^{\infty} f(x,y)\,dx, \quad -\infty < y < \infty. \tag{3.90}$$

3.6.12.4 Conditional distributions

Let X and Y be discrete random variables with joint distribution $p(x,y)$ and let $p_Y(y)$ be the marginal distribution for Y. The conditional distribution for X given $Y = y$ is

$$p(x\,|\,y) = \frac{p(x,y)}{p_Y(y)}, \qquad p_Y(y) \neq 0. \tag{3.91}$$

Let $p_X(x)$ be the conditional distribution for X. The conditional distribution for Y given $X = x$ is

$$p(y\,|\,x) = \frac{p(x,y)}{p_X(x)}, \qquad p_X(x) \neq 0. \tag{3.92}$$

Let X and Y be continuous random variables with joint distribution $f(x,y)$ and let $f_Y(y)$ be the marginal distribution for Y. The conditional distribution for X given $Y = y$ is

$$f(x\,|\,y) = \frac{f(x,y)}{f_Y(y)}, \qquad f_Y(y) \neq 0. \tag{3.93}$$

Let $f_X(x)$ be the conditional distribution for X. The conditional distribution for Y given $X = x$ is

$$f(y\,|\,x) = \frac{f(x,y)}{f_X(x)}, \qquad f_X(x) \neq 0. \tag{3.94}$$

3.6.12.5 *Conditional expectation*

Let X and Y be random variables and let $g(X)$ be a function of X. The conditional expectation of $g(X)$ given $Y = y$ is

$$E\left[g(X)\,|y\right] = \begin{cases} \displaystyle\sum_x g(x)p(x\,|y) & \text{if } X \text{ and } Y \text{ are discrete} \\[2mm] \displaystyle\int_{-\infty}^{\infty} g(x)f(x\,|y)\,dx & \text{if } X \text{ and } Y \text{ are continuous} \end{cases} \qquad (3.95)$$

Properties:

(1) The conditional mean, or conditional expectation, of X given $Y = y$ is

$$\mu_{X|y} = E\left[X\,|y\right] = \begin{cases} \displaystyle\sum_x xp(x\,|y) & \text{if } X \text{ and } Y \text{ are discrete} \\[2mm] \displaystyle\int_{-\infty}^{\infty} xf(x\,|y)\,dx & \text{if } X \text{ and } Y \text{ are continuous} \end{cases} \qquad (3.96)$$

(2) The conditional variance of X given $Y = y$ is

$$\sigma_{X|y}^2 = E\left[(X - \mu_{X|y})^2\,|y\right] = E\left[X^2\,|y\right] - \mu_{X|y}^2. \qquad (3.97)$$

(3) $E[X] = E[E[X\,|Y]]$

3.7 INEQUALITIES

1. *Bienaymé–Chebyshev's inequality*: If $E\left[|X|^r\right] < \infty$ for all $r > 0$ (r not necessarily an integer) then, for every $a > 0$

$$\text{Prob}\left[|X| \geq a\right] \leq \frac{E\left[|X|^r\right]}{a^r} \qquad (3.98)$$

2. *Bienaymé–Chebyshev's inequality (generalized)*: Let $g(x)$ be a nondecreasing nonnegative function defined on $(0, \infty)$. Then, for $a \geq 0$,

$$\text{Prob}\left[|X| \geq a\right] \leq \frac{E\left[g(|X|)\right]}{g(a)} \qquad (3.99)$$

3. *Cauchy–Schwartz inequality*: Let X and Y be random variables in which $E\left[Y^2\right]$ and $E\left[Z^2\right]$ exist, then

$$(E[YZ])^2 \leq E\left[Y^2\right] E\left[Z^2\right] \qquad (3.100)$$

4. *Chebyshev inequality*: Let c be any real number and let X be a random variable for which $E\left[(X - c)^2\right]$ is finite. Then for every $\epsilon > 0$ the following holds

$$\text{Prob}\left[|X - c| \geq \epsilon\right] \leq \frac{1}{\epsilon^2} E\left[(X - c)^2\right] \qquad (3.101)$$

5. *Chebyshev inequality* (*one-sided*):   Let X be a random variable with zero mean (i.e., $E[X] = 0$) and variance σ^2. Then for any positive a

$$\text{Prob}[X > a] \leq \frac{\sigma^2}{\sigma^2 + a^2} \tag{3.102}$$

6. *Chernoff bound*:   This bound is useful for sums of random variables. Let $Y_n = \sum_{i=1}^{n} X_i$ where each of the X_i is iid. Let $m_X(t) = E[e^{tX}]$ be the common moment generating function for the $\{X_i\}$, and define $c(t) = \log m_X(t)$. Then

$$\begin{aligned}
\text{Prob}[Y_n \geq nc'(t)] &\leq e^{-n[tc'(t)-c(t)]} \qquad \text{if } t \geq 0 \\
\text{Prob}[Y_n \leq nc'(t)] &\leq e^{-n[tc'(t)-c(t)]} \qquad \text{if } t \leq 0
\end{aligned} \tag{3.103}$$

7. *Jensen's inequality*:   If $E[X]$ exists, and if $f(x)$ is a convex $\cup$ ("convex cup") function, then

$$E[f(X)] \geq f(E[X]) \tag{3.104}$$

8. *Kolmogorov's inequality*:   Let $X_1, X_2, \ldots, X_n$ be n independent random variables such that $E[X_i] = 0$ and $\text{Var}(X_i) = \sigma_{X_i}^2$ is finite. Then, for all $a > 0$,

$$\text{Prob}\left[\max_{i=1,\ldots,n} |X_1 + X_2 + \cdots + X_i| > a\right] \leq \sum_{i=1}^{n} \frac{\sigma_i^2}{a^2} \tag{3.105}$$

9. *Kolmogorov's inequality*:   Let $X_1, X_2, \ldots, X_n$ be n mutually independent random variables with expectations $\mu_i = E[X_i]$ and variances σ_k^2. Define the sums $S_k = X_1 + \cdots + X_k$ so that $m_k = E[S_k] = \mu_1 + \cdots + \mu_k$ and $s_k^2 = \text{Var}[S_k] = \sigma_1^2 + \cdots + \sigma_k^2$. For every $t > 0$, the probability of the simultaneous realization of the n inequalities

$$|S_k - m_k| < t \, s_k \tag{3.106}$$

is at least $1 - t^{-2}$. (When $n = 1$ this is Chebyshev's inequality.)

10. *Markov's inequality*:   If X is random variable which takes only nonnegative values, then for any $a > 0$

$$\text{Prob}[X \geq a] \leq \frac{E[X]}{a}. \tag{3.107}$$

CHAPTER 4

Functions of Random Variables

Contents

Let $X_1, X_2, \ldots, X_n$ be a collection of random variables with joint probability mass function $p(x_1, x_2, \ldots, x_n)$ (if the collection is discrete) or joint density function $f(x_1, x_2, \ldots, x_n)$ (if the collection is continuous). Suppose the random variable $Y = Y(X_1, X_2, \ldots, X_n)$ is a function of $X_1, X_2, \ldots, X_n$. Methods for finding the distribution of Y are presented below and sampling distributions are discussed in the following section.

4.1   FINDING THE PROBABILITY DISTRIBUTION

The following techniques may be used to determine the probability distribution for $Y = Y(X_1, X_2, \ldots, X_n)$.

4.1.1   Method of distribution functions

Let $X_1, X_2, \ldots, X_n$ be a collection of continuous random variables.

(1) Determine the region $Y = y$.

(2) Determine the region $Y \leq y$.

(3) Compute $F(y) = \text{Prob}\,[Y \leq y]$ by integrating the joint density function $f(x_1, x_2, \ldots, x_n)$ over the region $Y \leq y$.

(4) Compute the probability density function for Y, $f(y)$, by differentiating $F(y)$:

$$f(y) = \frac{dF(y)}{dy}. \tag{4.1}$$

Example 4.22: Suppose the joint density function of X_1 and X_2 is given by

$$f(x_1, x_2) = \begin{cases} 4x_1 x_2 e^{-(x_1^2 + x_2^2)} & \text{for } x_1 > 0,\ x_2 > 0 \\ 0 & \text{elsewhere} \end{cases}$$

and $Y = \sqrt{X_1^2 + X_2^2}$. Find the cumulative distribution function for Y and the probability density function for Y.

Solution:

(S1) The region $Y \leq y$ is a quarter circle in quadrant I, shown shaded in Figure 4.1.

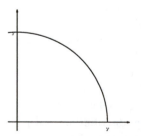

Figure 4.1: Integration region for example 4.22.

(S2) The cumulative distribution function for Y is given by

$$F(y) = \int_0^y \int_0^{\sqrt{y^2 - x_1^2}} 4x_1 x_2 e^{-(x_1^2 + x_2^2)} \, dx_2 \, dx_1$$

$$= \int_0^y 2x_1 (e^{-x_1^2} - e^{-y^2}) \, dx_1 \tag{4.2}$$

$$= 1 - (1 + y^2) e^{-y^2}$$

(S3) The probability density function for Y is given by

$$f(y) = F'(y) = -[(2y)e^{-y^2} + (1 + y^2)(-2y)e^{-y^2}]$$

$$= 2y^3 e^{-y^2}, \quad \text{when } y > 0 \tag{4.3}$$

4.1.2 Method of transformations (one variable)

Let X be a continuous random variable with probability density function $f_X(x)$. If $u(x)$ is differentiable and either increasing or decreasing, then $Y = u(X)$ has probability density function

$$f_Y(y) = f_X(w(y)) \cdot |w'(y)|, \qquad u'(x) \neq 0 \tag{4.4}$$

where $x = w(y) = u^{-1}(y)$.

Example 4.23: Let X be a standard normal random variable, and let $Y = X^2$. What is the distribution of Y?

Solution:

(S1) Since X can be both positive and negative, two regions of X correspond to the same value of Y.

(S2) The computation is

$$f_y(y) = \left[f_x(x) + f_x(-x) \right] \left| \frac{dy}{dx} \right|$$

$$= \left[\frac{e^{-x^2/2}}{\sqrt{2\pi}} + \frac{e^{-(-x)^2/2}}{\sqrt{2\pi}} + \right] \frac{1}{2\sqrt{y}}$$

$$= \frac{1}{\sqrt{2\pi y}} e^{-x^2/2} \tag{4.5}$$

$$= \frac{1}{\sqrt{2\pi y}} e^{-y/2}$$

which is the probability density function for a chi–square random variable with one degree of freedom.

Example 4.24: Given two independent random variables X and Y with joint probability density $f(x, y)$, let $U = X/Y$ be the ratio distribution. The probability density is:

$$f_U(u) = \int_{-\infty}^{\infty} |x| \, f(x, ux) \, dx \tag{4.6}$$

If X and Y are normally distributed, then U has a Cauchy distribution. If X and Y are uniformly distributed on $[0, 1]$, then

$$f_U(u) = \begin{cases} 0 & \text{for } u < 0 \\ \frac{1}{2} & \text{for } 0 \leq u \leq 1 \\ \frac{1}{2u^2} & \text{for } u > 1 \end{cases} \tag{4.7}$$

4.1.3   Method of transformations (two or more variables)

Let X_1 and X_2 be continuous random variables with joint density function $f(x_1, x_2)$. Let the functions $y_1 = u_1(x_1, x_2)$ and $y_2 = u_2(x_1, x_2)$ represent a one–to–one transformation from the x's to the y's and let the partial derivatives with respect to both x_1 and x_2 exist. The joint density function of $Y_1 = u_1(X_1, X_2)$ and $Y_2 = u_2(X_1, X_2)$ is

$$g(y_1, y_2) = f(w_1(y_1, y_2), w_2(y_1, y_2)) \cdot |J| \tag{4.8}$$

where $y_1 = u_1(x_1, x_2)$ and $y_2 = u_2(x_1, x_2)$ are uniquely solved for $x_1 = w_1(y_1, y_2)$ and $x_2 = w_2(y_1, y_2)$, and J is the determinant of the Jacobian

$$J = \begin{vmatrix} \dfrac{\partial x_1}{\partial y_1} & \dfrac{\partial x_1}{\partial y_2} \\[2ex] \dfrac{\partial x_2}{\partial y_1} & \dfrac{\partial x_2}{\partial y_2} \end{vmatrix}. \tag{4.9}$$

This method of transformations may be extended to functions of n random variables. Let $X_1, X_2, \ldots, X_n$ be continuous random variables with joint density function $f(x_1, x_2, \ldots, x_n)$. Let the functions $y_1 = u_1(x_1, x_2, \ldots, x_n)$, $y_2 = u_2(x_1, x_2, \ldots, x_n)$, $\ldots$, $y_n = u_n(x_1, x_2, \ldots, x_n)$ represent a one–to–one transformation from the x's to the y's and let the partial derivatives with respect to $x_1, x_2, \ldots, x_n$ exist. The joint density function of $Y_1 = u_1(X_1, X_2, \ldots, X_n)$, $Y_2 = u_2(X_1, X_2, \ldots, X_n)$, $\ldots$, $Y_n = u_n(X_1, X_2, \ldots, X_n)$ is

$$g(y_1, y_2, \ldots, y_n) = f(w_1(y_1, \ldots, y_n), \ldots, w_n(y_1, \ldots, y_n)) \cdot |J| \tag{4.10}$$

where the functions $y_1 = u_1(x_1, x_2, \ldots, x_n)$, $y_2 = u_2(x_1, x_2, \ldots, x_n)$, $\ldots$, $y_n = u_n(x_1, x_2, \ldots, x_n)$ are uniquely solved for $x_1 = w_1(y_1, y_2, \ldots, y_n)$, $x_2 = w_2(y_1, y_2, \ldots, y_n)$, $\ldots$, $x_n = w_n(y_1, y_2, \ldots, y_n)$ and J is the determinant of the Jacobian

$$J = \begin{vmatrix} \dfrac{\partial x_1}{\partial y_1} & \dfrac{\partial x_1}{\partial y_2} & \cdots & \dfrac{\partial x_1}{\partial y_n} \\[2ex] \dfrac{\partial x_2}{\partial y_1} & \dfrac{\partial x_2}{\partial y_2} & \cdots & \dfrac{\partial x_2}{\partial y_n} \\[2ex] \vdots & \vdots & \ddots & \vdots \\[2ex] \dfrac{\partial x_n}{\partial y_1} & \dfrac{\partial x_n}{\partial y_2} & \cdots & \dfrac{\partial x_n}{\partial y_n} \end{vmatrix} \tag{4.11}$$

Example 4.25: Suppose the random variables X and Y are independent with probability density functions $f_X(x)$ and $f_Y(y)$, then the probability density of their sum, $Z = X + Y$, is given by

$$f_Z(z) = \int_{-\infty}^{\infty} f_X(t) f_Y(z-t)\, dt \tag{4.12}$$

Example 4.26: Suppose the random variables X and Y are independent with probability density functions $f_X(x)$ and $f_Y(y)$, then the probability density of their product, $Z = XY$, is given by

$$f_Z(z) = \int_{-\infty}^{\infty} \frac{1}{|t|} f_X(t) f_Y\left(\frac{z}{t}\right) dt \tag{4.13}$$

Example 4.27: Two random variables X and Y have a joint normal distribution. The probability density is $f(x,y) = \dfrac{1}{2\pi\sigma^2} \exp\left(\dfrac{x^2+y^2}{2\sigma^2}\right)$. Find the probability density of the system (R, Φ) if

$$\begin{aligned} X &= R\cos\Phi \\ Y &= R\sin\Phi \end{aligned} \tag{4.14}$$

Solution:

(S1) Use $f(r,\phi) = f\left[x(r,\phi), y(r,\phi)\right] \left|\frac{\partial(x,y)}{\partial(r,\phi)}\right|$ and $\left|\frac{\partial(x,y)}{\partial(r,\phi)}\right| = r$.

(S2) Then

$$f(r,\phi) = \frac{r}{2\pi\sigma^2} \exp\left(-\frac{r^2\cos^2\phi + r^2\sin^2\phi}{2\sigma^2}\right)$$

$$= \underbrace{\frac{1}{2\pi}}_{f_\Phi(\phi)} \quad \underbrace{\frac{r}{\sigma^2} \exp\left(-\frac{r^2}{2\sigma^2}\right)}_{f_R(r)} \tag{4.15}$$

where $f_R(r)$ is a Rayleigh distribution and $f_\Phi(\phi)$ is a uniform distribution.

4.1.4   Method of moment generating functions

To determine the distribution of Y:

(1) Determine the moment generating function for Y, $m_Y(t)$.

(2) Compare $m_Y(t)$ with known moment generating functions. If $m_Y(t) = m_U(t)$ for all t, then Y and U have identical distributions.

Theorems:

(1) Let X and Y be random variables with moment generating functions $m_X(t)$ and $m_Y(t)$, respectively. If $m_X(t) = m_Y(t)$ for all t, then X and Y have the same probability distributions.

(2) Let $X_1, X_2, \ldots, X_n$ be independent random variables and let $Y = X_1 + X_2 + \cdots + X_n$, then

$$m_Y(t) = \prod_{i=1}^{n} m_{X_i}(t).$$ (4.16)

4.2 SUMS OF RANDOM VARIABLES

4.2.1 Deterministic sums of random variables

If $Y = X_1 + X_2 + \cdots + X_n$ and

(a) the $X_1, X_2, \ldots, X_n$ are independent random variables with factorial moment generating functions $P_{X_i}(t)$, then

$$P_Y(t) = \prod_{i=1}^{n} P_{X_i}(t)$$ (4.17)

(b) the $X_1, X_2, \ldots, X_n$ are independent random variables with the same factorial moment generating function $P_X(t)$, then

$$P_Y(t) = [P_X(t)]^n$$ (4.18)

(c) the $X_1, X_2, \ldots, X_n$ are independent random variables with characteristic functions $\phi_{X_i}(t)$, then

$$\phi_Y(t) = \prod_{i=1}^{n} \phi_{X_i}(t)$$ (4.19)

(d) the $X_1, X_2, \ldots, X_n$ are independent random variables with the same characteristic function $\phi_X(t)$, then

$$\phi_Y(t) = [\phi_X(t)]^n$$ (4.20)

Example 4.28: What is the distribution of the sum of two normal random variables?

Solution:

(S1) Let X_1 be $N(\mu_1, \sigma_1)$ and let X_2 be $N(\mu_2, \sigma_2)$.

(S2) The characteristic functions are (see page 148) $\phi_{X_1}(t) = \exp\left(\mu_1 it - \frac{\sigma_1^2 t}{2}\right)$ and $\phi_{X_2}(t) = \exp\left(\mu_2 it - \frac{\sigma_2^2 t}{2}\right)$.

(S3) From equation (4.19) the characteristic function for $Y = X_1 + X_2$ is

$$\phi_Y(t) = \phi_{X_1}(t) \cdot \phi_{X_2}(t) = \exp\left((\mu_1 + \mu_2)it - \frac{(\sigma_1^2 + \sigma_2^2)t}{2}\right)$$ (4.21)

(S4) This last expression is the characteristic function for a normal random variable with mean $\mu_Y = \mu_1 + \mu_2$ and variance of $\sigma_Y^2 = \sigma_1^2 + \sigma_2^2$.

(S5) Conclusion: the distribution of the sum of two normal random variables is normal; the means add and the variances add.

See section 3.6.11 for linear combinations of random variables.

4.2.2   Random sums of random variables

If $T = \sum_{i=1}^{N} X_i$ where N is an integer valued random variable with factorial generating function $P_N(t)$, the $\{X_i\}$ are discrete independent and identically distributed random variables with factorial generating function $P_X(t)$, and the $\{X_i\}$ are independent of N, then the factorial generating function for T is

$$P_T(t) = P_N(P_X(t)) \tag{4.22}$$

(If the $\{X_i\}$ are continuous random variables, then $\phi_T(t) = P_N(\phi_X(t))$.) Hence (using equation (3.44))

$$\mu_T = \mu_N \mu_X$$
$$\sigma_T^2 = \mu_N \sigma_X^2 + \mu_X \sigma_N^2 \tag{4.23}$$

Example 4.29:   A game is played as follows: There are two coins used to play the game. The probability of a head on the first coin is p_1 and the probability of a head on the second coin is p_2. The first coin is tossed. If the resulting toss is a head, the game is over. If the outcome is a tail, then the second coin is tossed. If the second coin lands head up, a \$1.00 payoff is made. There is no payoff for a tail. The first coin is tossed again and the game continues in this manner. What is the expected payoff for this game?

Solution:

(S1) In this game the number of rounds, N, has a geometric distribution, so that
$$P_N(t) = \frac{p_1 t}{1 - (1 - p_1)t}.$$

(S2) Let the random variable X be the payoff at each round. X has a Bernoulli distribution: $P_X(t) = (1 - p_2) + p_2 t$.

(S3) The generating function for the payoff is

$$P_T(t) = P_N(P_X(t)) = \frac{p_1[(1 - p_2) + p_2 t]}{1 - (1 - p_1)[(1 - p_2) + p_2 t]}. \tag{4.24}$$

(S4) Using $P_T(t)$ in equation (3.44) or using equation (4.23) (with $\mu_N = 1/p_1$ and $\mu_X = p_2$) results in $\mu_T = p_2/p_1$.

4.3   SAMPLING DISTRIBUTIONS

4.3.1   Definitions

(1) The random variables $X_1, X_2, \ldots, X_n$ are a random sample of size n from an infinite population if $X_1, X_2, \ldots, X_n$ are independent and identically distributed (iid).

(2) If $X_1, X_2, \ldots, X_n$ are a random sample, then the sample total and sample mean are

$$T = \sum_{i=1}^{n} X_i \quad \text{and} \quad \overline{X} = \frac{1}{n} \sum_{i=1}^{n} X_i, \tag{4.25}$$

respectively. The sample variance is

$$S^2 = \frac{1}{n-1} \sum_{i=1}^{n} (X_i - \overline{X})^2. \tag{4.26}$$

4.3.2 The sample mean

Consider an infinite population with mean μ, variance σ^2, skewness γ_1, and kurtosis γ_2. Using a sample of size n, the parameters describing the sample mean are:

$$\mu_{\overline{x}} = \mu$$

$$\sigma_{\overline{X}}^2 = \frac{\sigma^2}{n} \qquad \sigma_{\overline{X}} = \frac{\sigma}{\sqrt{n}} \tag{4.27}$$

$$\gamma_{1,\overline{X}} = \frac{\gamma_1}{\sqrt{n}} \qquad \gamma_{2,\overline{X}} = \frac{\gamma_2}{n}$$

When the population is finite and of size M,

$$\mu_{\overline{x}}^{(M)} = \mu$$

$$\sigma_{\overline{x}}^{2(M)} = \frac{\sigma^2}{N} \frac{M-N}{M-1} \tag{4.28}$$

If the underlying population is *normal*, then the sample mean $\overline{X}$ is normally distributed.

4.3.3 Central limit theorem

Let $X_1, X_2, \ldots, X_n$ be a random sample from an infinite population with mean μ and variance σ^2. The limiting distribution of

$$Z = \frac{\overline{X} - \mu}{\sigma/\sqrt{n}} \tag{4.29}$$

as $n \to \infty$ is the standard normal distribution. The limiting distribution of

$$T = \sum_{i=1}^{n} X_i \tag{4.30}$$

as $n \to \infty$ is normal with mean $n\mu$ and variance $n\sigma^2$.

4.3.4 The law of large numbers

Let $X_1, X_2, \ldots, X_n$ be a random sample from an infinite population with mean μ and variance σ^2. For any positive constant c, the probability the sample mean is within c units of μ is at least $1 - \dfrac{\sigma^2}{nc^2}$:

$$\text{Prob}\left[\mu - c < \overline{X} < \mu + c\right] \geq 1 - \frac{\sigma^2}{nc^2}. \tag{4.31}$$

As $n \to \infty$ the probability approaches 1. (See Chebyshev inequality on page 48.)

4.3.5 Laws of the iterated logarithm

Laws of the iterated logarithm (the following hold "a.s.", or "almost surely"):

$$\limsup_{t \downarrow 0} \frac{W_t}{\sqrt{2t \ln \ln(1/t)}} = 1 \qquad \limsup_{t \to \infty} \frac{W_t}{\sqrt{2t \ln \ln t}} = 1$$

$$\liminf_{t \downarrow 0} \frac{W_t}{\sqrt{2t \ln \ln(1/t)}} = -1 \qquad \liminf_{t \to \infty} \frac{W_t}{\sqrt{2t \ln \ln t}} = -1$$

where W is a Brownian motion.

4.4 FINITE POPULATION

Let $\{c_1, c_2, \ldots, c_N\}$ be a collection of numbers representing a finite population of size N and assume the sampling from this population is done without replacement. Let the random variable X_i be the ith observation selected from the population. The collection $X_1, X_2, \ldots, X_n$ is a random sample of size n from the finite population if the joint probability mass function for $X_1, X_2, \ldots, X_n$ is

$$p(x_1, x_2, \ldots, x_n) = \frac{1}{N(N-1)\cdots(N-n+1)}. \tag{4.32}$$

(1) The marginal probability distribution for the random variable X_i, $i = 1, 2, \ldots, n$ is

$$p_{X_i}(x_i) = \frac{1}{N} \quad \text{for} \quad x_i = c_1, c_2, \ldots, c_N. \tag{4.33}$$

(2) The mean and the variance of the finite population are

$$\mu = \sum_{i=1}^{N} c_i \frac{1}{N} \quad \text{and} \quad \sigma^2 = \sum_{i=1}^{N} (c_i - \mu)^2 \frac{1}{N}. \tag{4.34}$$

(3) The joint marginal probability mass function for any two random variables in the collection $X_1, X_2, \ldots, X_n$ is

$$p(x_i, x_j) = \frac{1}{N(N-1)}. \tag{4.35}$$

(4) The covariance between any two random variables in the collection $X_1, X_2, \ldots, X_n$ is

$$\text{Cov}[X_i, X_j] = -\frac{\sigma^2}{N-1}. \tag{4.36}$$

(5) Let $\overline{X}$ be the sample mean of the random sample of size n. The expected value and variance of $\overline{X}$ are

$$\mathrm{E}\left[\overline{X}\right] = \mu \quad \text{and} \quad \mathrm{Var}\left[\overline{X}\right] = \frac{\sigma^2}{n} \cdot \frac{N-n}{N-1}. \tag{4.37}$$

The quantity $(N-n)/(N-1)$ is the *finite population correction factor*.

4.5 THEOREMS

4.5.1 Theorems: the chi–square distribution

(1) Let Z be a standard normal random variable, then Z^2 has a chi–square distribution with 1 degree of freedom.

(2) Let $Z_1, Z_2, \ldots, Z_n$ be independent standard normal random variables. The random variable $Y = \sum_{i=1}^{n} Z_i^2$ has a chi–square distribution with n degrees of freedom.

(3) Let $X_1, X_2, \ldots, X_n$ be independent random variables such that X_i has a chi–square distribution with ν_i degrees of freedom. The random variable $Y = \sum_{i=1}^{n} X_i$ has a chi–square distribution with $\nu = \nu_1 + \nu_2 + \cdots + \nu_n$ degrees of freedom.

(4) Let U have a chi–square distribution with ν_1 degrees of freedom, U and V be independent, and $U + V$ have a chi–square distribution with $\nu > \nu_1$ degrees of freedom. The random variable V has a chi–sqaure distribution with $\nu - \nu_1$ degrees of freedom.

(5) Let $X_1, X_2, \ldots, X_n$ be a random sample from a normal population with mean μ and variance σ^2. Then

  (a) The sample mean, $\overline{X}$, and the sample variance, S^2, are independent, and

  (b) The random variable $\dfrac{(n-1)S^2}{\sigma^2}$ has a chi–square distribution with $n - 1$ degrees of freedom.

4.5.2 Theorems: the t distribution

(1) Let Z have a standard normal distribution, X have a chi–square distribution with ν degrees of freedom, and X and Z be independent. The random variable

$$T = \frac{Z}{\sqrt{X/\nu}} \tag{4.38}$$

has a t distribution with ν degrees of freedom.

(2) Let $X_1, X_2, \ldots, X_n$ be a random sample from a normal population with mean μ and variance σ^2. The random variable

$$T = \frac{\overline{X} - \mu}{S/\sqrt{n}} \qquad (4.39)$$

has a t distribution with $n - 1$ degrees of freedom.

4.5.3 Theorems: the F distribution

(1) Let U have a chi–square distribution with ν_1 degrees of freedom, V have a chi–square distribution with ν_2 degrees of freedom, and U and V be independent. The random variable

$$F = \frac{U/\nu_1}{V/\nu_2} \qquad (4.40)$$

has an F distribution with ν_1 and ν_2 degrees of freedom.

(2) Let $X_1, X_2, \ldots, X_m$ and $Y_1, Y_2, \ldots, Y_n$ be random samples from normal populations with variances σ_X^2 and σ_Y^2, respectively. The random variable

$$F = \frac{S_X^2/\sigma_X^2}{S_y^2/\sigma_Y^2} \qquad (4.41)$$

has an F distribution with $m - 1$ and $n - 1$ degrees of freedom.

(3) Let F_{α,ν_1,ν_2} be a critical value for the F distribution defined by $\text{Prob}\,[F \geq F_{\alpha,\nu_1,\nu_2}] = \alpha$. Then $F_{1-\alpha,\nu_1,\nu_2} = 1/F_{\alpha,\nu_2,\nu_1}$.

4.6 ORDER STATISTICS

4.6.1 Definition

Let $X_1, X_2, \ldots, X_n$ be independent continuous random variables with probability density function $f(x)$ and cumulative distribution function $F(x)$. The order statistic, $X_{(i)}$, $i = 1, 2, \ldots, n$, is a random variable defined to be the ith largest of the set $\{X_1, X_2, \ldots, X_n\}$. Therefore,

$$X_{(1)} \leq X_{(2)} \leq \cdots \leq X_{(n)} \qquad (4.42)$$

and in particular

$$\begin{aligned} X_{(1)} &= \min\{X_1, X_2, \ldots, X_n\} \quad \text{and} \\ X_{(n)} &= \max\{X_1, X_2, \ldots, X_n\}. \end{aligned} \qquad (4.43)$$

The cumulative distribution function for the ith order statistic is

$$F_{X_{(i)}}(x) = \text{Prob}\,[X_{(i)} \leq x] = \text{Prob}\,[i \text{ or more observations are} \leq x]$$

$$= \sum_{j=i}^{n} \binom{n}{j} [F(x)]^j [1 - F(x)]^{n-j} \qquad (4.44)$$

and the probability density function is

$$f_{X_{(i)}}(x) = n\binom{n-1}{i-1}[F(x)]^{i-1}[1 - F(x)]^{n-i}f(x)$$

$$= \frac{n!}{(i-1)!(n-i)!}[F(x)]^{i-1}f(x)[1 - F(x)]^{n-i}. \tag{4.45}$$

4.6.2 The first order statistic

The probability density function, $f_{X(1)}(x)$, and the cumulative distribution function, $F_{X(1)}(x)$, for $X_{(1)}$ are

$$f_{X_{(1)}}(x) = n[1 - F(x)]^{n-1}f(x) \qquad F_{X_{(1)}}(x) = 1 - [1 - F(x)]^n. \tag{4.46}$$

4.6.3 The nth order statistic

The probability density function, $f_{(n)}(x)$, and the cumulative distribution function, $F_{(n)}(x)$, for $X_{(n)}$ are

$$f_{X_{(n)}}(x) = n[F(x)]^{n-1}f(x) \qquad F_{X_{(n)}}(x) = [F(x)]^n. \tag{4.47}$$

4.6.4 The median

If the number of observations is odd, the median is the middle observation when the observations are in numerical order. If the number of observations is even, the median is (arbitrarily) defined as the average of the middle two of the ordered observations.

$$\text{median} = \begin{cases} X_{(k)} & \text{if } n \text{ is odd and } n = 2k - 1 \\ \frac{1}{2}[X_{(k)} + X_{(k+1)}] & \text{if } n \text{ is even and } n = 2k \end{cases} \tag{4.48}$$

4.6.5 Joint distributions

The joint density function for $X_{(1)}, X_{(2)}, \ldots, X_{(n)}$ is

$$g(x_1, x_2, \ldots, x_n) = n!f(x_1)f(x_2)\cdots f(x_n). \tag{4.49}$$

The joint density function for the ith and j^{th} $(i < j)$ order statistics is

$$f_{ij}(x, y) = \frac{n!}{(i-1)!(j-i-1)!(n-j)!}f(x)f(y)$$

$$\times [F(x)]^{i-1}[1 - F(y)]^{n-j}[F(y) - F(x)]^{j-i-1}. \tag{4.50}$$

The joint distribution function for $X_{(1)}$ and $X_{(n)}$ is

$$F_{1n}(x, y) = \text{Prob}\left[X_{(1)} \leq x \quad \text{and} \quad X_{(n)} \leq y\right]$$

$$= \begin{cases} [F(y)]^n - [F(y) - F(x)]^n & \text{if } x \leq y \\ [F(y)]^n & \text{if } x > y \end{cases} \tag{4.51}$$

and the joint density function is

$$f_{1n}(x,y) = \begin{cases} n(n-1)f(x)f(y)[F(y)-F(x)]^{n-2} & \text{if } x \leq y \\ 0 & \text{if } x > y \end{cases} \quad (4.52)$$

4.6.6 Midrange and range

The *midrange* is defined to be $A = \frac{1}{2}\left[X_{(1)} + X_{(n)}\right]$. Using $f_{1n}(x,y)$ for the joint density function of $X_{(1)}$ and $X_{(n)}$ results in

$$f_A(x) = 2 \int_{-\infty}^{x} f_{1n}(t, 2x - t)\, dt$$

$$= 2n(n-1) \int_{-\infty}^{x} f(t)f(2x-t)\left[F(2x-t)-F(t)\right]^{n-2} dt \quad (4.53)$$

The *range* is the difference between the largest and smallest observations: $R = X_{(n)} - X_{(1)}$. The random variable R is used in the construction of *tolerance intervals*.

$$f_R(r) = \int_{-\infty}^{\infty} f_{1n}(t, t + r)\, dt$$

$$= \begin{cases} n(n-1) \int_{-\infty}^{\infty} f(t)f(t+r)[F(t+r)-F(t)]^{n-2} dt & \text{if } r > 0 \\ 0 & \text{if } r \leq 0 \end{cases} \quad (4.54)$$

4.6.7 Uniform distribution: order statistics

If X is uniformly distributed on the interval $[0,1]$ then the density function for $X_{(i)}$ is

$$f_i(x) = n\binom{n-1}{i-1}x^{i-1}(1-x)^{n-i}, \quad 0 \leq x \leq 1 \quad (4.55)$$

which is a beta distribution with parameters i and $n - i + 1$.

(1) $\mathrm{E}\left[X_{(i)}\right] = \int_0^1 f_k(t)\, dt = \dfrac{i}{n+1}$.

(2) The expected value of the largest of n observations is $\dfrac{n}{n+1}$.

(3) The expected value of the smallest of n observations is $\dfrac{1}{n+1}$.

(4) The density function of the midrange is

$$f_A(x) = \begin{cases} n2^{n-1}x^{n-1} & \text{if } 0 < x \leq \frac{1}{2} \\ n2^{n-1}(1-x)^{n-1} & \text{if } \frac{1}{2} \leq x < 1 \end{cases} \quad (4.56)$$

(5) The density function of the range is

$$f_R(r) = \begin{cases} n(n-1)(1-r)r^{n-2} & \text{if } 0 < r < 1 \\ 0 & \text{otherwise} \end{cases} \qquad (4.57)$$

4.6.7.1   Tolerance intervals

In many applications, we need to estimate an interval in which a certain proportion of the population lies, with given probability. A tolerance interval may be constructed using the results relating to order statistics and the range. A table of required sample sizes for varying ranges and probabilities is in the following table.

Example 4.30: Assume a sample is drawn from a uniform population. Find a sample size n such that at least 99% of the sample population, with probability .95, lies between the smallest and largest observations. This problem may be written as a probability statement:

$$\begin{aligned} 0.95 &= \text{Prob}\left[F(Z_n) - F(Z_1) > 0.99\right] \\ &= \text{Prob}\left[R > 0.99\right] \\ &= n(n-1)\int_{0.99}^{1}(1-r)r^{n-2}\, dr \\ &= 1 - (0.99)^{n-1}(0.01n + 0.99) \end{aligned} \qquad (4.58)$$

Solving this results in the value $n \approx 473$.

Tolerance intervals, uniform distribution

Probability	This fraction of the total population is within the range							
	0.500	0.750	0.900	0.950	0.975	0.990	0.995	0.999
0.500	3	7	17	34	67	168	336	1679
0.750	5	10	26	53	107	269	538	2692
0.900	6	14	38	77	154	388	777	3889
0.950	8	17	46	93	188	473	947	4742
0.975	9	20	54	109	221	555	1112	5570
0.990	10	24	64	130	263	661	1325	6636
0.995	11	26	71	145	294	740	1483	7427

For tolerance intervals for normal samples, see section 7.3.

4.6.8   Normal distribution: order statistics

When the $\{X_i\}$ come from a standard normal distribution, the $\{X_{(i)}\}$ are called *standard order statistics*.

4.6.8.1   Expected value of normal order statistics

The tables on pages 65–66 gives expected values of standard order statistics

$$E\left[X_{(i)}\right] = n\binom{n-1}{i-1}\int_{-\infty}^{\infty} tf(t)[F(t)]^{i-1}[1 - F(t)]^{n-i}\, dt \qquad (4.59)$$

when $f(x) = \dfrac{e^{-x^2/2}}{\sqrt{2\pi}}$ and $F(x) = \displaystyle\int_{-\infty}^{x} \dfrac{e^{-t^2/2}}{\sqrt{2\pi}}\, dt$. Missing values (indicated by a dash) may be obtained from $\mathrm{E}\left[X_{(i)}\right] = -\mathrm{E}\left[X_{(n-i+1)}\right]$.

Example 4.31: If an average person takes five intelligence tests (each test having a normal distribution with a mean of 100 and a standard deviation of 20), what is the expected value of the largest score?

Solution:

(S1) We need to obtain the expected value of the largest normal order statistic when $n = 5$.

(S2) Using $n = 5$ and $i = 5$ in the table on page 65 yields (use $j = 1$) $\mathrm{E}\left[X_{(5)}\right]\big|_{n=5} = 1.1630$.

(S3) The expected value of the largest score is $100 + (1.1630)(20) \approx 123$.

Expected value of the ith normal order statistic (use $j = n - i + 1$)

j			$n = 2$	3	4	5	6	7	8	9
1			0.5642	0.8463	1.0294	1.1629	1.2672	1.3522	1.4236	1.4850
2			—	0.0000	0.2970	0.4950	0.6418	0.7574	0.8522	0.9323
3			—	—	—	0.0000	0.2015	0.3527	0.4728	0.5720
4			—	—	—	—	—	0.0000	0.1526	0.2745
5			—	—	—	—	—	—	—	0.0000

j	$n = 10$	11	12	13	14	15	16	17	18	19
1	1.5388	1.5865	1.6292	1.6680	1.7034	1.7359	1.7660	1.7939	1.8200	1.8445
2	1.0014	1.0619	1.1157	1.1641	1.2079	1.2479	1.2848	1.3188	1.3504	1.3800
3	0.6561	0.7288	0.7929	0.8498	0.9011	0.9477	0.9903	1.0295	1.0657	1.0995
4	0.3757	0.4619	0.5368	0.6028	0.6618	0.7149	0.7632	0.8074	0.8481	0.8859
5	0.1227	0.2249	0.3122	0.3883	0.4556	0.5157	0.5700	0.6195	0.6648	0.7066
6	—	0.0000	0.1025	0.1905	0.2672	0.3353	0.3962	0.4513	0.5016	0.5477
7	—	—	—	0.0000	0.0882	0.1653	0.2337	0.2952	0.3508	0.4016
8	—	—	—	—	—	0.0000	0.0772	0.1459	0.2077	0.2637
9	—	—	—	—	—	—	—	0.0000	0.0688	0.1307
10	—	—	—	—	—	—	—	—	—	0.0000

j	$n = 20$	21	22	23	24	25	26	27	28	29
1	1.8675	1.8892	1.9097	1.9292	1.9477	1.9653	1.9822	1.9983	2.0137	2.0285
2	1.4076	1.4336	1.4581	1.4813	1.5034	1.5243	1.5442	1.5632	1.5814	1.5988
3	1.1310	1.1605	1.1883	1.2145	1.2393	1.2628	1.2851	1.3064	1.3268	1.3462
4	0.9210	0.9538	0.9846	1.0136	1.0409	1.0668	1.0914	1.1147	1.1370	1.1582
5	0.7454	0.7816	0.8153	0.8470	0.8769	0.9051	0.9318	0.9571	0.9812	1.0042
6	0.5903	0.6298	0.6667	0.7012	0.7336	0.7641	0.7929	0.8202	0.8462	0.8709
7	0.4483	0.4915	0.5316	0.5690	0.6040	0.6369	0.6679	0.6973	0.7251	0.7515
8	0.3149	0.3620	0.4056	0.4461	0.4839	0.5193	0.5527	0.5841	0.6138	0.6420
9	0.1869	0.2384	0.2857	0.3296	0.3704	0.4086	0.4443	0.4780	0.5098	0.5398
10	0.0620	0.1183	0.1699	0.2175	0.2616	0.3026	0.3410	0.3770	0.4109	0.4430
11	—	0.0000	0.0564	0.1081	0.1558	0.2000	0.2413	0.2798	0.3160	0.3501
12	—	—	—	0.0000	0.0518	0.0995	0.1439	0.1852	0.2239	0.2602
13	—	—	—	—	—	0.0000	0.0478	0.0922	0.1336	0.1724
14	—	—	—	—	—	—	—	0.0000	0.0444	0.0859
15	—	—	—	—	—	—	—	—	—	0.0000

Expected value of the ith normal order statistic (use $j = n - i + 1$)

j	n = 30	31	32	33	34	35	36	37	38	39
1	2.0427	2.0564	2.0696	2.0824	2.0947	2.1066	2.1181	2.1292	2.1401	2.1505
2	1.6156	1.6316	1.6471	1.6620	1.6763	1.6902	1.7036	1.7165	1.7291	1.7413
3	1.3648	1.3827	1.3999	1.4164	1.4323	1.4476	1.4624	1.4768	1.4906	1.5040
4	1.1786	1.1980	1.2167	1.2347	1.2520	1.2686	1.2847	1.3002	1.3151	1.3296
5	1.0262	1.0472	1.0673	1.0866	1.1052	1.1230	1.1402	1.1568	1.1729	1.1884
6	0.8944	0.9169	0.9385	0.9591	0.9789	0.9979	1.0163	1.0339	1.0510	1.0674
7	0.7767	0.8007	0.8236	0.8456	0.8666	0.8868	0.9063	0.9250	0.9430	0.9604
8	0.6689	0.6944	0.7188	0.7420	0.7644	0.7857	0.8063	0.8261	0.8451	0.8634
9	0.5683	0.5954	0.6213	0.6460	0.6695	0.6921	0.7138	0.7346	0.7547	0.7740
10	0.4733	0.5020	0.5294	0.5555	0.5804	0.6043	0.6271	0.6490	0.6701	0.6904
11	0.3823	0.4129	0.4418	0.4694	0.4957	0.5208	0.5449	0.5679	0.5900	0.6113
12	0.2945	0.3268	0.3575	0.3867	0.4144	0.4409	0.4662	0.4904	0.5136	0.5359
13	0.2088	0.2432	0.2757	0.3065	0.3358	0.3637	0.3903	0.4157	0.4401	0.4635
14	0.1247	0.1613	0.1957	0.2283	0.2592	0.2886	0.3166	0.3433	0.3689	0.3934
15	0.0415	0.0804	0.1170	0.1515	0.1842	0.2151	0.2446	0.2727	0.2995	0.3252
16	—	0.0000	0.0389	0.0755	0.1101	0.1428	0.1739	0.2034	0.2316	0.2585
17	—	—	—	0.0000	0.0367	0.0713	0.1040	0.1351	0.1647	0.1929
18	—	—	—	—	—	0.0000	0.0346	0.0674	0.0986	0.1282
19	—	—	—	—	—	—	—	0.0000	0.0328	0.0640
20	—	—	—	—	—	—	—	—	—	0.0000

4.6.8.2   Variances and covariances of order statistics

Given n observations of independent standard normal variables, arrange the sample in ascending order of magnitude $X_{(1)}, X_{(2)}, \ldots, X_{(n)}$. The variances and covariances for expected values and product moments may be found from

$$E\left[X_{(i)}\right] = n\binom{n-1}{n-i} \int_{-\infty}^{\infty} t f(t) F^{i-1}(t)[1 - F(t)]^{n-i}\, dt$$

$$E\left[X_{(i)}^2\right] = n\binom{n-1}{n-i} \int_{-\infty}^{\infty} t^2 f(t) F^{i-1}(t)[1 - F(t)]^{n-i}\, dt$$

$$E\left[X_{(i)}X_{(j)}\right] = n\binom{n-1}{n-i} \int_{-\infty}^{\infty} \int_{-\infty}^{y} t y f(t) f(y)$$
$$\times\, [F(t)]^{i-1}[1 - F(y)]^{n-j}[F(y) - F(t)]^{j-i-1}\, dt\, dy$$

(4.60)

where $f(x) = \dfrac{e^{-x^2/2}}{\sqrt{2\pi}}$ and $F(x) = \displaystyle\int_{-\infty}^{x} \dfrac{e^{-x^2/2}}{\sqrt{2\pi}}\, dx$.

The following table gives the variances and covariances of order statistics in samples of sizes up to 10 from a standard normal distribution. Missing values may be obtained from $E\left[X_{(i)}X_{(j)}\right] = E\left[X_{(j)}X_{(i)}\right] = E\left[X_{(n-i+1)}X_{(n-j+1)}\right]$

See G. L. Tietjen, D. K. Kahaner, and R. J. Beckman, "Variances and covariances of the normal order statistics for samples sizes 2 to 50", *Selected Tables in Mathematical Statistics*, **5**, American Mathematical Society, Providence, RI, 1977.

Variances and covariances of normal order statistics

$\mathrm{E}\left[X_{(i)}X_{(j)}\right]$ is shown for samples of size n

(use $k = n - i + 1$ and $\ell = n - j + 1$)

n	k	ℓ	value	n	k	ℓ	value	n	k	ℓ	value
2	1	1	.6817	7	3	3	.2197	9	4	4	.1706
		2	.3183			4	.1656			5	.1370
	2	2	.6817			5	.1296			6	.1127
3	1	1	.5595		4	4	.2104		5	5	.1661
		2	.2757	8	1	1	.3729	10	1	1	.3443
		3	.1649			2	.1863			2	.1713
	2	2	.4487			3	.1260			3	.1163
4	1	1	.4917			4	.0947			4	.0882
		2	.2456			5	.0748			5	.0707
		3	.1580			6	.0602			6	.0584
		4	.1047			7	.0483			7	.0489
	2	2	.3605			8	.0368			8	.0411
		3	.2359		2	2	.2394			9	.0340
5	1	1	.4475			3	.1632			10	.0267
		2	.2243			4	.1233		2	2	.2145
		3	.1481			5	.0976			3	.1466
		4	.1058			6	.0787			4	.1117
		5	.0742			7	.0632			5	.0897
	2	2	.3115		3	3	.2008			6	.0742
		3	.2084			4	.1524			7	.0622
		4	.1499			5	.1210			8	.0523
	3	3	.2868			6	.0978			9	.0434
6	1	1	.4159		4	4	.1872		3	3	.1750
		2	.2085			5	.1492			4	.1338
		3	.1394	9	1	1	.3574			5	.1077
		4	.1024			2	.1781			6	.0892
		5	.0774			3	.1207			7	.0749
		6	.0563			4	.0913			8	.0630
	2	2	.2796			5	.0727		4	4	.1579
		3	.1890			6	.0595			5	.1275
		4	.1397			7	.0491			6	.1058
		5	.1059			8	.0401			7	.0889
	3	3	.2462			9	.0311		5	5	.1511
		4	.1833		2	2	.2257			6	.1256
7	1	1	.3919			3	.1541				
		2	.1962			4	.1170				
		3	.1321			5	.0934				
		4	.0985			6	.0765				
		5	.0766			7	.0632				
		6	.0599			8	.0517				
		7	.0448		3	3	.1864				
	2	2	.2567			4	.1421				
		3	.1745			5	.1138				
		4	.1307			6	.0934				
		5	.1020			7	.0772				
		6	.0800								

4.7   RANGE AND STUDENTIZED RANGE

4.7.1   Probability integral of the range

Let $\{X_1, X_2, \ldots, X_n\}$ denote a random sample of size n from a population with standard deviation σ, density function $f(x)$, and cumulative distribution function $F(x)$. Let $\{X_{(1)}, X_{(2)}, \ldots, X_{(n)}\}$ denote the same values in ascending order of magnitude. The sample range R is defined by

$$R = X_{(n)} - X_{(1)} \tag{4.61}$$

In standardized form

$$W = \frac{R}{\sigma} = \frac{X_{(n)} - X_{(1)}}{\sigma} \tag{4.62}$$

The probability that the range exceeds some value R, for a sample of size n, is (see equation (4.54))

$$\text{Prob} \begin{bmatrix} \text{range exceeds } R \text{ for} \\ \text{a sample of size } n \end{bmatrix} = \int_R^\infty f_R(r)\, dr$$

$$= n \int_{-\infty}^\infty \left[F(t + R) - F(t) \right]^{n-1} f(t)\, dt \tag{4.63}$$

The following tables provide values of this probability for the normal density function $f(x) = \dfrac{1}{\sqrt{2\pi}} e^{-x^2/2}$ for various values of n and W. (Note that since $\sigma = 1$ for this case $R = W$.)

Probability integral of the range

W	n = 2	3	4	5	6	7	8	9	10
0.00	0.0000	0.0000							
0.05	0.0282	0.0007	0.0000						
0.10	0.0564	0.0028	0.0001						
0.15	0.0845	0.0062	0.0004	0.0000					
0.20	0.1125	0.0110	0.0010	0.0001					
0.25	0.1403	0.0171	0.0020	0.0002	0.0000				
0.30	0.1680	0.0245	0.0034	0.0004	0.0001				
0.35	0.1955	0.0332	0.0053	0.0008	0.0001				
0.40	0.2227	0.0431	0.0079	0.0014	0.0002	0.0000			
0.45	0.2497	0.0543	0.0111	0.0022	0.0004	0.0001			
0.50	0.2763	0.0666	0.0152	0.0033	0.0007	0.0002	0.0000		
0.55	0.3027	0.0800	0.0200	0.0048	0.0011	0.0003	0.0001		
0.60	0.3286	0.0944	0.0257	0.0068	0.0017	0.0004	0.0001	0.0000	
0.65	0.3542	0.1099	0.0322	0.0092	0.0026	0.0007	0.0002	0.0001	
0.70	0.3794	0.1263	0.0398	0.0121	0.0036	0.0011	0.0003	0.0001	
0.75	0.4041	0.1436	0.0483	0.0157	0.0050	0.0016	0.0005	0.0002	0.0000
0.80	0.4284	0.1616	0.0578	0.0200	0.0068	0.0023	0.0008	0.0002	0.0001
0.85	0.4522	0.1805	0.0682	0.0250	0.0090	0.0032	0.0011	0.0004	0.0001
0.90	0.4755	0.2000	0.0797	0.0308	0.0117	0.0044	0.0016	0.0006	0.0002
0.95	0.4983	0.2201	0.0922	0.0375	0.0150	0.0059	0.0023	0.0009	0.0003
1.00	0.5205	0.2407	0.1057	0.0450	0.0188	0.0078	0.0032	0.0013	0.0005
1.05	0.5422	0.2618	0.1201	0.0535	0.0234	0.0101	0.0043	0.0018	0.0008
1.10	0.5633	0.2833	0.1355	0.0629	0.0287	0.0129	0.0058	0.0025	0.0011
1.15	0.5839	0.3052	0.1517	0.0733	0.0348	0.0163	0.0076	0.0035	0.0016
1.20	0.6039	0.3272	0.1688	0.0847	0.0417	0.0203	0.0098	0.0047	0.0022
1.25	0.6232	0.3495	0.1867	0.0970	0.0495	0.0249	0.0125	0.0062	0.0030
1.30	0.6420	0.3719	0.2054	0.1104	0.0583	0.0304	0.0157	0.0080	0.0041
1.35	0.6602	0.3943	0.2248	0.1247	0.0680	0.0366	0.0195	0.0103	0.0054
1.40	0.6778	0.4168	0.2448	0.1400	0.0787	0.0437	0.0240	0.0131	0.0071
1.45	0.6948	0.4392	0.2654	0.1562	0.0904	0.0516	0.0292	0.0164	0.0092
1.50	0.7112	0.4614	0.2865	0.1733	0.1031	0.0606	0.0353	0.0204	0.0117
1.55	0.7269	0.4835	0.3080	0.1913	0.1168	0.0705	0.0421	0.0250	0.0148
1.60	0.7421	0.5053	0.3299	0.2101	0.1315	0.0814	0.0499	0.0304	0.0184
1.65	0.7567	0.5269	0.3521	0.2296	0.1473	0.0934	0.0587	0.0366	0.0227
1.70	0.7707	0.5481	0.3745	0.2498	0.1639	0.1064	0.0684	0.0437	0.0278
1.75	0.7841	0.5690	0.3970	0.2706	0.1815	0.1204	0.0792	0.0517	0.0336
1.80	0.7969	0.5894	0.4197	0.2920	0.2000	0.1355	0.0910	0.0607	0.0403
1.85	0.8092	0.6094	0.4423	0.3138	0.2193	0.1516	0.1039	0.0707	0.0479
1.90	0.8209	0.6290	0.4649	0.3361	0.2394	0.1686	0.1178	0.0818	0.0565
1.95	0.8321	0.6480	0.4874	0.3587	0.2602	0.1867	0.1329	0.0939	0.0661
2.00	0.8427	0.6665	0.5096	0.3816	0.2816	0.2056	0.1489	0.1072	0.0768
2.05	0.8528	0.6845	0.5317	0.4046	0.3035	0.2254	0.1661	0.1216	0.0886
2.10	0.8624	0.7019	0.5534	0.4277	0.3260	0.2460	0.1842	0.1371	0.1015
2.15	0.8716	0.7187	0.5748	0.4508	0.3489	0.2673	0.2032	0.1536	0.1155
2.20	0.8802	0.7349	0.5957	0.4739	0.3720	0.2893	0.2232	0.1712	0.1307
2.25	0.8884	0.7505	0.6163	0.4969	0.3955	0.3118	0.2440	0.1899	0.1470

Probability integral of the range

W	$n = 11$	12	13	14	15	16	17	18	19	20
0.00										
0.05										
0.10										
0.15										
0.20										
0.25										
0.30										
0.35										
0.40										
0.45										
0.50										
0.55										
0.60										
0.65										
0.70										
0.75										
0.80										
0.85	0.0000									
0.90	0.0001									
0.95	0.0001	0.0000								
1.00	0.0002	0.0001	0.0000							
1.05	0.0003	0.0001	0.0001							
1.10	0.0005	0.0002	0.0001	0.0000						
1.15	0.0007	0.0003	0.0001	0.0001	0.0000					
1.20	0.0010	0.0005	0.0002	0.0001	0.0001					
1.25	0.0015	0.0007	0.0004	0.0002	0.0001	0.0000				
1.30	0.0021	0.0010	0.0005	0.0003	0.0001	0.0001	0.0000			
1.35	0.0028	0.0015	0.0008	0.0004	0.0002	0.0001	0.0001			
1.40	0.0038	0.0021	0.0011	0.0006	0.0003	0.0002	0.0001	0.0000		
1.45	0.0051	0.0028	0.0016	0.0009	0.0005	0.0003	0.0001	0.0001	0.0000	
1.50	0.0067	0.0038	0.0022	0.0012	0.0007	0.0004	0.0002	0.0001	0.0001	0.0000
1.55	0.0087	0.0051	0.0030	0.0017	0.0010	0.0006	0.0003	0.0002	0.0001	0.0001
1.60	0.0111	0.0067	0.0040	0.0024	0.0014	0.0008	0.0005	0.0003	0.0002	0.0001
1.65	0.0140	0.0086	0.0053	0.0032	0.0020	0.0012	0.0007	0.0004	0.0003	0.0002
1.70	0.0176	0.0111	0.0070	0.0044	0.0027	0.0017	0.0011	0.0007	0.0004	0.0003
1.75	0.0217	0.0140	0.0090	0.0058	0.0037	0.0023	0.0015	0.0010	0.0006	0.0004
1.80	0.0266	0.0175	0.0115	0.0075	0.0049	0.0032	0.0021	0.0014	0.0009	0.0006
1.85	0.0323	0.0217	0.0145	0.0097	0.0065	0.0043	0.0029	0.0019	0.0013	0.0008
1.90	0.0388	0.0266	0.0182	0.0124	0.0084	0.0057	0.0039	0.0026	0.0018	0.0012
1.95	0.0463	0.0323	0.0225	0.0156	0.0108	0.0075	0.0052	0.0036	0.0024	0.0017
2.00	0.0548	0.0389	0.0276	0.0195	0.0137	0.0097	0.0068	0.0048	0.0033	0.0023
2.05	0.0643	0.0465	0.0335	0.0241	0.0173	0.0124	0.0088	0.0063	0.0045	0.0032
2.10	0.0748	0.0550	0.0403	0.0295	0.0215	0.0156	0.0114	0.0082	0.0060	0.0043
2.15	0.0866	0.0646	0.0481	0.0357	0.0265	0.0196	0.0144	0.0106	0.0078	0.0058
2.20	0.0994	0.0753	0.0569	0.0429	0.0323	0.0242	0.0182	0.0136	0.0102	0.0076
2.25	0.1134	0.0872	0.0669	0.0511	0.0390	0.0297	0.0226	0.0172	0.0130	0.0099

Probability integral of the range

W	$n = 2$	3	4	5	6	7	8	9	10
2.25	0.8884	0.7505	0.6163	0.4969	0.3955	0.3118	0.2440	0.1899	0.1470
2.30	0.8961	0.7655	0.6363	0.5196	0.4190	0.3348	0.2656	0.2095	0.1645
2.35	0.9034	0.7799	0.6559	0.5421	0.4427	0.3582	0.2878	0.2300	0.1829
2.40	0.9103	0.7937	0.6748	0.5643	0.4663	0.3820	0.3107	0.2514	0.2025
2.45	0.9168	0.8069	0.6932	0.5861	0.4899	0.4059	0.3341	0.2735	0.2229
2.50	0.9229	0.8195	0.7110	0.6075	0.5132	0.4300	0.3579	0.2963	0.2443
2.55	0.9286	0.8315	0.7282	0.6283	0.5364	0.4541	0.3820	0.3198	0.2665
2.60	0.9340	0.8429	0.7448	0.6487	0.5592	0.4782	0.4064	0.3437	0.2894
2.65	0.9390	0.8537	0.7607	0.6685	0.5816	0.5022	0.4310	0.3680	0.3130
2.70	0.9438	0.8640	0.7759	0.6877	0.6036	0.5259	0.4555	0.3927	0.3372
2.75	0.9482	0.8737	0.7905	0.7063	0.6252	0.5494	0.4801	0.4175	0.3617
2.80	0.9523	0.8828	0.8045	0.7242	0.6461	0.5725	0.5045	0.4425	0.3867
2.85	0.9561	0.8915	0.8177	0.7415	0.6665	0.5952	0.5286	0.4675	0.4119
2.90	0.9597	0.8996	0.8304	0.7581	0.6863	0.6174	0.5525	0.4923	0.4372
2.95	0.9630	0.9073	0.8424	0.7739	0.7055	0.6391	0.5760	0.5171	0.4625
3.00	0.9661	0.9145	0.8537	0.7891	0.7239	0.6601	0.5991	0.5415	0.4878
3.05	0.9690	0.9212	0.8645	0.8036	0.7416	0.6806	0.6216	0.5656	0.5129
3.10	0.9716	0.9275	0.8746	0.8174	0.7587	0.7003	0.6436	0.5892	0.5378
3.15	0.9741	0.9334	0.8842	0.8305	0.7750	0.7194	0.6649	0.6124	0.5623
3.20	0.9763	0.9388	0.8931	0.8429	0.7905	0.7377	0.6856	0.6350	0.5864
3.25	0.9784	0.9439	0.9016	0.8546	0.8053	0.7553	0.7055	0.6569	0.6099
3.30	0.9804	0.9487	0.9095	0.8657	0.8194	0.7721	0.7248	0.6782	0.6329
3.35	0.9822	0.9531	0.9168	0.8761	0.8327	0.7881	0.7432	0.6988	0.6553
3.40	0.9838	0.9572	0.9237	0.8859	0.8454	0.8034	0.7609	0.7186	0.6769
3.45	0.9853	0.9610	0.9302	0.8951	0.8573	0.8179	0.7778	0.7376	0.6978
3.50	0.9867	0.9644	0.9361	0.9037	0.8685	0.8316	0.7938	0.7558	0.7180
3.55	0.9879	0.9677	0.9417	0.9117	0.8790	0.8446	0.8091	0.7732	0.7373
3.60	0.9891	0.9706	0.9468	0.9192	0.8889	0.8568	0.8236	0.7898	0.7558
3.65	0.9901	0.9734	0.9516	0.9261	0.8981	0.8683	0.8372	0.8055	0.7735
3.70	0.9911	0.9759	0.9560	0.9326	0.9067	0.8790	0.8501	0.8204	0.7903
3.75	0.9920	0.9782	0.9600	0.9386	0.9147	0.8891	0.8622	0.8345	0.8062
3.80	0.9928	0.9803	0.9637	0.9441	0.9222	0.8985	0.8736	0.8477	0.8212
3.85	0.9935	0.9822	0.9672	0.9493	0.9291	0.9073	0.8842	0.8602	0.8355
3.90	0.9942	0.9840	0.9703	0.9540	0.9355	0.9155	0.8941	0.8718	0.8488
3.95	0.9948	0.9856	0.9732	0.9583	0.9415	0.9230	0.9034	0.8827	0.8614
4.00	0.9953	0.9870	0.9758	0.9623	0.9469	0.9300	0.9120	0.8929	0.8731
4.05	0.9958	0.9883	0.9782	0.9660	0.9520	0.9365	0.9199	0.9024	0.8841
4.10	0.9963	0.9895	0.9804	0.9693	0.9566	0.9425	0.9273	0.9112	0.8943
4.15	0.9967	0.9906	0.9824	0.9724	0.9608	0.9480	0.9341	0.9193	0.9038
4.20	0.9970	0.9916	0.9842	0.9752	0.9647	0.9530	0.9404	0.9268	0.9126
4.25	0.9973	0.9925	0.9859	0.9777	0.9682	0.9576	0.9461	0.9338	0.9208
4.30	0.9976	0.9933	0.9874	0.9800	0.9715	0.9619	0.9514	0.9402	0.9283
4.35	0.9979	0.9941	0.9887	0.9821	0.9744	0.9657	0.9562	0.9460	0.9352
4.40	0.9981	0.9947	0.9899	0.9840	0.9771	0.9692	0.9607	0.9514	0.9416
4.45	0.9983	0.9953	0.9910	0.9857	0.9795	0.9724	0.9647	0.9563	0.9474
4.50	0.9985	0.9958	0.9920	0.9873	0.9817	0.9754	0.9684	0.9608	0.9527

Probability integral of the range

W	n = 11	12	13	14	15	16	17	18	19	20
2.25	0.1134	0.0872	0.0669	0.0511	0.0390	0.0297	0.0226	0.0172	0.0130	0.0099
2.30	0.1286	0.1003	0.0779	0.0604	0.0468	0.0361	0.0279	0.0214	0.0165	0.0127
2.35	0.1450	0.1145	0.0902	0.0709	0.0556	0.0435	0.0340	0.0265	0.0207	0.0161
2.40	0.1624	0.1299	0.1036	0.0825	0.0655	0.0519	0.0411	0.0325	0.0256	0.0202
2.45	0.1810	0.1466	0.1183	0.0953	0.0766	0.0615	0.0493	0.0394	0.0315	0.0251
2.50	0.2007	0.1643	0.1342	0.1094	0.0890	0.0722	0.0585	0.0474	0.0383	0.0309
2.55	0.2213	0.1833	0.1513	0.1247	0.1025	0.0842	0.0690	0.0565	0.0462	0.0377
2.60	0.2429	0.2032	0.1696	0.1413	0.1174	0.0974	0.0807	0.0668	0.0552	0.0455
2.65	0.2653	0.2243	0.1891	0.1590	0.1335	0.1119	0.0937	0.0783	0.0654	0.0545
2.70	0.2885	0.2462	0.2096	0.1780	0.1509	0.1278	0.1080	0.0911	0.0768	0.0647
2.75	0.3124	0.2690	0.2311	0.1981	0.1696	0.1449	0.1236	0.1053	0.0896	0.0761
2.80	0.3368	0.2926	0.2536	0.2194	0.1894	0.1632	0.1405	0.1208	0.1037	0.0889
2.85	0.3618	0.3169	0.2770	0.2416	0.2103	0.1828	0.1587	0.1376	0.1191	0.1031
2.90	0.3870	0.3417	0.3011	0.2647	0.2323	0.2036	0.1782	0.1557	0.1360	0.1186
2.95	0.4125	0.3670	0.3258	0.2887	0.2553	0.2255	0.1989	0.1752	0.1541	0.1355
3.00	0.4382	0.3927	0.3511	0.3134	0.2792	0.2484	0.2207	0.1959	0.1736	0.1537
3.05	0.4639	0.4186	0.3769	0.3387	0.3039	0.2723	0.2436	0.2177	0.1944	0.1733
3.10	0.4895	0.4446	0.4029	0.3645	0.3292	0.2969	0.2675	0.2407	0.2163	0.1942
3.15	0.5150	0.4706	0.4291	0.3907	0.3551	0.3223	0.2922	0.2646	0.2394	0.2163
3.20	0.5401	0.4965	0.4554	0.4171	0.3814	0.3483	0.3177	0.2894	0.2634	0.2395
3.25	0.5649	0.5222	0.4817	0.4437	0.4080	0.3748	0.3438	0.3151	0.2884	0.2638
3.30	0.5893	0.5475	0.5078	0.4703	0.4348	0.4016	0.3704	0.3413	0.3142	0.2890
3.35	0.6131	0.5725	0.5337	0.4967	0.4617	0.4286	0.3974	0.3681	0.3407	0.3150
3.40	0.6363	0.5970	0.5592	0.5230	0.4885	0.4557	0.4246	0.3953	0.3676	0.3416
3.45	0.6589	0.6209	0.5842	0.5489	0.5150	0.4827	0.4519	0.4227	0.3950	0.3688
3.50	0.6807	0.6442	0.6087	0.5744	0.5413	0.5096	0.4792	0.4502	0.4226	0.3964
3.55	0.7017	0.6668	0.6326	0.5994	0.5672	0.5362	0.5063	0.4777	0.4504	0.4242
3.60	0.7220	0.6886	0.6558	0.6237	0.5926	0.5624	0.5332	0.5051	0.4781	0.4522
3.65	0.7414	0.7096	0.6782	0.6474	0.6173	0.5881	0.5597	0.5322	0.5056	0.4801
3.70	0.7600	0.7298	0.6999	0.6704	0.6414	0.6132	0.5856	0.5588	0.5329	0.5078
3.75	0.7776	0.7491	0.7206	0.6925	0.6648	0.6376	0.6110	0.5850	0.5598	0.5352
3.80	0.7944	0.7675	0.7406	0.7138	0.6874	0.6613	0.6357	0.6106	0.5861	0.5622
3.85	0.8103	0.7850	0.7596	0.7342	0.7090	0.6842	0.6596	0.6355	0.6118	0.5887
3.90	0.8254	0.8016	0.7777	0.7537	0.7298	0.7062	0.6827	0.6596	0.6369	0.6145
3.95	0.8395	0.8173	0.7948	0.7723	0.7497	0.7273	0.7050	0.6829	0.6611	0.6397
4.00	0.8528	0.8321	0.8111	0.7899	0.7686	0.7474	0.7263	0.7053	0.6845	0.6640
4.05	0.8653	0.8460	0.8264	0.8066	0.7866	0.7666	0.7466	0.7268	0.7070	0.6874
4.10	0.8769	0.8590	0.8408	0.8223	0.8036	0.7848	0.7660	0.7472	0.7285	0.7099
4.15	0.8878	0.8712	0.8543	0.8371	0.8196	0.8021	0.7844	0.7667	0.7491	0.7315
4.20	0.8978	0.8826	0.8669	0.8509	0.8347	0.8183	0.8018	0.7852	0.7686	0.7520
4.25	0.9072	0.8931	0.8787	0.8639	0.8488	0.8336	0.8182	0.8027	0.7871	0.7715
4.30	0.9158	0.9029	0.8896	0.8760	0.8620	0.8479	0.8336	0.8191	0.8046	0.7899
4.35	0.9238	0.9120	0.8998	0.8872	0.8744	0.8613	0.8480	0.8346	0.8210	0.8074
4.40	0.9312	0.9204	0.9092	0.8976	0.8858	0.8737	0.8615	0.8490	0.8364	0.8237
4.45	0.9379	0.9281	0.9178	0.9073	0.8964	0.8853	0.8740	0.8625	0.8508	0.8391
4.50	0.9441	0.9352	0.9258	0.9162	0.9062	0.8960	0.8856	0.8750	0.8643	0.8534

Probability integral of the range

W	n = 2	3	4	5	6	7	8	9	10
4.50	0.9985	0.9958	0.9920	0.9873	0.9817	0.9754	0.9684	0.9608	0.9527
4.55	0.9987	0.9963	0.9929	0.9887	0.9837	0.9780	0.9717	0.9649	0.9576
4.60	0.9989	0.9967	0.9937	0.9899	0.9855	0.9804	0.9747	0.9686	0.9620
4.65	0.9990	0.9971	0.9944	0.9911	0.9871	0.9825	0.9775	0.9719	0.9660
4.70	0.9991	0.9974	0.9951	0.9921	0.9885	0.9845	0.9799	0.9750	0.9696
4.75	0.9992	0.9977	0.9956	0.9930	0.9898	0.9862	0.9822	0.9777	0.9729
4.80	0.9993	0.9980	0.9962	0.9938	0.9910	0.9878	0.9842	0.9802	0.9759
4.85	0.9994	0.9982	0.9966	0.9945	0.9920	0.9892	0.9860	0.9824	0.9786
4.90	0.9995	0.9985	0.9970	0.9952	0.9930	0.9904	0.9876	0.9844	0.9810
4.95	0.9995	0.9986	0.9974	0.9958	0.9938	0.9916	0.9890	0.9862	0.9832
5.00	0.9996	0.9988	0.9977	0.9963	0.9945	0.9926	0.9903	0.9878	0.9851
5.05	0.9996	0.9990	0.9980	0.9967	0.9952	0.9935	0.9915	0.9893	0.9869
5.10	0.9997	0.9991	0.9982	0.9971	0.9958	0.9942	0.9925	0.9906	0.9884
5.15	0.9997	0.9992	0.9985	0.9975	0.9963	0.9950	0.9934	0.9917	0.9898
5.20	0.9998	0.9993	0.9987	0.9978	0.9968	0.9956	0.9942	0.9927	0.9911
5.25	0.9998	0.9994	0.9988	0.9981	0.9972	0.9961	0.9949	0.9936	0.9922
5.30	0.9998	0.9995	0.9990	0.9983	0.9975	0.9966	0.9956	0.9944	0.9931
5.35	0.9998	0.9995	0.9991	0.9985	0.9979	0.9971	0.9961	0.9951	0.9940
5.40	0.9999	0.9996	0.9992	0.9987	0.9981	0.9974	0.9966	0.9957	0.9948
5.45	0.9999	0.9997	0.9993	0.9989	0.9984	0.9978	0.9971	0.9963	0.9954
5.50	0.9999	0.9997	0.9994	0.9990	0.9986	0.9981	0.9974	0.9968	0.9960
5.55	0.9999	0.9997	0.9995	0.9992	0.9988	0.9983	0.9978	0.9972	0.9965
5.60	0.9999	0.9998	0.9996	0.9993	0.9989	0.9985	0.9981	0.9976	0.9970
5.65	0.9999	0.9998	0.9996	0.9994	0.9991	0.9987	0.9983	0.9979	0.9974
5.70	0.9999	0.9998	0.9997	0.9995	0.9992	0.9989	0.9986	0.9982	0.9977
5.75	1.0000	0.9999	0.9997	0.9995	0.9993	0.9991	0.9988	0.9984	0.9980
5.80		0.9999	0.9998	0.9996	0.9994	0.9992	0.9989	0.9986	0.9983
5.85		0.9999	0.9998	0.9997	0.9995	0.9993	0.9991	0.9988	0.9985
5.90		0.9999	0.9998	0.9997	0.9996	0.9994	0.9992	0.9990	0.9988
5.95		0.9999	0.9998	0.9997	0.9996	0.9995	0.9993	0.9991	0.9989
6.00		0.9999	0.9999	0.9998	0.9997	0.9996	0.9994	0.9993	0.9991

Probability integral of the range

W	$n = 11$	12	13	14	15	16	17	18	19	20
4.50	0.9441	0.9352	0.9258	0.9162	0.9062	0.8960	0.8856	0.8750	0.8643	0.8534
4.55	0.9498	0.9417	0.9332	0.9244	0.9153	0.9060	0.8964	0.8867	0.8768	0.8667
4.60	0.9550	0.9476	0.9399	0.9319	0.9236	0.9151	0.9064	0.8975	0.8884	0.8791
4.65	0.9597	0.9530	0.9460	0.9388	0.9313	0.9235	0.9155	0.9074	0.8991	0.8906
4.70	0.9639	0.9579	0.9516	0.9451	0.9382	0.9312	0.9240	0.9165	0.9089	0.9012
4.75	0.9678	0.9624	0.9567	0.9508	0.9446	0.9383	0.9317	0.9249	0.9180	0.9110
4.80	0.9713	0.9665	0.9614	0.9560	0.9505	0.9447	0.9387	0.9326	0.9263	0.9199
4.85	0.9745	0.9702	0.9656	0.9608	0.9557	0.9505	0.9452	0.9396	0.9339	0.9281
4.90	0.9774	0.9735	0.9694	0.9650	0.9605	0.9559	0.9510	0.9460	0.9409	0.9356
4.95	0.9799	0.9765	0.9728	0.9689	0.9649	0.9607	0.9563	0.9518	0.9472	0.9424
5.00	0.9822	0.9791	0.9759	0.9724	0.9688	0.9650	0.9611	0.9571	0.9529	0.9486
5.05	0.9843	0.9816	0.9786	0.9756	0.9723	0.9690	0.9655	0.9618	0.9581	0.9543
5.10	0.9862	0.9837	0.9811	0.9784	0.9755	0.9725	0.9694	0.9661	0.9628	0.9593
5.15	0.9878	0.9856	0.9833	0.9809	0.9783	0.9757	0.9729	0.9700	0.9670	0.9639
5.20	0.9893	0.9874	0.9853	0.9832	0.9809	0.9785	0.9760	0.9735	0.9708	0.9681
5.25	0.9906	0.9889	0.9871	0.9852	0.9832	0.9811	0.9789	0.9766	0.9742	0.9718
5.30	0.9917	0.9903	0.9887	0.9870	0.9852	0.9833	0.9814	0.9794	0.9773	0.9751
5.35	0.9928	0.9915	0.9901	0.9886	0.9870	0.9854	0.9836	0.9819	0.9800	0.9781
5.40	0.9937	0.9925	0.9913	0.9900	0.9886	0.9872	0.9856	0.9841	0.9824	0.9807
5.45	0.9945	0.9935	0.9924	0.9913	0.9900	0.9888	0.9874	0.9860	0.9846	0.9831
5.50	0.9952	0.9943	0.9934	0.9924	0.9913	0.9902	0.9890	0.9878	0.9865	0.9852
5.55	0.9958	0.9951	0.9942	0.9933	0.9924	0.9914	0.9904	0.9893	0.9882	0.9870
5.60	0.9964	0.9957	0.9950	0.9942	0.9934	0.9925	0.9916	0.9907	0.9897	0.9887
5.65	0.9969	0.9963	0.9956	0.9950	0.9943	0.9935	0.9927	0.9919	0.9910	0.9901
5.70	0.9973	0.9968	0.9962	0.9956	0.9950	0.9944	0.9937	0.9930	0.9922	0.9914
5.75	0.9976	0.9972	0.9967	0.9962	0.9957	0.9951	0.9945	0.9939	0.9932	0.9925
5.80	0.9980	0.9976	0.9972	0.9967	0.9963	0.9958	0.9952	0.9947	0.9941	0.9935
5.85	0.9982	0.9979	0.9976	0.9972	0.9968	0.9963	0.9959	0.9954	0.9949	0.9944
5.90	0.9985	0.9982	0.9979	0.9976	0.9972	0.9968	0.9964	0.9960	0.9956	0.9952
5.95	0.9987	0.9985	0.9982	0.9979	0.9976	0.9973	0.9969	0.9966	0.9962	0.9958
6.00	0.9989	0.9987	0.9984	0.9982	0.9979	0.9977	0.9974	0.9971	0.9967	0.9964

Probability integral of the range

W	n = 2	3	4	5	6	7	8	9	10
6.00	1.0000	0.9999	0.9999	0.9998	0.9997	0.9996	0.9994	0.9993	0.9991
6.05		0.9999	0.9999	0.9998	0.9997	0.9996	0.9995	0.9994	0.9992
6.10		1.0000	0.9999	0.9998	0.9998	0.9997	0.9996	0.9995	0.9993
6.15			0.9999	0.9999	0.9998	0.9997	0.9996	0.9995	0.9994
6.20			0.9999	0.9999	0.9998	0.9998	0.9997	0.9996	0.9995
6.25			0.9999	0.9999	0.9999	0.9998	0.9997	0.9997	0.9996
6.30			1.0000	0.9999	0.9999	0.9998	0.9998	0.9997	0.9996
6.35				0.9999	0.9999	0.9999	0.9998	0.9998	0.9997
6.40				0.9999	0.9999	0.9999	0.9998	0.9998	0.9997
6.45				0.9999	0.9999	0.9999	0.9999	0.9998	0.9998
6.50				1.0000	0.9999	0.9999	0.9999	0.9999	0.9998
6.55					0.9999	0.9999	0.9999	0.9999	0.9998
6.60					1.0000	0.9999	0.9999	0.9999	0.9999
6.65						0.9999	0.9999	0.9999	0.9999
6.70						1.0000	0.9999	0.9999	0.9999
6.75							1.0000	0.9999	0.9999
6.80								0.9999	0.9999
6.85								1.0000	0.9999
6.90									1.0000
6.95									
7.00									
7.05									
7.10									
7.15									
7.20									
7.25									

Probability integral of the range

W	$n = 11$	12	13	14	15	16	17	18	19	20
6.00	0.9989	0.9987	0.9984	0.9982	0.9979	0.9977	0.9974	0.9971	0.9967	0.9964
6.05	0.9990	0.9989	0.9987	0.9985	0.9982	0.9980	0.9977	0.9975	0.9972	0.9969
6.10	0.9992	0.9990	0.9989	0.9987	0.9985	0.9983	0.9981	0.9978	0.9976	0.9973
6.15	0.9993	0.9992	0.9990	0.9989	0.9987	0.9985	0.9983	0.9981	0.9979	0.9977
6.20	0.9994	0.9993	0.9992	0.9990	0.9989	0.9987	0.9986	0.9984	0.9982	0.9980
6.25	0.9995	0.9994	0.9993	0.9992	0.9990	0.9989	0.9988	0.9986	0.9985	0.9983
6.30	0.9996	0.9995	0.9994	0.9993	0.9992	0.9991	0.9990	0.9988	0.9987	0.9986
6.35	0.9996	0.9996	0.9995	0.9994	0.9993	0.9992	0.9991	0.9990	0.9989	0.9988
6.40	0.9997	0.9996	0.9996	0.9995	0.9994	0.9993	0.9992	0.9992	0.9991	0.9990
6.45	0.9997	0.9997	0.9996	0.9996	0.9995	0.9994	0.9994	0.9993	0.9992	0.9991
6.50	0.9998	0.9997	0.9997	0.9996	0.9996	0.9995	0.9995	0.9994	0.9993	0.9993
6.55	0.9998	0.9998	0.9997	0.9997	0.9996	0.9996	0.9995	0.9995	0.9994	0.9994
6.60	0.9998	0.9998	0.9998	0.9997	0.9997	0.9997	0.9996	0.9996	0.9995	0.9995
6.65	0.9999	0.9998	0.9998	0.9998	0.9997	0.9997	0.9997	0.9996	0.9996	0.9995
6.70	0.9999	0.9999	0.9998	0.9998	0.9998	0.9998	0.9997	0.9997	0.9997	0.9996
6.75	0.9999	0.9999	0.9999	0.9998	0.9998	0.9998	0.9998	0.9997	0.9997	0.9997
6.80	0.9999	0.9999	0.9999	0.9999	0.9998	0.9998	0.9998	0.9998	0.9998	0.9997
6.85	0.9999	0.9999	0.9999	0.9999	0.9999	0.9999	0.9998	0.9998	0.9998	0.9998
6.90	0.9999	0.9999	0.9999	0.9999	0.9999	0.9999	0.9999	0.9998	0.9998	0.9998
6.95	1.0000	0.9999	0.9999	0.9999	0.9999	0.9999	0.9999	0.9999	0.9999	0.9998
7.00		1.0000	0.9999	0.9999	0.9999	0.9999	0.9999	0.9999	0.9999	0.9999
7.05			1.0000	0.9999	0.9999	0.9999	0.9999	0.9999	0.9999	0.9999
7.10				1.0000	0.9999	0.9999	0.9999	0.9999	0.9999	0.9999
7.15					1.0000	1.0000	0.9999	0.9999	0.9999	0.9999
7.20							1.0000	0.9999	0.9999	0.9999
7.25								1.0000	1.0000	0.9999

4.7.2 Percentage points, studentized range

The standardized range is $W = R/\sigma$ as defined in the previous section. If the population standard deviation σ is replaced by the sample standard deviation s (computed from another sample from the same population), then the studentized range Q is given by $Q = R/S$. Here, R is the range of the sample of size n and S is the independent of R and has ν degrees of freedom. The probability integral for the studentized range is given by

$$\text{Prob}\,[Q \le q] = \text{Prob}\left[\frac{R}{S} \le q\right] = \int_0^\infty \frac{2^{1-\nu/2}\nu^{\nu/2}s^{\nu-1}e^{-\nu s^2/2}f(qs)}{\Gamma(\nu/2)}\,ds \quad (4.64)$$

where f is the probability integral of the range for samples of size n.

The following tables provide values of the studentized range for the normal density function $f(x) = \frac{1}{\sqrt{2\pi}}e^{-x^2/2}$.

Upper 1% points of the studentized range

The entries are $q_{.01}$ where $\text{Prob}\,[Q < q_{.01}] = .99$.

$\nu/n =$	2	3	4	5	6	7	8	9	10	11	12	13	14	15	16	17	18	19	20
1	77.75	129.44	147.54	170.27	188.43	202.60	215.08	225.53	234.69	242.85	250.15	255.43	261.60	238.57	243.92	248.69	253.42	257.88	262.10
2	13.58	18.33	22.09	24.66	26.66	28.28	29.68	30.87	30.63	31.76	32.78	33.71	34.57	35.36	36.10	36.79	37.44	38.05	38.63
3	8.10	10.54	12.18	13.35	13.96	14.86	15.63	16.29	16.88	17.40	17.87	18.31	18.70	19.07	19.41	19.73	20.04	20.32	20.71
4	6.33	8.11	9.20	9.75	10.46	11.03	11.52	11.94	12.31	12.64	12.94	13.22	13.47	13.70	13.92	14.13	14.32	14.50	14.67
5	5.64	6.98	7.82	8.28	8.81	9.23	9.59	9.91	10.18	10.43	10.65	10.86	11.05	11.22	11.38	11.53	11.67	11.81	11.94
6	5.23	6.34	7.05	7.44	7.87	8.22	8.51	8.77	8.99	9.19	9.37	9.54	9.69	9.83	9.96	10.09	10.20	10.31	10.42
7	4.94	5.93	6.55	6.91	7.28	7.57	7.83	8.05	8.38	8.56	8.72	8.87	9.01	9.13	9.25	9.36	9.47	9.57	9.66
8	4.75	5.65	6.12	6.54	6.87	7.13	7.48	7.69	7.87	8.04	8.18	8.32	8.44	8.56	8.66	8.77	8.86	8.95	9.03
9	4.60	5.44	5.89	6.27	6.57	6.92	7.14	7.33	7.50	7.65	7.79	7.91	8.03	8.13	8.23	8.33	8.42	8.50	8.58
10	4.49	5.20	5.71	6.07	6.35	6.68	6.88	7.06	7.22	7.36	7.49	7.60	7.71	7.81	7.91	7.99	8.08	8.15	8.23
11	4.40	5.09	5.57	5.91	6.17	6.48	6.68	6.85	7.00	7.13	7.25	7.36	7.46	7.56	7.65	7.73	7.81	7.88	7.95
12	4.33	5.00	5.46	5.78	6.03	6.33	6.51	6.67	6.82	6.95	7.06	7.17	7.26	7.35	7.44	7.52	7.59	7.66	7.73
13	4.27	4.93	5.37	5.68	5.92	6.20	6.38	6.53	6.67	6.80	6.90	7.01	7.10	7.19	7.27	7.34	7.42	7.48	7.55
14	4.22	4.86	5.29	5.59	5.82	6.09	6.26	6.41	6.55	6.67	6.77	6.87	6.96	7.05	7.12	7.20	7.27	7.33	7.39
15	4.18	4.81	5.22	5.51	5.74	6.00	6.17	6.31	6.44	6.56	6.66	6.76	6.84	6.93	7.00	7.07	7.14	7.20	7.26
16	4.08	4.76	5.16	5.45	5.67	5.92	6.08	6.23	6.35	6.47	6.57	6.66	6.74	6.82	6.90	6.97	7.03	7.09	7.15
17	4.05	4.72	5.12	5.39	5.61	5.79	6.01	6.15	6.27	6.38	6.48	6.57	6.66	6.73	6.81	6.87	6.94	6.99	7.05
18	4.03	4.69	5.07	5.35	5.56	5.73	5.95	6.08	6.20	6.31	6.41	6.50	6.58	6.65	6.72	6.79	6.85	6.91	6.97
19	4.01	4.66	5.03	5.30	5.51	5.68	5.89	6.03	6.14	6.25	6.34	6.43	6.51	6.58	6.65	6.72	6.78	6.84	6.89
20	3.99	4.63	5.00	5.27	5.47	5.64	5.84	5.97	6.09	6.19	6.29	6.37	6.45	6.52	6.59	6.66	6.71	6.77	6.82
25	3.92	4.52	4.87	5.12	5.32	5.48	5.61	5.73	5.89	5.99	6.07	6.15	6.23	6.29	6.36	6.42	6.47	6.52	6.57
30	3.87	4.45	4.79	5.03	5.22	5.37	5.50	5.61	5.71	5.80	5.94	6.01	6.08	6.14	6.20	6.26	6.31	6.36	6.41
40	3.82	4.37	4.70	4.93	5.10	5.25	5.37	5.48	5.57	5.66	5.73	5.80	5.87	5.93	5.98	6.03	6.08	6.13	6.17
60	3.76	4.29	4.60	4.82	4.99	5.13	5.25	5.35	5.44	5.52	5.60	5.66	5.72	5.78	5.83	5.88	5.93	5.97	6.01
120	3.71	4.20	4.50	4.71	4.87	5.00	5.12	5.21	5.30	5.37	5.44	5.50	5.56	5.61	5.66	5.70	5.75	5.79	5.82
1000	3.64	4.13	4.42	4.62	4.78	4.91	5.02	5.11	5.20	5.27	5.34	5.40	5.46	5.51	5.56	5.61	5.65	5.69	5.73

Upper 5% points of the studentized range

The entries are $q_{.05}$ where Prob$[Q < q_{.05}] = .95$.

ν	$n=2$	3	4	5	6	7	8	9	10	11	12	13	14	15	16	17	18	19	20
1	17.79	26.70	32.79	37.07	39.84	42.67	45.05	47.10	48.89	50.49	51.91	53.14	54.28	55.36	56.34	57.22	58.05	58.83	59.56
2	6.10	8.31	9.81	10.89	11.70	12.52	13.25	13.87	14.42	14.91	15.35	15.75	16.13	16.47	16.79	17.09	17.38	17.64	17.90
3	4.50	5.91	6.83	7.46	8.03	8.50	8.90	9.24	9.56	9.84	10.09	10.31	10.52	10.72	10.90	11.06	11.22	11.37	11.52
4	3.93	5.04	5.76	6.26	6.68	7.03	7.32	7.58	7.80	8.00	8.19	8.35	8.50	8.65	8.78	8.90	9.02	9.12	9.23
5	3.63	4.60	5.22	5.65	6.04	6.33	6.58	6.80	7.00	7.17	7.33	7.47	7.60	7.72	7.83	7.94	8.03	8.12	8.21
6	3.46	4.34	4.88	5.28	5.63	5.90	6.12	6.32	6.49	6.65	6.79	6.92	7.04	7.14	7.24	7.34	7.43	7.51	7.59
7	3.35	4.15	4.67	5.05	5.36	5.61	5.82	6.00	6.16	6.30	6.43	6.55	6.66	6.76	6.85	6.94	7.02	7.10	7.17
8	3.26	4.03	4.53	4.88	5.15	5.40	5.60	5.77	5.92	6.06	6.18	6.29	6.39	6.48	6.57	6.65	6.73	6.80	6.87
9	3.20	3.95	4.42	4.75	5.01	5.25	5.43	5.60	5.74	5.87	5.98	6.09	6.19	6.28	6.36	6.44	6.51	6.58	6.64
10	3.15	3.88	4.33	4.65	4.90	5.11	5.31	5.46	5.60	5.72	5.83	5.94	6.03	6.12	6.19	6.27	6.34	6.41	6.47
11	3.11	3.83	4.27	4.58	4.82	5.02	5.20	5.35	5.49	5.61	5.71	5.81	5.90	5.99	6.06	6.14	6.20	6.27	6.33
12	3.08	3.78	4.21	4.52	4.75	4.94	5.11	5.27	5.40	5.51	5.62	5.71	5.80	5.88	5.95	6.02	6.09	6.15	6.21
13	3.06	3.75	4.17	4.46	4.69	4.88	5.04	5.18	5.32	5.43	5.53	5.63	5.71	5.79	5.86	5.93	6.00	6.06	6.11
14	3.04	3.72	4.13	4.42	4.65	4.83	4.99	5.12	5.24	5.35	5.46	5.56	5.64	5.71	5.79	5.85	5.92	5.97	6.03
15	3.02	3.69	4.10	4.38	4.61	4.79	4.94	5.08	5.19	5.30	5.39	5.48	5.56	5.64	5.70	5.79	5.85	5.90	5.96
16	3.00	3.67	4.07	4.35	4.57	4.75	4.90	5.03	5.15	5.26	5.35	5.44	5.51	5.59	5.66	5.72	5.78	5.84	5.89
17	2.99	3.65	4.05	4.33	4.54	4.72	4.87	5.00	5.11	5.22	5.31	5.39	5.47	5.55	5.61	5.68	5.73	5.79	5.84
18	2.97	3.63	4.02	4.30	4.52	4.69	4.84	4.97	5.08	5.18	5.28	5.36	5.44	5.51	5.57	5.64	5.70	5.75	5.80
19	2.95	3.62	4.00	4.28	4.49	4.67	4.81	4.94	5.05	5.15	5.24	5.33	5.40	5.48	5.54	5.60	5.66	5.72	5.77
20	2.94	3.60	3.99	4.26	4.47	4.65	4.79	4.92	5.03	5.13	5.22	5.30	5.38	5.45	5.52	5.58	5.63	5.69	5.74
25	2.91	3.55	3.92	4.19	4.39	4.56	4.71	4.83	4.94	5.03	5.12	5.20	5.28	5.35	5.41	5.47	5.53	5.58	5.63
30	2.88	3.51	3.88	4.14	4.34	4.51	4.65	4.77	4.88	4.97	5.06	5.14	5.21	5.28	5.34	5.40	5.46	5.51	5.56
40	2.86	3.47	3.83	4.09	4.28	4.44	4.58	4.70	4.80	4.90	4.98	5.06	5.13	5.20	5.26	5.30	5.35	5.40	5.45
60	2.83	3.43	3.78	4.03	4.22	4.38	4.52	4.63	4.73	4.81	4.89	4.97	5.04	5.10	5.16	5.22	5.27	5.32	5.36
120	2.80	3.36	3.68	3.91	4.10	4.25	4.38	4.49	4.59	4.67	4.75	4.83	4.89	4.95	5.01	5.07	5.12	5.16	5.20
1000	2.77	3.35	3.68	3.92	4.11	4.25	4.37	4.48	4.58	4.67	4.75	4.82	4.88	4.95	5.00	5.06	5.11	5.15	5.20

Upper 10% points of the studentized range

The entries are $q_{.10}$ where Prob $[Q < q_{.10}] = .90$.

ν	$n=2$	3	4	5	6	7	8	9	10	11	12	13	14	15	16	17	18	19	20
1	8.94	13.43	16.37	18.43	20.12	21.49	22.63	23.61	24.47	25.23	25.91	26.52	26.81	27.46	28.06	28.62	29.15	29.65	30.18
2	4.13	5.73	6.77	7.54	8.14	8.63	9.05	9.41	9.73	10.01	10.26	10.49	10.70	10.89	11.07	11.24	11.40	11.54	11.63
3	3.33	4.47	5.20	5.74	6.16	6.51	6.81	7.06	7.29	7.49	7.67	7.83	7.98	8.12	8.25	8.37	8.48	8.58	8.68
4	3.01	3.98	4.59	5.03	5.39	5.68	5.92	6.14	6.33	6.49	6.64	6.78	6.91	7.02	7.13	7.23	7.33	7.41	7.50
5	2.85	3.72	4.26	4.66	4.98	5.24	5.46	5.65	5.81	5.96	6.10	6.22	6.33	6.44	6.53	6.62	6.71	6.79	6.86
6	2.75	3.56	4.07	4.44	4.73	4.97	5.17	5.34	5.50	5.64	5.76	5.87	5.98	6.07	6.16	6.25	6.32	6.40	6.46
7	2.68	3.45	3.93	4.28	4.55	4.78	4.97	5.14	5.28	5.41	5.53	5.64	5.73	5.82	5.91	5.99	6.06	6.13	6.19
8	2.63	3.37	3.83	4.17	4.43	4.65	4.83	4.99	5.12	5.25	5.36	5.46	5.56	5.64	5.72	5.80	5.87	5.93	6.00
9	2.59	3.32	3.76	4.08	4.34	4.54	4.72	4.87	5.01	5.13	5.23	5.33	5.42	5.50	5.58	5.65	5.72	5.79	5.84
10	2.56	3.27	3.70	4.02	4.26	4.47	4.64	4.78	4.91	5.03	5.13	5.23	5.32	5.40	5.47	5.54	5.61	5.67	5.73
11	2.54	3.23	3.66	3.97	4.20	4.40	4.57	4.71	4.84	4.95	5.05	5.15	5.23	5.31	5.38	5.45	5.51	5.57	5.63
12	2.52	3.20	3.62	3.92	4.16	4.35	4.51	4.65	4.78	4.89	4.99	5.08	5.16	5.24	5.31	5.37	5.44	5.49	5.55
13	2.51	3.18	3.59	3.89	4.12	4.30	4.46	4.60	4.72	4.83	4.93	5.02	5.10	5.18	5.25	5.31	5.37	5.43	5.48
14	2.49	3.16	3.56	3.85	4.08	4.27	4.42	4.56	4.68	4.79	4.88	4.97	5.05	5.12	5.19	5.26	5.32	5.37	5.43
15	2.48	3.14	3.54	3.83	4.05	4.24	4.39	4.52	4.64	4.75	4.84	4.93	5.01	5.08	5.15	5.21	5.27	5.32	5.38
16	2.47	3.12	3.52	3.81	4.03	4.21	4.36	4.49	4.61	4.71	4.81	4.89	4.97	5.04	5.11	5.17	5.23	5.28	5.33
17	2.46	3.11	3.50	3.78	4.00	4.18	4.33	4.46	4.58	4.68	4.77	4.86	4.93	5.01	5.07	5.13	5.19	5.24	5.30
18	2.45	3.10	3.49	3.77	3.98	4.16	4.31	4.44	4.55	4.66	4.75	4.83	4.91	4.98	5.04	5.10	5.16	5.21	5.26
19	2.44	3.09	3.47	3.75	3.97	4.14	4.29	4.42	4.53	4.63	4.72	4.80	4.88	4.95	5.01	5.07	5.13	5.18	5.23
20	2.43	3.08	3.46	3.74	3.95	4.12	4.27	4.40	4.51	4.61	4.70	4.78	4.86	4.92	4.99	5.05	5.10	5.16	5.21
25	2.41	3.04	3.42	3.68	3.89	4.06	4.20	4.32	4.43	4.53	4.62	4.69	4.77	4.83	4.89	4.95	5.00	5.06	5.10
30	2.40	3.02	3.39	3.65	3.85	4.02	4.16	4.28	4.38	4.47	4.56	4.64	4.71	4.77	4.83	4.89	4.94	4.99	5.04
40	2.38	2.99	3.35	3.61	3.80	3.96	4.10	4.22	4.32	4.41	4.49	4.56	4.63	4.69	4.75	4.81	4.86	4.91	4.95
60	2.36	2.96	3.31	3.56	3.76	3.91	4.04	4.16	4.25	4.34	4.42	4.49	4.56	4.62	4.68	4.73	4.78	4.82	4.87
120	2.34	2.93	3.28	3.52	3.71	3.86	3.99	4.10	4.19	4.28	4.35	4.42	4.49	4.54	4.60	4.65	4.69	4.74	4.78
1000	2.33	2.90	3.24	3.48	3.66	3.81	3.93	4.04	4.13	4.21	4.28	4.35	4.41	4.48	4.53	4.59	4.64	4.68	4.73

CHAPTER 5

Discrete Probability Distributions

Contents

This chapter presents some common discrete probability distributions along with their properties. Relevant numerical tables are also included.

Notation used throughout this chapter:

Probability mass function (pmf)	$p(x) = \text{Prob}\,[X = x]$
Mean	$\mu = \text{E}\,[X]$
Variance	$\sigma^2 = \text{E}\,\big[(X - \mu)^2\big]$
Coefficient of skewness	$\beta_1 = \text{E}\,\big[(X - \mu)^3\big]$
Coefficient of kurtosis	$\beta_2 = \text{E}\,\big[(X - \mu)^4\big]$
Moment generating function (mgf)	$m(t) = \text{E}\,\big[e^{tX}\big]$
Characteristic function (char function)	$\phi(t) = \text{E}\,\big[e^{itX}\big]$
Factorial moment generating function (fact mgf)	$P(t) = \text{E}\,\big[t^X\big]$

5.1   BERNOULLI DISTRIBUTION

A Bernoulli distribution is used to describe an experiment in which there are only two possible outcomes, typically a *success* or a *failure*. This type of experiment is called a *Bernoulli trial*, or simply a *trial*. The probability of a success is p and a sequence of Bernoulli trials is referred to as *repeated trials*.

5.1.1   Properties

$$\text{pmf} \quad p(x) = \begin{cases} q & x = 0 \\ p & x = 1 \end{cases} \quad \text{(or } p^x q^{1-x} \text{ for } x = 0, 1)$$

$$0 \le p \le 1, \quad q = 1 - p$$

mean $\mu = p$

variance $\sigma^2 = pq$

skewness $\beta_1 = \dfrac{1 - 2p}{\sqrt{pq}}$

kurtosis $\beta_2 = 3 + \dfrac{1 - 6pq}{pq}$

mgf $m(t) = q + pe^t$

char function $\quad \phi(t) = q + pe^{it}$

fact mgf $\quad P(t) = q + pt$

5.1.2 Variates

(1) Let $X_1, X_2, \ldots, X_n$ be independent, identically distributed (iid) Bernoulli random variables with probability of a success p. The random variable $Y = X_1 + X_2 + \cdots + X_n$ has a binomial distribution with parameters n and p.

5.2 BETA BINOMIAL DISTRIBUTION

The beta binomial distribution is also known as the *negative hypergeometric distribution, inverse hypergeometric distribution, hypergeometric waiting–time distribution*, and *Markov–Pólya distribution*.

5.2.1 Properties

pmf $\quad p(x) = \dbinom{-a}{n} \dbinom{-b}{n-x} \bigg/ \dbinom{-a-b}{n} \qquad x = 0, 1, \ldots, n$

$$a, b, n > 0, n \text{ an integer}$$

mean $\qquad \mu = \dfrac{an}{a+b}$

variance $\qquad \sigma^2 = \dfrac{abn(a+b+n)}{(a+b)^2(a+b+1)}$

skewness $\qquad \beta_1 = \sqrt{\dfrac{(a+b+1)}{abn(a+b+n)}} \; \dfrac{(a-b)(a+b+2n)}{(a+b+2)}$

kurtosis $\qquad \beta_2 = \dfrac{(a+b)^2(a+b+1)}{abn(a+b+n)(a+b+2)(a+b+3)} \times$

$$\left[(a+b)(a+b+1+6n) + 3ab(n-2) + 6n^2 \right.$$

$$\left. - \dfrac{3abn(6-n)}{a+b} - \dfrac{18abn^2}{(a+b)^2} \right] + 3$$

mgf $\quad m(t) = {}_2F_1(a, -n; a+b; -e^t)$

char function $\quad \phi(t) = {}_2F_1(a, -n; a+b; e^{-it})$

fact mgf $\quad P(t) = {}_2F_1(a, -n; a+b; -t)$

where ${}_pF_q$ is the generalized hypergeometric function defined in Chapter 18 (see page 520).

5.2.2   Variates

Let X be a beta binomial random variable with parameters a, b, and n.

(1) If $a = b = 1$ and n is reduced by 1, then X is a rectangular (discrete uniform) random variable.

(2) As $n \to \infty$, X is approximately a binomial random variable with parameters n and $p = a/b$.

5.3   BETA PASCAL DISTRIBUTION

The beta Pascal distribution arises from a special case of the *urn scheme*.

5.3.1   Properties

$$\text{pmf} \quad p(x) = \frac{\Gamma(x)\Gamma(\nu)\Gamma(\rho + \nu)\Gamma(\nu + x - (\rho + r))}{\Gamma(r)\Gamma(x - r + 1)\Gamma(\rho)\Gamma(\nu - \rho)\Gamma(\nu + x)}$$

$$x = r, r + 1, \ldots, \quad r \in \mathcal{N}, \quad \nu > \rho > 0$$

$$\text{mean} \quad \mu = r\frac{\nu - 1}{\rho - 1}, \quad \rho > 1$$

$$\text{variance} \quad \sigma^2 = r(r + \rho - 1)\frac{(\nu - 1)(\nu - \rho)}{(\rho - 1)^2(\rho - 2)}, \quad \rho > 2$$

where $\Gamma(x)$ is the gamma function defined in Chapter 18 (see page 515).

5.4   BINOMIAL DISTRIBUTION

The binomial distribution is used to characterize the number of successes in n Bernoulli trials. It is used to model some very common experiments in which a sample of size n is taken from an infinite population such that each element is selected independently and has the same probability, p, of having a specified attribute.

5.4.1   Properties

$$\text{pmf} \quad p(x) = \binom{n}{x}p^x q^{n-x} \quad x = 0, 1, 2, \ldots, n$$

$$0 \le p \le 1, \quad q = 1 - p$$

$$\text{mean} \quad \mu = np$$

$$\text{variance} \quad \sigma^2 = npq$$

$$\text{skewness} \quad \beta_1 = \frac{1 - 2p}{\sqrt{npq}}$$

$$\text{kurtosis} \quad \beta_2 = 3 + \frac{1 - 6pq}{npq}$$

$$\text{mgf} \quad m(t) = (q + pe^t)^n$$

$$\text{char function} \quad \phi(t) = (q + pe^{it})^n$$

$$\text{fact mgf} \quad P(t) = (q + pt)^n$$

5.4.2 Variates

Let X be a binomial random variable with parameters n and p.

(1) If $n = 1$, then X is a Bernoulli random variable with probability of success p.

(2) As $n \to \infty$ if $np \geq 5$ and $n(1 - p) \geq 5$, then X is approximately normal with parameters $\mu = np$ and $\sigma^2 = np(1 - p)$.

(3) As $n \to \infty$ if $p < 0.1$ and $np < 10$, then X is approximately a Poisson random variable with parameter $\lambda = np$.

(4) Let $X_1, \ldots, X_k$ be independent, binomial random variables with parameters n_i and p, respectively. The random variable $Y = X_1 + X_2 + \cdots + X_k$ has a binomial distribution with parameters $n = n_1 + n_2 + \cdots + n_k$ and p.

5.4.3 Tables

The following tables only contain values of p up to $p = \frac{1}{2}$. By symmetry (replacing p with $1 - p$ and replacing x with $n - x$) values for $p > \frac{1}{2}$ can be reduced to the present tables.

Example 5.32: A biased coin has a probability of heads of .75; what is the probability of obtaining 5 or more heads in 8 flips?

Solution:

(S1) The answer is given by looking in cumulative distribution tables with $n = 8$, $x = 5$, and $p = 0.75$.

(S2) Making the substitutions mentioned above this is the same as $n = 8$, $x = 3$ and $p = 0.25$.

(S3) This value is in the tables and is equal to 0.8862. Hence, 89% of the time 5 or more heads would be likely to occur.

(S4) This result can be interpreted as the likelihood of flipping a biased coin that has a probability of tails equal to 25% and asking how likely it is to have 3 or fewer tails.

(S5) Note: The following tables are for the cumulative distribution function, not the probability mass function. The probability of obtaining exactly 5 heads in 8 flips is $\binom{8}{5}(0.75)^5(.25)^3 = .2076$.

Example 5.33: The probability a randomly selected home in Columbia County will lose power during a summer storm is .25. Suppose 14 homes in this county are selected at random. What is the probability exactly 4 homes will lose power, more than 6 will lose power, and between 2 and 7 (inclusive) will lose power?

Solution:

(S1) Let X be the number of homes (out of 14) that will lose power. The random variable X has a binomial distribution with parameters $n = 14$ and $p = 0.25$. Use the cumulative terms for the binomial distribution to answer each probability question.

(S2) $\operatorname{Prob}[X = 4] = \operatorname{Prob}[X \leq 4] - \operatorname{Prob}[X \leq 3]$
$$= 0.7415 - 0.5213 = 0.2202$$

(S3) $\operatorname{Prob}[X > 6] = 1 - \operatorname{Prob}[X \leq 6] = 1 - 0.9617 = 0.0383$

(S4) $\operatorname{Prob}[2 \leq X \leq 7] = \operatorname{Prob}[X \leq 7] - \operatorname{Prob}[X \leq 1]$
$$= 0.9897 - 0.1010 = 0.8887$$

Cumulative probability, Binomial distribution

n	x	$p=0.05$	0.10	0.15	0.20	0.25	0.30	0.40	0.50
2	0	0.9025	0.8100	0.7225	0.6400	0.5625	0.4900	0.3600	0.2500
	1	0.9975	0.9900	0.9775	0.9600	0.9375	0.9100	0.8400	0.7500
3	0	0.8574	0.7290	0.6141	0.5120	0.4219	0.3430	0.2160	0.1250
	1	0.9928	0.9720	0.9393	0.8960	0.8438	0.7840	0.6480	0.5000
	2	0.9999	0.9990	0.9966	0.9920	0.9844	0.9730	0.9360	0.8750
4	0	0.8145	0.6561	0.5220	0.4096	0.3164	0.2401	0.1296	0.0625
	1	0.9860	0.9477	0.8905	0.8192	0.7383	0.6517	0.4752	0.3125
	2	0.9995	0.9963	0.9880	0.9728	0.9492	0.9163	0.8208	0.6875
	3	1.0000	0.9999	0.9995	0.9984	0.9961	0.9919	0.9744	0.9375
5	0	0.7738	0.5905	0.4437	0.3277	0.2373	0.1681	0.0778	0.0313
	1	0.9774	0.9185	0.8352	0.7373	0.6328	0.5282	0.3370	0.1875
	2	0.9988	0.9914	0.9734	0.9421	0.8965	0.8369	0.6826	0.5000
	3	1.0000	0.9995	0.9978	0.9933	0.9844	0.9692	0.9130	0.8125
	4	1.0000	1.0000	0.9999	0.9997	0.9990	0.9976	0.9898	0.9688
6	0	0.7351	0.5314	0.3771	0.2621	0.1780	0.1177	0.0467	0.0156
	1	0.9672	0.8857	0.7765	0.6554	0.5339	0.4202	0.2333	0.1094
	2	0.9978	0.9841	0.9527	0.9011	0.8306	0.7443	0.5443	0.3438
	3	0.9999	0.9987	0.9941	0.9830	0.9624	0.9295	0.8208	0.6563
	4	1.0000	1.0000	0.9996	0.9984	0.9954	0.9891	0.9590	0.8906
	5	1.0000	1.0000	1.0000	0.9999	0.9998	0.9993	0.9959	0.9844
7	0	0.6983	0.4783	0.3206	0.2097	0.1335	0.0824	0.0280	0.0078
	1	0.9556	0.8503	0.7166	0.5767	0.4450	0.3294	0.1586	0.0625
	2	0.9962	0.9743	0.9262	0.8520	0.7564	0.6471	0.4199	0.2266
	3	0.9998	0.9973	0.9879	0.9667	0.9294	0.8740	0.7102	0.5000
	4	1.0000	0.9998	0.9988	0.9953	0.9871	0.9712	0.9037	0.7734
	5	1.0000	1.0000	0.9999	0.9996	0.9987	0.9962	0.9812	0.9375
	6	1.0000	1.0000	1.0000	1.0000	0.9999	0.9998	0.9984	0.9922
8	0	0.6634	0.4305	0.2725	0.1678	0.1001	0.0576	0.0168	0.0039
	1	0.9428	0.8131	0.6572	0.5033	0.3671	0.2553	0.1064	0.0352
	2	0.9942	0.9619	0.8948	0.7969	0.6785	0.5518	0.3154	0.1445
	3	0.9996	0.9950	0.9787	0.9437	0.8862	0.8059	0.5941	0.3633
	4	1.0000	0.9996	0.9971	0.9896	0.9727	0.9420	0.8263	0.6367
	5	1.0000	1.0000	0.9998	0.9988	0.9958	0.9887	0.9502	0.8555
	6	1.0000	1.0000	1.0000	0.9999	0.9996	0.9987	0.9915	0.9648
	7	1.0000	1.0000	1.0000	1.0000	1.0000	0.9999	0.9993	0.9961
9	0	0.6302	0.3874	0.2316	0.1342	0.0751	0.0403	0.0101	0.0019
	1	0.9288	0.7748	0.5995	0.4362	0.3003	0.1960	0.0705	0.0195
	2	0.9916	0.9470	0.8591	0.7382	0.6007	0.4628	0.2318	0.0898
	3	0.9994	0.9917	0.9661	0.9144	0.8343	0.7297	0.4826	0.2539
	4	1.0000	0.9991	0.9944	0.9804	0.9511	0.9012	0.7334	0.5000
	5	1.0000	0.9999	0.9994	0.9969	0.9900	0.9747	0.9006	0.7461
	6	1.0000	1.0000	1.0000	0.9997	0.9987	0.9957	0.9750	0.9102
	7	1.0000	1.0000	1.0000	1.0000	0.9999	0.9996	0.9962	0.9805
	8	1.0000	1.0000	1.0000	1.0000	1.0000	1.0000	0.9997	0.9980

Cumulative probability, Binomial distribution

n	x	p =0.05	0.10	0.15	0.20	0.25	0.30	0.40	0.50
10	0	0.5987	0.3487	0.1969	0.1074	0.0563	0.0283	0.0060	0.0010
	1	0.9139	0.7361	0.5443	0.3758	0.2440	0.1493	0.0464	0.0107
	2	0.9885	0.9298	0.8202	0.6778	0.5256	0.3828	0.1673	0.0547
	3	0.9990	0.9872	0.9500	0.8791	0.7759	0.6496	0.3823	0.1719
	4	0.9999	0.9984	0.9901	0.9672	0.9219	0.8497	0.6331	0.3770
	5	1.0000	0.9999	0.9986	0.9936	0.9803	0.9526	0.8338	0.6230
	6	1.0000	1.0000	0.9999	0.9991	0.9965	0.9894	0.9452	0.8281
	7	1.0000	1.0000	1.0000	0.9999	0.9996	0.9984	0.9877	0.9453
	8	1.0000	1.0000	1.0000	1.0000	1.0000	0.9999	0.9983	0.9893
	9	1.0000	1.0000	1.0000	1.0000	1.0000	1.0000	0.9999	0.9990
11	0	0.5688	0.3138	0.1673	0.0859	0.0422	0.0198	0.0036	0.0005
	1	0.8981	0.6974	0.4922	0.3221	0.1971	0.1130	0.0302	0.0059
	2	0.9848	0.9104	0.7788	0.6174	0.4552	0.3127	0.1189	0.0327
	3	0.9984	0.9815	0.9306	0.8389	0.7133	0.5696	0.2963	0.1133
	4	0.9999	0.9972	0.9841	0.9496	0.8854	0.7897	0.5328	0.2744
	5	1.0000	0.9997	0.9973	0.9883	0.9657	0.9218	0.7535	0.5000
	6	1.0000	1.0000	0.9997	0.9980	0.9924	0.9784	0.9006	0.7256
	7	1.0000	1.0000	1.0000	0.9998	0.9988	0.9957	0.9707	0.8867
	8	1.0000	1.0000	1.0000	1.0000	0.9999	0.9994	0.9941	0.9673
	9	1.0000	1.0000	1.0000	1.0000	1.0000	1.0000	0.9993	0.9941
	10	1.0000	1.0000	1.0000	1.0000	1.0000	1.0000	1.0000	0.9995
12	0	0.5404	0.2824	0.1422	0.0687	0.0317	0.0138	0.0022	0.0002
	1	0.8816	0.6590	0.4435	0.2749	0.1584	0.0850	0.0196	0.0032
	2	0.9804	0.8891	0.7358	0.5584	0.3907	0.2528	0.0834	0.0193
	3	0.9978	0.9744	0.9078	0.7946	0.6488	0.4925	0.2253	0.0730
	4	0.9998	0.9957	0.9761	0.9274	0.8424	0.7237	0.4382	0.1938
	5	1.0000	0.9995	0.9954	0.9806	0.9456	0.8821	0.6652	0.3872
	6	1.0000	1.0000	0.9993	0.9961	0.9858	0.9614	0.8418	0.6128
	7	1.0000	1.0000	0.9999	0.9994	0.9972	0.9905	0.9427	0.8062
	8	1.0000	1.0000	1.0000	0.9999	0.9996	0.9983	0.9847	0.9270
	9	1.0000	1.0000	1.0000	1.0000	1.0000	0.9998	0.9972	0.9807
	10	1.0000	1.0000	1.0000	1.0000	1.0000	1.0000	0.9997	0.9968
	11	1.0000	1.0000	1.0000	1.0000	1.0000	1.0000	1.0000	0.9998
13	0	0.5133	0.2542	0.1209	0.0550	0.0238	0.0097	0.0013	0.0001
	1	0.8646	0.6213	0.3983	0.2336	0.1267	0.0637	0.0126	0.0017
	2	0.9755	0.8661	0.6920	0.5017	0.3326	0.2025	0.0579	0.0112
	3	0.9969	0.9658	0.8820	0.7473	0.5843	0.4206	0.1686	0.0461
	4	0.9997	0.9935	0.9658	0.9009	0.7940	0.6543	0.3530	0.1334
	5	1.0000	0.9991	0.9925	0.9700	0.9198	0.8346	0.5744	0.2905
	6	1.0000	0.9999	0.9987	0.9930	0.9757	0.9376	0.7712	0.5000
	7	1.0000	1.0000	0.9998	0.9988	0.9943	0.9818	0.9023	0.7095
	8	1.0000	1.0000	1.0000	0.9998	0.9990	0.9960	0.9679	0.8666
	9	1.0000	1.0000	1.0000	1.0000	0.9999	0.9993	0.9922	0.9539
	10	1.0000	1.0000	1.0000	1.0000	1.0000	0.9999	0.9987	0.9888
	11	1.0000	1.0000	1.0000	1.0000	1.0000	1.0000	0.9999	0.9983
	12	1.0000	1.0000	1.0000	1.0000	1.0000	1.0000	1.0000	0.9999

Cumulative probability, Binomial distribution

n	x	$p=0.05$	0.10	0.15	0.20	0.25	0.30	0.40	0.50
14	0	0.4877	0.2288	0.1028	0.0440	0.0178	0.0068	0.0008	0.0001
	1	0.8470	0.5846	0.3567	0.1979	0.1010	0.0475	0.0081	0.0009
	2	0.9699	0.8416	0.6479	0.4481	0.2811	0.1608	0.0398	0.0065
	3	0.9958	0.9559	0.8535	0.6982	0.5213	0.3552	0.1243	0.0287
	4	0.9996	0.9908	0.9533	0.8702	0.7415	0.5842	0.2793	0.0898
	5	1.0000	0.9985	0.9885	0.9562	0.8883	0.7805	0.4859	0.2120
	6	1.0000	0.9998	0.9978	0.9884	0.9617	0.9067	0.6925	0.3953
	7	1.0000	1.0000	0.9997	0.9976	0.9897	0.9685	0.8499	0.6047
	8	1.0000	1.0000	1.0000	0.9996	0.9979	0.9917	0.9417	0.7880
	9	1.0000	1.0000	1.0000	1.0000	0.9997	0.9983	0.9825	0.9102
	10	1.0000	1.0000	1.0000	1.0000	1.0000	0.9998	0.9961	0.9713
	11	1.0000	1.0000	1.0000	1.0000	1.0000	1.0000	0.9994	0.9935
	12	1.0000	1.0000	1.0000	1.0000	1.0000	1.0000	0.9999	0.9991
	13	1.0000	1.0000	1.0000	1.0000	1.0000	1.0000	1.0000	0.9999
15	0	0.4633	0.2059	0.0873	0.0352	0.0134	0.0047	0.0005	0.0000
	1	0.8290	0.5490	0.3186	0.1671	0.0802	0.0353	0.0052	0.0005
	2	0.9638	0.8159	0.6042	0.3980	0.2361	0.1268	0.0271	0.0037
	3	0.9945	0.9444	0.8227	0.6482	0.4613	0.2969	0.0905	0.0176
	4	0.9994	0.9873	0.9383	0.8358	0.6865	0.5155	0.2173	0.0592
	5	1.0000	0.9978	0.9832	0.9389	0.8516	0.7216	0.4032	0.1509
	6	1.0000	0.9997	0.9964	0.9819	0.9434	0.8689	0.6098	0.3036
	7	1.0000	1.0000	0.9994	0.9958	0.9827	0.9500	0.7869	0.5000
	8	1.0000	1.0000	0.9999	0.9992	0.9958	0.9848	0.9050	0.6964
	9	1.0000	1.0000	1.0000	0.9999	0.9992	0.9963	0.9662	0.8491
	10	1.0000	1.0000	1.0000	1.0000	0.9999	0.9993	0.9907	0.9408
	11	1.0000	1.0000	1.0000	1.0000	1.0000	0.9999	0.9981	0.9824
	12	1.0000	1.0000	1.0000	1.0000	1.0000	1.0000	0.9997	0.9963
	13	1.0000	1.0000	1.0000	1.0000	1.0000	1.0000	1.0000	0.9995
	14	1.0000	1.0000	1.0000	1.0000	1.0000	1.0000	1.0000	1.0000
16	0	0.4401	0.1853	0.0742	0.0282	0.0100	0.0033	0.0003	0.0000
	1	0.8108	0.5147	0.2839	0.1407	0.0635	0.0261	0.0033	0.0003
	2	0.9571	0.7893	0.5614	0.3518	0.1971	0.0994	0.0183	0.0021
	3	0.9930	0.9316	0.7899	0.5981	0.4050	0.2459	0.0651	0.0106
	4	0.9991	0.9830	0.9210	0.7983	0.6302	0.4499	0.1666	0.0384
	5	0.9999	0.9967	0.9765	0.9183	0.8104	0.6598	0.3288	0.1051
	6	1.0000	0.9995	0.9944	0.9733	0.9204	0.8247	0.5272	0.2273
	7	1.0000	0.9999	0.9989	0.9930	0.9729	0.9256	0.7161	0.4018
	8	1.0000	1.0000	0.9998	0.9985	0.9925	0.9743	0.8577	0.5982
	9	1.0000	1.0000	1.0000	0.9998	0.9984	0.9929	0.9417	0.7728
	10	1.0000	1.0000	1.0000	1.0000	0.9997	0.9984	0.9809	0.8949
	11	1.0000	1.0000	1.0000	1.0000	1.0000	0.9997	0.9951	0.9616
	12	1.0000	1.0000	1.0000	1.0000	1.0000	1.0000	0.9991	0.9894
	13	1.0000	1.0000	1.0000	1.0000	1.0000	1.0000	0.9999	0.9979
	14	1.0000	1.0000	1.0000	1.0000	1.0000	1.0000	1.0000	0.9997
	15	1.0000	1.0000	1.0000	1.0000	1.0000	1.0000	1.0000	1.0000

Cumulative probability, Binomial distribution

n	x	p = 0.05	0.10	0.15	0.20	0.25	0.30	0.40	0.50
17	0	0.4181	0.1668	0.0631	0.0225	0.0075	0.0023	0.0002	0.0000
	1	0.7922	0.4818	0.2525	0.1182	0.0501	0.0193	0.0021	0.0001
	2	0.9497	0.7618	0.5198	0.3096	0.1637	0.0774	0.0123	0.0012
	3	0.9912	0.9174	0.7556	0.5489	0.3530	0.2019	0.0464	0.0064
	4	0.9988	0.9779	0.9013	0.7582	0.5739	0.3887	0.1260	0.0245
	5	0.9999	0.9953	0.9681	0.8943	0.7653	0.5968	0.2639	0.0717
	6	1.0000	0.9992	0.9917	0.9623	0.8929	0.7752	0.4478	0.1661
	7	1.0000	0.9999	0.9983	0.9891	0.9598	0.8954	0.6405	0.3145
	8	1.0000	1.0000	0.9997	0.9974	0.9876	0.9597	0.8011	0.5000
	9	1.0000	1.0000	1.0000	0.9995	0.9969	0.9873	0.9081	0.6855
	10	1.0000	1.0000	1.0000	0.9999	0.9994	0.9968	0.9652	0.8338
	11	1.0000	1.0000	1.0000	1.0000	0.9999	0.9993	0.9894	0.9283
	12	1.0000	1.0000	1.0000	1.0000	1.0000	0.9999	0.9975	0.9755
	13	1.0000	1.0000	1.0000	1.0000	1.0000	1.0000	0.9996	0.9936
	14	1.0000	1.0000	1.0000	1.0000	1.0000	1.0000	0.9999	0.9988
	15	1.0000	1.0000	1.0000	1.0000	1.0000	1.0000	1.0000	0.9999
	16	1.0000	1.0000	1.0000	1.0000	1.0000	1.0000	1.0000	1.0000
18	0	0.3972	0.1501	0.0537	0.0180	0.0056	0.0016	0.0001	0.0000
	1	0.7735	0.4503	0.2240	0.0991	0.0395	0.0142	0.0013	0.0001
	2	0.9419	0.7338	0.4797	0.2713	0.1353	0.0600	0.0082	0.0007
	3	0.9891	0.9018	0.7202	0.5010	0.3057	0.1646	0.0328	0.0038
	4	0.9984	0.9718	0.8794	0.7164	0.5187	0.3327	0.0942	0.0154
	5	0.9998	0.9936	0.9581	0.8671	0.7175	0.5344	0.2088	0.0481
	6	1.0000	0.9988	0.9882	0.9487	0.8610	0.7217	0.3743	0.1189
	7	1.0000	0.9998	0.9973	0.9837	0.9431	0.8593	0.5634	0.2403
	8	1.0000	1.0000	0.9995	0.9958	0.9807	0.9404	0.7368	0.4073
	9	1.0000	1.0000	0.9999	0.9991	0.9946	0.9790	0.8653	0.5927
	10	1.0000	1.0000	1.0000	0.9998	0.9988	0.9939	0.9424	0.7597
	11	1.0000	1.0000	1.0000	1.0000	0.9998	0.9986	0.9797	0.8811
	12	1.0000	1.0000	1.0000	1.0000	1.0000	0.9997	0.9942	0.9519
	13	1.0000	1.0000	1.0000	1.0000	1.0000	1.0000	0.9987	0.9846
	14	1.0000	1.0000	1.0000	1.0000	1.0000	1.0000	0.9998	0.9962
	15	1.0000	1.0000	1.0000	1.0000	1.0000	1.0000	1.0000	0.9993
	16	1.0000	1.0000	1.0000	1.0000	1.0000	1.0000	1.0000	0.9999
	17	1.0000	1.0000	1.0000	1.0000	1.0000	1.0000	1.0000	1.0000

Cumulative probability, Binomial distribution

n	x	p =0.05	0.10	0.15	0.20	0.25	0.30	0.40	0.50
19	0	0.3774	0.1351	0.0456	0.0144	0.0042	0.0011	0.0001	0.0000
	1	0.7547	0.4203	0.1985	0.0829	0.0310	0.0104	0.0008	0.0000
	2	0.9335	0.7054	0.4413	0.2369	0.1113	0.0462	0.0055	0.0004
	3	0.9868	0.8850	0.6842	0.4551	0.2631	0.1332	0.0230	0.0022
	4	0.9980	0.9648	0.8556	0.6733	0.4654	0.2822	0.0696	0.0096
	5	0.9998	0.9914	0.9463	0.8369	0.6678	0.4739	0.1629	0.0318
	6	1.0000	0.9983	0.9837	0.9324	0.8251	0.6655	0.3081	0.0835
	7	1.0000	0.9997	0.9959	0.9767	0.9225	0.8180	0.4878	0.1796
	8	1.0000	1.0000	0.9992	0.9933	0.9712	0.9161	0.6675	0.3238
	9	1.0000	1.0000	0.9999	0.9984	0.9911	0.9675	0.8139	0.5000
	10	1.0000	1.0000	1.0000	0.9997	0.9977	0.9895	0.9115	0.6762
	11	1.0000	1.0000	1.0000	1.0000	0.9995	0.9972	0.9648	0.8204
	12	1.0000	1.0000	1.0000	1.0000	0.9999	0.9994	0.9884	0.9165
	13	1.0000	1.0000	1.0000	1.0000	1.0000	0.9999	0.9969	0.9682
	14	1.0000	1.0000	1.0000	1.0000	1.0000	1.0000	0.9994	0.9904
	15	1.0000	1.0000	1.0000	1.0000	1.0000	1.0000	0.9999	0.9978
	16	1.0000	1.0000	1.0000	1.0000	1.0000	1.0000	1.0000	0.9996
	17	1.0000	1.0000	1.0000	1.0000	1.0000	1.0000	1.0000	1.0000
	18	1.0000	1.0000	1.0000	1.0000	1.0000	1.0000	1.0000	1.0000
20	0	0.3585	0.1216	0.0388	0.0115	0.0032	0.0008	0.0000	0.0000
	1	0.7358	0.3917	0.1756	0.0692	0.0243	0.0076	0.0005	0.0000
	2	0.9245	0.6769	0.4049	0.2061	0.0913	0.0355	0.0036	0.0002
	3	0.9841	0.8670	0.6477	0.4114	0.2252	0.1071	0.0160	0.0013
	4	0.9974	0.9568	0.8298	0.6297	0.4148	0.2375	0.0510	0.0059
	5	0.9997	0.9888	0.9327	0.8042	0.6172	0.4164	0.1256	0.0207
	6	1.0000	0.9976	0.9781	0.9133	0.7858	0.6080	0.2500	0.0577
	7	1.0000	0.9996	0.9941	0.9679	0.8982	0.7723	0.4159	0.1316
	8	1.0000	0.9999	0.9987	0.9900	0.9591	0.8867	0.5956	0.2517
	9	1.0000	1.0000	0.9998	0.9974	0.9861	0.9520	0.7553	0.4119
	10	1.0000	1.0000	1.0000	0.9994	0.9961	0.9829	0.8725	0.5881
	11	1.0000	1.0000	1.0000	0.9999	0.9991	0.9949	0.9435	0.7483
	12	1.0000	1.0000	1.0000	1.0000	0.9998	0.9987	0.9790	0.8684
	13	1.0000	1.0000	1.0000	1.0000	1.0000	0.9997	0.9935	0.9423
	14	1.0000	1.0000	1.0000	1.0000	1.0000	1.0000	0.9984	0.9793
	15	1.0000	1.0000	1.0000	1.0000	1.0000	1.0000	0.9997	0.9941
	16	1.0000	1.0000	1.0000	1.0000	1.0000	1.0000	1.0000	0.9987
	17	1.0000	1.0000	1.0000	1.0000	1.0000	1.0000	1.0000	0.9998
	18	1.0000	1.0000	1.0000	1.0000	1.0000	1.0000	1.0000	1.0000
	19	1.0000	1.0000	1.0000	1.0000	1.0000	1.0000	1.0000	1.0000

5.5   GEOMETRIC DISTRIBUTION

In a series of Bernoulli trials with probability of success p, a geometric random variable is the number of the trial on which the first success occurs. Hence, this is a waiting time distribution. The geometric distribution, sometimes called the *Pascal distribution*, is often thought of as the discrete version of an exponential distribution.

5.5.1   Properties

$$\text{pmf}\quad p(x) = pq^{x-1}\quad x = 1, 2, 3, \ldots, \quad 0 \le p \le 1, \quad q = 1 - p$$

$$\text{mean}\qquad \mu = 1/p$$

$$\text{variance}\qquad \sigma^2 = q/p^2$$

$$\text{skewness}\qquad \beta_1 = \frac{2 - p}{\sqrt{q}}$$

$$\text{kurtosis}\qquad \beta_2 = \frac{p^2 + 6q}{q}$$

$$\text{mgf}\quad m(t) = \frac{pe^t}{1 - qe^t}$$

$$\text{char function}\quad \phi(t) = \frac{pe^{it}}{1 - qe^{it}}$$

$$\text{fact mgf}\quad P(t) = \frac{p^t}{1 - qt}$$

5.5.2   Variates

Let $X_1, X_2, \ldots, X_n$ be independent, identically distributed geometric random variables with parameter p.

(1) The random variable $Y = X_1 + X_2 + \cdots + X_n$ has a negative binomial distribution with parameters n and p.

(2) The random variable $Y = \min(X_1, X_2, \ldots, X_n)$ has a geometric distribution with parameter p.

5.5.3   Tables

Example 5.34:   When flipping a biased coin (so that heads occur only 30% of the time), what is the probability that the first head occurs on the 10th flip?

Solution:

(S1) Using the probability mass table below with $x = 10$ and $p = 0.3$ results in 0.0121.

(S2) Hence, this is likely to occur only about 1% of the time.

Example 5.35: The probability a randomly selected customer has the correct change when making a purchase at the local donut shop is 0.1. What is the probability the first person to have correct change will be the fifth customer? What is the probability the first person with correct change will be at least the sixth customer?

Solution:

(S1) Let X be the number of the first customer with correct change. The random variable X has a geometric distribution with parameter $p = 0.1$. Use the table for cumulative terms of the geometric probabilities to answer each question.

(S2) $\text{Prob}\,[X = 5] = 0.0656$

(S3) $\text{Prob}\,[X \geq 6] = 1 - \text{Prob}\,[X \leq 5]$
$$= 1 - (0.1000 + 0.0900 + 0.0810 + 0.0729 + 0.656)$$
$$= 1 - 0.4095 = 0.5905$$

Probability mass, Geometric distribution

x	$p = 0.1$	0.2	0.3	0.4	0.5	0.6	0.7	0.8	0.9
1	0.1000	0.2000	0.3000	0.4000	0.5000	0.6000	0.7000	0.8000	0.9000
2	0.0900	0.1600	0.2100	0.2400	0.2500	0.2400	0.2100	0.1600	0.0900
3	0.0810	0.1280	0.1470	0.1440	0.1250	0.0960	0.0630	0.0320	0.0090
4	0.0729	0.1024	0.1029	0.0864	0.0625	0.0384	0.0189	0.0064	0.0009
5	0.0656	0.0819	0.0720	0.0518	0.0313	0.0154	0.0057	0.0013	0.0001
6	0.0590	0.0655	0.0504	0.0311	0.0156	0.0061	0.0017	0.0003	
7	0.0531	0.0524	0.0353	0.0187	0.0078	0.0025	0.0005	0.0001	
8	0.0478	0.0419	0.0247	0.0112	0.0039	0.0010	0.0002		
9	0.0430	0.0336	0.0173	0.0067	0.0020	0.0004			
10	0.0387	0.0268	0.0121	0.0040	0.0010	0.0002			
11	0.0349	0.0215	0.0085	0.0024	0.0005	0.0001			
12	0.0314	0.0172	0.0059	0.0015	0.0002				
13	0.0282	0.0137	0.0042	0.0009	0.0001				
14	0.0254	0.0110	0.0029	0.0005	0.0001				
15	0.0229	0.0088	0.0020	0.0003					
20	0.0135	0.0029	0.0003						

Cumulative probability, Geometric distribution

x	$p = 0.1$	0.2	0.3	0.4	0.5	0.6	0.7	0.8	0.9
1	0.1000	0.2000	0.3000	0.4000	0.5000	0.6000	0.7000	0.8000	0.9000
2	0.1900	0.3600	0.5100	0.6400	0.7500	0.8400	0.9100	0.9600	0.9900
3	0.2710	0.4880	0.6570	0.7840	0.8750	0.9360	0.9730	0.9920	0.9990
4	0.3439	0.5904	0.7599	0.8704	0.9375	0.9744	0.9919	0.9984	0.9999
5	0.4095	0.6723	0.8319	0.9222	0.9688	0.9898	0.9976	0.9997	1
6	0.4686	0.7379	0.8824	0.9533	0.9844	0.9959	0.9993	0.9999	1
7	0.5217	0.7903	0.9176	0.9720	0.9922	0.9984	0.9998	1	
8	0.5695	0.8322	0.9424	0.9832	0.9961	0.9993	0.9999	1	
9	0.6126	0.8658	0.9596	0.9899	0.9980	0.9997	1		
10	0.6513	0.8926	0.9718	0.9940	0.9990	0.9999	1		
11	0.6862	0.9141	0.9802	0.9964	0.9995	1			
12	0.7176	0.9313	0.9862	0.9978	0.9998	1			
13	0.7458	0.9450	0.9903	0.9987	0.9999	1			
14	0.7712	0.9560	0.9932	0.9992	0.9999	1			
15	0.7941	0.9648	0.9953	0.9995	1				
20	0.8784	0.9885	0.9992	1					

5.6   HYPERGEOMETRIC DISTRIBUTION

In a finite population of size N suppose there are M successes (and $N - M$ failures). The hypergeometric distribution is used to describe the number of successes, X, in n trials (n observations drawn from the population). Unlike a binomial distribution, the probability of a success does not remain constant from trial to trial.

5.6.1   Properties

$$\text{pmf} \quad p(x) = \frac{\binom{M}{x}\binom{N-M}{n-x}}{\binom{N}{n}} \quad x = 0, 1, \ldots, n \quad x \leq M$$

$$n - x \leq N - M, \quad n, M, N \in \mathcal{N}, \quad 1 \leq n \leq N$$

$$1 \leq M \leq N, \quad N = 1, 2, \ldots$$

$$\text{mean} \quad \mu = n\frac{M}{N}$$

$$\text{variance} \quad \sigma^2 = \left(\frac{N-n}{N-1}\right) n\frac{M}{N}\left(1 - \frac{M}{N}\right)$$

$$\text{skewness} \quad \beta_1 = \frac{(N - 2M)(N - 2n)\sqrt{N-1}}{(N-2)\sqrt{nM(N-M)(N-n)}}$$

$$\text{kurtosis} \quad \beta_2 = \frac{N^2(N-1)}{(N-2)(N-3)nM(N-m)(N-n)} \times$$

$$\left[N(N+1) - 6n(N-n) + 3\frac{M}{N^2}(N-M)\times\right.$$

$$\left.[N^2(n-2) - Nn^2 + 6n(N-n)]\right]$$

$$\text{mgf} \quad m(t) = {}_2F_1(-n, -M; -N; 1 - e^t)$$

$$\text{char function} \quad \phi(t) = {}_2F_1(-n, -M; -N; 1 - e^{it})$$

$$\text{fact mgf} \quad P(t) = {}_2F_1(-n, -M; -N; 1 - t)$$

where $_pF_q$ is the generalized hypergeometric function defined in Chapter 18 (see page 520).

5.6.2   Variates

Let X be a hypergeometric random variable with parameters n, m, and N.

(1) As $N \to \infty$ if $n/N < 0.1$ then X is approximately a binomial random variable with parameters n and $p = M/N$.

(2) As n, M, and N all tend to infinity, if M/N is small then X has approximately a Poisson distribution with parameter $\lambda = nM/N$.

5.6.3 Tables

Let X be a hypergeometric random variable with parameters n, M, and N. The probability mass function $p(x) = f(x; n, M, N)$ is the probability of exactly x successes in a sample of n items. The cumulative distribution function

$$F(x; n, M, N) = \sum_{r=0}^{x} f(r; n, M, N) = \sum_{r=0}^{x} \frac{\binom{M}{r}\binom{N-M}{n-r}}{\binom{N}{n}} \tag{5.1}$$

is the probability of x or fewer successes in the sample of n items. The following table contains values for $f(x; n, M, N)$ and $F(x; n, M, N)$ for various values of x, n, M, and N.

Example 5.36: A New York City transportation company has 10 taxis, 3 of which have broken meters. Suppose 4 taxis are selected at random. What is the probability exactly 1 will have a broken meter, fewer than 2 will have a broken meter, all will have working meters?

Solution:

(S1) Let X be the number of taxis selected with broken meters. The random variable X has a hypergeometric distribution with $N = 10$, $n = 4$, and $M = 3$.

(S2) $\mathrm{Prob}\,[X = 1] = \dfrac{\binom{M}{1}\binom{N-M}{n-1}}{\binom{N}{n}} = \dfrac{\binom{3}{1}\binom{7}{3}}{\binom{10}{4}} = \dfrac{3 \cdot 35}{210} = 0.5$

(S3) $\mathrm{Prob}\,[X = 0] = \dfrac{\binom{M}{0}\binom{N-M}{n}}{\binom{N}{n}} = \dfrac{\binom{3}{0}\binom{7}{4}}{\binom{10}{4}} = \dfrac{1 \cdot 35}{210} = 0.16667$

(S4) $\mathrm{Prob}\,[X < 2] = \mathrm{Prob}\,[X \leq 1] = \mathrm{Prob}\,[X = 0] + \mathrm{Prob}\,[X = 1] = 0.66667$

Hypergeometric probability and distribution functions

N	n	M	x	$F(x)$	$f(x)$	N	n	M	x	$F(x)$	$f(x)$
2	1	1	0	0.50000	0.50000	6	2	2	2	1.00000	0.06667
2	1	1	1	1.00000	0.50000	6	3	1	0	0.50000	0.50000
3	1	1	0	0.66667	0.66667	6	3	1	1	1.00000	0.50000
3	1	1	1	1.00000	0.33333	6	3	2	0	0.20000	0.20000
3	2	1	0	0.33333	0.33333	6	3	2	1	0.80000	0.60000
3	2	1	1	1.00000	0.66667	6	3	2	2	1.00000	0.20000
3	2	2	1	0.66667	0.66667	6	3	3	0	0.05000	0.05000
3	2	2	2	1.00000	0.33333	6	3	3	1	0.50000	0.45000
4	1	1	0	0.75000	0.75000	6	3	3	2	0.95000	0.45000
4	1	1	1	1.00000	0.25000	6	3	3	3	1.00000	0.05000
4	2	1	0	0.50000	0.50000	6	4	1	0	0.33333	0.33333
4	2	1	1	1.00000	0.50000	6	4	1	1	1.00000	0.66667
4	2	2	0	0.16667	0.16667	6	4	2	0	0.06667	0.06667
4	2	2	1	0.83333	0.66667	6	4	2	1	0.60000	0.53333
4	2	2	2	1.00000	0.16667	6	4	2	2	1.00000	0.40000
4	3	1	0	0.25000	0.25000	6	4	3	1	0.20000	0.20000
4	3	1	1	1.00000	0.75000	6	4	3	2	0.80000	0.60000
4	3	2	1	0.50000	0.50000	6	4	3	3	1.00000	0.20000
4	3	2	2	1.00000	0.50000	6	4	4	2	0.40000	0.40000
4	3	3	2	0.75000	0.75000	6	4	4	3	0.93333	0.53333
4	3	3	3	1.00000	0.25000	6	4	4	4	1.00000	0.06667
5	1	1	0	0.80000	0.80000	6	5	1	0	0.16667	0.16667
5	1	1	1	1.00000	0.20000	6	5	1	1	1.00000	0.83333
5	2	1	0	0.60000	0.60000	6	5	2	1	0.33333	0.33333
5	2	1	1	1.00000	0.40000	6	5	2	2	1.00000	0.66667
5	2	2	0	0.30000	0.30000	6	5	3	2	0.50000	0.50000
5	2	2	1	0.90000	0.60000	6	5	3	3	1.00000	0.50000
5	2	2	2	1.00000	0.10000	6	5	4	3	0.66667	0.66667
5	3	1	0	0.40000	0.40000	6	5	4	4	1.00000	0.33333
5	3	1	1	1.00000	0.60000	6	5	5	4	0.83333	0.83333
5	3	2	0	0.10000	0.10000	6	5	5	5	1.00000	0.16667
5	3	2	1	0.70000	0.60000	7	1	1	0	0.85714	0.85714
5	3	2	2	1.00000	0.30000	7	1	1	1	1.00000	0.14286
5	3	3	1	0.30000	0.30000	7	2	1	0	0.71429	0.71429
5	3	3	2	0.90000	0.60000	7	2	1	1	1.00000	0.28571
5	3	3	3	1.00000	0.10000	7	2	2	0	0.47619	0.47619
5	4	1	0	0.20000	0.20000	7	2	2	1	0.95238	0.47619
5	4	1	1	1.00000	0.80000	7	2	2	2	1.00000	0.04762
5	4	2	1	0.40000	0.40000	7	3	1	0	0.57143	0.57143
5	4	2	2	1.00000	0.60000	7	3	1	1	1.00000	0.42857
5	4	3	2	0.60000	0.60000	7	3	2	0	0.28571	0.28571
5	4	3	3	1.00000	0.40000	7	3	2	1	0.85714	0.57143
5	4	4	3	0.80000	0.80000	7	3	2	2	1.00000	0.14286
5	4	4	4	1.00000	0.20000	7	3	3	0	0.11429	0.11429
6	1	1	0	0.83333	0.83333	7	3	3	1	0.62857	0.51429
6	1	1	1	1.00000	0.16667	7	3	3	2	0.97143	0.34286
6	2	1	0	0.66667	0.66667	7	3	3	3	1.00000	0.02857
6	2	1	1	1.00000	0.33333	7	4	1	0	0.42857	0.42857
6	2	2	0	0.40000	0.40000	7	4	1	1	1.00000	0.57143
6	2	2	1	0.93333	0.53333	7	4	2	0	0.14286	0.14286

Hypergeometric probability and distribution functions

N	n	M	x	$F(x)$	$f(x)$	N	n	M	x	$F(x)$	$f(x)$
7	4	2	1	0.71429	0.57143	8	3	3	2	0.98214	0.26786
7	4	2	2	1.00000	0.28571	8	3	3	3	1.00000	0.01786
7	4	3	0	0.02857	0.02857	8	4	1	0	0.50000	0.50000
7	4	3	1	0.37143	0.34286	8	4	1	1	1.00000	0.50000
7	4	3	2	0.88571	0.51429	8	4	2	0	0.21429	0.21429
7	4	3	3	1.00000	0.11429	8	4	2	1	0.78571	0.57143
7	4	4	1	0.11429	0.11429	8	4	2	2	1.00000	0.21429
7	4	4	2	0.62857	0.51429	8	4	3	0	0.07143	0.07143
7	4	4	3	0.97143	0.34286	8	4	3	1	0.50000	0.42857
7	4	4	4	1.00000	0.02857	8	4	3	2	0.92857	0.42857
7	5	1	0	0.28571	0.28571	8	4	3	3	1.00000	0.07143
7	5	1	1	1.00000	0.71429	8	4	4	0	0.01429	0.01429
7	5	2	0	0.04762	0.04762	8	4	4	1	0.24286	0.22857
7	5	2	1	0.52381	0.47619	8	4	4	2	0.75714	0.51429
7	5	2	2	1.00000	0.47619	8	4	4	3	0.98571	0.22857
7	5	3	1	0.14286	0.14286	8	4	4	4	1.00000	0.01429
7	5	3	2	0.71429	0.57143	8	5	1	0	0.37500	0.37500
7	5	3	3	1.00000	0.28571	8	5	1	1	1.00000	0.62500
7	5	4	2	0.28571	0.28571	8	5	2	0	0.10714	0.10714
7	5	4	3	0.85714	0.57143	8	5	2	1	0.64286	0.53571
7	5	4	4	1.00000	0.14286	8	5	2	2	1.00000	0.35714
7	5	5	3	0.47619	0.47619	8	5	3	0	0.01786	0.01786
7	5	5	4	0.95238	0.47619	8	5	3	1	0.28571	0.26786
7	5	5	5	1.00000	0.04762	8	5	3	2	0.82143	0.53571
7	6	1	0	0.14286	0.14286	8	5	3	3	1.00000	0.17857
7	6	1	1	1.00000	0.85714	8	5	4	1	0.07143	0.07143
7	6	2	1	0.28571	0.28571	8	5	4	2	0.50000	0.42857
7	6	2	2	1.00000	0.71429	8	5	4	3	0.92857	0.42857
7	6	3	2	0.42857	0.42857	8	5	4	4	1.00000	0.07143
7	6	3	3	1.00000	0.57143	8	5	5	2	0.17857	0.17857
7	6	4	3	0.57143	0.57143	8	5	5	3	0.71429	0.53571
7	6	4	4	1.00000	0.42857	8	5	5	4	0.98214	0.26786
7	6	5	4	0.71429	0.71429	8	5	5	5	1.00000	0.01786
7	6	5	5	1.00000	0.28571	8	6	1	0	0.25000	0.25000
7	6	6	5	0.85714	0.85714	8	6	1	1	1.00000	0.75000
7	6	6	6	1.00000	0.14286	8	6	2	0	0.03571	0.03571
8	1	1	0	0.87500	0.87500	8	6	2	1	0.46429	0.42857
8	1	1	1	1.00000	0.12500	8	6	2	2	1.00000	0.53571
8	2	1	0	0.75000	0.75000	8	6	3	1	0.10714	0.10714
8	2	1	1	1.00000	0.25000	8	6	3	2	0.64286	0.53571
8	2	2	0	0.53571	0.53571	8	6	3	3	1.00000	0.35714
8	2	2	1	0.96429	0.42857	8	6	4	2	0.21429	0.21429
8	2	2	2	1.00000	0.03571	8	6	4	3	0.78571	0.57143
8	3	1	0	0.62500	0.62500	8	6	4	4	1.00000	0.21429
8	3	1	1	1.00000	0.37500	8	6	5	3	0.35714	0.35714
8	3	2	0	0.35714	0.35714	8	6	5	4	0.89286	0.53571
8	3	2	1	0.89286	0.53571	8	6	5	5	1.00000	0.10714
8	3	2	2	1.00000	0.10714	8	6	6	4	0.53571	0.53571
8	3	3	0	0.17857	0.17857	8	6	6	5	0.96429	0.42857
8	3	3	1	0.71429	0.53571	8	6	6	6	1.00000	0.03571

Hypergeometric probability and distribution functions

N	n	M	x	$F(x)$	$f(x)$	N	n	M	x	$F(x)$	$f(x)$
8	7	1	0	0.12500	0.12500	9	5	3	1	0.40476	0.35714
8	7	1	1	1.00000	0.87500	9	5	3	2	0.88095	0.47619
8	7	2	1	0.25000	0.25000	9	5	3	3	1.00000	0.11905
8	7	2	2	1.00000	0.75000	9	5	4	0	0.00794	0.00794
8	7	3	2	0.37500	0.37500	9	5	4	1	0.16667	0.15873
8	7	3	3	1.00000	0.62500	9	5	4	2	0.64286	0.47619
8	7	4	3	0.50000	0.50000	9	5	4	3	0.96032	0.31746
8	7	4	4	1.00000	0.50000	9	5	4	4	1.00000	0.03968
8	7	5	4	0.62500	0.62500	9	5	5	1	0.03968	0.03968
8	7	5	5	1.00000	0.37500	9	5	5	2	0.35714	0.31746
8	7	6	5	0.75000	0.75000	9	5	5	3	0.83333	0.47619
8	7	6	6	1.00000	0.25000	9	5	5	4	0.99206	0.15873
8	7	7	6	0.87500	0.87500	9	5	5	5	1.00000	0.00794
8	7	7	7	1.00000	0.12500	9	6	1	0	0.33333	0.33333
9	1	1	0	0.88889	0.88889	9	6	1	1	1.00000	0.66667
9	1	1	1	1.00000	0.11111	9	6	2	0	0.08333	0.08333
9	2	1	0	0.77778	0.77778	9	6	2	1	0.58333	0.50000
9	2	1	1	1.00000	0.22222	9	6	2	2	1.00000	0.41667
9	2	2	0	0.58333	0.58333	9	6	3	0	0.01191	0.01191
9	2	2	1	0.97222	0.38889	9	6	3	1	0.22619	0.21429
9	2	2	2	1.00000	0.02778	9	6	3	2	0.76191	0.53571
9	3	1	0	0.66667	0.66667	9	6	3	3	1.00000	0.23810
9	3	1	1	1.00000	0.33333	9	6	4	1	0.04762	0.04762
9	3	2	0	0.41667	0.41667	9	6	4	2	0.40476	0.35714
9	3	2	1	0.91667	0.50000	9	6	4	3	0.88095	0.47619
9	3	2	2	1.00000	0.08333	9	6	4	4	1.00000	0.11905
9	3	3	0	0.23810	0.23810	9	6	5	2	0.11905	0.11905
9	3	3	1	0.77381	0.53571	9	6	5	3	0.59524	0.47619
9	3	3	2	0.98809	0.21429	9	6	5	4	0.95238	0.35714
9	3	3	3	1.00000	0.01191	9	6	5	5	1.00000	0.04762
9	4	1	0	0.55556	0.55556	9	6	6	3	0.23810	0.23810
9	4	1	1	1.00000	0.44444	9	6	6	4	0.77381	0.53571
9	4	2	0	0.27778	0.27778	9	6	6	5	0.98809	0.21429
9	4	2	1	0.83333	0.55556	9	6	6	6	1.00000	0.01191
9	4	2	2	1.00000	0.16667	9	7	1	0	0.22222	0.22222
9	4	3	0	0.11905	0.11905	9	7	1	1	1.00000	0.77778
9	4	3	1	0.59524	0.47619	9	7	2	0	0.02778	0.02778
9	4	3	2	0.95238	0.35714	9	7	2	1	0.41667	0.38889
9	4	3	3	1.00000	0.04762	9	7	2	2	1.00000	0.58333
9	4	4	0	0.03968	0.03968	9	7	3	1	0.08333	0.08333
9	4	4	1	0.35714	0.31746	9	7	3	2	0.58333	0.50000
9	4	4	2	0.83333	0.47619	9	7	3	3	1.00000	0.41667
9	4	4	3	0.99206	0.15873	9	7	4	2	0.16667	0.16667
9	4	4	4	1.00000	0.00794	9	7	4	3	0.72222	0.55556
9	5	1	0	0.44444	0.44444	9	7	4	4	1.00000	0.27778
9	5	1	1	1.00000	0.55556	9	7	5	3	0.27778	0.27778
9	5	2	0	0.16667	0.16667	9	7	5	4	0.83333	0.55556
9	5	2	1	0.72222	0.55556	9	7	5	5	1.00000	0.16667
9	5	2	2	1.00000	0.27778	9	7	6	4	0.41667	0.41667
9	5	3	0	0.04762	0.04762	9	7	6	5	0.91667	0.50000

Hypergeometric probability and distribution functions

N	n	M	x	$F(x)$	$f(x)$	N	n	M	x	$F(x)$	$f(x)$
9	7	6	6	1.00000	0.08333	10	5	1	0	0.50000	0.50000
9	7	7	5	0.58333	0.58333	10	5	1	1	1.00000	0.50000
9	7	7	6	0.97222	0.38889	10	5	2	0	0.22222	0.22222
9	7	7	7	1.00000	0.02778	10	5	2	1	0.77778	0.55556
9	8	1	0	0.11111	0.11111	10	5	2	2	1.00000	0.22222
9	8	1	1	1.00000	0.88889	10	5	3	0	0.08333	0.08333
9	8	2	1	0.22222	0.22222	10	5	3	1	0.50000	0.41667
9	8	2	2	1.00000	0.77778	10	5	3	2	0.91667	0.41667
9	8	3	2	0.33333	0.33333	10	5	3	3	1.00000	0.08333
9	8	3	3	1.00000	0.66667	10	5	4	0	0.02381	0.02381
9	8	4	3	0.44444	0.44444	10	5	4	1	0.26190	0.23810
9	8	4	4	1.00000	0.55556	10	5	4	2	0.73809	0.47619
9	8	5	4	0.55556	0.55556	10	5	4	3	0.97619	0.23810
9	8	5	5	1.00000	0.44444	10	5	4	4	1.00000	0.02381
9	8	6	5	0.66667	0.66667	10	5	5	0	0.00397	0.00397
9	8	6	6	1.00000	0.33333	10	5	5	1	0.10318	0.09921
9	8	7	6	0.77778	0.77778	10	5	5	2	0.50000	0.39682
9	8	7	7	1.00000	0.22222	10	5	5	3	0.89682	0.39682
9	8	8	7	0.88889	0.88889	10	5	5	4	0.99603	0.09921
9	8	8	8	1.00000	0.11111	10	5	5	5	1.00000	0.00397
10	1	1	0	0.90000	0.90000	10	6	1	0	0.40000	0.40000
10	1	1	1	1.00000	0.10000	10	6	1	1	1.00000	0.60000
10	2	1	0	0.80000	0.80000	10	6	2	0	0.13333	0.13333
10	2	1	1	1.00000	0.20000	10	6	2	1	0.66667	0.53333
10	2	2	0	0.62222	0.62222	10	6	2	2	1.00000	0.33333
10	2	2	1	0.97778	0.35556	10	6	3	0	0.03333	0.03333
10	2	2	2	1.00000	0.02222	10	6	3	1	0.33333	0.30000
10	3	1	0	0.70000	0.70000	10	6	3	2	0.83333	0.50000
10	3	1	1	1.00000	0.30000	10	6	3	3	1.00000	0.16667
10	3	2	0	0.46667	0.46667	10	6	4	0	0.00476	0.00476
10	3	2	1	0.93333	0.46667	10	6	4	1	0.11905	0.11429
10	3	2	2	1.00000	0.06667	10	6	4	2	0.54762	0.42857
10	3	3	0	0.29167	0.29167	10	6	4	3	0.92857	0.38095
10	3	3	1	0.81667	0.52500	10	6	4	4	1.00000	0.07143
10	3	3	2	0.99167	0.17500	10	6	5	1	0.02381	0.02381
10	3	3	3	1.00000	0.00833	10	6	5	2	0.26190	0.23810
10	4	1	0	0.60000	0.60000	10	6	5	3	0.73809	0.47619
10	4	1	1	1.00000	0.40000	10	6	5	4	0.97619	0.23810
10	4	2	0	0.33333	0.33333	10	6	5	5	1.00000	0.02381
10	4	2	1	0.86667	0.53333	10	6	6	2	0.07143	0.07143
10	4	2	2	1.00000	0.13333	10	6	6	3	0.45238	0.38095
10	4	3	0	0.16667	0.16667	10	6	6	4	0.88095	0.42857
10	4	3	1	0.66667	0.50000	10	6	6	5	0.99524	0.11429
10	4	3	2	0.96667	0.30000	10	6	6	6	1.00000	0.00476
10	4	3	3	1.00000	0.03333	10	7	1	0	0.30000	0.30000
10	4	4	0	0.07143	0.07143	10	7	1	1	1.00000	0.70000
10	4	4	1	0.45238	0.38095	10	7	2	0	0.06667	0.06667
10	4	4	2	0.88095	0.42857	10	7	2	1	0.53333	0.46667
10	4	4	3	0.99524	0.11429	10	7	2	2	1.00000	0.46667
10	4	4	4	1.00000	0.00476	10	7	3	0	0.00833	0.00833

Hypergeometric probability and distribution functions

N	n	M	x	$F(x)$	$f(x)$	N	n	M	x	$F(x)$	$f(x)$
10	7	3	1	0.18333	0.17500	10	9	5	4	0.50000	0.50000
10	7	3	2	0.70833	0.52500	10	9	5	5	1.00000	0.50000
10	7	3	3	1.00000	0.29167	10	9	6	5	0.60000	0.60000
10	7	4	1	0.03333	0.03333	10	9	6	6	1.00000	0.40000
10	7	4	2	0.33333	0.30000	10	9	7	6	0.70000	0.70000
10	7	4	3	0.83333	0.50000	10	9	7	7	1.00000	0.30000
10	7	4	4	1.00000	0.16667	10	9	8	7	0.80000	0.80000
10	7	5	2	0.08333	0.08333	10	9	8	8	1.00000	0.20000
10	7	5	3	0.50000	0.41667	10	9	9	8	0.90000	0.90000
10	7	5	4	0.91667	0.41667	10	9	9	9	1.00000	0.10000
10	7	5	5	1.00000	0.08333	11	1	1	0	0.90909	0.90909
10	7	6	3	0.16667	0.16667	11	1	1	1	1.00000	0.09091
10	7	6	4	0.66667	0.50000	11	2	1	0	0.81818	0.81818
10	7	6	5	0.96667	0.30000	11	2	1	1	1.00000	0.18182
10	7	6	6	1.00000	0.03333	11	2	2	0	0.65455	0.65455
10	7	7	4	0.29167	0.29167	11	2	2	1	0.98182	0.32727
10	7	7	5	0.81667	0.52500	11	2	2	2	1.00000	0.01818
10	7	7	6	0.99167	0.17500	11	3	1	0	0.72727	0.72727
10	7	7	7	1.00000	0.00833	11	3	1	1	1.00000	0.27273
10	8	1	0	0.20000	0.20000	11	3	2	0	0.50909	0.50909
10	8	1	1	1.00000	0.80000	11	3	2	1	0.94546	0.43636
10	8	2	0	0.02222	0.02222	11	3	2	2	1.00000	0.05455
10	8	2	1	0.37778	0.35556	11	3	3	0	0.33939	0.33939
10	8	2	2	1.00000	0.62222	11	3	3	1	0.84849	0.50909
10	8	3	1	0.06667	0.06667	11	3	3	2	0.99394	0.14546
10	8	3	2	0.53333	0.46667	11	3	3	3	1.00000	0.00606
10	8	3	3	1.00000	0.46667	11	4	1	0	0.63636	0.63636
10	8	4	2	0.13333	0.13333	11	4	1	1	1.00000	0.36364
10	8	4	3	0.66667	0.53333	11	4	2	0	0.38182	0.38182
10	8	4	4	1.00000	0.33333	11	4	2	1	0.89091	0.50909
10	8	5	3	0.22222	0.22222	11	4	2	2	1.00000	0.10909
10	8	5	4	0.77778	0.55556	11	4	3	0	0.21212	0.21212
10	8	5	5	1.00000	0.22222	11	4	3	1	0.72121	0.50909
10	8	6	4	0.33333	0.33333	11	4	3	2	0.97576	0.25455
10	8	6	5	0.86667	0.53333	11	4	3	3	1.00000	0.02424
10	8	6	6	1.00000	0.13333	11	4	4	0	0.10606	0.10606
10	8	7	5	0.46667	0.46667	11	4	4	1	0.53030	0.42424
10	8	7	6	0.93333	0.46667	11	4	4	2	0.91212	0.38182
10	8	7	7	1.00000	0.06667	11	4	4	3	0.99697	0.08485
10	8	8	6	0.62222	0.62222	11	4	4	4	1.00000	0.00303
10	8	8	7	0.97778	0.35556	11	5	1	0	0.54546	0.54546
10	8	8	8	1.00000	0.02222	11	5	1	1	1.00000	0.45454
10	9	1	0	0.10000	0.10000	11	5	2	0	0.27273	0.27273
10	9	1	1	1.00000	0.90000	11	5	2	1	0.81818	0.54546
10	9	2	1	0.20000	0.20000	11	5	2	2	1.00000	0.18182
10	9	2	2	1.00000	0.80000	11	5	3	0	0.12121	0.12121
10	9	3	2	0.30000	0.30000	11	5	3	1	0.57576	0.45454
10	9	3	3	1.00000	0.70000	11	5	3	2	0.93939	0.36364
10	9	4	3	0.40000	0.40000	11	5	3	3	1.00000	0.06061
10	9	4	4	1.00000	0.60000	11	5	4	0	0.04546	0.04546

5.7   MULTINOMIAL DISTRIBUTION

The multinomial distribution is a generalization of the binomial distribution. Suppose there are n independent trials, and each trial results in exactly one of k possible distinct outcomes. For $i = 1, 2, \ldots, k$ let p_i be the probability that outcome i occurs on any given trial (with $\sum_{i=1}^{k} p_i = 1$). The multinomial random variable is the random vector $\mathbf{X} = [X_1, X_2, \ldots, X_k]^{\mathrm{T}}$ where X_i is the number of times outcome i occurs.

5.7.1   Properties

$$\text{pmf}\quad p(x_1, x_2, \ldots, x_k) = n! \prod_{i=1}^{k} \frac{p_i^{x_i}}{x_i!}, \quad \sum_{i=1}^{k} x_i = n$$

$$\text{mean of } X_i \quad \mu_i = np_i$$

$$\text{variance of } X_i \quad \sigma_i^2 = np_i(1 - p_i)$$

$$\text{Cov}[X_i, X_j] \quad \sigma_{ij} = -np_i p_j, \quad i \neq j$$

$$\text{joint mgf}\quad m(t_1, t_2, \ldots, t_k) = (p_1 e^{t_1} + p_2 e^{t_2} + \cdots + p_k e^{t_k})^n$$

$$\text{joint char function}\quad \phi(t_1, t_2, \ldots, t_k) = (p_1 e^{it_1} + p_2 e^{it_2} + \cdots + p_k e^{it_k})^n$$

$$\text{joint fact mgf}\quad P(t_1, t_2, \ldots, t_k) = (p_1 t_1 + p_2 t_2 + \cdots + p_k t_k)^n$$

5.7.2   Variates

Let $\mathbf{X}$ be a multinomial random variable with parameters n (number of trials) and $p_1, p_2, \ldots, p_k$.

(1) The marginal distribution of X_i is binomial with parameters n and p_i.

(2) If $k = 2$ and $p_1 = p$, then the multinomial random variable corresponds to the binomial random variable with parameters n and p.

5.8   NEGATIVE BINOMIAL DISTRIBUTION

Consider a sequence of Bernoulli trials with probability of success p. The negative binomial distribution is used to describe the number of failures, X, before the nth success.

5.8.1   Properties

$$\text{pmf}\quad p(x) = \binom{x + n - 1}{n - 1} p^n q^x \quad x = 0, 1, 2, \ldots, \quad n = 1, 2, \ldots$$

$$0 \leq p \leq 1, \quad q = 1 - p$$

$$\text{mean}\quad \mu = \frac{nq}{p}$$

$$\text{variance} \quad \sigma^2 = \frac{nq}{p^2}$$

$$\text{skewness} \quad \beta_1 = \frac{2-p}{\sqrt{nq}}$$

$$\text{kurtosis} \quad \beta_2 = 3 + \frac{p^2 + 6q}{nq}$$

$$\text{mgf} \quad m(t) = \left(\frac{p}{1-qe^t}\right)^n$$

$$\text{char function} \quad \phi(t) = \left(\frac{p}{1-qe^{it}}\right)^n$$

$$\text{fact mgf} \quad P(t) = \left(\frac{p}{1-qt}\right)^n$$

Using $p = k/(m+k)$ and $n = k$, there is the following alternative characterization.

5.8.1.1 Alternative characterization

$$\text{pmf} \quad p(x) = \frac{\Gamma(k+x)}{x!\,\Gamma(k)} \left(\frac{k}{m+k}\right)^k \left(\frac{m}{m+k}\right)^x$$

$$x = 0, 1, 2, \ldots, \quad m, k > 0$$

$$\text{mean} \quad \mu = m$$

$$\text{variance} \quad \sigma^2 = m + m^2/k$$

$$\text{skewness} \quad \beta_1 = \frac{2m+k}{\sqrt{mk(m+k)}}$$

$$\text{kurtosis} \quad \beta_2 = 3 + \frac{6m^2 + 6mk + k^2}{mk(m+k)}$$

$$\text{mgf} \quad m(t) = \left(1 - \frac{m}{k}(e^t - 1)\right)^{-k}$$

$$\text{char function} \quad \phi(t) = \left(1 - \frac{m}{k}(e^{it} - 1)\right)^{-k}$$

$$\text{fact mgf} \quad P(t) = \left(1 - \frac{m}{k}(t - 1)\right)^{-k}$$

where $\Gamma(x)$ is the gamma function defined in Chapter 18 (see page 515).

5.8.2 Variates

Let X be a negative binomial random variable with parameters n and p.

(1) If $n = 1$ then X is a geometric random variable with probability of success p.

(2) As $n \to \infty$ and $p \to 1$ with $n(1-p)$ held constant, X is approximately a Poisson random variable with $\lambda = n(1-p)$.

(3) Let $X_1, X_2, \ldots, X_k$ be independent negative binomial random variables with parameters n_i and p, respectively. The random variable $Y = X_1 + X_2 + \cdots + X_k$ has a negative binomial distribution with parameters $n = n_1 + n_2 + \cdots + n_k$ and p.

5.8.3 Tables

Example 5.37: Suppose a biased coin has probability of heads 0.3. What is the probability that the 5th head occurs after the 8th tail?

Solution:

(S1) Recognizing that $n = 5$ and $x = 8$ with $p = 0.3$ and $q = 1 - p = 0.7$, the probability is

$$\text{Prob}\,[X = 8] = \binom{5+8-1}{5-1}(0.3)^5(0.7)^8 = 495(0.3)^5(0.7)^8 = 0.0693$$

(S2) This value is in the table below with $n = 5$, $x = 8$, and $p = 0.3$.

Probability mass, Negative binomial distribution

(n,x)	$p = 0.1$	0.2	0.3	0.4	0.5	0.6	0.7	0.8	0.9
(1,2)	0.0810	0.1280	0.1470	0.1440	0.1250	0.0960	0.0630	0.0320	0.00900
(1,5)	0.0590	0.0655	0.0504	0.0311	0.0156	0.0061	0.0017	0.0003	
(1,8)	0.0430	0.0336	0.0173	0.0067	0.0020	0.0004			
(1,10)	0.0349	0.0215	0.0085	0.0024	0.0005	0.0001			
(3,2)	0.0049	0.0307	0.0794	0.1382	0.1875	0.2074	0.1852	0.1229	0.04370
(3,5)	0.0124	0.0551	0.0953	0.1045	0.0820	0.0464	0.0175	0.0034	0.00020
(3,8)	0.0194	0.0604	0.0700	0.0484	0.0220	0.0064	0.0010	0.0001	
(3,10)	0.0230	0.0567	0.0503	0.0255	0.0081	0.0015	0.0001		
(5,2)	0.0001	0.0031	0.0179	0.0553	0.1172	0.1866	0.2269	0.1966	0.08860
(5,5)	0.0007	0.0132	0.0515	0.1003	0.1230	0.1003	0.0515	0.0132	0.00070
(5,8)	0.0021	0.0266	0.0693	0.0851	0.0604	0.0252	0.0055	0.0004	
(5,10)	0.0035	0.0344	0.0687	0.0620	0.0305	0.0082	0.0010		
(8,2)		0.0001	0.0012	0.0085	0.0352	0.0967	0.1868	0.2416	0.15500
(8,5)		0.0007	0.0087	0.0404	0.0967	0.1362	0.1109	0.0425	0.00340
(8,8)		0.0028	0.0243	0.0708	0.0982	0.0708	0.0243	0.0028	
(8,10)	0.0001	0.0053	0.0360	0.0771	0.0742	0.0343	0.0066	0.0003	

5.9 POISSON DISTRIBUTION

The Poisson, or *rare event*, distribution is completely described by a single parameter, λ. This distribution is used to model the number of successes, X, in a specified time interval or given region. It is assumed the numbers of successes occurring in different time intervals or regions are independent, the probability of a success in a time interval or region is very small and proportional to the length of the time interval or the size of the region, and the probability of more than one success during any one time interval or region is negligible.

5.9.1 Properties

$$
\begin{aligned}
\text{pmf} \quad & p(x) = \frac{e^{-\lambda}\lambda^x}{x!} \quad x = 0, 1, 2, \ldots, \quad \lambda > 0 \\
\text{mean} \quad & \mu = \lambda \\
\text{variance} \quad & \sigma^2 = \lambda \\
\text{skewness} \quad & \beta_1 = 1/\sqrt{\lambda} \\
\text{kurtosis} \quad & \beta_2 = 3 + (1/\lambda) \\
\text{mgf} \quad & m(t) = \exp[\lambda(e^t - 1)] \\
\text{char function} \quad & \phi(t) = \exp[\lambda(e^{it} - 1)] \\
\text{fact mgf} \quad & P(t) = \exp[\lambda(t - 1)]
\end{aligned}
$$

Note that the waiting time between Poisson arrivals is exponentially distributed.

5.9.2 Variates

Let X be a Poisson random variable with parameter λ.

(1) As $\lambda \to \infty$, X is approximately normal with parameters $\mu = \lambda$ and $\sigma^2 = \lambda$.

(2) Let $X_1, X_2, \ldots, X_n$ be independent Poisson random variables with parameters λ_i, respectively. The random variable $Y = X_1 + X_2 + \cdots + X_n$ has a Poisson distribution with parameter $\lambda = \lambda_1 + \lambda_2 + \cdots + \lambda_n$.

5.9.3 Tables

Example 5.38:   The number of black bear sightings in Northeastern Pennsylvania during a given week has a Poisson distribution with $\lambda = 3$. For a randomly selected week, what is the probability of exactly 2 sightings, more than 5 sightings, between 4 and 7 sightings (inclusive)?

Solution:

(S1) Let X be the random variable representing the number of black bear sightings during any given week; X is Poisson with $\lambda = 3$. Use the table below to answer the probability questions.

(S2) $\text{Prob}[X = 2] = \text{Prob}[X \leq 2] - \text{Prob}[X \leq 1] = 0.423 - 0.199 = 0.224$

(S3) $\text{Prob}[X > 5] = 1 - \text{Prob}[X \leq 4] = 1 - 0.815 = 0.185$

(S4) $\text{Prob}[4 \leq X \leq 7] = \text{Prob}[X \leq 7] - \text{Prob}[X \leq 3] = .988 - .647 = .341$

Cumulative probability, Poisson distribution

λ \ x =0	1	2	3	4	5	6	7	8	9
0.02	0.980	1.000								
0.04	0.961	0.999	1.000							
0.06	0.942	0.998	1.000							
0.08	0.923	0.997	1.000							
0.10	0.905	0.995	1.000							
0.15	0.861	0.990	1.000	1.000						
0.20	0.819	0.983	0.999	1.000						
0.25	0.779	0.974	0.998	1.000						
0.30	0.741	0.963	0.996	1.000						
0.35	0.705	0.951	0.995	1.000						
0.40	0.670	0.938	0.992	0.999	1.000					
0.45	0.638	0.925	0.989	0.999	1.000					
0.50	0.607	0.910	0.986	0.998	1.000					
0.55	0.577	0.894	0.982	0.998	1.000					
0.60	0.549	0.878	0.977	0.997	1.000					
0.65	0.522	0.861	0.972	0.996	0.999	1.000				
0.70	0.497	0.844	0.966	0.994	0.999	1.000				
0.75	0.472	0.827	0.960	0.993	0.999	1.000				
0.80	0.449	0.809	0.953	0.991	0.999	1.000				
0.85	0.427	0.791	0.945	0.989	0.998	1.000				
0.90	0.407	0.772	0.937	0.987	0.998	1.000				
0.95	0.387	0.754	0.929	0.984	0.997	1.000				
1.00	0.368	0.736	0.920	0.981	0.996	0.999	1.000			
1.1	0.333	0.699	0.900	0.974	0.995	0.999	1.000			
1.2	0.301	0.663	0.879	0.966	0.992	0.999	1.000			
1.3	0.273	0.627	0.857	0.957	0.989	0.998	1.000			
1.4	0.247	0.592	0.834	0.946	0.986	0.997	0.999	1.000		
1.5	0.223	0.558	0.809	0.934	0.981	0.996	0.999	1.000		
1.6	0.202	0.525	0.783	0.921	0.976	0.994	0.999	1.000		
1.7	0.183	0.493	0.757	0.907	0.970	0.992	0.998	1.000		
1.8	0.165	0.463	0.731	0.891	0.964	0.990	0.997	0.999	1.000	
1.9	0.150	0.434	0.704	0.875	0.956	0.987	0.997	0.999	1.000	
2.0	0.135	0.406	0.677	0.857	0.947	0.983	0.996	0.999	1.000	
2.2	0.111	0.355	0.623	0.819	0.927	0.975	0.993	0.998	1.000	
2.4	0.091	0.308	0.570	0.779	0.904	0.964	0.988	0.997	0.999	1.000
2.6	0.074	0.267	0.518	0.736	0.877	0.951	0.983	0.995	0.999	1.000
2.8	0.061	0.231	0.469	0.692	0.848	0.935	0.976	0.992	0.998	0.999
3.0	0.050	0.199	0.423	0.647	0.815	0.916	0.967	0.988	0.996	0.999
3.2	0.041	0.171	0.380	0.603	0.781	0.895	0.955	0.983	0.994	0.998
3.4	0.033	0.147	0.340	0.558	0.744	0.871	0.942	0.977	0.992	0.997
3.6	0.027	0.126	0.303	0.515	0.706	0.844	0.927	0.969	0.988	0.996
3.8	0.022	0.107	0.269	0.473	0.668	0.816	0.909	0.960	0.984	0.994

continued on next page

continued from previous page

λ	x =0	1	2	3	4	5	6	7	8	9
4.0	0.018	0.092	0.238	0.433	0.629	0.785	0.889	0.949	0.979	0.992
4.2	0.015	0.078	0.210	0.395	0.590	0.753	0.868	0.936	0.972	0.989
4.4	0.012	0.066	0.185	0.359	0.551	0.720	0.844	0.921	0.964	0.985
4.6	0.010	0.056	0.163	0.326	0.513	0.686	0.818	0.905	0.955	0.981
4.8	0.008	0.048	0.142	0.294	0.476	0.651	0.791	0.887	0.944	0.975
5.0	0.007	0.040	0.125	0.265	0.441	0.616	0.762	0.867	0.932	0.968
5.2	0.005	0.034	0.109	0.238	0.406	0.581	0.732	0.845	0.918	0.960
5.4	0.004	0.029	0.095	0.213	0.373	0.546	0.702	0.822	0.903	0.951
5.6	0.004	0.024	0.082	0.191	0.342	0.512	0.670	0.797	0.886	0.941
5.8	0.003	0.021	0.071	0.170	0.313	0.478	0.638	0.771	0.867	0.929
6.0	0.003	0.017	0.062	0.151	0.285	0.446	0.606	0.744	0.847	0.916
6.2	0.002	0.015	0.054	0.134	0.259	0.414	0.574	0.716	0.826	0.902
6.4	0.002	0.012	0.046	0.119	0.235	0.384	0.542	0.687	0.803	0.886
6.6	0.001	0.010	0.040	0.105	0.213	0.355	0.511	0.658	0.780	0.869
6.8	0.001	0.009	0.034	0.093	0.192	0.327	0.480	0.628	0.755	0.850
7.0	0.001	0.007	0.030	0.082	0.173	0.301	0.450	0.599	0.729	0.831
7.2	0.001	0.006	0.025	0.072	0.155	0.276	0.420	0.569	0.703	0.810
7.4	0.001	0.005	0.022	0.063	0.140	0.253	0.392	0.539	0.676	0.788
7.6	0.001	0.004	0.019	0.055	0.125	0.231	0.365	0.510	0.648	0.765
7.8	0.000	0.004	0.016	0.049	0.112	0.210	0.338	0.481	0.620	0.741
8.0	0.000	0.003	0.014	0.042	0.100	0.191	0.313	0.453	0.593	0.717
8.5	0.000	0.002	0.009	0.030	0.074	0.150	0.256	0.386	0.523	0.653
9.0	0.000	0.001	0.006	0.021	0.055	0.116	0.207	0.324	0.456	0.587
9.5	0.000	0.001	0.004	0.015	0.040	0.088	0.165	0.269	0.392	0.522
10.0	0.000	0.001	0.003	0.010	0.029	0.067	0.130	0.220	0.333	0.458
10.5	0.000	0.000	0.002	0.007	0.021	0.050	0.102	0.178	0.279	0.397
11.0	0.000	0.000	0.001	0.005	0.015	0.037	0.079	0.143	0.232	0.341
11.5	0.000	0.000	0.001	0.003	0.011	0.028	0.060	0.114	0.191	0.289
12.0	0.000	0.000	0.001	0.002	0.008	0.020	0.046	0.089	0.155	0.242
12.5	0.000	0.000	0.000	0.002	0.005	0.015	0.035	0.070	0.125	0.201
13.0	0.000	0.000	0.000	0.001	0.004	0.011	0.026	0.054	0.100	0.166
13.5	0.000	0.000	0.000	0.001	0.003	0.008	0.019	0.042	0.079	0.135
14.0	0.000	0.000	0.000	0.001	0.002	0.005	0.014	0.032	0.062	0.109
14.5	0.000	0.000	0.000	0.000	0.001	0.004	0.011	0.024	0.048	0.088
15.0	0.000	0.000	0.000	0.000	0.001	0.003	0.008	0.018	0.037	0.070

Cumulative probability, Poisson distribution

λ	x =10	11	12	13	14	15	16	17	18	19
2.8	1.000									
3.0	1.000									
3.2	1.000									
3.4	0.999	1.000								
3.6	0.999	1.000								
3.8	0.998	0.999	1.000							
4.0	0.997	0.999	1.000							

continued on next page

λ	$x =10$	11	12	13	14	15	16	17	18	19
continued from previous page										
4.2	0.996	0.999	1.000							
4.4	0.994	0.998	0.999	1.000						
4.6	0.992	0.997	0.999	1.000						
4.8	0.990	0.996	0.999	1.000						
5.0	0.986	0.995	0.998	0.999	1.000					
5.2	0.982	0.993	0.997	0.999	1.000					
5.4	0.978	0.990	0.996	0.999	1.000					
5.6	0.972	0.988	0.995	0.998	0.999	1.000				
5.8	0.965	0.984	0.993	0.997	0.999	1.000				
6.0	0.957	0.980	0.991	0.996	0.999	1.000	1.000			
6.2	0.949	0.975	0.989	0.995	0.998	0.999	1.000			
6.4	0.939	0.969	0.986	0.994	0.997	0.999	1.000			
6.6	0.927	0.963	0.982	0.992	0.997	0.999	1.000	1.000		
6.8	0.915	0.955	0.978	0.990	0.996	0.998	0.999	1.000		
7.0	0.901	0.947	0.973	0.987	0.994	0.998	0.999	1.000		
7.2	0.887	0.937	0.967	0.984	0.993	0.997	0.999	1.000		
7.4	0.871	0.926	0.961	0.981	0.991	0.996	0.998	0.999	1.000	
7.6	0.854	0.915	0.954	0.976	0.989	0.995	0.998	0.999	1.000	
7.8	0.835	0.902	0.945	0.971	0.986	0.993	0.997	0.999	1.000	
8.0	0.816	0.888	0.936	0.966	0.983	0.992	0.996	0.998	0.999	1.000
8.5	0.763	0.849	0.909	0.949	0.973	0.986	0.993	0.997	0.999	1.000
9.0	0.706	0.803	0.876	0.926	0.959	0.978	0.989	0.995	0.998	0.999
9.5	0.645	0.752	0.836	0.898	0.940	0.967	0.982	0.991	0.996	0.998
10.0	0.583	0.697	0.792	0.865	0.916	0.951	0.973	0.986	0.993	0.997
10.5	0.521	0.639	0.742	0.825	0.888	0.932	0.960	0.978	0.989	0.994
11.0	0.460	0.579	0.689	0.781	0.854	0.907	0.944	0.968	0.982	0.991
11.5	0.402	0.520	0.633	0.733	0.815	0.878	0.924	0.954	0.974	0.986
12.0	0.347	0.462	0.576	0.681	0.772	0.844	0.899	0.937	0.963	0.979
12.5	0.297	0.406	0.519	0.628	0.725	0.806	0.869	0.916	0.948	0.969
13.0	0.252	0.353	0.463	0.573	0.675	0.764	0.836	0.890	0.930	0.957
13.5	0.211	0.304	0.409	0.518	0.623	0.718	0.797	0.861	0.908	0.942
14.0	0.176	0.260	0.358	0.464	0.570	0.669	0.756	0.827	0.883	0.923
14.5	0.145	0.220	0.311	0.412	0.518	0.619	0.711	0.790	0.853	0.901
15.0	0.118	0.185	0.268	0.363	0.466	0.568	0.664	0.749	0.820	0.875

Cumulative probability, Poisson distribution

λ	$x =20$	21	22	23	24	25	26	27	28	29
8.5	1.000									
9.0	1.000									
9.5	0.999	1.000								
10.0	0.998	0.999	1.000							
10.5	0.997	0.999	0.999	1.000						
11.0	0.995	0.998	0.999	1.000						
11.5	0.993	0.996	0.998	0.999	1.000					
12.0	0.988	0.994	0.997	0.999	0.999	1.000				

continued on next page

λ	$x = 20$	21	22	23	24	25	26	27	28	29
										continued from previous page
12.5	0.983	0.991	0.995	0.998	0.999	0.999	1.000			
13.0	0.975	0.986	0.992	0.996	0.998	0.999	1.000			
13.5	0.965	0.980	0.989	0.994	0.997	0.998	0.999	1.000		
14.0	0.952	0.971	0.983	0.991	0.995	0.997	0.999	0.999	1.000	
14.5	0.936	0.960	0.976	0.986	0.992	0.996	0.998	0.999	1.000	1.000
15.0	0.917	0.947	0.967	0.981	0.989	0.994	0.997	0.998	0.999	1.000

Cumulative probability, Poisson distribution

λ	$x = 5$	6	7	8	9	10	11	12	13	14
16	0.001	0.004	0.010	0.022	0.043	0.077	0.127	0.193	0.275	0.367
17	0.001	0.002	0.005	0.013	0.026	0.049	0.085	0.135	0.201	0.281
18	0.000	0.001	0.003	0.007	0.015	0.030	0.055	0.092	0.143	0.208
19	0.000	0.001	0.002	0.004	0.009	0.018	0.035	0.061	0.098	0.150
20	0.000	0.000	0.001	0.002	0.005	0.011	0.021	0.039	0.066	0.105
21	0.000	0.000	0.000	0.001	0.003	0.006	0.013	0.025	0.043	0.072
22	0.000	0.000	0.000	0.001	0.002	0.004	0.008	0.015	0.028	0.048
23	0.000	0.000	0.000	0.000	0.001	0.002	0.004	0.009	0.017	0.031
24	0.000	0.000	0.000	0.000	0.000	0.001	0.003	0.005	0.011	0.020
25	0.000	0.000	0.000	0.000	0.000	0.001	0.001	0.003	0.006	0.012
26	0.000	0.000	0.000	0.000	0.000	0.000	0.001	0.002	0.004	0.008
27	0.000	0.000	0.000	0.000	0.000	0.000	0.000	0.001	0.002	0.005
28	0.000	0.000	0.000	0.000	0.000	0.000	0.000	0.001	0.001	0.003
29	0.000	0.000	0.000	0.000	0.000	0.000	0.000	0.000	0.001	0.002
30	0.000	0.000	0.000	0.000	0.000	0.000	0.000	0.000	0.000	0.001

Cumulative probability, Poisson distribution

λ	$x = 15$	16	17	18	19	20	21	22	23	24
16	0.467	0.566	0.659	0.742	0.812	0.868	0.911	0.942	0.963	0.978
17	0.371	0.468	0.564	0.655	0.736	0.805	0.862	0.905	0.937	0.959
18	0.287	0.375	0.469	0.562	0.651	0.731	0.799	0.855	0.899	0.932
19	0.215	0.292	0.378	0.469	0.561	0.647	0.726	0.793	0.849	0.893
20	0.157	0.221	0.297	0.381	0.470	0.559	0.644	0.721	0.787	0.843
21	0.111	0.163	0.227	0.302	0.384	0.471	0.558	0.640	0.716	0.782
22	0.077	0.117	0.169	0.233	0.306	0.387	0.472	0.556	0.637	0.712
23	0.052	0.082	0.123	0.175	0.238	0.310	0.389	0.472	0.555	0.635
24	0.034	0.056	0.087	0.128	0.180	0.243	0.314	0.392	0.473	0.554
25	0.022	0.038	0.060	0.092	0.134	0.185	0.247	0.318	0.394	0.473
26	0.014	0.025	0.041	0.065	0.097	0.139	0.191	0.252	0.321	0.396
27	0.009	0.016	0.027	0.044	0.069	0.102	0.144	0.195	0.256	0.324
28	0.005	0.010	0.018	0.030	0.048	0.073	0.106	0.148	0.200	0.260
29	0.003	0.006	0.011	0.020	0.033	0.051	0.077	0.110	0.153	0.204
30	0.002	0.004	0.007	0.013	0.022	0.035	0.054	0.081	0.115	0.157

Cumulative probability, Poisson distribution

λ	x =25	26	27	28	29	30	31	32	33	34
16	0.987	0.993	0.996	0.998	0.999	0.999	1.000			
17	0.975	0.985	0.991	0.995	0.997	0.999	0.999	1.000		
18	0.955	0.972	0.983	0.990	0.994	0.997	0.998	0.999	1.000	
19	0.927	0.951	0.969	0.981	0.988	0.993	0.996	0.998	0.999	0.999
20	0.888	0.922	0.948	0.966	0.978	0.987	0.992	0.995	0.997	0.999
21	0.838	0.883	0.917	0.944	0.963	0.976	0.985	0.991	0.995	0.997
22	0.777	0.832	0.877	0.913	0.940	0.960	0.974	0.983	0.990	0.994
23	0.708	0.772	0.827	0.873	0.908	0.936	0.956	0.971	0.981	0.988
24	0.632	0.704	0.768	0.823	0.868	0.904	0.932	0.953	0.969	0.979
25	0.553	0.629	0.700	0.763	0.818	0.863	0.900	0.928	0.950	0.966
26	0.474	0.552	0.627	0.697	0.759	0.813	0.859	0.896	0.925	0.947
27	0.398	0.474	0.551	0.625	0.694	0.755	0.809	0.855	0.892	0.921
28	0.327	0.400	0.475	0.550	0.623	0.690	0.751	0.805	0.851	0.888
29	0.264	0.330	0.401	0.475	0.549	0.621	0.687	0.748	0.801	0.847
30	0.208	0.267	0.333	0.403	0.476	0.548	0.619	0.684	0.744	0.797

Cumulative probability, Poisson distribution

λ	x =35	36	37	38	39	40	41	42	43	44
16	1.000									
17	1.000									
18	1.000									
19	1.000									
20	0.999	1.000								
21	0.998	0.999	1.000	1.000						
22	0.996	0.998	0.999	0.999	1.000					
23	0.993	0.996	0.997	0.999	0.999	1.000				
24	0.987	0.992	0.995	0.997	0.998	0.999	1.000	1.000		
25	0.978	0.985	0.991	0.994	0.997	0.998	0.999	0.999	1.000	
26	0.964	0.976	0.984	0.990	0.994	0.996	0.998	0.999	0.999	1.000
27	0.944	0.961	0.974	0.983	0.989	0.993	0.996	0.997	0.998	0.999
28	0.918	0.941	0.959	0.972	0.981	0.988	0.992	0.995	0.997	0.998
29	0.884	0.914	0.938	0.956	0.970	0.980	0.986	0.991	0.994	0.997
30	0.843	0.880	0.911	0.935	0.954	0.968	0.978	0.985	0.990	0.994

5.10   RECTANGULAR (DISCRETE UNIFORM) DISTRIBUTION

A general rectangular distribution is used to describe a random variable, X, that can assume n different values with equal probabilities. In the special case presented here, we assume the random variable can assume the first n positive integers.

5.10.1   Properties

pmf	$p(x) = 1/n, \quad x = 1, 2, \ldots, n, \quad n \in \mathcal{N}$
mean	$\mu = (n+1)/2$
variance	$\sigma^2 = (n^2 - 1)/12$
skewness	$\beta_1 = 0$
kurtosis	$\beta_2 = \dfrac{3}{5}\left(3 - \dfrac{4}{n^2 - 1}\right)$
mgf	$m(t) = \dfrac{e^t(1 - e^{nt})}{n(1 - e^t)}$
char function	$\phi(t) = \dfrac{e^{it}(1 - e^{nit})}{n(1 - e^{it})}$
fact mgf	$P(t) = \dfrac{t(1 - t^n)}{n(1 - t)}$

Example 5.39:   A new family game has a special 12-sided numbered die, manufactured so that each side is equally likely to occur. Find the mean and variance of the number rolled, and the probability of rolling a 2, 3, or 12.

Solution:

(S1) Let X be the number on the side facing up; X has a discrete uniform distribution with $n = 12$.

(S2) Using the properties given above:

$$\mu = (n+1)/2 = (12+1)/2 = {}^{13}\!/_2 = 6.5$$
$$\sigma^2 = (n^2 - 1)/12 = (12^2 - 1)/12 = {}^{143}\!/_{12} = 11.9167$$

(S3) $\text{Prob}\,[X = 2, 3, 12] = \dfrac{1}{12} + \dfrac{1}{12} + \dfrac{1}{12} = \dfrac{3}{12} = 0.25$

CHAPTER 6

Continuous Probability Distributions

Contents

This chapter presents some common continuous probability distributions along with their properties. Relevant numerical tables are also included.

Notation used throughout this chapter:

Probability density function $f(x)$ $\quad\text{Prob}\,[a \leq X \leq b] = \int_a^b f(x)\,dx$
(pdf)

Cumulative distrib function $\quad F(x) = \text{Prob}\,[X \leq x] = \int_{-\infty}^x f(x)\,dx$
(cdf)

$$\text{Mean} \qquad \mu = \text{E}\,[X]$$

$$\text{Variance} \qquad \sigma^2 = \text{E}\,[(X - \mu)^2]$$

$$\text{Coefficient of skewness} \qquad \beta_1 = \text{E}\,[(X - \mu)^3]/\sigma^3$$

$$\text{Coefficient of kurtosis} \qquad \beta_2 = \text{E}\,[(X - \mu)^4]/\sigma^4$$

$$\text{Moment generating function} \qquad m(t) = \text{E}\,[e^{tX}]$$
(mgf)

$$\text{Characteristic function} \qquad \phi(t) = \text{E}\,[e^{itX}]$$
(char function)

6.1   ARCSIN DISTRIBUTION

6.1.1   Properties

$$\text{pdf} \quad f(x) = \frac{1}{\pi\sqrt{x(1 - x)}}, \quad 0 < x < 1$$

$$\text{mean} \quad \mu = 1/2$$

$$\text{variance} \quad \sigma^2 = 1/8$$

$$\text{skewness} \quad \beta_1 = 0$$

$$\text{kurtosis} \quad \beta_2 = 3/2$$

$$\text{mgf} \quad m(t) = e^{t/2} I_0(t/2)$$

$$\text{char function} \quad \phi(t) = J_0(t/2)\cos(t/2) + i J_0(t/2)\sin(t/2)$$

where $J_n(x)$ is the Bessel function of the first kind and $I_p(x)$ is the modified Bessel function of the first kind defined in Chapter 18 (see page 506).

6.1.2   Probability density function

The probability density function is "U" shaped. As $x \to 0^+$ and as $x \to 1^-$, $f(x) \to \infty$.

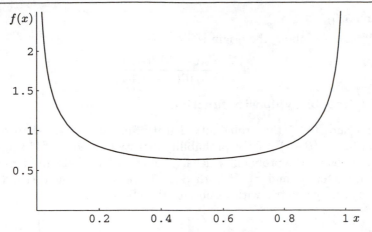

Figure 6.1: Probability density function for an arcsin random variable.

6.2 BETA DISTRIBUTION

6.2.1 Properties

$$\text{pdf} \quad f(x) = \frac{\Gamma(\alpha + \beta)}{\Gamma(\alpha)\Gamma(\beta)} x^{\alpha-1}(1-x)^{\beta-1} = \frac{x^{\alpha-1}(1-x)^{\beta-1}}{B(\alpha, \beta)}$$

$$0 \le x \le 1, \quad \alpha, \ \beta > 0$$

$$\text{mean} \quad \mu = \frac{\alpha}{\alpha + \beta}$$

$$\text{variance} \quad \sigma^2 = \frac{\alpha\beta}{(\alpha + \beta)^2(\alpha + \beta + 1)}$$

$$\text{skewness} \quad \beta_1 = \frac{2(\beta - \alpha)\sqrt{\alpha + \beta + 1}}{\sqrt{\alpha\beta}(\alpha + \beta + 2)}$$

$$\text{kurtosis} \quad \beta_2 = \frac{3(\alpha + \beta + 1)[2(\alpha + \beta)^2 + \alpha\beta(\alpha + \beta - 6)]}{\alpha\beta(\alpha + \beta + 2)(\alpha + \beta + 3)}$$

$$\text{mgf} \quad m(t) = {}_1F_1(\alpha; \beta; t)$$

char function $\quad \phi(t) =$

$$\frac{1}{\alpha + \beta}\left(iat \ {}_2F_3\left[\left\{\frac{1}{2} + \frac{\alpha}{2}, 1 + \frac{\alpha}{2}\right\}; \left\{\frac{3}{2}, \frac{1}{2} + \frac{\alpha}{2} + \frac{\beta}{2}, 1 + \frac{\alpha}{2} + \frac{\beta}{2}\right\}; -\frac{t^2}{4}\right]\right)$$

$$+ {}_2F_3\left[\left\{\frac{1}{2} + \frac{\alpha}{2}, \frac{\alpha}{2}\right\}; \left\{\frac{1}{2}, \frac{\alpha}{2} + \frac{\beta}{2}, \frac{1}{2} + \frac{\alpha}{2} + \frac{\beta}{2}\right\}; -\frac{t^2}{4}\right]$$

where $\Gamma(x)$ is the gamma function, $B(a, b)$ is the beta function, and ${}_pF_q$ is the generalized hypergeometric function defined in Chapter 18 (see pages 515,

511, and 520).

The rth moment about the origin is

$$\mu'_r = \frac{\Gamma(\alpha+\beta)\Gamma(\alpha+r)}{\Gamma(\alpha)\Gamma(\alpha+\beta+r)} \tag{6.1}$$

6.2.2 Probability density function

If $\alpha < 1$ and $\beta < 1$ the probability density function is "U" shaped. If the product $(\alpha-1)(\beta-1) < 0$ the probability density function is "J" shaped. Let $f(x;\alpha,\beta)$ denote the probability density function for a beta random variable with parameters α and β. If both $\alpha > 1$ and $\beta > 1$ then $f(x;\alpha,\beta)$ and $f(x;\beta,\alpha)$ are symmetric with respect to the line $x = .5$.

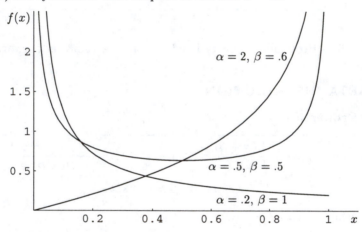

Figure 6.2: Probability density functions for a beta random variable, various shape parameters.

6.2.3 Related distributions

Let X be a beta random variable with parameters α and β.

(1) If $\alpha = \beta = 1/2$, then X is an arcsin random variable.

(2) If $\alpha = \beta = 1$, then X is a uniform random variable with parameters $a = 0$ and $b = 1$.

(3) If $\beta = 1$, then X is a power function random variable with parameters $b = 1$ and $c = \alpha$.

(4) As α and β tend to infinity such that α/β is constant, X tends to a standard normal random variable.

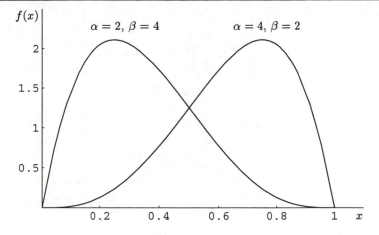

Figure 6.3: Probability density functions for a beta random variable, example of symmetry.

6.3 CAUCHY DISTRIBUTION

6.3.1 Properties

$$\text{pdf} \quad f(x) = \frac{1}{b\pi \left(1 + \left(\frac{x-a}{b}\right)^2\right)}, \quad x \in \mathcal{R}, \ a \in \mathcal{R}, \ b > 0$$

mean	μ	= does not exist
variance	σ^2	= does not exist
skewness	β_1	= does not exist
kurtosis	β_2	= does not exist
mgf	$m(t)$	= does not exist
char function	$\phi(t)$	= $e^{ait - b|t|}$

6.3.2 Probability density function

The probability density function for a Cauchy random variable is unimodal and symmetric about the parameter a. The tails are heavier than those of a normal random variable.

6.3.3 Related distributions

Let X be a Cauchy random variable with parameters a and b.

(1) If $a = 0$ and $b = 1$ then X is a standard Cauchy random variable.

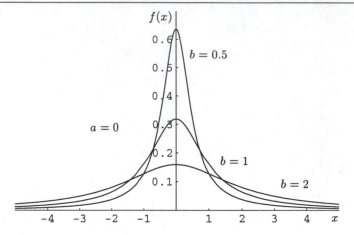

Figure 6.4: Probability density functions for a Cauchy random variable.

(2) The random variable $1/X$ is also a Cauchy random variable with parameters $a/(a^2 + b^2)$ and $b/(a^2 + b^2)$.

(3) Let X_i (for $i = 1, 2, \ldots, n$) be independent, Cauchy random variables with parameters a_i and b_i, respectively. The random variable $Y = X_1 + X_2 + \cdots + X_n$ has a Cauchy distribution with parameters $a = a_1 + a_2 + \cdots + a_n$ and $b = b_1 + b_2 + \cdots + b_n$.

6.4   CHI–SQUARE DISTRIBUTION

6.4.1   Properties

$$\text{pdf} \quad f(x) = \frac{e^{-x/2} x^{(\nu/2)-1}}{2^{\nu/2} \Gamma(\nu/2)}, \quad x \geq 0, \; \nu \in \mathcal{N}$$

$$\text{mean} \quad \mu = \nu$$

$$\text{variance} \quad \sigma^2 = 2\nu$$

$$\text{skewness} \quad \beta_1 = 2\sqrt{2/\nu}$$

$$\text{kurtosis} \quad \beta_2 = 3 + \frac{12}{\nu}$$

$$\text{mgf} \quad m(t) = (1 - 2t)^{-\nu/2}, \quad t < 1/2$$

$$\text{char function} \quad \phi(t) = (1 - 2it)^{-\nu/2}$$

---

where $\Gamma(x)$ is the gamma function (see page 515).

A chi–square(χ^2) distribution is completely characterized by the parameter ν, the *degrees of freedom*.

6.4.2 Probability density function

The probability density function for a chi–square random variable is positively skewed. As ν tends to infinity, the density function becomes more bell–shaped and symmetric.

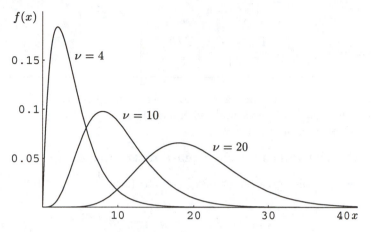

Figure 6.5: Probability density functions for a chi–square random variable.

6.4.3 Related distributions

(1) If X is a chi–square random variable with $\nu = 2$, then X is an exponential random variable with $\lambda = 1/2$.

(2) If X_1 and X_2 are independent chi–square random variables with parameters ν_1 and ν_2, then the random variable $(X_1/\nu_1)/(X_2/\nu_2)$ has an F distribution with ν_1 and ν_2 degrees of freedom.

(3) If X_1 and X_2 are independent chi–square random variables with parameters $\nu_1 = \nu_2 = \nu$, the random variable

$$Y = \frac{\sqrt{\nu}}{2} \frac{X_1 - X_2}{\sqrt{X_1 X_2}} \tag{6.2}$$

has a t distribution with ν degrees of freedom.

(4) Let X_i (for $i = 1, 2, \ldots, n$) be independent chi–square random variables with parameters ν_i. The random variable $Y = X_1 + X_2 + \cdots + X_n$ has a chi–square distribution with $\nu = \nu_1 + \nu_2 + \cdots + \nu_n$ degrees of freedom.

(5) If X is a chi–square random variable with ν degrees of freedom, the random variable $\sqrt{X}$ has a **chi distribution** with parameter ν. Properties of a chi random variable:

---

pdf $f(x) = \dfrac{x^{n-1}e^{-x^2/2}}{2^{(n/2)-1}\Gamma(n/2)}, \quad x \geq 0, \; n \in \mathcal{N}$

mean $\mu = \dfrac{\Gamma\left(\frac{n+1}{2}\right)}{\Gamma\left(\frac{n}{2}\right)}$

variance $\sigma^2 = \dfrac{\Gamma\left(\frac{n+2}{2}\right)}{\Gamma\left(\frac{n}{2}\right)} - \left[\dfrac{\Gamma\left(\frac{n+1}{2}\right)}{\Gamma\left(\frac{n}{2}\right)}\right]^2$

---

where $\Gamma(x)$ is the gamma function (see page 515).
If X is a chi random variable with parameter $n = 2$, then X is a Rayleigh random variable with $\sigma = 1$.

6.4.4   Critical values for chi–square distribution

The following tables give values of $\chi^2_{\alpha,\nu}$ such that

$$1 - \alpha = F\left(\chi^2_{\alpha,\nu}\right) = \int_0^{\chi^2_{\alpha,\nu}} \frac{1}{2^{\nu/2}\,\Gamma(\nu/2)} x^{(\nu-2)/2} e^{-x/2}\, dx \qquad (6.3)$$

where ν, the number of degrees of freedom, varies from 1 to 10,000 and α varies from 0.0001 to 0.9999.

(a) For $\nu > 30$, the expression $\sqrt{2\chi^2} - \sqrt{2\nu - 1}$ is approximately a standard normal distribution. Hence, $\chi^2_{\alpha,\nu}$ is approximately

$$\chi^2_{\alpha,\nu} \approx \frac{1}{2}\left[z_\alpha + \sqrt{2\nu - 1}\right]^2 \qquad \text{for } \nu \gg 1 \qquad (6.4)$$

(b) For even values of ν, $F\left(\chi^2_{\alpha,\nu}\right)$ can be written as

$$1 - F\left(\chi^2_{\alpha,\nu}\right) = \sum_{x=0}^{x'-1} \frac{e^{-\lambda}\lambda^x}{x!} \qquad (6.5)$$

with $\lambda = \chi^2_{\alpha,\nu}/2$ and $x' = \nu/2$. Hence, the cumulative chi–square distribution is related to the cumulative Poisson distribution.

Example 6.40:   Use the table on page 121 to find the values $\chi^2_{.99,36}$ and $\chi^2_{.05,20}$.

Solution:

(S1) The left–hand column of the table on page 121 contains entries for the number of degrees of freedom and the top row lists values for α. The intersection of the ν degrees of freedom row and the α column contains $\chi^2_{\alpha,\nu}$ such that $\text{Prob}\left[\chi^2 \geq \chi^2_{\alpha,\nu}\right] = \alpha$.

(S2) $\chi^2_{.99,36} = 19.2327 \implies \text{Prob}\left[\chi^2 \geq 19.2327\right] = .99$

     $\chi^2_{.05,20} = 31.4104 \implies \text{Prob}\left[\chi^2 \geq 31.4104\right] = .05$

Critical values for the chi–square distribution $\chi^2_{\alpha,\nu}$.

ν	.9999	.9995	.999	.995	.99	.975	.95	.90
1	$.0^7157$	$.0^6393$	$.0^5157$	$.0^4393$	.0002	.0010	.0039	.0158
2	.0002	.0010	.0020	.0100	.0201	.0506	.1026	.2107
3	.0052	.0153	.0243	.0717	.1148	.2158	.3518	.5844
4	.0284	.0639	.0908	.2070	.2971	.4844	.7107	1.0636
5	.0822	.1581	.2102	.4117	.5543	.8312	1.1455	1.6103
6	.1724	.2994	.3811	.6757	.8721	1.2373	1.6354	2.2041
7	.3000	.4849	.5985	.9893	1.2390	1.6899	2.1673	2.8331
8	.4636	.7104	.8571	1.3444	1.6465	2.1797	2.7326	3.4895
9	.6608	.9717	1.1519	1.7349	2.0879	2.7004	3.3251	4.1682
10	.8889	1.2650	1.4787	2.1559	2.5582	3.2470	3.9403	4.8652
11	1.1453	1.5868	1.8339	2.6032	3.0535	3.8157	4.5748	5.5778
12	1.4275	1.9344	2.2142	3.0738	3.5706	4.4038	5.2260	6.3038
13	1.7333	2.3051	2.6172	3.5650	4.1069	5.0088	5.8919	7.0415
14	2.0608	2.6967	3.0407	4.0747	4.6604	5.6287	6.5706	7.7895
15	2.4082	3.1075	3.4827	4.6009	5.2293	6.2621	7.2609	8.5468
16	2.7739	3.5358	3.9416	5.1422	5.8122	6.9077	7.9616	9.3122
17	3.1567	3.9802	4.4161	5.6972	6.4078	7.5642	8.6718	10.0852
18	3.5552	4.4394	4.9048	6.2648	7.0149	8.2307	9.3905	10.8649
19	3.9683	4.9123	5.4068	6.8440	7.6327	8.9065	10.1170	11.6509
20	4.3952	5.3981	5.9210	7.4338	8.2604	9.5908	10.8508	12.4426
21	4.8348	5.8957	6.4467	8.0337	8.8972	10.2829	11.5913	13.2396
22	5.2865	6.4045	6.9830	8.6427	9.5425	10.9823	12.3380	14.0415
23	5.7494	6.9237	7.5292	9.2604	10.1957	11.6886	13.0905	14.8480
24	6.2230	7.4527	8.0849	9.8862	10.8564	12.4012	13.8484	15.6587
25	6.7066	7.9910	8.6493	10.5197	11.5240	13.1197	14.6114	16.4734
26	7.1998	8.5379	9.2221	11.1602	12.1981	13.8439	15.3792	17.2919
27	7.7019	9.0932	9.8028	11.8076	12.8785	14.5734	16.1514	18.1139
28	8.2126	9.6563	10.3909	12.4613	13.5647	15.3079	16.9279	18.9392
29	8.7315	10.2268	10.9861	13.1211	14.2565	16.0471	17.7084	19.7677
30	9.2581	10.8044	11.5880	13.7867	14.9535	16.7908	18.4927	20.5992
31	9.7921	11.3887	12.1963	14.4578	15.6555	17.5387	19.2806	21.4336
32	10.3331	11.9794	12.8107	15.1340	16.3622	18.2908	20.0719	22.2706
33	10.8810	12.5763	13.4309	15.8153	17.0735	19.0467	20.8665	23.1102
34	11.4352	13.1791	14.0567	16.5013	17.7891	19.8063	21.6643	23.9523
35	11.9957	13.7875	14.6878	17.1918	18.5089	20.5694	22.4650	24.7967
36	12.5622	14.4012	15.3241	17.8867	19.2327	21.3359	23.2686	25.6433
37	13.1343	15.0202	15.9653	18.5858	19.9602	22.1056	24.0749	26.4921
38	13.7120	15.6441	16.6112	19.2889	20.6914	22.8785	24.8839	27.3430
39	14.2950	16.2729	17.2616	19.9959	21.4262	23.6543	25.6954	28.1958
40	14.8831	16.9062	17.9164	20.7065	22.1643	24.4330	26.5093	29.0505

Critical values for the chi–square distribution $\chi^2_{\alpha,\nu}$.

ν	α .9999	.9995	.999	.995	.99	.975	.95	.90
41	15.48	17.54	18.58	21.42	22.91	25.21	27.33	29.91
42	16.07	18.19	19.24	22.14	23.65	26.00	28.14	30.77
43	16.68	18.83	19.91	22.86	24.40	26.79	28.96	31.63
44	17.28	19.48	20.58	23.58	25.15	27.57	29.79	32.49
45	17.89	20.14	21.25	24.31	25.90	28.37	30.61	33.35
46	18.51	20.79	21.93	25.04	26.66	29.16	31.44	34.22
47	19.13	21.46	22.61	25.77	27.42	29.96	32.27	35.08
48	19.75	22.12	23.29	26.51	28.18	30.75	33.10	35.95
49	20.38	22.79	23.98	27.25	28.94	31.55	33.93	36.82
50	21.01	23.46	24.67	27.99	29.71	32.36	34.76	37.69
60	27.50	30.34	31.74	35.53	37.48	40.48	43.19	46.46
70	34.26	37.47	39.04	43.28	45.44	48.76	51.74	55.33
80	41.24	44.79	46.52	51.17	53.54	57.15	60.39	64.28
90	48.41	52.28	54.16	59.20	61.75	65.65	69.13	73.29
100	55.72	59.90	61.92	67.33	70.06	74.22	77.93	82.36
200	134.02	140.66	143.84	152.24	156.43	162.73	168.28	174.84
300	217.33	225.89	229.96	240.66	245.97	253.91	260.88	269.07
400	303.26	313.43	318.26	330.90	337.16	346.48	354.64	364.21
500	390.85	402.45	407.95	422.30	429.39	439.94	449.15	459.93
600	479.64	492.52	498.62	514.53	522.37	534.02	544.18	556.06
700	569.32	583.39	590.05	607.38	615.91	628.58	639.61	652.50
800	659.72	674.89	682.07	700.73	709.90	723.51	735.36	749.19
900	750.70	766.91	774.57	794.47	804.25	818.76	831.37	846.07
1000	842.17	859.36	867.48	888.56	898.91	914.26	927.59	943.13
1500	1304.80	1326.30	1336.42	1362.67	1375.53	1394.56	1411.06	1430.25
2000	1773.30	1798.42	1810.24	1840.85	1855.82	1877.95	1897.12	1919.39
2500	2245.54	2273.86	2287.17	2321.62	2338.45	2363.31	2384.84	2409.82
3000	2720.44	2751.65	2766.32	2804.23	2822.75	2850.08	2873.74	2901.17
3500	3197.36	3231.23	3247.14	3288.25	3308.31	3337.92	3363.53	3393.22
4000	3675.88	3712.22	3729.29	3773.37	3794.87	3826.60	3854.03	3885.81
4500	4155.71	4194.37	4212.52	4259.39	4282.25	4315.96	4345.10	4378.86
5000	4636.62	4677.48	4696.67	4746.17	4770.31	4805.90	4836.66	4872.28
5500	5118.47	5161.42	5181.58	5233.60	5258.96	5296.34	5328.63	5366.03
6000	5601.13	5646.08	5667.17	5721.59	5748.11	5787.20	5820.96	5860.05
6500	6084.50	6131.36	6153.35	6210.07	6237.70	6278.43	6313.60	6354.32
7000	6568.49	6617.20	6640.05	6698.98	6727.69	6769.99	6806.52	6848.80
7500	7053.05	7103.53	7127.22	7188.28	7218.03	7261.85	7299.69	7343.48
8000	7538.11	7590.32	7614.81	7677.94	7708.68	7753.98	7793.08	7838.33
8500	8023.63	8077.51	8102.78	8167.91	8199.63	8246.35	8286.68	8333.34
9000	8509.57	8565.07	8591.09	8658.17	8690.83	8738.94	8780.46	8828.50
9500	8995.90	9052.97	9079.73	9148.70	9182.28	9231.74	9274.42	9323.78
10000	9482.59	9541.19	9568.67	9639.48	9673.95	9724.72	9768.53	9819.19

Critical values for the chi–square distribution $\chi^2_{\alpha,\nu}$.

ν	.10	.05	.025	.01	.005	.001	.0005	.0001
1	2.7055	3.8415	5.0239	6.6349	7.8794	10.8276	12.1157	15.1367
2	4.6052	5.9915	7.3778	9.2103	10.5966	13.8155	15.2018	18.4207
3	6.2514	7.8147	9.3484	11.3449	12.8382	16.2662	17.7300	21.1075
4	7.7794	9.4877	11.1433	13.2767	14.8603	18.4668	19.9974	23.5127
5	9.2364	11.0705	12.8325	15.0863	16.7496	20.5150	22.1053	25.7448
6	10.6446	12.5916	14.4494	16.8119	18.5476	22.4577	24.1028	27.8563
7	12.0170	14.0671	16.0128	18.4753	20.2777	24.3219	26.0178	29.8775
8	13.3616	15.5073	17.5345	20.0902	21.9550	26.1245	27.8680	31.8276
9	14.6837	16.9190	19.0228	21.6660	23.5894	27.8772	29.6658	33.7199
10	15.9872	18.3070	20.4832	23.2093	25.1882	29.5883	31.4198	35.5640
11	17.2750	19.6751	21.9200	24.7250	26.7568	31.2641	33.1366	37.3670
12	18.5493	21.0261	23.3367	26.2170	28.2995	32.9095	34.8213	39.1344
13	19.8119	22.3620	24.7356	27.6882	29.8195	34.5282	36.4778	40.8707
14	21.0641	23.6848	26.1189	29.1412	31.3193	36.1233	38.1094	42.5793
15	22.3071	24.9958	27.4884	30.5779	32.8013	37.6973	39.7188	44.2632
16	23.5418	26.2962	28.8454	31.9999	34.2672	39.2524	41.3081	45.9249
17	24.7690	27.5871	30.1910	33.4087	35.7185	40.7902	42.8792	47.5664
18	25.9894	28.8693	31.5264	34.8053	37.1565	42.3124	44.4338	49.1894
19	27.2036	30.1435	32.8523	36.1909	38.5823	43.8202	45.9731	50.7955
20	28.4120	31.4104	34.1696	37.5662	39.9968	45.3147	47.4985	52.3860
21	29.6151	32.6706	35.4789	38.9322	41.4011	46.7970	49.0108	53.9620
22	30.8133	33.9244	36.7807	40.2894	42.7957	48.2679	50.5111	55.5246
23	32.0069	35.1725	38.0756	41.6384	44.1813	49.7282	52.0002	57.0746
24	33.1962	36.4150	39.3641	42.9798	45.5585	51.1786	53.4788	58.6130
25	34.3816	37.6525	40.6465	44.3141	46.9279	52.6197	54.9475	60.1403
26	35.5632	38.8851	41.9232	45.6417	48.2899	54.0520	56.4069	61.6573
27	36.7412	40.1133	43.1945	46.9629	49.6449	55.4760	57.8576	63.1645
28	37.9159	41.3371	44.4608	48.2782	50.9934	56.8923	59.3000	64.6624
29	39.0875	42.5570	45.7223	49.5879	52.3356	58.3012	60.7346	66.1517
30	40.2560	43.7730	46.9792	50.8922	53.6720	59.7031	62.1619	67.6326
31	41.4217	44.9853	48.2319	52.1914	55.0027	61.0983	63.5820	69.1057
32	42.5847	46.1943	49.4804	53.4858	56.3281	62.4872	64.9955	70.5712
33	43.7452	47.3999	50.7251	54.7755	57.6484	63.8701	66.4025	72.0296
34	44.9032	48.6024	51.9660	56.0609	58.9639	65.2472	67.8035	73.4812
35	46.0588	49.8018	53.2033	57.3421	60.2748	66.6188	69.1986	74.9262
36	47.2122	50.9985	54.4373	58.6192	61.5812	67.9852	70.5881	76.3650
37	48.3634	52.1923	55.6680	59.8925	62.8833	69.3465	71.9722	77.7977
38	49.5126	53.3835	56.8955	61.1621	64.1814	70.7029	73.3512	79.2247
39	50.6598	54.5722	58.1201	62.4281	65.4756	72.0547	74.7253	80.6462
40	51.8051	55.7585	59.3417	63.6907	66.7660	73.4020	76.0946	82.0623

Critical values for the chi–square distribution $\chi^2_{\alpha,\nu}$.

ν	.10	.05	.025	.01	.005	.001	.0005	.0001
41	52.95	56.94	60.56	64.95	68.05	74.74	77.46	83.47
42	54.09	58.12	61.78	66.21	69.34	76.08	78.82	84.88
43	55.23	59.30	62.99	67.46	70.62	77.42	80.18	86.28
44	56.37	60.48	64.20	68.71	71.89	78.75	81.53	87.68
45	57.51	61.66	65.41	69.96	73.17	80.08	82.88	89.07
46	58.64	62.83	66.62	71.20	74.44	81.40	84.22	90.46
47	59.77	64.00	67.82	72.44	75.70	82.72	85.56	91.84
48	60.91	65.17	69.02	73.68	76.97	84.04	86.90	93.22
49	62.04	66.34	70.22	74.92	78.23	85.35	88.23	94.60
50	63.17	67.50	71.42	76.15	79.49	86.66	89.56	95.97
60	74.40	79.08	83.30	88.38	91.95	99.61	102.69	109.50
70	85.53	90.53	95.02	100.43	104.21	112.32	115.58	122.75
80	96.58	101.88	106.63	112.33	116.32	124.84	128.26	135.78
90	107.57	113.15	118.14	124.12	128.30	137.21	140.78	148.63
100	118.50	124.34	129.56	135.81	140.17	149.45	153.17	161.32
200	226.02	233.99	241.06	249.45	255.26	267.54	272.42	283.06
300	331.79	341.40	349.87	359.91	366.84	381.43	387.20	399.76
400	436.65	447.63	457.31	468.72	476.61	493.13	499.67	513.84
500	540.93	553.13	563.85	576.49	585.21	603.45	610.65	626.24
600	644.80	658.09	669.77	683.52	692.98	712.77	720.58	737.46
700	748.36	762.66	775.21	789.97	800.13	821.35	829.71	847.78
800	851.67	866.91	880.28	895.98	906.79	929.33	938.21	957.38
900	954.78	970.90	985.03	1001.63	1013.04	1036.83	1046.19	1066.40
1000	1057.72	1074.68	1089.53	1106.97	1118.95	1143.92	1153.74	1174.93
1500	1570.61	1591.21	1609.23	1630.35	1644.84	1674.97	1686.81	1712.30
2000	2081.47	2105.15	2125.84	2150.07	2166.66	2201.16	2214.68	2243.81
2500	2591.04	2617.43	2640.47	2667.43	2685.89	2724.22	2739.25	2771.57
3000	3099.69	3128.54	3153.70	3183.13	3203.28	3245.08	3261.45	3296.66
3500	3607.64	3638.75	3665.87	3697.57	3719.26	3764.26	3781.87	3819.74
4000	4115.05	4148.25	4177.19	4211.01	4234.14	4282.11	4300.88	4341.22
4500	4622.00	4657.17	4687.83	4723.63	4748.12	4798.87	4818.73	4861.40
5000	5128.58	5165.61	5197.88	5235.57	5261.34	5314.73	5335.62	5380.48
5500	5634.83	5673.64	5707.45	5746.93	5773.91	5829.81	5851.68	5898.63
6000	6140.81	6181.31	6216.59	6257.78	6285.92	6344.23	6367.02	6415.98
6500	6646.54	6688.67	6725.36	6768.18	6797.45	6858.05	6881.74	6932.61
7000	7152.06	7195.75	7233.79	7278.19	7308.53	7371.35	7395.90	7448.62
7500	7657.38	7702.58	7741.93	7787.86	7819.23	7884.18	7909.57	7964.06
8000	8162.53	8209.19	8249.81	8297.20	8329.58	8396.59	8422.78	8479.00
8500	8667.52	8715.59	8757.44	8806.26	8839.60	8908.62	8935.59	8993.48
9000	9172.36	9221.81	9264.85	9315.05	9349.34	9420.30	9448.03	9507.53
9500	9677.07	9727.86	9772.05	9823.60	9858.81	9931.67	9960.13	10021.21
10000	10181.66	10233.75	10279.07	10331.93	10368.03	10442.73	10471.91	10534.52

6.4.5   Percentage points, chi–square over degrees of freedom distribution

The following table gives the percentage points of the sampling distribution of s^2/σ^2, referred to as the percentage points of the $\chi^2/\text{d.f.}$ distribution (read "chi–square over degrees of freedom"). These percentage points are a function of the sample size.

ν	0.05	0.1	0.5	1.0	2.5	5.0	95	97.5	99	99.5	99.9	99.95
	Probability in percent						Probability in percent					
1	0.000	0.000	0.000	0.000	0.001	0.004	3.841	5.024	6.635	7.879	10.828	12.116
2	0.001	0.001	0.005	0.010	0.025	0.051	2.996	3.689	4.605	5.298	6.908	7.601
3	0.005	0.008	0.024	0.038	0.072	0.117	2.605	3.116	3.782	4.279	5.422	5.910
4	0.016	0.023	0.052	0.074	0.121	0.178	2.372	2.786	3.319	3.715	4.617	4.999
5	0.032	0.042	0.082	0.111	0.166	0.229	2.214	2.567	3.017	3.350	4.103	4.421
6	0.050	0.064	0.113	0.145	0.206	0.273	2.099	2.408	2.802	3.091	3.743	4.017
7	0.069	0.085	0.141	0.177	0.241	0.310	2.010	2.288	2.639	2.897	3.475	3.717
8	0.089	0.107	0.168	0.206	0.272	0.342	1.938	2.192	2.511	2.744	3.266	3.484
9	0.108	0.128	0.193	0.232	0.300	0.369	1.880	2.114	2.407	2.621	3.097	3.296
10	0.126	0.148	0.216	0.256	0.325	0.394	1.831	2.048	2.321	2.519	2.959	3.142
15	0.207	0.232	0.307	0.349	0.417	0.484	1.666	1.833	2.039	2.187	2.513	2.648
20	0.270	0.296	0.372	0.413	0.480	0.543	1.571	1.708	1.878	2.000	2.266	2.375
25	0.320	0.346	0.421	0.461	0.525	0.584	1.506	1.626	1.773	1.877	2.105	2.198
30	0.360	0.386	0.460	0.498	0.560	0.616	1.459	1.566	1.696	1.789	1.990	2.072
40	0.423	0.448	0.518	0.554	0.611	0.663	1.394	1.484	1.592	1.669	1.835	1.902
50	0.469	0.493	0.560	0.594	0.647	0.695	1.350	1.428	1.523	1.590	1.733	1.791
75	0.548	0.570	0.629	0.660	0.706	0.747	1.283	1.345	1.419	1.470	1.581	1.626
100	0.599	0.619	0.673	0.701	0.742	0.779	1.243	1.296	1.358	1.402	1.494	1.532
150	0.663	0.681	0.728	0.751	0.787	0.818	1.197	1.239	1.288	1.322	1.395	1.424
200	0.703	0.719	0.761	0.782	0.814	0.841	1.170	1.205	1.247	1.276	1.338	1.362
500	0.810	0.816	0.845	0.859	0.880	0.898	1.111	1.128	1.153	1.170	1.207	1.221
1000	0.859	0.868	0.889	0.899	0.914	0.928	1.075	1.090	1.107	1.119	1.144	1.154

6.5   ERLANG DISTRIBUTION

6.5.1   Properties

$$\text{pdf} \quad f(x) = \frac{x^{n-1}\, e^{-x/\beta}}{\beta^n (n-1)!}, \quad x \geq 0,\ \beta > 0,\ n \in \mathcal{N}$$

$$\text{mean} \quad \mu = n\beta$$

$$\text{variance} \quad \sigma^2 = n\beta^2$$

$$\text{skewness} \quad \beta_1 = 2/\sqrt{n}$$

$$\text{kurtosis} \quad \beta_2 = 3 + \frac{6}{n}$$

$$\text{mgf} \quad m(t) = (1 - \beta t)^{-n}$$

$$\text{char function} \quad \phi(t) = (1 - \beta i t)^{-n}$$

6.5.2   Probability density function

The probability density function is skewed to the right with n as the *shape* parameter.

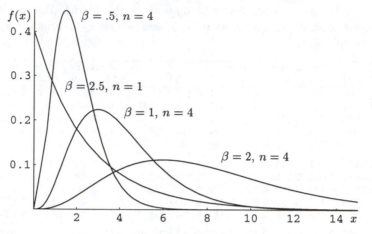

Figure 6.6: Probability density functions for an Erlang random variable.

6.5.3   Related distributions

If X is an Erlang random variable with parameters β and $n = 1$, then X is an exponential random variable with parameter $\lambda = 1/\beta$.

6.6   EXPONENTIAL DISTRIBUTION

6.6.1   Properties

$$\text{pdf}\quad f(x) = \lambda e^{-\lambda x}, \quad x \geq 0,\ \lambda > 0$$

$$\text{mean}\quad \mu = 1/\lambda$$

$$\text{variance}\quad \sigma^2 = 1/\lambda^2$$

$$\text{skewness}\quad \beta_1 = 2$$

$$\text{kurtosis}\quad \beta_2 = 9$$

$$\text{mgf}\quad m(t) = \frac{\lambda}{\lambda - t}$$

$$\text{char function}\quad \phi(t) = \frac{\lambda}{\lambda - it}$$

6.6.2   Probability density function

The probability density function is skewed to the right. The tail of the distribution is heavier for larger values of λ.

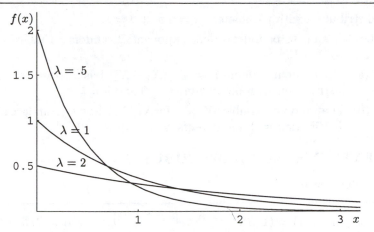

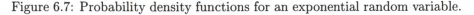

Figure 6.7: Probability density functions for an exponential random variable.

6.6.3 Related distributions

Let X be an exponential random variable with parameter λ.

(1) If $\lambda = 1/2$, then X is a chi–square random variable with $\nu = 2$.

(2) The random variable $\sqrt{X}$ has a Rayleigh distribution with parameter $\sigma = \sqrt{1/(2\lambda)}$.

(3) The random variable $Y = X^{1/\alpha}$ has a Weibull distribution with parameters α and $\lambda^{-1/\alpha}$.

(4) The random variable $Y = e^{-X}$ has a power function distribution with parameters $b = 1$ and $c = \lambda$.

(5) The random variable $Y = ae^{X}$ has a Pareto distribution with parameters a and $\theta = \lambda$.

(6) The random variable $Y = \alpha - \ln X$ has an extreme–value distribution with parameters α and $\beta = 1/\lambda$.

(7) Let $X_1, X_2, \ldots, X_n$ be independent exponential random variables each with parameter λ.

  (a) The random variable $Y = \min(X_1, X_2, \ldots, X_n)$ has an exponential distribution with parameter $n\lambda$.

  (b) The random variable $Y = X_1 + X_2 + \cdots + X_n$ has an Erlang distribution with parameters $\beta = 1/\lambda$ and n.

(8) Let X_1 and X_2 be independent exponential random variables each with parameter λ. The random variable $Y = X_1 - X_2$ has a Laplace distribution with parameters 0 and $1/\lambda$.

(9) Let X be an exponential random variable with parameter $\lambda = 1$. The random variable $Y = -\ln[e^{-X}/(1 + e^{-X})]$ has a (standard) logistic

distribution with parameters $\alpha = 0$ and $\beta = 1$.

(10) Let X_1 and X_2 be independent exponential random variables with parameter $\lambda = 1$.

    (a) The random variable $Y = X_1/(X_1 + X_2)$ has a (standard) uniform distribution with parameters $a = 0$ and $b = 1$.

    (b) The random variable $W = -\ln(X_1/X_2)$ has a (standard) logistic distribution with parameters $\alpha = 0$ and $\beta = 1$.

6.7 EXTREME–VALUE DISTRIBUTION

6.7.1 Properties

$$\text{pdf}\quad f(x) = (1/\beta)e^{-(x-\alpha)/\beta}e^{\left[-e^{-(x-\alpha)/\beta}\right]}\qquad x, \alpha \in \mathcal{R}, \beta > 0$$

$$\text{mean}\qquad \mu = \alpha + \gamma\beta,\quad \gamma = 0.5772156649\ldots(\text{Euler's constant})$$

$$\text{variance}\qquad \sigma^2 = \frac{\pi^2\beta^2}{6}$$

$$\text{skewness}\qquad \beta_1 = -\frac{6\sqrt{6}\,\varphi''(1)}{\pi^3}$$

$$\text{kurtosis}\qquad \beta_2 = {}^{27}\!/_5 = 5.4$$

$$\text{mgf}\quad m(t) = e^{\alpha t}\Gamma(1 - \beta t),\quad t < 1/\beta$$

$$\text{char function}\quad \phi(t) = e^{\alpha it}\Gamma(1 - \beta it)$$

where $\Gamma(x)$ is the gamma function and $\varphi(x)$ is the digamma function (see pages 515 and 518).

6.7.2 Probability density function

The probability density function is skewed slightly to the right with *location* parameter α.

6.7.3 Related distributions

(1) The standard extreme–value distribution has $\alpha = 0$ and $\beta = 1$.

(2) If X is an extreme–value random variable with parameters α and β, then the random variable $Y = (X - \alpha)/\beta$ has a (standard) extreme–value distribution with parameters 0 and 1.

(3) If X is a (standard) extreme–value random variable with parameters $\alpha = 0$ and $\beta = 1$, then the random variable $Y = e^{\left(-e^{-X/c}\right)}$ has a power function distribution with parameters $b = 0$ and c.

(4) If X is a extreme–value random variable with parameters $\alpha = 0$ and $\beta = 1$, then the random variable $Y = a\left[1 - e^{\left(-e^{-x}\right)}\right]^{1/\theta}$ has a Pareto distribution with parameters a and θ.

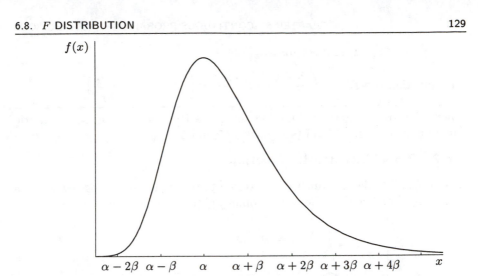

Figure 6.8: Probability density functions for an extreme–value random variable.

(5) Let X_1 and X_2 be independent extreme–value random variables with parameters α and β. The random variable $Y = X_1 - X_2$ has a logistic distribution with parameters 0 and β.

6.8 *F* DISTRIBUTION

6.8.1 Properties

$$\text{pdf} \quad f(x) = \frac{\Gamma\left(\frac{\nu_1+\nu_2}{2}\right)\nu_1^{\frac{\nu_1}{2}}\nu_2^{\frac{\nu_2}{2}}}{\Gamma(\nu_1/2)\Gamma(\nu_2/2)}\, x^{(\nu_1/2)-1}\,(\nu_2 + \nu_1 x)^{-(\nu_1+\nu_2)/2}$$

$$x > 0,\ \nu_1,\ \nu_2 > 0$$

$$\text{mean} \quad \mu = \frac{\nu_2}{\nu_2 - 2},\quad \nu_2 \geq 3$$

$$\text{variance} \quad \sigma^2 = \frac{2\nu_2^2(\nu_1 + \nu_2 - 2)}{\nu_1(\nu_2 - 2)^2(\nu_2 - 4)},\quad \nu_2 \geq 5$$

$$\text{skewness} \quad \beta_1 = \frac{(2\nu_1 + \nu_2 - 2)\sqrt{8(\nu_2 - 4)}}{\sqrt{\nu_1}(\nu_2 - 6)\sqrt{\nu_1 + \nu_2 - 2}},\quad \nu_2 \geq 7$$

$$\text{kurtosis} \quad \beta_2 = 3 +$$

$$\frac{12[(\nu_2 - 2)^2(\nu_2 - 4) + \nu_1(\nu_1 + \nu_2 - 2)(5\nu_2 - 22)]}{\nu_1(\nu_2 - 6)(\nu_2 - 8)(\nu_1 + \nu_2 - 2)}$$

$$\nu_2 \geq 9$$

mgf $m(t) = $ does not exist

char function $\phi(t) = \Gamma\left(\dfrac{\nu_1 + \nu_2}{2}\right) \Gamma\left(\dfrac{\nu_2}{2}\right) \psi\left(\dfrac{\nu_1}{2}, 1 - \dfrac{\nu_2}{2}; \dfrac{it\nu_2}{\nu_1}\right)$

where $\Gamma(x)$ is the gamma function and ψ is the confluent hypergeometric function of the second kind (see pages 515 and 521).

6.8.2   Probability density function

The probability density function is skewed to the right with *shape* parameters ν_1 and ν_2. For fixed ν_2, the tail becomes lighter as ν_1 increases.

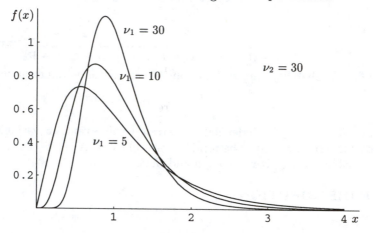

Figure 6.9: Probability density functions for an F random variable.

6.8.3   Related distributions

(1) If X has an F distribution with ν_1 and ν_2 degrees of freedom, then the random variable $Y = 1/X$ has an F distribution with ν_2 and ν_1 degrees of freedom.

(2) If X has an F distribution with ν_1 and ν_2 degrees of freedom, the random variable $\nu_1 X$ tends to a chi–square distribution with ν_1 degrees of freedom as $\nu_2 \to \infty$.

(3) Let X_1 and X_2 be independent F random variables with $\nu_1 = \nu_2 = \nu$ degrees of freedom. The random variable

$$Y = \frac{\sqrt{\nu}}{2}\left(\sqrt{X_1} - \sqrt{X_2}\right) \tag{6.6}$$

has a t distribution with ν degrees of freedom.

(4) If X has an F distribution with parameters ν_1 and ν_2, the random variable

$$Y = \frac{\nu_1 X/\nu_2}{1 + \dfrac{\nu_1 X}{\nu_2}} \tag{6.7}$$

has a beta distribution with parameters $\alpha = \nu_2/2$ and $\beta = \nu_1/2$.

6.8.4   Critical values for the *F* distribution

Given values of ν_1, ν_2, and α, the tables on pages 132–137 contain values of F_{α,ν_1,ν_2} such that

$$
\begin{aligned}
1 - \alpha &= \int_0^{F_{\alpha,\nu_1,\nu_2}} f(x)\, dx \\
&= \int_0^{F_{\alpha,\nu_1,\nu_2}} \frac{\Gamma\left(\frac{\nu_1+\nu_2}{2}\right) \nu_1^{\frac{\nu_1}{2}} \nu_2^{\frac{\nu_2}{2}}}{\Gamma(\nu_1/2)\Gamma(\nu_2/2)}\, x^{(\nu_1/2)-1} \left(\nu_2 + \nu_1 x\right)^{-(\nu_1+\nu_2)/2}\, dx
\end{aligned} \tag{6.8}
$$

Note that $F_{1-\alpha}$ for ν_1 and ν_2 degrees of freedom is the reciprocal of F_α for ν_2 and ν_1 degrees of freedom. For example,

$$F_{.05,4,7} = \frac{1}{F_{.95,7,4}} = \frac{1}{6.09} = .164 \tag{6.9}$$

Example 6.41:   Use the following tables to find the values $F_{.1,4,9}$ and $F_{.95,12,15}$.

Solution:

(S1) The top rows of the tables on pages 132–137 contain entries for the numerator degrees of freedom and the left–hand column contains the denominator degrees of freedom. The intersection of the ν_1 degrees of freedom column and the ν_2 row may be used to find critical values of the form F_{α,ν_1,ν_2} such that $\text{Prob}\left[F \geq F_{\alpha,\nu_1,\nu_2}\right] = \alpha$.

(S2) $F_{.1,4,9} = 2.69 \implies \text{Prob}\left[F \geq 2.69\right] = .1$

$$F_{.95,12,15} = \frac{1}{F_{.05,15,12}} = \frac{1}{2.62} = .3817 \implies \text{Prob}\left[F \geq .3817\right] = .95$$

(S3) Illustrations:

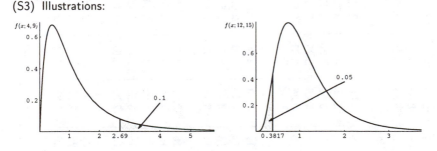

Critical values for the F distribution

For given values of ν_1 and ν_2, the following table contains values of $F_{0.1,\nu_1,\nu_2}$; defined by $\text{Prob}\,[F \geq F_{0.1,\nu_1,\nu_2}] = \alpha = 0.1$.

ν_2	$\nu_1=1$	2	3	4	5	6	7	8	9	10	50	100	∞
1	39.86	49.50	53.59	55.83	57.24	58.20	58.91	59.44	59.86	60.19	62.69	63.01	63.33
2	8.53	9.00	9.16	9.24	9.29	9.33	9.35	9.37	9.38	9.39	9.47	9.48	9.49
3	5.54	5.46	5.39	5.34	5.31	5.28	5.27	5.25	5.24	5.23	5.15	5.14	5.13
4	4.54	4.32	4.19	4.11	4.05	4.01	3.98	3.95	3.94	3.92	3.80	3.78	3.76
5	4.06	3.78	3.62	3.52	3.45	3.40	3.37	3.34	3.32	3.30	3.15	3.13	3.10
6	3.78	3.46	3.29	3.18	3.11	3.05	3.01	2.98	2.96	2.94	2.77	2.75	2.72
7	3.59	3.26	3.07	2.96	2.88	2.83	2.78	2.75	2.72	2.70	2.52	2.50	2.47
8	3.46	3.11	2.92	2.81	2.73	2.67	2.62	2.59	2.56	2.54	2.35	2.32	2.29
9	3.36	3.01	2.81	2.69	2.61	2.55	2.51	2.47	2.44	2.42	2.22	2.19	2.16
10	3.29	2.92	2.73	2.61	2.52	2.46	2.41	2.38	2.35	2.32	2.12	2.09	2.06
11	3.23	2.86	2.66	2.54	2.45	2.39	2.34	2.30	2.27	2.25	2.04	2.01	1.97
12	3.18	2.81	2.61	2.48	2.39	2.33	2.28	2.24	2.21	2.19	1.97	1.94	1.90
13	3.14	2.76	2.56	2.43	2.35	2.28	2.23	2.20	2.16	2.14	1.92	1.88	1.85
14	3.10	2.73	2.52	2.39	2.31	2.24	2.19	2.15	2.12	2.10	1.87	1.83	1.80
15	3.07	2.70	2.49	2.36	2.27	2.21	2.16	2.12	2.09	2.06	1.83	1.79	1.76
16	3.05	2.67	2.46	2.33	2.24	2.18	2.13	2.09	2.06	2.03	1.79	1.76	1.72
17	3.03	2.64	2.44	2.31	2.22	2.15	2.10	2.06	2.03	2.00	1.76	1.73	1.69
18	3.01	2.62	2.42	2.29	2.20	2.13	2.08	2.04	2.00	1.98	1.74	1.70	1.66
19	2.99	2.61	2.40	2.27	2.18	2.11	2.06	2.02	1.98	1.96	1.71	1.67	1.63
20	2.97	2.59	2.38	2.25	2.16	2.09	2.04	2.00	1.96	1.94	1.69	1.65	1.61
25	2.92	2.53	2.32	2.18	2.09	2.02	1.97	1.93	1.89	1.87	1.61	1.56	1.52
50	2.81	2.41	2.20	2.06	1.97	1.90	1.84	1.80	1.76	1.73	1.44	1.39	1.34
100	2.76	2.36	2.14	2.00	1.91	1.83	1.78	1.73	1.69	1.66	1.35	1.29	1.20
∞	2.71	2.30	2.08	1.94	1.85	1.77	1.72	1.67	1.63	1.60	1.24	1.17	1.00

Critical values for the *F* distribution

For given values of ν_1 and ν_2, the following table contains values of $F_{0.05,\nu_1,\nu_2}$; defined by $\text{Prob}\left[F \geq F_{0.05,\nu_1,\nu_2}\right] = \alpha = 0.05$.

ν_2	$\nu_1=1$	2	3	4	5	6	7	8	9	10	50	100	∞
1	161.4	199.5	215.7	224.6	230.2	234.0	236.8	238.9	240.5	241.9	251.8	253.0	254.3
2	18.51	19.00	19.16	19.25	19.30	19.33	19.35	19.37	19.38	19.40	19.48	19.49	19.50
3	10.13	9.55	9.28	9.12	9.01	8.94	8.89	8.85	8.81	8.79	8.58	8.55	8.53
4	7.71	6.94	6.59	6.39	6.26	6.16	6.09	6.04	6.00	5.96	5.70	5.66	5.63
5	6.61	5.79	5.41	5.19	5.05	4.95	4.88	4.82	4.77	4.74	4.44	4.41	4.36
6	5.99	5.14	4.76	4.53	4.39	4.28	4.21	4.15	4.10	4.06	3.75	3.71	3.67
7	5.59	4.74	4.35	4.12	3.97	3.87	3.79	3.73	3.68	3.64	3.32	3.27	3.23
8	5.32	4.46	4.07	3.84	3.69	3.58	3.50	3.44	3.39	3.35	3.02	2.97	2.93
9	5.12	4.26	3.86	3.63	3.48	3.37	3.29	3.23	3.18	3.14	2.80	2.76	2.71
10	4.96	4.10	3.71	3.48	3.33	3.22	3.14	3.07	3.02	2.98	2.64	2.59	2.54
11	4.84	3.98	3.59	3.36	3.20	3.09	3.01	2.95	2.90	2.85	2.51	2.46	2.40
12	4.75	3.89	3.49	3.26	3.11	3.00	2.91	2.85	2.80	2.75	2.40	2.35	2.30
13	4.67	3.81	3.41	3.18	3.03	2.92	2.83	2.77	2.71	2.67	2.31	2.26	2.21
14	4.60	3.74	3.34	3.11	2.96	2.85	2.76	2.70	2.65	2.60	2.24	2.19	2.13
15	4.54	3.68	3.29	3.06	2.90	2.79	2.71	2.64	2.59	2.54	2.18	2.12	2.07
16	4.49	3.63	3.24	3.01	2.85	2.74	2.66	2.59	2.54	2.49	2.12	2.07	2.01
17	4.45	3.59	3.20	2.96	2.81	2.70	2.61	2.55	2.49	2.45	2.08	2.02	1.96
18	4.41	3.55	3.16	2.93	2.77	2.66	2.58	2.51	2.46	2.41	2.04	1.98	1.92
19	4.38	3.52	3.13	2.90	2.74	2.63	2.54	2.48	2.42	2.38	2.00	1.94	1.88
20	4.35	3.49	3.10	2.87	2.71	2.60	2.51	2.45	2.39	2.35	1.97	1.91	1.84
25	4.24	3.39	2.99	2.76	2.60	2.49	2.40	2.34	2.28	2.24	1.84	1.78	1.71
50	4.03	3.18	2.79	2.56	2.40	2.29	2.20	2.13	2.07	2.03	1.60	1.52	1.45
100	3.94	3.09	2.70	2.46	2.31	2.19	2.10	2.03	1.97	1.93	1.48	1.39	1.28
∞	3.84	3.00	2.60	2.37	2.21	2.10	2.01	1.94	1.88	1.83	1.35	1.25	1.00

Critical values for the F distribution

For given values of ν_1 and ν_2, the following table contains values of $F_{0.025,\nu_1,\nu_2}$; defined by $\text{Prob}\,[F \geq F_{0.025,\nu_1,\nu_2}] = \alpha = 0.025$.

ν_2	$\nu_1=1$	2	3	4	5	6	7	8	9	10	50	100	∞
1	647.8	799.5	864.2	899.6	921.8	937.1	948.2	956.7	963.3	968.6	1008	1013	1018
2	38.51	39.00	39.17	39.25	39.30	39.33	39.36	39.37	39.39	39.40	39.48	39.49	39.50
3	17.44	16.04	15.44	15.10	14.88	14.73	14.62	14.54	14.47	14.42	14.01	13.96	13.90
4	12.22	10.65	9.98	9.60	9.36	9.20	9.07	8.98	8.90	8.84	8.38	8.32	8.26
5	10.01	8.43	7.76	7.39	7.15	6.98	6.85	6.76	6.68	6.62	6.14	6.08	6.02
6	8.81	7.26	6.60	6.23	5.99	5.82	5.70	5.60	5.52	5.46	4.98	4.92	4.85
7	8.07	6.54	5.89	5.52	5.29	5.12	4.99	4.90	4.82	4.76	4.28	4.21	4.14
8	7.57	6.06	5.42	5.05	4.82	4.65	4.53	4.43	4.36	4.30	3.81	3.74	3.67
9	7.21	5.71	5.08	4.72	4.48	4.32	4.20	4.10	4.03	3.96	3.47	3.40	3.33
10	6.94	5.46	4.83	4.47	4.24	4.07	3.95	3.85	3.78	3.72	3.22	3.15	3.08
11	6.72	5.26	4.63	4.28	4.04	3.88	3.76	3.66	3.59	3.53	3.03	2.96	2.88
12	6.55	5.10	4.47	4.12	3.89	3.73	3.61	3.51	3.44	3.37	2.87	2.80	2.72
13	6.41	4.97	4.35	4.00	3.77	3.60	3.48	3.39	3.31	3.25	2.74	2.67	2.60
14	6.30	4.86	4.24	3.89	3.66	3.50	3.38	3.29	3.21	3.15	2.64	2.56	2.49
15	6.20	4.77	4.15	3.80	3.58	3.41	3.29	3.20	3.12	3.06	2.55	2.47	2.40
16	6.12	4.69	4.08	3.73	3.50	3.34	3.22	3.12	3.05	2.99	2.47	2.40	2.32
17	6.04	4.62	4.01	3.66	3.44	3.28	3.16	3.06	2.98	2.92	2.41	2.33	2.25
18	5.98	4.56	3.95	3.61	3.38	3.22	3.10	3.01	2.93	2.87	2.35	2.27	2.19
19	5.92	4.51	3.90	3.56	3.33	3.17	3.05	2.96	2.88	2.82	2.30	2.22	2.13
20	5.87	4.46	3.86	3.51	3.29	3.13	3.01	2.91	2.84	2.77	2.25	2.17	2.09
25	5.69	4.29	3.69	3.35	3.13	2.97	2.85	2.75	2.68	2.61	2.08	2.00	1.91
50	5.34	3.97	3.39	3.05	2.83	2.67	2.55	2.46	2.38	2.32	1.75	1.66	1.54
100	5.18	3.83	3.25	2.92	2.70	2.54	2.42	2.32	2.24	2.18	1.59	1.48	1.37
∞	5.02	3.69	3.12	2.79	2.57	2.41	2.29	2.19	2.11	2.05	1.43	1.27	1.00

Critical values for the F distribution

For given values of ν_1 and ν_2, the following table contains values of $F_{0.01,\nu_1,\nu_2}$; defined by $\text{Prob}\,[F \geq F_{0.01,\nu_1,\nu_2}] = \alpha = 0.01$.

ν_2	$\nu_1=1$	2	3	4	5	6	7	8	9	10	50	100	∞
1	4052	5000	5403	5625	5764	5859	5928	5981	6022	6056	6303	6334	6336
2	98.50	99.00	99.17	99.25	99.30	99.33	99.36	99.37	99.39	99.40	99.48	99.49	99.50
3	34.12	30.82	29.46	28.71	28.24	27.91	27.67	27.49	27.35	27.23	26.35	26.24	26.13
4	21.20	18.00	16.69	15.98	15.52	15.21	14.98	14.80	14.66	14.55	13.69	13.58	13.46
5	16.26	13.27	12.06	11.39	10.97	10.67	10.46	10.29	10.16	10.05	9.24	9.13	9.02
6	13.75	10.92	9.78	9.15	8.75	8.47	8.26	8.10	7.98	7.87	7.09	6.99	6.88
7	12.25	9.55	8.45	7.85	7.46	7.19	6.99	6.84	6.72	6.62	5.86	5.75	5.65
8	11.26	8.65	7.59	7.01	6.63	6.37	6.18	6.03	5.91	5.81	5.07	4.96	4.86
9	10.56	8.02	6.99	6.42	6.06	5.80	5.61	5.47	5.35	5.26	4.52	4.41	4.31
10	10.04	7.56	6.55	5.99	5.64	5.39	5.20	5.06	4.94	4.85	4.12	4.01	3.91
11	9.65	7.21	6.22	5.67	5.32	5.07	4.89	4.74	4.63	4.54	3.81	3.71	3.60
12	9.33	6.93	5.95	5.41	5.06	4.82	4.64	4.50	4.39	4.30	3.57	3.47	3.36
13	9.07	6.70	5.74	5.21	4.86	4.62	4.44	4.30	4.19	4.10	3.38	3.27	3.17
14	8.86	6.51	5.56	5.04	4.69	4.46	4.28	4.14	4.03	3.94	3.22	3.11	3.00
15	8.68	6.36	5.42	4.89	4.56	4.32	4.14	4.00	3.89	3.80	3.08	2.98	2.87
16	8.53	6.23	5.29	4.77	4.44	4.20	4.03	3.89	3.78	3.69	2.97	2.86	2.75
17	8.40	6.11	5.18	4.67	4.34	4.10	3.93	3.79	3.68	3.59	2.87	2.76	2.65
18	8.29	6.01	5.09	4.58	4.25	4.01	3.84	3.71	3.60	3.51	2.78	2.68	2.57
19	8.18	5.93	5.01	4.50	4.17	3.94	3.77	3.63	3.52	3.43	2.71	2.60	2.49
20	8.10	5.85	4.94	4.43	4.10	3.87	3.70	3.56	3.46	3.37	2.64	2.54	2.42
25	7.77	5.57	4.68	4.18	3.85	3.63	3.46	3.32	3.22	3.13	2.40	2.29	2.17
50	7.17	5.06	4.20	3.72	3.41	3.19	3.02	2.89	2.78	2.70	1.95	1.82	1.70
100	6.90	4.82	3.98	3.51	3.21	2.99	2.82	2.69	2.59	2.50	1.74	1.60	1.45
∞	6.63	4.61	3.78	3.32	3.02	2.80	2.64	2.51	2.41	2.32	1.53	1.32	1.00

Critical values for the F distribution

For given values of ν_1 and ν_2, the following table contains values of $F_{0.005,\nu_1,\nu_2}$; defined by $\text{Prob}\,[F \geq F_{0.005,\nu_1,\nu_2}] = \alpha = 0.005$.

ν_2	$\nu_1=1$	2	3	4	5	6	7	8	9	10	50	100	∞
1	16211	20000	21615	22500	23056	23437	23715	23925	24091	24224	25211	25337	25465
2	198.5	199.0	199.2	199.2	199.3	199.3	199.4	199.4	199.4	199.4	199.5	199.5	199.5
3	55.55	49.80	47.47	46.19	45.39	44.84	44.43	44.13	43.88	43.69	42.21	42.02	41.83
4	31.33	26.28	24.26	23.15	22.46	21.97	21.62	21.35	21.14	20.97	19.67	19.50	19.32
5	22.78	18.31	16.53	15.56	14.94	14.51	14.20	13.96	13.77	13.62	12.45	12.30	12.14
6	18.63	14.54	12.92	12.03	11.46	11.07	10.79	10.57	10.39	10.25	9.17	9.03	8.88
7	16.24	12.40	10.88	10.05	9.52	9.16	8.89	8.68	8.51	8.38	7.35	7.22	7.08
8	14.69	11.04	9.60	8.81	8.30	7.95	7.69	7.50	7.34	7.21	6.22	6.09	5.95
9	13.61	10.11	8.72	7.96	7.47	7.13	6.88	6.69	6.54	6.42	5.45	5.32	5.19
10	12.83	9.43	8.08	7.34	6.87	6.54	6.30	6.12	5.97	5.85	4.90	4.77	4.64
11	12.23	8.91	7.60	6.88	6.42	6.10	5.86	5.68	5.54	5.42	4.49	4.36	4.23
12	11.75	8.51	7.23	6.52	6.07	5.76	5.52	5.35	5.20	5.09	4.17	4.04	3.90
13	11.37	8.19	6.93	6.23	5.79	5.48	5.25	5.08	4.94	4.82	3.91	3.78	3.65
14	11.06	7.92	6.68	6.00	5.56	5.26	5.03	4.86	4.72	4.60	3.70	3.57	3.44
15	10.80	7.70	6.48	5.80	5.37	5.07	4.85	4.67	4.54	4.42	3.52	3.39	3.26
16	10.58	7.51	6.30	5.64	5.21	4.91	4.69	4.52	4.38	4.27	3.37	3.25	3.11
17	10.38	7.35	6.16	5.50	5.07	4.78	4.56	4.39	4.25	4.14	3.25	3.12	2.98
18	10.22	7.21	6.03	5.37	4.96	4.66	4.44	4.28	4.14	4.03	3.14	3.01	2.87
19	10.07	7.09	5.92	5.27	4.85	4.56	4.34	4.18	4.04	3.93	3.04	2.91	2.78
20	9.94	6.99	5.82	5.17	4.76	4.47	4.26	4.09	3.96	3.85	2.96	2.83	2.69
25	9.48	6.60	5.46	4.84	4.43	4.15	3.94	3.78	3.64	3.54	2.65	2.52	2.38
50	8.63	5.90	4.83	4.23	3.85	3.58	3.38	3.22	3.09	2.99	2.10	1.95	1.81
100	8.24	5.59	4.54	3.96	3.59	3.33	3.13	2.97	2.85	2.74	1.84	1.68	1.51
∞	7.88	5.30	4.28	3.72	3.35	3.09	2.90	2.74	2.62	2.52	1.60	1.36	1.00

Critical values for the *F* distribution

For given values of ν_1 and ν_2, the following table contains values of $F_{0.001,\nu_1,\nu_2}$; defined by $\mathrm{Prob}\,[F \geq F_{0.001,\nu_1,\nu_2}] = \alpha = 0.001$.

ν_2	$\nu_1=1$	2	3	4	5	6	7	8	9	10	50	100	∞
2	998.5	999.0	999.2	999.2	999.3	999.3	999.4	999.4	999.4	999.4	999.5	999.5	999.5
3	167.0	148.5	141.1	137.1	134.6	132.8	131.6	130.6	129.9	129.2	124.7	124.1	123.5
4	74.14	61.25	56.18	53.44	51.71	50.53	49.66	49.00	48.47	48.05	44.88	44.47	44.05
5	47.18	37.12	33.20	31.09	29.75	28.83	28.16	27.65	27.24	26.92	24.44	24.12	23.79
6	35.51	27.00	23.70	21.92	20.80	20.03	19.46	19.03	18.69	18.41	16.31	16.03	15.75
7	29.25	21.69	18.77	17.20	16.21	15.52	15.02	14.63	14.33	14.08	12.20	11.95	11.70
8	25.41	18.49	15.83	14.39	13.48	12.86	12.40	12.05	11.77	11.54	9.80	9.57	9.33
9	22.86	16.39	13.90	12.56	11.71	11.13	10.70	10.37	10.11	9.89	8.26	8.04	7.81
10	21.04	14.91	12.55	11.28	10.48	9.93	9.52	9.20	8.96	8.75	7.19	6.98	6.76
11	19.69	13.81	11.56	10.35	9.58	9.05	8.66	8.35	8.12	7.92	6.42	6.21	6.00
12	18.64	12.97	10.80	9.63	8.89	8.38	8.00	7.71	7.48	7.29	5.83	5.63	5.42
13	17.82	12.31	10.21	9.07	8.35	7.86	7.49	7.21	6.98	6.80	5.37	5.17	4.97
14	17.14	11.78	9.73	8.62	7.92	7.44	7.08	6.80	6.58	6.40	5.00	4.81	4.60
15	16.59	11.34	9.34	8.25	7.57	7.09	6.74	6.47	6.26	6.08	4.70	4.51	4.31
16	16.12	10.97	9.01	7.94	7.27	6.80	6.46	6.19	5.98	5.81	4.45	4.26	4.06
17	15.72	10.66	8.73	7.68	7.02	6.56	6.22	5.96	5.75	5.58	4.24	4.05	3.85
18	15.38	10.39	8.49	7.46	6.81	6.35	6.02	5.76	5.56	5.39	4.06	3.87	3.67
19	15.08	10.16	8.28	7.27	6.62	6.18	5.85	5.59	5.39	5.22	3.90	3.71	3.51
20	14.82	9.95	8.10	7.10	6.46	6.02	5.69	5.44	5.24	5.08	3.77	3.58	3.38
25	13.88	9.22	7.45	6.49	5.89	5.46	5.15	4.91	4.71	4.56	3.28	3.09	2.89
50	12.22	7.96	6.34	5.46	4.90	4.51	4.22	4.00	3.82	3.67	2.44	2.25	2.06
100	11.50	7.41	5.86	5.02	4.48	4.11	3.83	3.61	3.44	3.30	2.08	1.87	1.65
∞	10.83	6.91	5.42	4.62	4.10	3.74	3.47	3.27	3.10	2.96	1.75	1.45	1.00

6.9   GAMMA DISTRIBUTION

6.9.1   Properties

$$\text{pdf} \quad f(x) = \frac{x^{\alpha-1}\, e^{-x/\beta}}{\beta^{\alpha}\Gamma(\alpha)}$$

$$\text{mean} \quad \mu = \alpha\beta$$

$$\text{variance} \quad \sigma^2 = \alpha\beta^2$$

$$\text{skewness} \quad \beta_1 = 2/\sqrt{\alpha}$$

$$\text{kurtosis} \quad \beta_2 = 3\left(1 + \frac{2}{\alpha}\right)$$

$$\text{mgf} \quad m(t) = (1 - \beta t)^{-\alpha}$$

$$\text{char function} \quad \phi(t) = (1 - i\beta t)^{-\alpha}$$

---

where $\Gamma(x)$ is the gamma function (see page 515).

6.9.2   Probability density function

The probability density function is skewed to the right. For fixed β the tail becomes heavier as α increases.

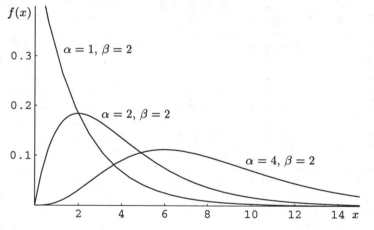

Figure 6.10: Probability density functions for a gamma random variable.

6.9.3   Related distributions

Let X be a gamma random variable with parameters α and β.

(1) The random variable X has a standard gamma distribution if $\alpha = 1$.

(2) If $\alpha = 1$ and $\beta = 1/\lambda$, then X has an exponential distribution with parameter λ.

(3) If $\alpha = \nu/2$ and $\beta = 2$, then X has a chi–square distribution with ν degrees of freedom.

(4) If $\alpha = n$ is an integer, then X has an Erlang distribution with parameters β and n.

(5) If $\alpha = \nu/2$ and $\beta = 1$, then the random variable $Y = 2X$ has a chi–square distribution with ν degrees of freedom.

(6) As $\alpha \to \infty$, X tends to a normal distribution with parameters $\mu = \alpha\beta$ and $\sigma^2 = \alpha\beta^2$.

(7) Suppose X_1 is a gamma random variable with parameters $\alpha = 1$ and $\beta = \beta_1$, X_2 is a gamma random variable with parameters $\alpha = 1$ and $\beta = \beta_2$, and X_1 and X_2 are independent. The random variable $Y = X_1/(X_1 + X_2)$ has a beta distribution with parameters β_1 and β_2.

(8) Let $X_1, X_2, \ldots, X_n$ be independent gamma random variables with parameters α and β_i for $i = 1, 2, \ldots, n$. The random variable $Y = X_1 + X_2 + \cdots + X_n$ has a gamma distribution with parameters α and $\beta = \beta_1 + \beta_2 + \cdots + \beta_n$.

6.10 HALF–NORMAL DISTRIBUTION

6.10.1 Properties

$$\text{pdf} \quad f(x) = \frac{2\theta}{\pi} \exp\left(-\frac{\theta^2 x^2}{\pi^2}\right), \quad x \geq 0, \; \theta > 0$$

$$\text{mean} \quad \mu = 1/\theta$$

$$\text{variance} \quad \sigma^2 = \frac{\pi - 2}{2\theta^2}$$

$$\text{skewness} \quad \beta_1 = \frac{\sqrt{2}(4 - \pi)}{(\pi - 2)^{3/2}}$$

$$\text{kurtosis} \quad \beta_2 = \frac{3\pi^2 - 4\pi - 12}{(\pi - 2)^2}$$

$$\text{mgf} \quad m(t) = \exp\left(\frac{\pi t^2}{4\theta^2}\right)\left(1 + \text{erf}\left[\frac{\sqrt{\pi}t}{2\theta}\right]\right)$$

$$\text{char function} \quad \phi(t) = \exp\left(-\frac{\pi t^2}{4\theta^2}\right)\left(1 + \text{erf}\left[\frac{\sqrt{\pi}it}{2\theta}\right]\right)$$

where $\text{erf}(x)$ is the error function (see page 512).

6.10.2 Probability density function

The probability density function is skewed to the right. As θ increases the tail becomes lighter.

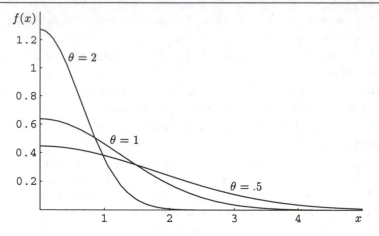

Figure 6.11: Probability density functions for a half–normal random variable.

6.11   INVERSE GAUSSIAN (WALD) DISTRIBUTION

6.11.1   Properties

$$\text{pdf} \quad f(x) = \sqrt{\frac{\lambda}{2\pi x^3}} \; \exp\left(\frac{-\lambda(x-\mu)^2}{2\mu^2 x}\right) \qquad x, \mu, \lambda > 0$$

$$\text{mean} \qquad \mu = \mu$$

$$\text{variance} \qquad \sigma^2 = \mu^3/\lambda$$

$$\text{skewness} \qquad \beta_1 = 3\sqrt{\mu/\lambda}$$

$$\text{kurtosis} \qquad \beta_2 = 3 + \frac{15\mu}{\lambda}$$

$$\text{mgf} \quad m(t) = \exp\left(\frac{\lambda}{\mu}\left[1 - \sqrt{1 - \frac{2\mu^2 t}{\lambda}}\right]\right)$$

$$\text{char function} \quad \phi(t) = \exp\left(\frac{\lambda}{\mu}\left[1 - \sqrt{1 - \frac{2\mu^2 it}{\lambda}}\right]\right)$$

6.11.2   Probability density function

The probability density function is skewed right. For fixed μ the probability density function becomes more bell–shaped as λ increases.

6.11.3   Related distributions

If X is an inverse Gaussian random variable with parameters μ and λ, the random variable $Y = \dfrac{\lambda(X-\mu)^2}{\mu^2 X}$ has a chi–square distribution with 1 degree

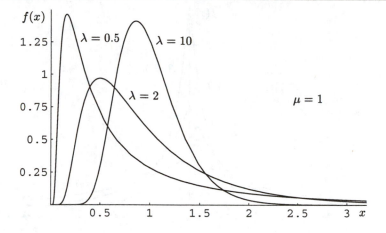

Figure 6.12: Probability density functions for an inverse Gaussian random variable.

of freedom.

6.12 LAPLACE DISTRIBUTION

6.12.1 Properties

$$\text{pdf} \quad f(x) = \frac{1}{2\beta} \exp\left(-\frac{|x - \alpha|}{\beta}\right), \quad x \in \mathcal{R}, \ \alpha \in \mathcal{R}, \ \beta > 0$$

mean $\quad \mu = \alpha$

variance $\quad \sigma^2 = 2\beta^2$

skewness $\quad \beta_1 = 0$

kurtosis $\quad \beta_2 = 6$

$$\text{mgf} \quad m(t) = \frac{e^{\alpha t}}{1 - \beta^2 t^2}$$

$$\text{char function} \quad \phi(t) = \frac{e^{\alpha i t}}{1 + \beta^2 t^2}$$

6.12.2 Probability density function

The probability density function is symmetric about the parameter α. For fixed α the tails become heavier as β increases.

6.12.3 Related distributions

(1) Let X be a Laplace random variable with parameters α and β. The random variable $Y = |X - \alpha|$ has an exponential distribution with pa-

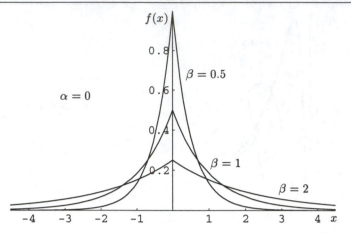

Figure 6.13: Probability density functions for a Laplace random variable.

rameter $\lambda = \beta$. The random variable $W = |X - \alpha|/\beta$ has an exponential distribution with parameter $\lambda = 1$.

(2) Let X_1 and X_2 be independent Laplace random variables with parameters $\alpha = 0$, and β_1 and β_2, respectively. The random variable $Y = |X_1/X_2|$ has an F distribution with parameters $\nu_1 = \nu_2 = 2$.

6.13 LOGISTIC DISTRIBUTION

6.13.1 Properties

$$\text{pdf} \quad f(x) = \frac{e^{-(x-\alpha)/\beta}}{\beta(1 + e^{-(x-\alpha)/\beta})^2}, \quad x \in \mathcal{R}, \ \alpha \in \mathcal{R}, \ \beta \in \mathcal{R}$$

$$\text{mean} \quad \mu = \alpha$$

$$\text{variance} \quad \sigma^2 = \beta^2 \pi^2 / 3$$

$$\text{skewness} \quad \beta_1 = 0$$

$$\text{kurtosis} \quad \beta_2 = 21/5$$

$$\text{mgf} \quad m(t) = \pi \beta t e^{\alpha t} / \sin(\pi \beta t)$$

$$\text{char function} \quad \phi(t) = \pi \beta t e^{i\alpha t} / \sinh(\pi \beta t)$$

6.13.2 Probability density function

The probability density function is symmetric about the parameter α. For fixed α the tails become heavier as β increases.

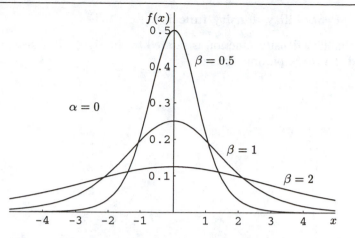

Figure 6.14: Probability density functions for a logistic random variable.

6.13.3 Related distributions

(1) The random variable X has a standard logistic distribution if $\alpha = 0$ and $\beta = 1$.

(2) If X is a logistic random variable with parameters α and β, then the random variable $Y = (X - \alpha)/\beta$ has a (standard) logistic distribution with parameters 0 and 1.

6.14 LOGNORMAL DISTRIBUTION

6.14.1 Properties

$$\text{pdf} \quad f(x) = \frac{1}{\sqrt{2\pi}\,\sigma x} \exp\left(-\frac{1}{2\sigma^2}(\ln x - \mu)^2\right)$$

$$x > 0, \ \mu \in \mathcal{R}, \ \sigma > 0$$

$$\text{mean} \quad \mu = e^{\mu + \sigma^2/2}$$

$$\text{variance} \quad \sigma^2 = e^{2\mu + \sigma^2}(e^{\sigma^2} - 1)$$

$$\text{skewness} \quad \beta_1 = (e^{\sigma^2} + 2)\sqrt{e^{\sigma^2} - 1}$$

$$\text{kurtosis} \quad \beta_2 = e^{4\sigma^2} + 2e^{3\sigma^2} + 3e^{2\sigma^2}$$

$$\text{mgf} \quad m(t) = \text{does not exist}$$

$$\text{char function} \quad \phi(t) = \text{does not exist}$$

6.14.2   Probability density function

The probability density function is skewed to the right. The *scale* parameter is μ and the *shape* parameter is σ.

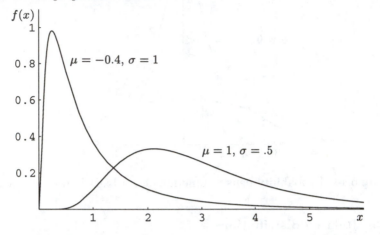

Figure 6.15: Probability density functions for a lognormal random variable.

6.14.3   Related distributions

(1) If X is a lognormal random variable with parameters μ and σ, then the random variable $Y = \ln X$ has a normal distribution with mean μ and variance σ^2.

(2) If X is a lognormal random variable with parameters μ and σ and a and b are constants, then the random variable $Y = e^a X^b$ has a lognormal distribution with parameters $a + b\mu$ and $b\sigma$.

(3) Let X_1 and X_2 be independent lognormal random variables with parameters μ_1, σ_1 and μ_2, σ_2, respectively. The random variable $Y = X_1/X_2$ has a lognormal distribution with parameters $\mu_1 - \mu_2$ and $\sigma_1 + \sigma_2$.

(4) Let $X_1, X_2, \ldots, X_n$ be independent lognormal random variables with parameters μ_i and σ_i for $i = 1, 2, \ldots, n$. The random variable $Y = X_1 \cdot X_2 \cdots X_n$ has a lognormal distribution with parameters $\mu = \mu_1 + \mu_2 + \cdots + \mu_n$ and $\sigma = \sigma_1 + \sigma_2 + \cdots + \sigma_n$.

(5) Let $X_1, X_2, \ldots, X_n$ be independent lognormal random variables with parameters μ and σ. The random variable $Y = \sqrt[n]{X_1 \cdots X_n}$ has a lognormal distribution with parameters μ and σ/n.

6.15 NONCENTRAL CHI–SQUARE DISTRIBUTION

6.15.1 Properties

$$\text{pdf}\quad f(x) = \frac{e^{\left[\frac{1}{2}(x+\lambda)\right]}}{2^{\nu/2}} \sum_{j=1}^{\infty} \frac{x^{(\nu/2)+j-1}\lambda^j}{\Gamma\left(\frac{\nu}{2}+j\right) 2^{2j} j!} \qquad x,\lambda > 0,\ \nu \in \mathcal{N}$$

$$\text{mean}\quad \mu = \nu + \lambda$$

$$\text{variance}\quad \sigma^2 = 2\nu + 4\lambda$$

$$\text{skewness}\quad \beta_1 = \frac{2\sqrt{2}(\nu + 3\lambda)}{(\nu + 2\lambda)^{3/2}}$$

$$\text{kurtosis}\quad \beta_2 = 3 + \frac{12(\nu + 4\lambda)}{(\nu + 2\lambda)^2}$$

$$\text{mgf}\quad m(t) = (1 - 2t)^{-\nu/2} \exp\left(\frac{\lambda t}{1 - 2t}\right)$$

$$\text{char function}\quad \phi(t) = (1 - 2it)^{-\nu/2} \exp\left(\frac{\lambda it}{1 - 2it}\right)$$

where $\Gamma(x)$ is the gamma function (see page 515).

6.15.2 Probability density function

The probability density function is skewed to the right. For fixed ν the tail becomes heavier as the *noncentrality* parameter λ increases.

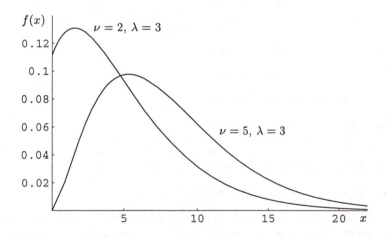

Figure 6.16: Probability density functions for a noncentral chi–square random variable.

6.15.3 Related distributions

(1) If X is a noncentral chi–square random variable with parameters ν and $\lambda = 0$, then X is a chi–square random variable with ν degrees of freedom.

(2) If X_1 is a noncentral chi–square random variable with parameters ν_1 and λ, X_2 is a chi–square random variable with parameter ν_2, and X_1 and X_2 are independent, then the random variable $Y = (X_1/\nu_1)/(X_2/\nu_2)$ has a noncentral F distribution with parameters ν_1, ν_2, and λ.

(3) Let $X_1, X_2, \ldots, X_n$ be independent noncentral chi–square random variables with parameters ν_i and λ_i (for $i = 1, 2, \ldots, n$). The random variable $Y = X_1 + X_2 + \cdots + X_n$ has a noncentral chi–square distribution with parameters $\nu = \nu_1 + \nu_2 + \cdots + \nu_n$ and $\lambda = \lambda_1 + \lambda_2 + \cdots + \lambda_n$.

6.16   NONCENTRAL F DISTRIBUTION

6.16.1 Properties

$$
\text{pdf} \quad f(x) = \frac{e^{-\lambda/2}\nu_1^{\nu_1/2}\nu_2^{\nu_2/2}x^{\frac{1}{2}(\nu_1-2)}(\nu_1 x + \nu_2)^{-\frac{1}{2}(\nu_1+\nu_2)}}{B(\nu_1/2, \nu_2/2)} \times
$$

$$
{}_1F_1\left[\frac{\nu_1 + \nu_2}{2}, \frac{\nu_1}{2}, \frac{\nu_1 \lambda x}{2(\nu_1 x + \nu_2)}\right]
$$

$$
x > 0, \ \nu_1, \nu_2 \in \mathcal{N}, \ \lambda > 0
$$

$$
\text{mean} \quad \mu = \frac{\nu_2(\nu_1 + \lambda)}{\nu_1(\nu_2 - 2)}, \quad \nu_2 > 2
$$

$$
\text{variance} \quad \sigma^2 = \frac{2\nu_2^2\left((\nu_1 + \lambda) + (\nu_2 - 2)(\nu_1 + 2\lambda)\right)}{\nu_1^2(\nu_2 - 4)(\nu_2 - 2)^2}, \quad \nu_2 > 4
$$

skewness $\quad \beta_1 = $ does not exist

kurtosis $\quad \beta_2 = $ does not exist

mgf $\quad m(t) = $ does not exist

char function $\quad \phi(t) = $ does not exist

where $B(a, b)$ is the beta function and ${}_pF_q$ is the generalized hypergeometric function defined in Chapter 18 (see pages 511 and 520).

6.16.2 Probability density function

The probability density function is skewed to the right. The parameters ν_1 and ν_2 are the *shape* parameters and λ is the *noncentrality* parameter.

6.16.3 Related distributions

If X has a noncentral F distribution with parameters ν_1, ν_2, and λ, then the random variable X tends to an F distribution as $\lambda \to 0$.

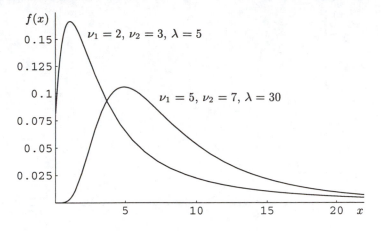

Figure 6.17: Probability density functions for a noncentral F random variable.

6.17   NONCENTRAL t DISTRIBUTION

6.17.1   Properties

pdf $f(x) = \dfrac{\nu^{\nu/2} e^{-\lambda^2/2}}{\sqrt{\pi}\,\Gamma(\nu/2)(\nu+x^2)^{(\nu+1)/2}} \times$

$$\sum_{j=0}^{\infty} \Gamma\left(\frac{\nu+j+1}{2}\right)\left(\frac{\lambda^j}{j!}\right)\left(\frac{2x^2}{\nu+x^2}\right)^{j/2} \qquad x, \lambda \in \mathcal{R},\ \nu \in \mathcal{N}$$

where $\Gamma(x)$ is the gamma function (see page 515).

The moments about the origin are

$$\mu'_r = c_r \frac{\nu^{r/2}\Gamma[(\nu-r)/2]}{2^{r/2}\Gamma(\nu/2)}, \qquad \nu > r \tag{6.10}$$

where

$$c_{2r-1} = \sum_{j=1}^{r} \frac{(2r-1)!\lambda^{2r-1}}{(2j-1)(r-j)!2^{r-j}}, \qquad r = 1, 2, 3, \ldots$$

$$c_{2r} = \sum_{j=0}^{r} \frac{(2r)!\lambda^{2j}}{(2j)!(r-j)!2^{r-j}}, \qquad r = 1, 2, 3, \ldots \tag{6.11}$$

6.17.2   Probability density function

The probability density function is skewed to the right. The *shape* parameter is ν and the *noncentrality* parameter is λ. For fixed ν the tail becomes heavier as λ increases. For large values of ν, the probability density function is approximately symmetric.

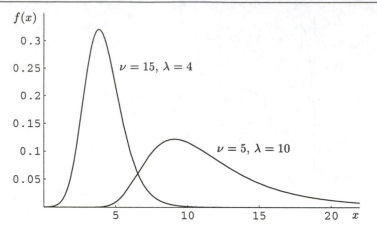

Figure 6.18: Probability density functions for a noncentral t random variable.

6.17.3   Related distributions

If X has a noncentral t distribution with parameters ν and $\lambda = 0$, then X has a t distribution with ν degrees of freedom.

6.18   NORMAL DISTRIBUTION

6.18.1   Properties

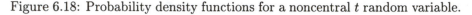

$$\text{pdf} \quad f(x) = \frac{1}{\sigma\sqrt{2\pi}}\, e^{-(x-\mu)^2/2\sigma^2}, \quad x \in \mathcal{R},\ \mu \in \mathcal{R},\ \sigma > 0$$

$$\text{mean} \quad \mu = \mu$$

$$\text{variance} \quad \sigma^2 = \sigma^2$$

$$\text{skewness} \quad \beta_1 = 0$$

$$\text{kurtosis} \quad \beta_2 = 3$$

$$\text{mgf} \quad m(t) = \exp\left(\mu t + \frac{\sigma^2 t^2}{2}\right)$$

$$\text{char function} \quad \phi(t) = \exp\left(\mu i t - \frac{\sigma^2 t^2}{2}\right)$$

See Chapter 7 for more details.

6.18.2   Probability density function

The probability density function is symmetric and bell–shaped about the *location* parameter μ. For small values of the *scale* parameter σ the probability density function is more compact.

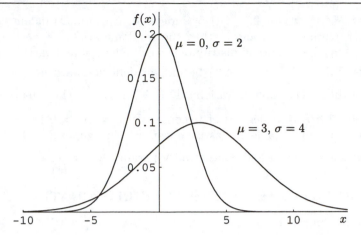

Figure 6.19: Probability density functions for a normal random variable.

6.18.3 Related distributions

(1) The random variable X has a standard normal distribution if $\mu = 0$ and $\sigma = 1$.

(2) If X is a normal random variable with parameters μ and σ, the random variable $Y = (X - \mu)/\sigma$ has a (standard) normal distribution with parameters 0 and 1.

(3) If X is a normal random variable with parameters μ and σ, the random variable $Y = e^X$ has a lognormal distribution with parameters μ and σ.

(4) If X is a normal random variable with parameters $\mu = 0$ and $\sigma = 1$, then the random variable $Y = e^{\mu + \sigma X}$ has a lognormal distribution with parameters μ and σ.

(5) If X is a normal random variable with parameters μ and σ, and a and b are constants, then the random variable $Y = a + bX$ has a normal distribution with parameters $a + b\mu$ and $b\sigma$.

(6) If X_1 and X_2 are independent standard normal random variables, the random variable $Y = X_1/X_2$ has a Cauchy distribution with parameters $a = 0$ and $b = 1$.

(7) If X_1 and X_2 are independent normal random variables with parameters $\mu = 0$ and σ, then the random variable $Y = \sqrt{X_1^2 + X_2^2}$ has a Rayleigh distribution with parameter σ.

(8) Let X_i (for $i = 1, 2, \ldots, n$) be independent, normal random variables with parameters μ_i and σ_i, and let c_i be any constants. The random variable $Y = \sum_{i=1}^{n} c_i X_i$ has a normal distribution with parameters $\mu = \sum_{i=1}^{n} c_i \mu_i$ and $\sigma^2 = \sum_{i=1}^{n} c_i^2 \sigma_i^2$.

(9) Let X_i (for $i = 1, 2, \ldots, n$) be independent, normal random variables with parameters μ and σ, then the random variable $Y = X_1 + X_2 + \cdots + X_n$ has a normal distribution with mean $n\mu$ and variance $n\sigma^2$.

(10) Let X_i (for $i = 1, 2, \ldots, n$) be independent standard normal random variables. The random variable $Y = \sum\limits_{i=1}^{n} X_i^2$ has a chi–square distribution with $\nu = n$ degrees of freedom. If $\mu_i = \lambda_i > 0$ ($\sigma_i = 1$), then the random variable Y has a noncentral chi–square distribution with parameters $\nu = n$ and noncentrality parameter $\lambda = \sum\limits_{i=1}^{n} \lambda_i^2$.

6.19   NORMAL DISTRIBUTION: MULTIVARIATE

6.19.1   Properties

$$\text{pdf} \quad f(\mathbf{x}) = \frac{1}{(2\pi)^{n/2}\sqrt{\det(\Sigma)}} \exp\left[-\frac{(\mathbf{x} - \boldsymbol{\mu})^{\mathrm{T}}\Sigma^{-1}(\mathbf{x} - \boldsymbol{\mu})}{2}\right]$$

mean $\qquad \boldsymbol{\mu}$

covariance matrix $\qquad \Sigma$

$$\text{char function} \quad \phi(\mathbf{t}) = \exp\left(i\mathbf{t}^{\mathrm{T}}\boldsymbol{\mu} - \frac{1}{2}\mathbf{t}^{\mathrm{T}}\Sigma\mathbf{t}\right)$$

---

where $\mathbf{x} = \begin{bmatrix} x_1, & x_2, & \ldots, & x_n \end{bmatrix}^{\mathrm{T}}$ (with $x_i \in \mathcal{R}$) and Σ is a positive semi-definite matrix.

Section 7.6 discusses the bivariate normal.

6.19.2   Probability density function

The probability density function is smooth and unimodal. Figure 6.20 shows two views of a bivariate normal with $\boldsymbol{\mu} = \begin{bmatrix} 1 & 0 \end{bmatrix}^{\mathrm{T}}$ and $\Sigma = \begin{bmatrix} 1 & 2 \\ 0 & 4 \end{bmatrix}$.

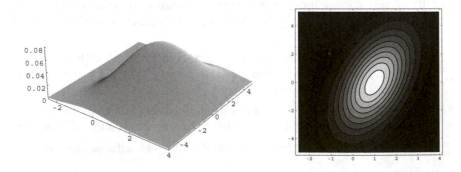

Figure 6.20: Two views of the probability density for a bivariate normal.

6.20 PARETO DISTRIBUTION

6.20.1 Properties

$$\text{pdf} \quad f(x) = \frac{\theta a^\theta}{x^{\theta+1}}, \quad x \geq a, \ \theta > 0, \ a > 0$$

$$\text{mean} \quad \mu = \frac{a\theta}{\theta - 1}, \quad \theta > 1$$

$$\text{variance} \quad \sigma^2 = \frac{a^2 \theta}{(\theta - 1)^2 (\theta - 2)}, \quad \theta > 2$$

$$\text{skewness} \quad \beta_1 = \frac{2(\theta + 1)\sqrt{\theta - 2}}{(\theta - 3)\sqrt{\theta}}, \quad \theta > 3$$

$$\text{kurtosis} \quad \beta_2 = \frac{3(\theta - 2)(3\theta^2 + \theta + 2)}{\theta(\theta - 3)(\theta - 4)}, \quad \theta > 4$$

$$\text{mgf} \quad m(t) = \text{does not exist}$$

$$\text{char function} \quad \phi(t) = -a^\theta t^\theta \cos(\pi\theta/2)\Gamma(1 - \theta) +$$
$$_1F_2\left[\left\{-\tfrac{\theta}{2}\right\}, \left\{\tfrac{1}{2}, 1 - \tfrac{\theta}{2}\right\}, -\tfrac{1}{4}a^2 t^2\right] -$$
$$\tfrac{1}{1-\theta}\left(ati\theta\, _1F_2\left[\left\{\tfrac{1}{2} - \tfrac{\theta}{2}\right\}, \left\{\tfrac{3}{2}, \tfrac{3}{2} - \tfrac{\theta}{2}\right\}, -\tfrac{1}{4}a^2 t^2\right] \text{sgn}(t)\right) +$$
$$ia^\theta t^\theta \Gamma(1 - \theta)\,\text{sgn}(t)\sin(\pi\theta/2)$$

where $_pF_q$ is the generalized hypergeometric function and $\text{sgn}(t)$ is the signum function defined in Chapter 18 (see pages 520 and 523).

6.20.2 Probability density function

The probability density function is skewed to the right. The *shape* parameter is θ and the *location* parameter is a.

6.20.3 Related distributions

(1) Let X be a Pareto random variable with parameters a and θ.

    (a) The random variable $Y = \ln(X/a)$ has an exponential distribution with parameter $\lambda = 1/\theta$.

    (b) The random variable $Y = 1/X$ has a power function distribution with parameters $1/a$ and θ.

    (c) The random variable $Y = -\ln\left[(X/a)^\theta - 1\right]$ has a logistic distribution with parameters $\alpha = 0$ and $\beta = 1$.

(2) Let X_i (for $i = 1, 2, \ldots, n$) be independent Pareto random variables with parameters a and θ. The random variable $Y = 2a \sum_{i=1}^{n} \ln(X_i/\theta)$ has a chi–square distribution with $\nu = 2n$.

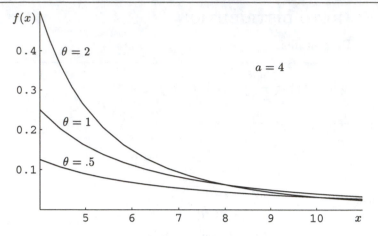

Figure 6.21: Probability density functions for a Pareto random variable.

6.21   POWER FUNCTION DISTRIBUTION

6.21.1   Properties

$$\text{pdf} \quad f(x) = \frac{cx^{c-1}}{b^c}, \quad 0 \le x \le b, \ b > 0, \ c > 0$$

$$\text{mean} \quad \mu = \frac{bc}{c+1}$$

$$\text{variance} \quad \sigma^2 = \frac{b^2 c}{(c+1)^2(c+2)}$$

$$\text{skewness} \quad \beta_1 = \frac{2(1-c)\sqrt{c+2}}{(c+3)\sqrt{c}}$$

$$\text{kurtosis} \quad \beta_2 = \frac{3(c+2)(3c^2 - c + 2)}{c(c+3)(c+4)}$$

$$\text{mgf} \quad m(t) = \text{does not exist}$$

$$\text{char function} \quad \phi(t) = {}_1F_2\left[\left\{\tfrac{c}{2}\right\}, \left\{\tfrac{1}{2}, 1 + \tfrac{c}{2}\right\}, -\tfrac{1}{4}b^2 t^2\right] +$$

$$\tfrac{1}{c+1}\left(ibct\, {}_1F_2\left[\left\{\tfrac{1}{2} + \tfrac{c}{2}\right\}, \left\{\tfrac{3}{2}, \tfrac{3}{2} + \tfrac{c}{2}\right\}, -\tfrac{1}{4}b^2 t^2\right]\text{sgn}(t)\right)$$

where ${}_pF_q$ is the generalized hypergeometric function and $\text{sgn}(t)$ is the signum function defined in Chapter 18 (see pages 520 and 523).

6.21.2   Probability density function

The probability density function is "J" shaped for $c < 1$ and is skewed left for $c > 1$.

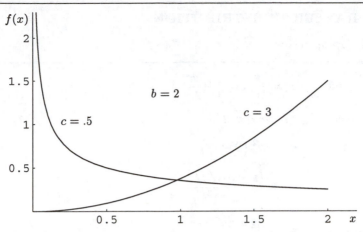

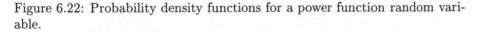

Figure 6.22: Probability density functions for a power function random variable.

6.21.3 Related distributions

(1) Let X be a power function random variable with parameters b and c.

   (a) The random variable $Y = 1/X$ has a power function distribution with parameters $1/b$ and c.

   (b) If $b = 1$:

      (1) The random variable X has a beta distribution with parameters $\alpha = c$ and $\beta = 1$.

      (2) The random variable $Y = -\ln X$ has an exponential distribution with parameter $\lambda = c$.

      (3) The random variable $Y = 1/X$ has a Pareto distribution with parameters $a = 0$ and $\theta = c$.

      (4) The random variable $Y = -\ln(X^{-c} - 1)$ has a logistic distribution with parameters $\alpha = 0$ and $\beta = 1$.

      (5) The random variable $Y = (-\ln X^c)^{1/k}$ has a Weibull distribution with parameters $\alpha = k$ and $\beta = 1$.

      (6) The random variable $Y = -\ln(-c\ln X)$ has an extreme–value distribution with parameters $\alpha = 0$ and $\beta = 1$.

   (c) If $c = 1$ then X has a uniform distribution with parameters $a = 0$ and b.

(7) Let X_1, X_2 be independent power function random variables with parameters $b = 1$ and c. The random variable $Y = -c\ln(X_1/X_2)$ has a Laplace distribution with parameters $\alpha = 0$ and $\beta = 1$.

6.22   RAYLEIGH DISTRIBUTION

6.22.1   Properties

$$\text{pdf} \quad f(x) = \frac{x}{\sigma^2} \exp\left(-\frac{x^2}{2\sigma^2}\right), \quad x \geq 0, \ \sigma > 0$$

$$\text{mean} \quad \mu = \sigma\sqrt{\pi/2}$$

$$\text{variance} \quad \sigma^2 = \sigma^2\left(2 - \frac{\pi}{2}\right)$$

$$\text{skewness} \quad \beta_1 = \frac{(\pi - 3)\sqrt{\pi/2}}{\left(2 - \frac{\pi}{2}\right)^{3/2}}$$

$$\text{kurtosis} \quad \beta_2 = \frac{32 - 3\pi^2}{(4 - \pi)^2}$$

$$\text{mgf} \quad m(t) = \frac{1}{2}\left(2 + \sqrt{2\pi}\,\sigma\,t\,e^{\sigma^2 t^2/2}\left[1 + \text{erf}\left(\frac{\sigma t}{\sqrt{2}}\right)\right]\right)$$

$$\text{char function} \quad \phi(t) = 1 + ie^{-\sigma^2 t^2/2}\sqrt{\frac{\pi}{2}}\,\sigma\,t\left[1 - \text{erf}\left(-\frac{i\sigma t}{\sqrt{2}}\right)\right]$$

where $\text{erf}(x)$ is the error function (see page 512).

6.22.2   Probability density function

The probability density function is skewed to the right. For large values of σ the tail is heavier.

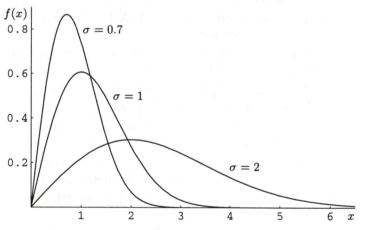

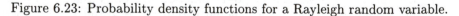

Figure 6.23: Probability density functions for a Rayleigh random variable.

6.22.3 Related distributions

(1) If X is a Rayleigh random variable with parameter $\sigma = 1$, then X is a chi random variable with parameter $n = 2$.

(2) If X is a Rayleigh random variable with parameter σ, then the random variable $Y = X^2$ has an exponential distribution with parameter $\lambda = 1/(2\sigma^2)$.

6.23 *t* DISTRIBUTION

6.23.1 Properties

$$\text{pdf} \quad f(x) = \frac{1}{\sqrt{\pi \nu}} \frac{\Gamma\left(\frac{\nu+1}{2}\right)}{\Gamma\left(\frac{\nu}{2}\right)} \left(1 + \frac{x^2}{\nu}\right)^{-(\nu+1)/2} \qquad x \in \mathcal{R},\ \nu \in \mathcal{N}$$

$$\text{mean} \quad \mu = 0, \quad \nu \geq 2$$

$$\text{variance} \quad \sigma^2 = \frac{\nu}{\nu - 2}, \quad \nu \geq 3$$

$$\text{skewness} \quad \beta_1 = 0, \quad \nu \geq 4$$

$$\text{kurtosis} \quad \beta_2 = 3 + \frac{6}{\nu - 4}, \quad \nu \geq 5$$

$$\text{mgf} \quad m(t) = \text{does not exist}$$

$$\text{char function} \quad \phi(t) = \frac{2^{1 - \frac{\nu}{2}} \nu^{\nu/4} |t|^{\nu/2} K_{\nu/2}(\sqrt{\nu}|t|)}{\Gamma(\nu/2)}$$

where $K_n(x)$ is a modified Bessel function and $\Gamma(x)$ is the gamma function defined in Chapter 18 (see pages 506 and 18.8).

6.23.2 Probability density function

The probability density function is symmetric and bell–shaped centered about 0. As the degrees of freedom, ν, increases the distribution becomes more compact.

6.23.3 Related distributions

(1) If X is a *t* random variable with parameter ν, then the random variable $Y = X^2$ has an F distribution with 1 and ν degrees of freedom.

(2) If X is a *t* random variable with parameter $\nu = 1$, then X has a Cauchy distribution with parameters $a = 0$ and $b = 1$.

(3) If X is a *t* random variable with parameter ν, as ν tends to infinity X tends to a standard normal distribution. The approximation is reasonable for $\nu \geq 30$.

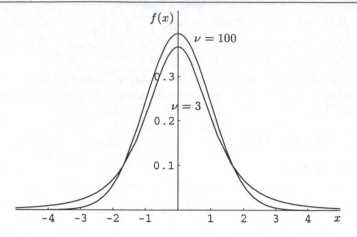

Figure 6.24: Probability density functions for a t random variable.

6.23.4   Critical values for the t distribution

For a given value of ν, the number of degrees of freedom, the table on page 157 contains values of $t_{\alpha,\nu}$ such that

$$\text{Prob}\,[t \geq t_{\alpha,\nu}] = \alpha \tag{6.12}$$

Example 6.42:   Use the table on page 157 to find the values $t_{.05,11}$ and $-t_{.01,24}$.

Solution:

(S1) The top row of the following table contains cumulative probability and the left–hand column contains the degrees of freedom. The values in the body of the table may be used to find critical values.

(S2) $t_{.05,11} = 1.7959$ since $F(1.7959; 11) = .95$ $\implies$ $\text{Prob}\,[t \geq 1.7959] = .05$

(S3) $-t_{.01,24} = -2.4922$ since $F(2.4922; 24) = .99$
      $\implies$ $\text{Prob}\,[t \leq -2.4922] = .01$

(S4) Illustrations:

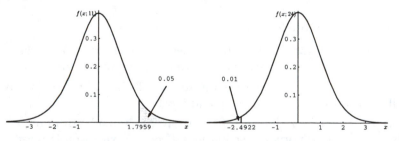

Critical values for the *t* distribution.

ν	$\alpha = 0.1$	0.05	0.025	0.01	0.005	0.0025	0.001
1	3.078	6.314	12.706	31.821	63.657	318.309	636.619
2	1.886	2.920	4.303	6.965	9.925	22.327	31.599
3	1.638	2.353	3.182	4.541	5.841	10.215	12.924
4	1.533	2.132	2.776	3.747	4.604	7.173	8.610
5	1.476	2.015	2.571	3.365	4.032	5.893	6.869
6	1.440	1.943	2.447	3.143	3.707	5.208	5.959
7	1.415	1.895	2.365	2.998	3.499	4.785	5.408
8	1.397	1.860	2.306	2.896	3.355	4.501	5.041
9	1.383	1.833	2.262	2.821	3.250	4.297	4.781
10	1.372	1.812	2.228	2.764	3.169	4.144	4.587
11	1.363	1.796	2.201	2.718	3.106	4.025	4.437
12	1.356	1.782	2.179	2.681	3.055	3.930	4.318
13	1.350	1.771	2.160	2.650	3.012	3.852	4.221
14	1.345	1.761	2.145	2.624	2.977	3.787	4.140
15	1.341	1.753	2.131	2.602	2.947	3.733	4.073
16	1.337	1.746	2.120	2.583	2.921	3.686	4.015
17	1.333	1.740	2.110	2.567	2.898	3.646	3.965
18	1.330	1.734	2.101	2.552	2.878	3.610	3.922
19	1.328	1.729	2.093	2.539	2.861	3.579	3.883
20	1.325	1.725	2.086	2.528	2.845	3.552	3.850
21	1.323	1.721	2.080	2.518	2.831	3.527	3.819
22	1.321	1.717	2.074	2.508	2.819	3.505	3.792
23	1.319	1.714	2.069	2.500	2.807	3.485	3.768
24	1.318	1.711	2.064	2.492	2.797	3.467	3.745
25	1.316	1.708	2.060	2.485	2.787	3.450	3.725
26	1.315	1.706	2.056	2.479	2.779	3.435	3.707
27	1.314	1.703	2.052	2.473	2.771	3.421	3.69o
28	1.313	1.701	2.048	2.467	2.763	3.408	3.674
29	1.311	1.699	2.045	2.462	2.756	3.396	3.659
30	1.310	1.697	2.042	2.457	2.750	3.385	3.646
35	1.306	1.690	2.030	2.438	2.724	3.340	3.591
40	1.303	1.684	2.021	2.423	2.704	3.307	3.551
45	1.301	1.679	2.014	2.412	2.690	3.281	3.520
50	1.299	1.676	2.009	2.403	2.678	3.261	3.496
100	0.290	1.660	1.984	2.364	2.626	3.174	3.390
∞	1.282	1.645	1.960	2.326	2.576	3.091	3.291

6.24   TRIANGULAR DISTRIBUTION

6.24.1   Properties

$$\text{pdf} \quad f(x) = \begin{cases} 0 & x \le a \\ 4(x-a)/(b-a)^2 & a < x \le (a+b)/2 \\ 4(b-x)/(b-a)^2 & (a+b)/2 < x < b \\ 0 & x \ge b \end{cases}$$

$$a < b \in \mathcal{R}$$

$$\text{mean} \quad \mu = \frac{a+b}{2}$$

$$\text{variance} \quad \sigma^2 = \frac{(b-a)^2}{24}$$

$$\text{skewness} \quad \beta_1 = 0$$

$$\text{kurtosis} \quad \beta_2 = 12/5$$

$$\text{mgf} \quad m(t) = \frac{4(e^{at/2} - e^{bt/2})^2}{(b-a)^2 t^2}$$

$$\text{char function} \quad \phi(t) = -\frac{4(e^{ait/2} - e^{bit/2})^2}{(b-a)^2 t^2}$$

6.24.2   Probability density function

The probability density function is symmetric about the mean and consists of two line segments.

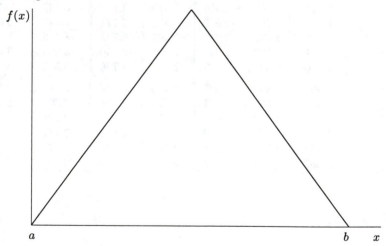

Figure 6.25: Probability density function for a triangular random variable.

6.25 UNIFORM DISTRIBUTION

6.25.1 Properties

$$\text{pdf} \quad f(x) = \frac{1}{b-a}, \quad a \le x \le b,\ a < b \in \mathcal{R}$$

$$\text{mean} \quad \mu = \frac{a+b}{2}$$

$$\text{variance} \quad \sigma^2 = \frac{(b-a)^2}{12}$$

$$\text{skewness} \quad \beta_1 = 0$$

$$\text{kurtosis} \quad \beta_2 = 9/5$$

$$\text{mgf} \quad m(t) = \frac{e^{bt} - e^{at}}{(b-a)t}$$

$$\text{char function} \quad \phi(t) = \frac{e^{bit} - e^{ait}}{(b-a)it}$$

6.25.2 Probability density function

The probability density function is a horizontal line segment between a and b at $1/(b-a)$.

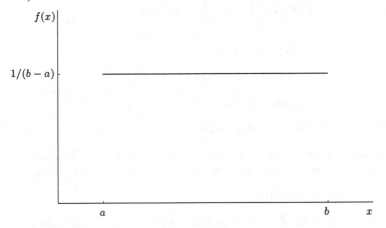

Figure 6.26: Probability density functions for a uniform random variable.

6.25.3 Related distributions

(1) The random variable X has a standard uniform distribution if $a = 0$ and $b = 1$.

(2) If X is a uniform random variable with parameters $a = 0$ and $b = 1$, the random variable $Y = -(\ln X)/\lambda$ has an exponential distribution with parameter λ.

(3) Let X_1 and X_2 be independent uniform random variables with parameters $a = 0$ and $b = 1$. The random variable $Y = (X_1 + X_2)/2$ has a triangular distribution with parameters 0 and 1.

(4) If X is a uniform random variable with parameters $a = -\pi/2$ and $b = \pi/2$, then the random variable $Y = \tan X$ has a Cauchy distribution with parameters $a = 0$ and $b = 1$.

6.26   WEIBULL DISTRIBUTION

6.26.1   Properties

$$\text{pdf} \quad \dot{f}(x) = \frac{\alpha}{\beta^\alpha} x^{\alpha-1} e^{-(x/\beta)^\alpha}$$

$$\text{mean} \quad \mu = \beta \Gamma\left(1 + \frac{1}{\alpha}\right)$$

$$\text{variance} \quad \sigma^2 = \beta^2 \left[\Gamma\left(1 + \frac{2}{\alpha}\right) - \Gamma^2\left(1 + \frac{1}{\alpha}\right)\right]$$

$$\text{skewness} \quad \beta_1 = \frac{2\Gamma^3(1 + \frac{1}{\alpha}) - 3\Gamma(1 + \frac{1}{\alpha})\,\Gamma(1 + \frac{2}{\alpha}) + \Gamma(1 + \frac{3}{\alpha})}{\left[\Gamma(1 + \frac{2}{\alpha}) - \Gamma^2(1 + \frac{1}{\alpha})\right]^{3/2}}$$

$$\text{kurtosis} \quad \beta_2 =$$

$$\frac{-3\Gamma^4(1 + \frac{1}{\alpha}) + 6\Gamma^2(1 + \frac{1}{\alpha})\,\Gamma(1 + \frac{2}{\alpha}) - 4\Gamma(1 + \frac{1}{\alpha})\,\Gamma(1 + \frac{3}{\alpha}) + \Gamma(1 + \frac{4}{\alpha})}{\left[\Gamma(1 + \frac{2}{\alpha}) - \Gamma^2(1 + \frac{1}{\alpha})\right]^2}$$

$$\text{mgf} \quad m(t) = \text{does not exist}$$

$$\text{char function} \quad \phi(t) = \text{does not exist}$$

---

where $\Gamma(x)$ is the gamma function (see page 515).

6.26.2   Probability density function

The probability density function is skewed to the right. For fixed β the tail becomes lighter and the distribution becomes more bell–shaped as α increases.

6.26.3   Related distributions

Suppose X is a Weibull random variable with parameters α and β.

(1) The random variable X has a standard Weibull distribution if $\beta = 1$.

(2) If $\alpha = 1$ then X has an exponential distribution with parameter $\lambda = 1/\beta$.

(3) The random variable $Y = X^\alpha$ has an exponential distribution with parameter $\lambda = \beta$.

(4) If $\alpha = 2$ then X has a Rayleigh distribution with parameter $\sigma = \beta/\sqrt{2}$.

(5) The random variable $Y = -\alpha \ln(X/\beta)$ has a (standard) extreme–value distribution with parameters $\alpha = 0$ and $\beta = 1$.

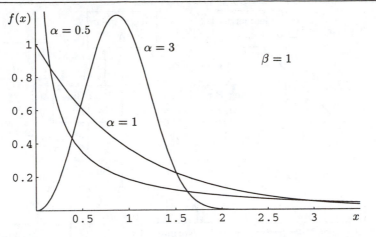

Figure 6.27: Probability density functions for a Weibull random variable.

6.27 RELATIONSHIPS AMONG DISTRIBUTIONS

Figure 6.27 presents some of the relationships among common univariate distributions. The first line of each box is the name of the distribution and the second line lists the parameters that characterize the distribution. The random variable X is used to represent each distribution. The three types of relationships presented in the figure are transformations (independent random variables are assumed) and special cases (both indicated with a solid arrow), and limiting distributions (indicated with a dashed arrow).

6.27.1 Other relationships among distributions

(1) If X_1 has a standard normal distribution, X_2 has a chi–square distribution with ν degrees of freedom, and X_1 and X_2 are independent, then the random variable

$$Y = \frac{X_1}{\sqrt{X_2/\nu}} \tag{6.13}$$

has a t distribution with ν degrees of freedom.

(2) Let $X_1, X_2, \ldots, X_n$ be independent normal random variables with parameters μ and σ, and define

$$\overline{X} = \frac{1}{n} \sum_{i=1}^{n} X_i \quad \text{and} \quad S^2 = \frac{1}{n} \sum_{i=1}^{n} (X_i - \overline{X})^2. \tag{6.14}$$

(a) The random variable $Y = nS^2/\sigma^2$ has a chi–square distribution with $n - 1$ degrees of freedom.

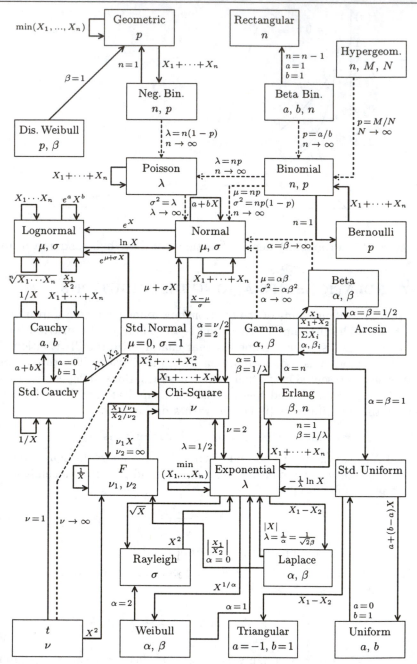

Figure 6.28: Relationships among distributions (see page 161).

(b) The random variable

$$W = \frac{\overline{X} - \mu}{S/\sqrt{n-1}} \tag{6.15}$$

has a t distribution with $n - 1$ degrees of freedom.

(3) Let $X_1, X_2, \ldots, X_n$ be independent normal random variables with parameters μ and σ, and define

$$\overline{X} = \frac{1}{n}\sum_{i=1}^{n} X_i \quad \text{and} \quad S^2 = \frac{1}{n-1}\sum_{i=1}^{n}(X_i - \overline{X})^2. \tag{6.16}$$

The random variable

$$Y = \frac{\overline{X} - \mu}{S/\sqrt{n}} \tag{6.17}$$

has a t distribution with $n - 1$ degrees of freedom.

(4) Let $X_1, X_2, \ldots, X_{n_1}$ be independent normal random variables with parameters μ_1 and σ, and $Y_1, Y_2, \ldots, Y_{n_2}$ be independent normal random variables with parameters μ_2 and σ. Define

$$\overline{X} = \frac{1}{n_1}\sum_{i=1}^{n_1} X_i \qquad S_1^2 = \frac{1}{n_1}\sum_{i=1}^{n_1}(X_i - \overline{X})^2$$

$$\overline{Y} = \frac{1}{n_2}\sum_{i=1}^{n_2} Y_i \qquad S_2^2 = \frac{1}{n_2}\sum_{i=1}^{n_2}(Y_i - \overline{Y})^2 \tag{6.18}$$

(a) The random variable $Y = (n_1 S_1^2 + n_2 S_2^2)/\sigma^2$ has a chi–square distribution with $n_1 + n_2 - 2$ degrees of freedom.

(b) The random variable

$$W = \frac{(\overline{X} - \overline{Y}) - (\mu_1 - \mu_2)}{\sqrt{\frac{1}{n_1} + \frac{1}{n_2}}\sqrt{\frac{n_1 S_1^2 + n_2 S_2^2}{n_1 + n_2 - 2}}} \tag{6.19}$$

has a t distribution with $n_1 + n_2 - 2$ degrees of freedom.

(5) Let $X_1, X_2, \ldots, X_{n_1}$ be independent normal random variables with parameters μ_1 and σ_1, and $Y_1, Y_2, \ldots, Y_{n_2}$ be independent normal random variables with parameters μ_2 and σ_2. Define

$$\overline{X} = \frac{1}{n_1}\sum_{i=1}^{n_1} X_i \qquad S_1^2 = \frac{1}{n_1}\sum_{i=1}^{n_1}(X_i - \overline{X})^2$$

$$\overline{Y} = \frac{1}{n_2}\sum_{i=1}^{n_2} Y_i \qquad S_2^2 = \frac{1}{n_2}\sum_{i=1}^{n_2}(Y_i - \overline{Y})^2 \tag{6.20}$$

The random variable

$$Y = \frac{n_1 S_1^2}{(n_1 - 1)\sigma_1^2} \left/ \frac{n_2 S_2^2}{(n_2 - 1)\sigma_2^2} \right. \tag{6.21}$$

has an F distribution with n_1 and n_2 degrees of freedom.

(6) Let X_1 be a normal random variable with parameters $\mu = \lambda$ and $\sigma = 1$, X_2 a chi–square random variable with parameter ν, and X_1 and X_2 be independent. The random variable $Y = X_1/\sqrt{X_2/\nu}$ has a noncentral t distribution with parameters ν and λ.

(7) Let X be a continuous random variable with cumulative distribution function $F(x)$.

    (a) The random variable $Y = F(X)$ has a (standard) uniform distribution with parameters $a = 0$ and $b = 1$.

    (b) The random variable $Y = -\ln[1 - F(X)]$ has a (standard) exponential distribution with parameter $\lambda = 1$.

Let X be a continuous random variable with probability density function $f(x)$. The random variable $Y = |X|$ has probability density function $g(y)$ given by

$$g(y) = \begin{cases} f(y) + f(-y) & \text{if } y > 0 \\ 0 & \text{elsewhere} \end{cases} \tag{6.22}$$

If X has a standard normal distribution ($\mu = 0$, $\sigma = 1$) then $g(y) = 2f(y)$.

CHAPTER 7

Standard Normal Distribution

Contents

7.1 THE PROBABILITY DENSITY FUNCTION AND RELATED FUNCTIONS

Let Z be a standard normal random variable ($\mu = 0$, $\sigma = 1$). The probability density function is given by

$$f(z) = \frac{1}{\sqrt{2\pi}}\, e^{-z^2/2}. \tag{7.1}$$

The following tables contain values for:

(1) $f(z)$

(2) $F(z) = \text{Prob}\,[Z \le z] = \int_{-\infty}^{z} \frac{1}{\sqrt{2\pi}}\, e^{-t^2/2}\, dt.$

$\quad$ = the cumulative distribution function

165

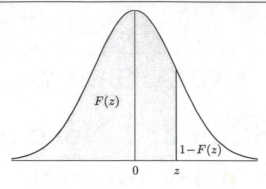

Figure 7.1: Cumulative distribution function for a standard normal random variable.

Note:

(1) For all z, $f(-z) = f(z)$.

(2) For all z, $F(-z) = 1 - F(z)$

(3) For all z, $\text{Prob}\,[|Z| \leq z] = F(z) - F(-z)$

(4) For all z, $\text{Prob}\,[|Z| \geq z] = 1 - F(z) + F(-z)$

(5) The function $\Phi(z) = F(z)$ is often used to represent the normal distribution function.

(6) $f'(x) = -\frac{x}{\sqrt{2\pi}}e^{-x^2/2} = -x\,f(x)$

(7) $f''(x) = \left(x^2 - 1\right) f(x)$

(8) $f'''(x) = \left(3x - x^3\right) f(x)$

(9) $f^{(4)}(x) = \left(x^4 - 6x^2 + 3\right) f(x)$

(10) For large values of x:

$$\left[\frac{e^{-x^2/2}}{\sqrt{2\pi}} \left(\frac{1}{x} - \frac{1}{x^3}\right)\right] < 1 - \Phi(x) < \left[\frac{e^{-x^2/2}}{\sqrt{2\pi}} \left(\frac{1}{x}\right)\right] \qquad (7.2)$$

(11) Order statistics for the normal distribution may be found in section 4.6.8 on page 64.

Normal distribution

z	$f(z)$	$F(z)$	$1 - F(z)$	z	$f(z)$	$F(z)$	$1 - F(z)$
−4.00	0.0001	0.0000	1.0000	−3.50	0.0009	0.0002	0.9998
−3.99	0.0001	0.0000	1.0000	−3.49	0.0009	0.0002	0.9998
−3.98	0.0001	0.0000	1.0000	−3.48	0.0009	0.0003	0.9998
−3.97	0.0001	0.0000	1.0000	−3.47	0.0010	0.0003	0.9997
−3.96	0.0002	0.0000	1.0000	−3.46	0.0010	0.0003	0.9997
−3.95	0.0002	0.0000	1.0000	−3.45	0.0010	0.0003	0.9997
−3.94	0.0002	0.0000	1.0000	−3.44	0.0011	0.0003	0.9997
−3.93	0.0002	0.0000	1.0000	−3.43	0.0011	0.0003	0.9997
−3.92	0.0002	0.0000	1.0000	−3.42	0.0011	0.0003	0.9997
−3.91	0.0002	0.0001	1.0000	−3.41	0.0012	0.0003	0.9997
−3.90	0.0002	0.0001	1.0000	−3.40	0.0012	0.0003	0.9997
−3.89	0.0002	0.0001	1.0000	−3.39	0.0013	0.0003	0.9997
−3.88	0.0002	0.0001	1.0000	−3.38	0.0013	0.0004	0.9996
−3.87	0.0002	0.0001	1.0000	−3.37	0.0014	0.0004	0.9996
−3.86	0.0002	0.0001	0.9999	−3.36	0.0014	0.0004	0.9996
−3.85	0.0002	0.0001	0.9999	−3.35	0.0015	0.0004	0.9996
−3.84	0.0003	0.0001	0.9999	−3.34	0.0015	0.0004	0.9996
−3.83	0.0003	0.0001	0.9999	−3.33	0.0016	0.0004	0.9996
−3.82	0.0003	0.0001	0.9999	−3.32	0.0016	0.0004	0.9996
−3.81	0.0003	0.0001	0.9999	−3.31	0.0017	0.0005	0.9995
−3.80	0.0003	0.0001	0.9999	−3.30	0.0017	0.0005	0.9995
−3.79	0.0003	0.0001	0.9999	−3.29	0.0018	0.0005	0.9995
−3.78	0.0003	0.0001	0.9999	−3.28	0.0018	0.0005	0.9995
−3.77	0.0003	0.0001	0.9999	−3.27	0.0019	0.0005	0.9995
−3.76	0.0003	0.0001	0.9999	−3.26	0.0020	0.0006	0.9994
−3.75	0.0003	0.0001	0.9999	−3.25	0.0020	0.0006	0.9994
−3.74	0.0004	0.0001	0.9999	−3.24	0.0021	0.0006	0.9994
−3.73	0.0004	0.0001	0.9999	−3.23	0.0022	0.0006	0.9994
−3.72	0.0004	0.0001	0.9999	−3.22	0.0022	0.0006	0.9994
−3.71	0.0004	0.0001	0.9999	−3.21	0.0023	0.0007	0.9993
−3.70	0.0004	0.0001	0.9999	−3.20	0.0024	0.0007	0.9993
−3.69	0.0004	0.0001	0.9999	−3.19	0.0025	0.0007	0.9993
−3.68	0.0005	0.0001	0.9999	−3.18	0.0025	0.0007	0.9993
−3.67	0.0005	0.0001	0.9999	−3.17	0.0026	0.0008	0.9992
−3.66	0.0005	0.0001	0.9999	−3.16	0.0027	0.0008	0.9992
−3.65	0.0005	0.0001	0.9999	−3.15	0.0028	0.0008	0.9992
−3.64	0.0005	0.0001	0.9999	−3.14	0.0029	0.0008	0.9992
−3.63	0.0006	0.0001	0.9999	−3.13	0.0030	0.0009	0.9991
−3.62	0.0006	0.0001	0.9999	−3.12	0.0031	0.0009	0.9991
−3.61	0.0006	0.0001	0.9999	−3.11	0.0032	0.0009	0.9991
−3.60	0.0006	0.0002	0.9998	−3.10	0.0033	0.0010	0.9990
−3.59	0.0006	0.0002	0.9998	−3.09	0.0034	0.0010	0.9990
−3.58	0.0007	0.0002	0.9998	−3.08	0.0035	0.0010	0.9990
−3.57	0.0007	0.0002	0.9998	−3.07	0.0036	0.0011	0.9989
−3.56	0.0007	0.0002	0.9998	−3.06	0.0037	0.0011	0.9989
−3.55	0.0007	0.0002	0.9998	−3.05	0.0038	0.0011	0.9989
−3.54	0.0008	0.0002	0.9998	−3.04	0.0039	0.0012	0.9988
−3.53	0.0008	0.0002	0.9998	−3.03	0.0040	0.0012	0.9988
−3.52	0.0008	0.0002	0.9998	−3.02	0.0042	0.0013	0.9987
−3.51	0.0008	0.0002	0.9998	−3.01	0.0043	0.0013	0.9987
−3.50	0.0009	0.0002	0.9998	−3.00	0.0044	0.0014	0.9987

Normal distribution

z	$f(z)$	$F(z)$	$1 - F(z)$	z	$f(z)$	$F(z)$	$1 - F(z)$
−3.00	0.0044	0.0014	0.9987	−2.50	0.0175	0.0062	0.9938
−2.99	0.0046	0.0014	0.9986	−2.49	0.0180	0.0064	0.9936
−2.98	0.0047	0.0014	0.9986	−2.48	0.0184	0.0066	0.9934
−2.97	0.0049	0.0015	0.9985	−2.47	0.0189	0.0068	0.9932
−2.96	0.0050	0.0015	0.9985	−2.46	0.0194	0.0069	0.9930
−2.95	0.0051	0.0016	0.9984	−2.45	0.0198	0.0071	0.9929
−2.94	0.0053	0.0016	0.9984	−2.44	0.0203	0.0073	0.9927
−2.93	0.0054	0.0017	0.9983	−2.43	0.0208	0.0076	0.9925
−2.92	0.0056	0.0018	0.9982	−2.42	0.0213	0.0078	0.9922
−2.91	0.0058	0.0018	0.9982	−2.41	0.0219	0.0080	0.9920
−2.90	0.0060	0.0019	0.9981	−2.40	0.0224	0.0082	0.9918
−2.89	0.0061	0.0019	0.9981	−2.39	0.0229	0.0084	0.9916
−2.88	0.0063	0.0020	0.9980	−2.38	0.0235	0.0087	0.9913
−2.87	0.0065	0.0021	0.9980	−2.37	0.0241	0.0089	0.9911
−2.86	0.0067	0.0021	0.9979	−2.36	0.0246	0.0091	0.9909
−2.85	0.0069	0.0022	0.9978	−2.35	0.0252	0.0094	0.9906
−2.84	0.0071	0.0023	0.9977	−2.34	0.0258	0.0096	0.9904
−2.83	0.0073	0.0023	0.9977	−2.33	0.0264	0.0099	0.9901
−2.82	0.0075	0.0024	0.9976	−2.32	0.0271	0.0102	0.9898
−2.81	0.0077	0.0025	0.9975	−2.31	0.0277	0.0104	0.9896
−2.80	0.0079	0.0026	0.9974	−2.30	0.0283	0.0107	0.9893
−2.79	0.0081	0.0026	0.9974	−2.29	0.0290	0.0110	0.9890
−2.78	0.0084	0.0027	0.9973	−2.28	0.0296	0.0113	0.9887
−2.77	0.0086	0.0028	0.9972	−2.27	0.0303	0.0116	0.9884
−2.76	0.0089	0.0029	0.9971	−2.26	0.0310	0.0119	0.9881
−2.75	0.0091	0.0030	0.9970	−2.25	0.0317	0.0122	0.9878
−2.74	0.0094	0.0031	0.9969	−2.24	0.0325	0.0126	0.9875
−2.73	0.0096	0.0032	0.9968	−2.23	0.0332	0.0129	0.9871
−2.72	0.0099	0.0033	0.9967	−2.22	0.0339	0.0132	0.9868
−2.71	0.0101	0.0034	0.9966	−2.21	0.0347	0.0135	0.9865
−2.70	0.0104	0.0035	0.9965	−2.20	0.0355	0.0139	0.9861
−2.69	0.0107	0.0036	0.9964	−2.19	0.0363	0.0143	0.9857
−2.68	0.0110	0.0037	0.9963	−2.18	0.0371	0.0146	0.9854
−2.67	0.0113	0.0038	0.9962	−2.17	0.0379	0.0150	0.9850
−2.66	0.0116	0.0039	0.9961	−2.16	0.0387	0.0154	0.9846
−2.65	0.0119	0.0040	0.9960	−2.15	0.0396	0.0158	0.9842
−2.64	0.0122	0.0042	0.9959	−2.14	0.0404	0.0162	0.9838
−2.63	0.0126	0.0043	0.9957	−2.13	0.0413	0.0166	0.9834
−2.62	0.0129	0.0044	0.9956	−2.12	0.0422	0.0170	0.9830
−2.61	0.0132	0.0045	0.9955	−2.11	0.0431	0.0174	0.9826
−2.60	0.0136	0.0047	0.9953	−2.10	0.0440	0.0179	0.9821
−2.59	0.0139	0.0048	0.9952	−2.09	0.0449	0.0183	0.9817
−2.58	0.0143	0.0049	0.9951	−2.08	0.0459	0.0188	0.9812
−2.57	0.0147	0.0051	0.9949	−2.07	0.0468	0.0192	0.9808
−2.56	0.0151	0.0052	0.9948	−2.06	0.0478	0.0197	0.9803
−2.55	0.0155	0.0054	0.9946	−2.05	0.0488	0.0202	0.9798
−2.54	0.0158	0.0055	0.9945	−2.04	0.0498	0.0207	0.9793
−2.53	0.0163	0.0057	0.9943	−2.03	0.0508	0.0212	0.9788
−2.52	0.0167	0.0059	0.9941	−2.02	0.0519	0.0217	0.9783
−2.51	0.0171	0.0060	0.9940	−2.01	0.0529	0.0222	0.9778
−2.50	0.0175	0.0062	0.9938	−2.00	0.0540	0.0227	0.9772

Normal distribution

z	$f(z)$	$F(z)$	$1 - F(z)$	z	$f(z)$	$F(z)$	$1 - F(z)$
−2.00	0.0540	0.0227	0.9772	−1.50	0.1295	0.0668	0.9332
−1.99	0.0551	0.0233	0.9767	−1.49	0.1315	0.0681	0.9319
−1.98	0.0562	0.0238	0.9761	−1.48	0.1334	0.0694	0.9306
−1.97	0.0573	0.0244	0.9756	−1.47	0.1354	0.0708	0.9292
−1.96	0.0584	0.0250	0.9750	−1.46	0.1374	0.0722	0.9278
−1.95	0.0596	0.0256	0.9744	−1.45	0.1394	0.0735	0.9265
−1.94	0.0608	0.0262	0.9738	−1.44	0.1415	0.0749	0.9251
−1.93	0.0619	0.0268	0.9732	−1.43	0.1435	0.0764	0.9236
−1.92	0.0632	0.0274	0.9726	−1.42	0.1456	0.0778	0.9222
−1.91	0.0644	0.0281	0.9719	−1.41	0.1476	0.0793	0.9207
−1.90	0.0656	0.0287	0.9713	−1.40	0.1497	0.0808	0.9192
−1.89	0.0669	0.0294	0.9706	−1.39	0.1518	0.0823	0.9177
−1.88	0.0681	0.0301	0.9699	−1.38	0.1540	0.0838	0.9162
−1.87	0.0694	0.0307	0.9693	−1.37	0.1561	0.0853	0.9147
−1.86	0.0707	0.0314	0.9686	−1.36	0.1582	0.0869	0.9131
−1.85	0.0721	0.0322	0.9678	−1.35	0.1604	0.0885	0.9115
−1.84	0.0734	0.0329	0.9671	−1.34	0.1626	0.0901	0.9099
−1.83	0.0748	0.0336	0.9664	−1.33	0.1647	0.0918	0.9082
−1.82	0.0761	0.0344	0.9656	−1.32	0.1669	0.0934	0.9066
−1.81	0.0775	0.0352	0.9648	−1.31	0.1691	0.0951	0.9049
−1.80	0.0790	0.0359	0.9641	−1.30	0.1714	0.0968	0.9032
−1.79	0.0804	0.0367	0.9633	−1.29	0.1736	0.0985	0.9015
−1.78	0.0818	0.0375	0.9625	−1.28	0.1759	0.1003	0.8997
−1.77	0.0833	0.0384	0.9616	−1.27	0.1781	0.1020	0.8980
−1.76	0.0848	0.0392	0.9608	−1.26	0.1804	0.1038	0.8962
−1.75	0.0863	0.0401	0.9599	−1.25	0.1827	0.1056	0.8943
−1.74	0.0878	0.0409	0.9591	−1.24	0.1849	0.1075	0.8925
−1.73	0.0893	0.0418	0.9582	−1.23	0.1872	0.1094	0.8907
−1.72	0.0909	0.0427	0.9573	−1.22	0.1895	0.1112	0.8888
−1.71	0.0925	0.0436	0.9564	−1.21	0.1919	0.1131	0.8869
−1.70	0.0940	0.0446	0.9554	−1.20	0.1942	0.1151	0.8849
−1.69	0.0957	0.0455	0.9545	−1.19	0.1965	0.1170	0.8830
−1.68	0.0973	0.0465	0.9535	−1.18	0.1989	0.1190	0.8810
−1.67	0.0989	0.0475	0.9525	−1.17	0.2012	0.1210	0.8790
−1.66	0.1006	0.0485	0.9515	−1.16	0.2036	0.1230	0.8770
−1.65	0.1023	0.0495	0.9505	−1.15	0.2059	0.1251	0.8749
−1.64	0.1040	0.0505	0.9495	−1.14	0.2083	0.1271	0.8729
−1.63	0.1057	0.0515	0.9485	−1.13	0.2107	0.1292	0.8708
−1.62	0.1074	0.0526	0.9474	−1.12	0.2131	0.1314	0.8686
−1.61	0.1091	0.0537	0.9463	−1.11	0.2155	0.1335	0.8665
−1.60	0.1109	0.0548	0.9452	−1.10	0.2178	0.1357	0.8643
−1.59	0.1127	0.0559	0.9441	−1.09	0.2203	0.1379	0.8621
−1.58	0.1145	0.0570	0.9429	−1.08	0.2226	0.1401	0.8599
−1.57	0.1163	0.0582	0.9418	−1.07	0.2251	0.1423	0.8577
−1.56	0.1182	0.0594	0.9406	−1.06	0.2275	0.1446	0.8554
−1.55	0.1200	0.0606	0.9394	−1.05	0.2299	0.1469	0.8531
−1.54	0.1219	0.0618	0.9382	−1.04	0.2323	0.1492	0.8508
−1.53	0.1238	0.0630	0.9370	−1.03	0.2347	0.1515	0.8485
−1.52	0.1257	0.0643	0.9357	−1.02	0.2371	0.1539	0.8461
−1.51	0.1276	0.0655	0.9345	−1.01	0.2396	0.1563	0.8438
−1.50	0.1295	0.0668	0.9332	−1.00	0.2420	0.1587	0.8413

Normal distribution

z	$f(z)$	$F(z)$	$1 - F(z)$	z	$f(z)$	$F(z)$	$1 - F(z)$
−1.00	0.2420	0.1587	0.8413	−0.50	0.3521	0.3085	0.6915
−0.99	0.2444	0.1611	0.8389	−0.49	0.3538	0.3121	0.6879
−0.98	0.2468	0.1635	0.8365	−0.48	0.3555	0.3156	0.6844
−0.97	0.2492	0.1660	0.8340	−0.47	0.3572	0.3192	0.6808
−0.96	0.2516	0.1685	0.8315	−0.46	0.3589	0.3228	0.6772
−0.95	0.2541	0.1711	0.8289	−0.45	0.3605	0.3264	0.6736
−0.94	0.2565	0.1736	0.8264	−0.44	0.3621	0.3300	0.6700
−0.93	0.2589	0.1762	0.8238	−0.43	0.3637	0.3336	0.6664
−0.92	0.2613	0.1788	0.8212	−0.42	0.3653	0.3372	0.6628
−0.91	0.2637	0.1814	0.8186	−0.41	0.3668	0.3409	0.6591
−0.90	0.2661	0.1841	0.8159	−0.40	0.3683	0.3446	0.6554
−0.89	0.2685	0.1867	0.8133	−0.39	0.3697	0.3483	0.6517
−0.88	0.2709	0.1894	0.8106	−0.38	0.3711	0.3520	0.6480
−0.87	0.2732	0.1921	0.8078	−0.37	0.3725	0.3557	0.6443
−0.86	0.2756	0.1949	0.8051	−0.36	0.3739	0.3594	0.6406
−0.85	0.2780	0.1977	0.8023	−0.35	0.3752	0.3632	0.6368
−0.84	0.2803	0.2004	0.7995	−0.34	0.3765	0.3669	0.6331
−0.83	0.2827	0.2033	0.7967	−0.33	0.3778	0.3707	0.6293
−0.82	0.2850	0.2061	0.7939	−0.32	0.3790	0.3745	0.6255
−0.81	0.2874	0.2090	0.7910	−0.31	0.3802	0.3783	0.6217
−0.80	0.2897	0.2119	0.7881	−0.30	0.3814	0.3821	0.6179
−0.79	0.2920	0.2148	0.7852	−0.29	0.3825	0.3859	0.6141
−0.78	0.2943	0.2177	0.7823	−0.28	0.3836	0.3897	0.6103
−0.77	0.2966	0.2207	0.7793	−0.27	0.3847	0.3936	0.6064
−0.76	0.2989	0.2236	0.7764	−0.26	0.3857	0.3974	0.6026
−0.75	0.3011	0.2266	0.7734	−0.25	0.3867	0.4013	0.5987
−0.74	0.3034	0.2296	0.7703	−0.24	0.3876	0.4052	0.5948
−0.73	0.3056	0.2327	0.7673	−0.23	0.3885	0.4091	0.5909
−0.72	0.3079	0.2358	0.7642	−0.22	0.3894	0.4129	0.5871
−0.71	0.3101	0.2389	0.7611	−0.21	0.3902	0.4168	0.5832
−0.70	0.3123	0.2420	0.7580	−0.20	0.3910	0.4207	0.5793
−0.69	0.3144	0.2451	0.7549	−0.19	0.3918	0.4247	0.5754
−0.68	0.3166	0.2482	0.7518	−0.18	0.3925	0.4286	0.5714
−0.67	0.3187	0.2514	0.7486	−0.17	0.3932	0.4325	0.5675
−0.66	0.3209	0.2546	0.7454	−0.16	0.3939	0.4364	0.5636
−0.65	0.3230	0.2579	0.7421	−0.15	0.3945	0.4404	0.5596
−0.64	0.3251	0.2611	0.7389	−0.14	0.3951	0.4443	0.5557
−0.63	0.3271	0.2643	0.7357	−0.13	0.3956	0.4483	0.5517
−0.62	0.3292	0.2676	0.7324	−0.12	0.3961	0.4522	0.5478
−0.61	0.3312	0.2709	0.7291	−0.11	0.3965	0.4562	0.5438
−0.60	0.3332	0.2742	0.7258	−0.10	0.3970	0.4602	0.5398
−0.59	0.3352	0.2776	0.7224	−0.09	0.3973	0.4641	0.5359
−0.58	0.3372	0.2810	0.7190	−0.08	0.3977	0.4681	0.5319
−0.57	0.3391	0.2843	0.7157	−0.07	0.3980	0.4721	0.5279
−0.56	0.3411	0.2877	0.7123	−0.06	0.3982	0.4761	0.5239
−0.55	0.3429	0.2912	0.7088	−0.05	0.3984	0.4801	0.5199
−0.54	0.3448	0.2946	0.7054	−0.04	0.3986	0.4840	0.5160
−0.53	0.3467	0.2981	0.7019	−0.03	0.3988	0.4880	0.5120
−0.52	0.3485	0.3015	0.6985	−0.02	0.3989	0.4920	0.5080
−0.51	0.3503	0.3050	0.6950	−0.01	0.3989	0.4960	0.5040
−0.50	0.3521	0.3085	0.6915	0.00	0.3989	0.5000	0.5000

Normal distribution

z	$f(z)$	$F(z)$	$1 - F(z)$	z	$f(z)$	$F(z)$	$1 - F(z)$
0.00	0.3989	0.5000	0.5000	0.50	0.3521	0.6915	0.3085
0.01	0.3989	0.5040	0.4960	0.51	0.3503	0.6950	0.3050
0.02	0.3989	0.5080	0.4920	0.52	0.3485	0.6985	0.3015
0.03	0.3988	0.5120	0.4880	0.53	0.3467	0.7019	0.2981
0.04	0.3986	0.5160	0.4840	0.54	0.3448	0.7054	0.2946
0.05	0.3984	0.5199	0.4801	0.55	0.3429	0.7088	0.2912
0.06	0.3982	0.5239	0.4761	0.56	0.3411	0.7123	0.2877
0.07	0.3980	0.5279	0.4721	0.57	0.3391	0.7157	0.2843
0.08	0.3977	0.5319	0.4681	0.58	0.3372	0.7190	0.2810
0.09	0.3973	0.5359	0.4641	0.59	0.3352	0.7224	0.2776
0.10	0.3970	0.5398	0.4602	0.60	0.3332	0.7258	0.2742
0.11	0.3965	0.5438	0.4562	0.61	0.3312	0.7291	0.2709
0.12	0.3961	0.5478	0.4522	0.62	0.3292	0.7324	0.2676
0.13	0.3956	0.5517	0.4483	0.63	0.3271	0.7357	0.2643
0.14	0.3951	0.5557	0.4443	0.64	0.3251	0.7389	0.2611
0.15	0.3945	0.5596	0.4404	0.65	0.3230	0.7421	0.2579
0.16	0.3939	0.5636	0.4364	0.66	0.3209	0.7454	0.2546
0.17	0.3932	0.5675	0.4325	0.67	0.3187	0.7486	0.2514
0.18	0.3925	0.5714	0.4286	0.68	0.3166	0.7518	0.2482
0.19	0.3918	0.5754	0.4247	0.69	0.3144	0.7549	0.2451
0.20	0.3910	0.5793	0.4207	0.70	0.3123	0.7580	0.2420
0.21	0.3902	0.5832	0.4168	0.71	0.3101	0.7611	0.2389
0.22	0.3894	0.5871	0.4129	0.72	0.3079	0.7642	0.2358
0.23	0.3885	0.5909	0.4091	0.73	0.3056	0.7673	0.2327
0.24	0.3876	0.5948	0.4052	0.74	0.3034	0.7703	0.2296
0.25	0.3867	0.5987	0.4013	0.75	0.3011	0.7734	0.2266
0.26	0.3857	0.6026	0.3974	0.76	0.2989	0.7764	0.2236
0.27	0.3847	0.6064	0.3936	0.77	0.2966	0.7793	0.2207
0.28	0.3836	0.6103	0.3897	0.78	0.2943	0.7823	0.2177
0.29	0.3825	0.6141	0.3859	0.79	0.2920	0.7852	0.2148
0.30	0.3814	0.6179	0.3821	0.80	0.2897	0.7881	0.2119
0.31	0.3802	0.6217	0.3783	0.81	0.2874	0.7910	0.2090
0.32	0.3790	0.6255	0.3745	0.82	0.2850	0.7939	0.2061
0.33	0.3778	0.6293	0.3707	0.83	0.2827	0.7967	0.2033
0.34	0.3765	0.6331	0.3669	0.84	0.2803	0.7995	0.2004
0.35	0.3752	0.6368	0.3632	0.85	0.2780	0.8023	0.1977
0.36	0.3739	0.6406	0.3594	0.86	0.2756	0.8051	0.1949
0.37	0.3725	0.6443	0.3557	0.87	0.2732	0.8078	0.1921
0.38	0.3711	0.6480	0.3520	0.88	0.2709	0.8106	0.1894
0.39	0.3697	0.6517	0.3483	0.89	0.2685	0.8133	0.1867
0.40	0.3683	0.6554	0.3446	0.90	0.2661	0.8159	0.1841
0.41	0.3668	0.6591	0.3409	0.91	0.2637	0.8186	0.1814
0.42	0.3653	0.6628	0.3372	0.92	0.2613	0.8212	0.1788
0.43	0.3637	0.6664	0.3336	0.93	0.2589	0.8238	0.1762
0.44	0.3621	0.6700	0.3300	0.94	0.2565	0.8264	0.1736
0.45	0.3605	0.6736	0.3264	0.95	0.2541	0.8289	0.1711
0.46	0.3589	0.6772	0.3228	0.96	0.2516	0.8315	0.1685
0.47	0.3572	0.6808	0.3192	0.97	0.2492	0.8340	0.1660
0.48	0.3555	0.6844	0.3156	0.98	0.2468	0.8365	0.1635
0.49	0.3538	0.6879	0.3121	0.99	0.2444	0.8389	0.1611
0.50	0.3521	0.6915	0.3085	1.00	0.2420	0.8413	0.1587

Normal distribution

z	$f(z)$	$F(z)$	$1-F(z)$	z	$f(z)$	$F(z)$	$1-F(z)$
1.00	0.2420	0.8413	0.1587	1.50	0.1295	0.9332	0.0668
1.01	0.2396	0.8438	0.1563	1.51	0.1276	0.9345	0.0655
1.02	0.2371	0.8461	0.1539	1.52	0.1257	0.9357	0.0643
1.03	0.2347	0.8485	0.1515	1.53	0.1238	0.9370	0.0630
1.04	0.2323	0.8508	0.1492	1.54	0.1219	0.9382	0.0618
1.05	0.2299	0.8531	0.1469	1.55	0.1200	0.9394	0.0606
1.06	0.2275	0.8554	0.1446	1.56	0.1182	0.9406	0.0594
1.07	0.2251	0.8577	0.1423	1.57	0.1163	0.9418	0.0582
1.08	0.2226	0.8599	0.1401	1.58	0.1145	0.9429	0.0570
1.09	0.2203	0.8621	0.1379	1.59	0.1127	0.9441	0.0559
1.10	0.2178	0.8643	0.1357	1.60	0.1109	0.9452	0.0548
1.11	0.2155	0.8665	0.1335	1.61	0.1091	0.9463	0.0537
1.12	0.2131	0.8686	0.1314	1.62	0.1074	0.9474	0.0526
1.13	0.2107	0.8708	0.1292	1.63	0.1057	0.9485	0.0515
1.14	0.2083	0.8729	0.1271	1.64	0.1040	0.9495	0.0505
1.15	0.2059	0.8749	0.1251	1.65	0.1023	0.9505	0.0495
1.16	0.2036	0.8770	0.1230	1.66	0.1006	0.9515	0.0485
1.17	0.2012	0.8790	0.1210	1.67	0.0989	0.9525	0.0475
1.18	0.1989	0.8810	0.1190	1.68	0.0973	0.9535	0.0465
1.19	0.1965	0.8830	0.1170	1.69	0.0957	0.9545	0.0455
1.20	0.1942	0.8849	0.1151	1.70	0.0940	0.9554	0.0446
1.21	0.1919	0.8869	0.1131	1.71	0.0925	0.9564	0.0436
1.22	0.1895	0.8888	0.1112	1.72	0.0909	0.9573	0.0427
1.23	0.1872	0.8907	0.1094	1.73	0.0893	0.9582	0.0418
1.24	0.1849	0.8925	0.1075	1.74	0.0878	0.9591	0.0409
1.25	0.1827	0.8943	0.1056	1.75	0.0863	0.9599	0.0401
1.26	0.1804	0.8962	0.1038	1.76	0.0848	0.9608	0.0392
1.27	0.1781	0.8980	0.1020	1.77	0.0833	0.9616	0.0384
1.28	0.1759	0.8997	0.1003	1.78	0.0818	0.9625	0.0375
1.29	0.1736	0.9015	0.0985	1.79	0.0804	0.9633	0.0367
1.30	0.1714	0.9032	0.0968	1.80	0.0790	0.9641	0.0359
1.31	0.1691	0.9049	0.0951	1.81	0.0775	0.9648	0.0352
1.32	0.1669	0.9066	0.0934	1.82	0.0761	0.9656	0.0344
1.33	0.1647	0.9082	0.0918	1.83	0.0748	0.9664	0.0336
1.34	0.1626	0.9099	0.0901	1.84	0.0734	0.9671	0.0329
1.35	0.1604	0.9115	0.0885	1.85	0.0721	0.9678	0.0322
1.36	0.1582	0.9131	0.0869	1.86	0.0707	0.9686	0.0314
1.37	0.1561	0.9147	0.0853	1.87	0.0694	0.9693	0.0307
1.38	0.1540	0.9162	0.0838	1.88	0.0681	0.9699	0.0301
1.39	0.1518	0.9177	0.0823	1.89	0.0669	0.9706	0.0294
1.40	0.1497	0.9192	0.0808	1.90	0.0656	0.9713	0.0287
1.41	0.1476	0.9207	0.0793	1.91	0.0644	0.9719	0.0281
1.42	0.1456	0.9222	0.0778	1.92	0.0632	0.9726	0.0274
1.43	0.1435	0.9236	0.0764	1.93	0.0619	0.9732	0.0268
1.44	0.1415	0.9251	0.0749	1.94	0.0608	0.9738	0.0262
1.45	0.1394	0.9265	0.0735	1.95	0.0596	0.9744	0.0256
1.46	0.1374	0.9278	0.0722	1.96	0.0584	0.9750	0.0250
1.47	0.1354	0.9292	0.0708	1.97	0.0573	0.9756	0.0244
1.48	0.1334	0.9306	0.0694	1.98	0.0562	0.9761	0.0238
1.49	0.1315	0.9319	0.0681	1.99	0.0551	0.9767	0.0233
1.50	0.1295	0.9332	0.0668	2.00	0.0540	0.9772	0.0227

Normal distribution

z	$f(z)$	$F(z)$	$1-F(z)$	z	$f(z)$	$F(z)$	$1-F(z)$
2.00	0.0540	0.9772	0.0227	2.50	0.0175	0.9938	0.0062
2.01	0.0529	0.9778	0.0222	2.51	0.0171	0.9940	0.0060
2.02	0.0519	0.9783	0.0217	2.52	0.0167	0.9941	0.0059
2.03	0.0508	0.9788	0.0212	2.53	0.0163	0.9943	0.0057
2.04	0.0498	0.9793	0.0207	2.54	0.0158	0.9945	0.0055
2.05	0.0488	0.9798	0.0202	2.55	0.0155	0.9946	0.0054
2.06	0.0478	0.9803	0.0197	2.56	0.0151	0.9948	0.0052
2.07	0.0468	0.9808	0.0192	2.57	0.0147	0.9949	0.0051
2.08	0.0459	0.9812	0.0188	2.58	0.0143	0.9951	0.0049
2.09	0.0449	0.9817	0.0183	2.59	0.0139	0.9952	0.0048
2.10	0.0440	0.9821	0.0179	2.60	0.0136	0.9953	0.0047
2.11	0.0431	0.9826	0.0174	2.61	0.0132	0.9955	0.0045
2.12	0.0422	0.9830	0.0170	2.62	0.0129	0.9956	0.0044
2.13	0.0413	0.9834	0.0166	2.63	0.0126	0.9957	0.0043
2.14	0.0404	0.9838	0.0162	2.64	0.0122	0.9959	0.0042
2.15	0.0396	0.9842	0.0158	2.65	0.0119	0.9960	0.0040
2.16	0.0387	0.9846	0.0154	2.66	0.0116	0.9961	0.0039
2.17	0.0379	0.9850	0.0150	2.67	0.0113	0.9962	0.0038
2.18	0.0371	0.9854	0.0146	2.68	0.0110	0.9963	0.0037
2.19	0.0363	0.9857	0.0143	2.69	0.0107	0.9964	0.0036
2.20	0.0355	0.9861	0.0139	2.70	0.0104	0.9965	0.0035
2.21	0.0347	0.9865	0.0135	2.71	0.0101	0.9966	0.0034
2.22	0.0339	0.9868	0.0132	2.72	0.0099	0.9967	0.0033
2.23	0.0332	0.9871	0.0129	2.73	0.0096	0.9968	0.0032
2.24	0.0325	0.9875	0.0126	2.74	0.0094	0.9969	0.0031
2.25	0.0317	0.9878	0.0122	2.75	0.0091	0.9970	0.0030
2.26	0.0310	0.9881	0.0119	2.76	0.0089	0.9971	0.0029
2.27	0.0303	0.9884	0.0116	2.77	0.0086	0.9972	0.0028
2.28	0.0296	0.9887	0.0113	2.78	0.0084	0.9973	0.0027
2.29	0.0290	0.9890	0.0110	2.79	0.0081	0.9974	0.0026
2.30	0.0283	0.9893	0.0107	2.80	0.0079	0.9974	0.0026
2.31	0.0277	0.9896	0.0104	2.81	0.0077	0.9975	0.0025
2.32	0.0271	0.9898	0.0102	2.82	0.0075	0.9976	0.0024
2.33	0.0264	0.9901	0.0099	2.83	0.0073	0.9977	0.0023
2.34	0.0258	0.9904	0.0096	2.84	0.0071	0.9977	0.0023
2.35	0.0252	0.9906	0.0094	2.85	0.0069	0.9978	0.0022
2.36	0.0246	0.9909	0.0091	2.86	0.0067	0.9979	0.0021
2.37	0.0241	0.9911	0.0089	2.87	0.0065	0.9980	0.0021
2.38	0.0235	0.9913	0.0087	2.88	0.0063	0.9980	0.0020
2.39	0.0229	0.9916	0.0084	2.89	0.0061	0.9981	0.0019
2.40	0.0224	0.9918	0.0082	2.90	0.0060	0.9981	0.0019
2.41	0.0219	0.9920	0.0080	2.91	0.0058	0.9982	0.0018
2.42	0.0213	0.9922	0.0078	2.92	0.0056	0.9982	0.0018
2.43	0.0208	0.9925	0.0076	2.93	0.0054	0.9983	0.0017
2.44	0.0203	0.9927	0.0073	2.94	0.0053	0.9984	0.0016
2.45	0.0198	0.9929	0.0071	2.95	0.0051	0.9984	0.0016
2.46	0.0194	0.9930	0.0069	2.96	0.0050	0.9985	0.0015
2.47	0.0189	0.9932	0.0068	2.97	0.0049	0.9985	0.0015
2.48	0.0184	0.9934	0.0066	2.98	0.0047	0.9986	0.0014
2.49	0.0180	0.9936	0.0064	2.99	0.0046	0.9986	0.0014
2.50	0.0175	0.9938	0.0062	3.00	0.0044	0.9987	0.0014

Normal distribution

z	$f(z)$	$F(z)$	$1 - F(z)$	z	$f(z)$	$F(z)$	$1 - F(z)$
3.00	0.0044	0.9987	0.0014	3.50	0.0009	0.9998	0.0002
3.01	0.0043	0.9987	0.0013	3.51	0.0008	0.9998	0.0002
3.02	0.0042	0.9987	0.0013	3.52	0.0008	0.9998	0.0002
3.03	0.0040	0.9988	0.0012	3.53	0.0008	0.9998	0.0002
3.04	0.0039	0.9988	0.0012	3.54	0.0008	0.9998	0.0002
3.05	0.0038	0.9989	0.0011	3.55	0.0007	0.9998	0.0002
3.06	0.0037	0.9989	0.0011	3.56	0.0007	0.9998	0.0002
3.07	0.0036	0.9989	0.0011	3.57	0.0007	0.9998	0.0002
3.08	0.0035	0.9990	0.0010	3.58	0.0007	0.9998	0.0002
3.09	0.0034	0.9990	0.0010	3.59	0.0006	0.9998	0.0002
3.10	0.0033	0.9990	0.0010	3.60	0.0006	0.9998	0.0002
3.11	0.0032	0.9991	0.0009	3.61	0.0006	0.9999	0.0001
3.12	0.0031	0.9991	0.0009	3.62	0.0006	0.9999	0.0001
3.13	0.0030	0.9991	0.0009	3.63	0.0006	0.9999	0.0001
3.14	0.0029	0.9992	0.0008	3.64	0.0005	0.9999	0.0001
3.15	0.0028	0.9992	0.0008	3.65	0.0005	0.9999	0.0001
3.16	0.0027	0.9992	0.0008	3.66	0.0005	0.9999	0.0001
3.17	0.0026	0.9992	0.0008	3.67	0.0005	0.9999	0.0001
3.18	0.0025	0.9993	0.0007	3.68	0.0005	0.9999	0.0001
3.19	0.0025	0.9993	0.0007	3.69	0.0004	0.9999	0.0001
3.20	0.0024	0.9993	0.0007	3.70	0.0004	0.9999	0.0001
3.21	0.0023	0.9993	0.0007	3.71	0.0004	0.9999	0.0001
3.22	0.0022	0.9994	0.0006	3.72	0.0004	0.9999	0.0001
3.23	0.0022	0.9994	0.0006	3.73	0.0004	0.9999	0.0001
3.24	0.0021	0.9994	0.0006	3.74	0.0004	0.9999	0.0001
3.25	0.0020	0.9994	0.0006	3.75	0.0003	0.9999	0.0001
3.26	0.0020	0.9994	0.0006	3.76	0.0003	0.9999	0.0001
3.27	0.0019	0.9995	0.0005	3.77	0.0003	0.9999	0.0001
3.28	0.0018	0.9995	0.0005	3.78	0.0003	0.9999	0.0001
3.29	0.0018	0.9995	0.0005	3.79	0.0003	0.9999	0.0001
3.30	0.0017	0.9995	0.0005	3.80	0.0003	0.9999	0.0001
3.31	0.0017	0.9995	0.0005	3.81	0.0003	0.9999	0.0001
3.32	0.0016	0.9996	0.0004	3.82	0.0003	0.9999	0.0001
3.33	0.0016	0.9996	0.0004	3.83	0.0003	0.9999	0.0001
3.34	0.0015	0.9996	0.0004	3.84	0.0003	0.9999	0.0001
3.35	0.0015	0.9996	0.0004	3.85	0.0002	0.9999	0.0001
3.36	0.0014	0.9996	0.0004	3.86	0.0002	0.9999	0.0001
3.37	0.0014	0.9996	0.0004	3.87	0.0002	1.0000	0.0001
3.38	0.0013	0.9996	0.0004	3.88	0.0002	1.0000	0.0001
3.39	0.0013	0.9997	0.0003	3.89	0.0002	1.0000	0.0001
3.40	0.0012	0.9997	0.0003	3.90	0.0002	1.0000	0.0001
3.41	0.0012	0.9997	0.0003	3.91	0.0002	1.0000	0.0001
3.42	0.0011	0.9997	0.0003	3.92	0.0002	1.0000	0.0000
3.43	0.0011	0.9997	0.0003	3.93	0.0002	1.0000	0.0000
3.44	0.0011	0.9997	0.0003	3.94	0.0002	1.0000	0.0000
3.45	0.0010	0.9997	0.0003	3.95	0.0002	1.0000	0.0000
3.46	0.0010	0.9997	0.0003	3.96	0.0002	1.0000	0.0000
3.47	0.0010	0.9997	0.0003	3.97	0.0001	1.0000	0.0000
3.48	0.0009	0.9998	0.0003	3.98	0.0001	1.0000	0.0000
3.49	0.0009	0.9998	0.0002	3.99	0.0001	1.0000	0.0000
3.50	0.0009	0.9998	0.0002	4.00	0.0001	1.0000	0.0000

7.2 CRITICAL VALUES

Table 7.1 lists common critical values for a standard normal random variable, z_α, defined by (see Figure 7.2):

$$\text{Prob}\,[Z \geq z_\alpha] = \alpha. \tag{7.3}$$

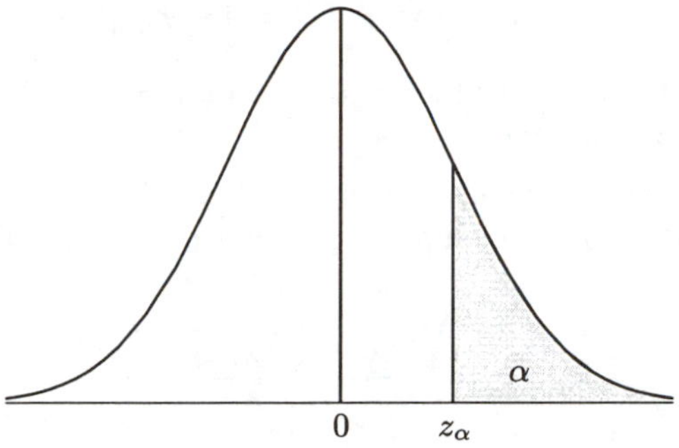

Figure 7.2: Critical values for a normal random variable.

α	z_α	α	z_α	α	z_α
.10	1.2816	.00009	3.7455	.000001	4.75
.05	1.6449	.00008	3.7750	.0000001	5.20
.025	1.9600	.00007	3.8082	.00000001	5.61
.01	2.3263	.00006	3.8461	.000000001	6.00
.005	2.5758	.00005	3.8906	.0000000001	6.36
.0025	2.8070	.00004	3.9444		
.001	3.0902	.00003	4.0128		
.0005	3.2905	.00002	4.1075		
.0001	3.7190	.00001	4.2649		

Table 7.1: Common critical values

7.3 TOLERANCE FACTORS FOR NORMAL DISTRIBUTIONS

Suppose $X_1, X_2, \ldots, X_n$ is a random sample of size n from a normal population with mean μ and standard deviation σ. Using the summary statistics $\bar{x}$ and s, a tolerance interval $[L, U]$ may be constructed to capture $100P\%$ of the population with probability $1 - \alpha$. The following procedures may be used.

(1) Two-sided tolerance interval: A $100(1 - \alpha)\%$ tolerance interval that captures $100P\%$ of the population has as endpoints

$$[L, U] = \bar{x} \pm K_{\alpha,n,P} \cdot s \tag{7.4}$$

(2) One-sided tolerance interval, upper tailed: A $100(1 - \alpha)\%$ tolerance interval bounded below has

$$L = \bar{x} - k_{\alpha,n,P} \cdot s \qquad U = \infty \qquad\qquad (7.5)$$

(3) One-sided tolerance interval, lower tailed: A $100(1 - \alpha)\%$ tolerance interval bounded above has

$$L = -\infty \qquad U = \bar{x} + k_{\alpha,n,P} \cdot s \qquad\qquad (7.6)$$

where $K_{\alpha,n,P}$ is the tolerance factor given in section 7.3.1 and $k_{\alpha,n,P}$ is computed using the formula below.

Values of $K_{\alpha,n,P}$ are given in section 7.3.1 for $P = 0.75,\ 0.90,\ 0.95,\ 0.99,\ 0.999$, $\alpha = 0.75, 0.90, 0.95, 0.99$, and various values of n. The value of $k_{\alpha,n,P}$ is given by

$$k_{\alpha,n,P} = \frac{z_{1-P} + \sqrt{(z_{1-P})^2 - ab}}{a}$$

$$a = 1 - \frac{(z_\alpha)^2}{2(n-1)} \qquad\qquad (7.7)$$

$$b = z_{1-P}^2 - \frac{z_\alpha^2}{n}$$

where z_{1-P} and z_α are critical values for a standard normal random variable (see page 175).

Example 7.43:   Suppose a sample of size $n = 30$ from a normal distribution has $\bar{x} = 10.02$ and $s = 0.13$. Find tolerance intervals with a confidence level 95% ($\alpha = .05$) and $P = .90$.

Solution:

(S1) Two-sided interval:

  1. From the tables in section 7.3.1 we find $K_{.05,30,.90} = 2.413$.
  2. The interval is $\bar{x} \pm K \cdot s = 10.02 \pm 0.31$; or $I = [9.71, 10.33]$.
  3. We conclude: in each sample of size 30, at least 90% of the normal population being sampled will be in the interval I, with probability 95%.

(S2) One-sided intervals:

  1. The critical values used in equation (7.7) are $z_{1-P} = z_{.10} = 1.282$ and $z_\alpha = z_{.05} = 1.645$. Using this equation: $a = 1 - \frac{(1.645)^2}{2(29)} = 0.9533$, $b = (1.282)^2 - \frac{(1.645)^2}{30} = 1.553$, and $k_{.05,30,.90} = 1.768$
  2. The lower bound is $L = \bar{x} - k \cdot s = 9.79$.
  3. The upper bound is $U = \bar{x} + k \cdot s = 10.25$.
  4. We conclude:
      (a) In each sample of size 30, at least 90% of the normal population being sampled will be greater than L, with probability 95%.
      (b) In each sample of size 30, at least 90% of the normal population being sampled will be smaller than U, with probability 95%.

7.3.1 Tables of tolerance intervals for normal distributions

Tolerance factors for normal distributions

			$P = .90$						
n	$\alpha = .10$	.05	.01	.001	n	$\alpha = .10$	.05	.01	.001
2	15.978	18.800	24.167	30.227	20	2.152	2.564	3.368	4.300
3	5.847	6.919	8.974	11.309	25	2.077	2.474	3.251	4.151
4	4.166	4.943	6.440	8.149	30	2.025	2.413	3.170	4.049
5	3.494	4.152	5.423	6.879	40	1.959	2.334	3.066	3.917
6	3.131	3.723	4.870	6.188	50	1.916	2.284	3.001	3.833
7	2.902	3.452	4.521	5.750	75	1.856	2.211	2.906	3.712
8	2.743	3.264	4.278	5.446	100	1.822	2.172	2.854	3.646
9	2.626	3.125	4.098	5.220	500	1.717	2.046	2.689	3.434
10	2.535	3.018	3.959	5.046	1000	1.695	2.019	2.654	3.390
15	2.278	2.713	3.562	4.545	∞	1.645	1.960	2.576	3.291

Tolerance factors for normal distributions

			$P = .95$						
n	$\alpha = .10$	.05	.01	.001	n	$\alpha = .10$	.05	.01	.001
2	32.019	37.674	48.430	60.573	20	2.310	2.752	3.615	4.614
3	8.380	9.916	12.861	16.208	25	2.208	2.631	3.457	4.413
4	5.369	6.370	8.299	10.502	30	2.140	2.549	3.350	4.278
5	4.275	5.079	6.634	8.415	40	2.052	2.445	3.213	4.104
6	3.712	4.414	5.775	7.337	50	1.996	2.379	3.126	3.993
7	3.369	4.007	5.248	6.676	75	1.917	2.285	3.002	3.835
8	3.136	3.732	4.891	6.226	100	1.874	2.233	2.934	3.748
9	2.967	3.532	4.631	5.899	500	1.737	2.070	2.721	3.475
10	2.839	3.379	4.433	5.649	1000	1.709	2.036	2.676	3.418
15	2.480	2.954	3.878	4.949	∞	1.645	1.960	2.576	3.291

Tolerance factors for normal distributions

			$P = .99$						
n	$\alpha = .10$	.05	.01	.001	n	$\alpha = .10$	.05	.01	.001
2	160.193	188.491	242.300	303.054	20	2.659	3.168	4.161	5.312
3	18.930	22.401	29.055	36.616	25	2.494	2.972	3.904	4.985
4	9.398	11.150	14.527	18.383	30	2.385	2.841	3.733	4.768
5	6.612	7.855	10.260	13.015	40	2.247	2.677	3.518	4.493
6	5.337	6.345	8.301	10.548	50	2.162	2.576	3.385	4.323
7	4.613	5.488	7.187	9.142	75	2.042	2.433	3.197	4.084
8	4.147	4.936	6.468	8.234	100	1.977	2.355	3.096	3.954
9	3.822	4.550	5.966	7.600	500	1.777	2.117	2.783	3.555
10	3.582	4.265	5.594	7.129	1000	1.736	2.068	2.718	3.472
15	2.945	3.507	4.605	5.876	∞	1.645	1.960	2.576	3.291

7.4   OPERATING CHARACTERISTIC CURVES

7.4.1   One-sample Z test

Consider a one-sample hypothesis test on a population mean of a normal distribution with known standard deviation σ (see section 10.2). The general form of the hypothesis test (for each possible alternative hypothesis) is:

$H_0 : \mu = \mu_0$

$H_a : \mu > \mu_0, \quad \mu < \mu_0, \quad \mu \neq \mu_0$

TS: $Z = \dfrac{\bar{X} - \mu_0}{\sigma/\sqrt{n}}$

RR: $Z \geq z_\alpha, \quad Z \leq -z_\alpha, \quad |Z| \geq z_{\alpha/2}$

Let α be the probability of a Type I error, β the probability of a Type II error, and μ_a an alternative mean. For $\Delta = |\mu_a - \mu_0|/\sigma$ the operating characteristic curve returns the probability of not rejecting the null hypothesis given $\mu = \mu_a$. The curves may be used to determine the appropriate sample size for given values of α, β, and Δ.

7.4.2   Two-sample Z test

Consider a two-sample hypothesis test for comparing population means from normal distributions with known standard deviations σ_1 and σ_2 (see section 10.3). The general form of the hypothesis test for testing the equality of means (for each possible alternative hypothesis) is:

$H_0 : \mu_1 - \mu_2 = 0$

$H_a : \mu_1 - \mu_2 > 0, \quad \mu_1 - \mu_2 < 0, \quad \mu_1 - \mu_2 \neq 0$

TS: $Z = \dfrac{\bar{X}_1 - \bar{X}_2}{\sqrt{\dfrac{\sigma_1^2}{n_1} + \dfrac{\sigma_2^2}{n_2}}}$

RR: $Z \geq z_\alpha, \quad Z \leq -z_\alpha, \quad |Z| \geq z_{\alpha/2}$

Let α be the probability of a Type I error and β the probability of a Type II error. For given values of α, and $\Delta = \dfrac{|\mu_1 - \mu_2|}{\sqrt{\sigma_1^2 + \sigma_2^2}}$ the operating characteristic curve returns the probability of not rejecting the null hypothesis. The curves may be used to determine an appropriate sample size ($n = n_1 = n_2$) for desired levels of α, β, and Δ.

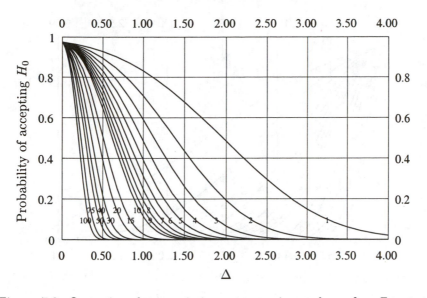

Figure 7.3: Operating characteristic curves, various values of n, Z test, two-sided alternative, $\alpha = .05$.

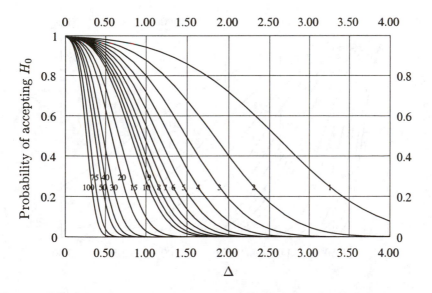

Figure 7.4: Operating characteristic curves, various values of n, Z test, two-sided alternative, $\alpha = .01$.

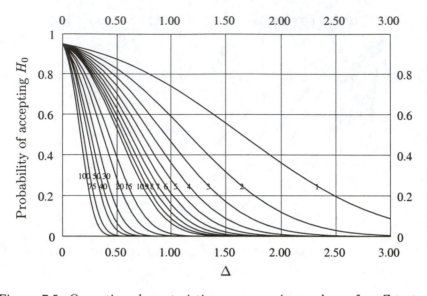

Figure 7.5: Operating characteristic curves, various values of n, Z test, one-sided alternative, $\alpha = .05$.

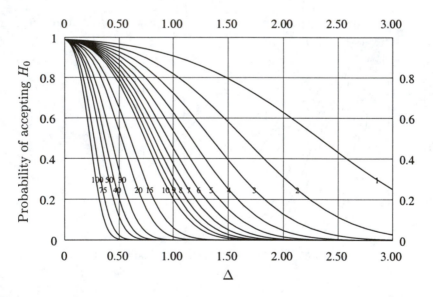

Figure 7.6: Operating characteristic curves, various values of n, Z test, one-sided alternative, $\alpha = .01$.

7.5 MULTIVARIATE NORMAL DISTRIBUTION

Let each $\{X_i\}$ (for $i = 1, \ldots, n$) be a normal random variable with mean μ_i and variance σ_{ii}. If the covariance of X_i and X_j is σ_{ij}, then the joint probability density of the $\{X_i\}$ is:

$$f(\mathbf{x}) = \frac{1}{(2\pi)^{n/2}\sqrt{\det(\mathbf{\Sigma})}} \exp\left[-\frac{1}{2}(\mathbf{x} - \boldsymbol{\mu})^{\mathrm{T}}\mathbf{\Sigma}^{-1}(\mathbf{x} - \boldsymbol{\mu})\right] \tag{7.8}$$

where

(a) $\mathbf{x} = \begin{bmatrix} x_1 & x_2 & \ldots & x_n \end{bmatrix}^{\mathrm{T}}$

(b) $\boldsymbol{\mu} = \begin{bmatrix} \mu_1 & \mu_2 & \ldots & \mu_n \end{bmatrix}^{\mathrm{T}}$

(c) $\mathbf{\Sigma}$ is an $n \times n$ matrix with elements σ_{ij}

The corresponding characteristic function is

$$\phi(\mathbf{t}) = \exp\left[i\boldsymbol{\mu}^{\mathrm{T}}\mathbf{t} - \frac{1}{2}\mathbf{t}^{\mathrm{T}}\mathbf{\Sigma}\mathbf{t}\right] \tag{7.9}$$

The form of the characteristic function implies that all cumulants of higher order than 2 vanish (see Marcienkiewicz's theorem). Therefore, all moments of order higher than 2 may be expressed in terms of those of order 1 and 2. If $\boldsymbol{\mu} = \mathbf{0}$ then the odd moments vanish and the $(2n)^{\mathrm{th}}$ moment satisfies

$$\mathrm{E}\left[\underbrace{X_i X_j X_k X_l \cdots}_{2n \text{ terms}}\right] = \frac{(2n)!}{n!2^n}\{\sigma_{ij}\sigma_{kl}\cdots\}_{\mathrm{sym}} \tag{7.10}$$

where the subscript "sym" means the symmetrized form of the product of the σ's.

Example 7.44: For $n = 2$ we can compute fourth moments

$$\mathrm{E}\left[X_1 X_2 X_3 X_4\right] = \frac{4!}{2! \cdot 2^2}\left[\frac{1}{3}\left(\sigma_{12}\sigma_{34} + \sigma_{41}\sigma_{23} + \sigma_{13}\sigma_{24}\right)\right]$$

$$= \sigma_{12}\sigma_{34} + \sigma_{41}\sigma_{23} + \sigma_{13}\sigma_{24} \tag{7.11}$$

$$\mathrm{E}\left[X_1^4\right] = \frac{4!}{2! \cdot 2^2}\left[\sigma_{11}^2\right] = 3\sigma_{11}^2$$

See C. W. Gardiner *Handbook of Stochastic Methods*, Springer–Verlag, New York, 1985, pages 36–37.

7.6 DISTRIBUTION OF THE CORRELATION COEFFICIENT FOR A BIVARIATE NORMAL

The bivariate normal probability function is given by

$$f(x,y) = \frac{1}{2\pi\sigma_x\sigma_y\sqrt{1-\rho^2}} \exp\left(-\frac{1}{2(1-\rho^2)}\right. \tag{7.12}$$

$$\times \left. \left[\left(\frac{x-\mu_x}{\sigma_x}\right)^2 - 2\rho\left(\frac{x-\mu_x}{\sigma_x}\right)\left(\frac{y-\mu_y}{\sigma_y}\right) + \left(\frac{y-\mu_y}{\sigma_y}\right)^2\right]\right)$$

where μ_x = mean of x

μ_y = mean of y

σ_x = standard deviation of x

σ_y = standard deviation of y

ρ = correlation coefficient between x and y

Given a sample $\{(x_1, y_1), \ldots, (x_n, y_n)\}$ of size n, the sample correlation coefficient, an estimate of ρ, is

$$r = \frac{\sum\limits_{i=1}^{n}(x_i - \bar{x})(y_i - \bar{y})}{\sqrt{\left(\sum\limits_{i=1}^{n}(x_i - \bar{x})^2\right)\left(\sum\limits_{i=1}^{n}(y_i - \bar{y})^2\right)}} \tag{7.13}$$

where $\bar{x} = \left(\sum_{i=1}^{n} x_i\right)/n$ and $\bar{y} = \left(\sum_{i=1}^{n} y_i\right)/n$. For given n, the distribution of r is independent of $\{\mu_x, \mu_y, \sigma_x, \sigma_y\}$, but depends on ρ. For $-1 \leq \rho \leq 1$, the density function for r is $f_n(r; \rho)$:

$$f_n(r; \rho) = \frac{1}{\pi}(n-2)(1-r^2)^{(n-4)/2}\left(1-\rho^2\right)^{(n-1)/2}\int_0^\infty \frac{dz}{(\cosh z - \rho r)^{n-1}}$$

$$= \frac{1}{\pi}(n-2)(1-r^2)^{(n-4)/2}\left(1-\rho^2\right)^{(n-1)/2}\sqrt{\frac{\pi}{2}}\frac{\Gamma(n-1)}{\Gamma(n-1/2)}$$

$$\times (1-\rho r)^{-(n-3/2)} {}_2F_1\left(\frac{1}{2}, \frac{1}{2}; \frac{2n-1}{2}; \frac{\rho r + 1}{2}\right) \tag{7.14}$$

where $\Gamma(x)$ is the gamma function and $_2F_1$ is the hypergeometric function defined in Chapter 18 (see pages 515 and 520). The moments are given by

$$\mu_r = \rho - \frac{\rho(1 - \rho^2)}{(n+1)}$$

$$\sigma_r^2 = \frac{(1 - \rho^2)^2}{n+1}\left(1 + \frac{11\rho^2}{2(n+1)} + \cdots\right)$$

$$\gamma_1 = \frac{6\rho}{\sqrt{n+1}}\left(1 + \frac{77\rho^2 - 30}{12(n+1)} + \cdots\right)$$

$$\gamma_2 = \frac{6}{n+1}\left(12\rho^2 - 1\right) + \cdots$$

(7.15)

7.6.1 Normal approximation

If r is the sample correlation coefficient (defined in equation (7.13)), the random variable

$$Z = \tanh^{-1} r = \frac{1}{2}\ln\frac{1+r}{1-r} \tag{7.16}$$

is approximately normally distributed with parameters

$$\mu_Z = \frac{1}{2}\ln\left[\frac{1+\rho}{1-\rho}\right] = \tanh^{-1}\rho \quad \text{and} \quad \sigma_Z^2 = \frac{1}{n-3} \tag{7.17}$$

7.6.2 Zero correlation coefficient for bivariate normal

In the special case where $\rho = 0$, the density function of r becomes

$$f_n(r;0) = \frac{1}{\sqrt{\pi}}\frac{\Gamma((n-1)/2)}{\Gamma((n-2)/2)}\left(1 - r^2\right)^{(n-4)/2} \tag{7.18}$$

Under the transformation

$$r^2 = \frac{t^2}{t^2 + \nu} \tag{7.19}$$

$f_n(r;0)$, as given by equation (7.18), has a t-distribution with $\nu = n-1$ degrees of freedom. The following table gives percentage points of the distribution of the correlation coefficient when $\rho = 0$.

Percentage points of the correlation coefficient, when $\rho = 0$
Prob $[r \leq$ tabulated value$] = 1 - \alpha$

$\alpha =$	0.05	0.025	0.01	0.005	0.0025	0.0005
$2\alpha =$	0.1	0.05	0.02	0.01	0.005	0.001
$\nu = 1$	0.988	0.997	0.9^3507	0.9^3877	0.9^469	0.9^6
2	0.900	0.950	0.980	0.990	0.995	0.999
3	0.805	0.878	0.934	0.959	0.974	0.991
4	0.729	0.811	0.882	0.917	0.942	0.974
5	0.669	0.754	0.833	0.875	0.906	0.951
6	0.621	0.707	0.789	0.834	0.870	0.925
7	0.582	0.666	0.750	0.798	0.836	0.898
8	0.549	0.632	0.715	0.765	0.805	0.872
9	0.521	0.602	0.685	0.735	0.776	0.847
10	0.497	0.576	0.658	0.708	0.750	0.823
11	0.476	0.553	0.634	0.684	0.726	0.801
12	0.458	0.532	0.612	0.661	0.703	0.780
13	0.441	0.514	0.592	0.641	0.683	0.760
14	0.426	0.497	0.574	0.623	0.664	0.742
15	0.412	0.482	0.558	0.606	0.647	0.725
16	0.400	0.468	0.543	0.590	0.631	0.708
17	0.389	0.456	0.529	0.575	0.616	0.693
18	0.378	0.444	0.516	0.561	0.602	0.679
19	0.369	0.433	0.503	0.549	0.589	0.665
20	0.360	0.423	0.492	0.537	0.576	0.652
25	0.323	0.381	0.445	0.487	0.524	0.597
30	0.296	0.349	0.409	0.449	0.484	0.554
35	0.275	0.325	0.381	0.418	0.452	0.519
40	0.257	0.304	0.358	0.393	0.425	0.490
45	0.243	0.288	0.338	0.372	0.403	0.465
50	0.231	0.273	0.322	0.354	0.384	0.443
60	0.211	0.250	0.295	0.325	0.352	0.408
70	0.195	0.232	0.274	0.302	0.327	0.380
80	0.183	0.217	0.257	0.283	0.307	0.357
90	0.173	0.205	0.242	0.267	0.290	0.338
100	0.164	0.195	0.230	0.254	0.276	0.321

Use the α value for a single-tail test. For a two-tail test, use the 2α value. If r is computed from n paired observations, enter the table with $\nu = n - 2$. For partial correlations, enter the table with $\nu = n - 2 - k$, where k is the number of variables held constant.

7.7 CIRCULAR NORMAL PROBABILITIES

The joint probability density of two independent and normally distributed random variables X and Y (each of mean zero and variance σ^2) is

$$f(x,y) = \frac{1}{2\pi\sigma^2} \exp\left[-\frac{1}{2\sigma^2}\left(x^2 + y^2\right)\right] \qquad (7.20)$$

The following table gives the probability that a sample value of X and Y is obtained inside a circle C of radius R at a distance d from the origin:

$$F\left(\frac{d}{\sigma}, \frac{R}{\sigma}\right) = \frac{1}{2\pi\sigma^2} \iint_C \exp\left[-\frac{1}{2\sigma^2}\left(x^2 + y^2\right)\right] dx\, dy \qquad (7.21)$$

Circular normal probabilities

d/σ	R/σ								
	0.2	0.4	0.6	0.8	1.0	1.5	2.0	2.5	3.0
0.2	0.020	0.077	0.164	0.273	0.392	0.674	0.863	0.955	0.989
0.3	0.019	0.075	0.162	0.269	0.387	0.668	0.859	0.953	0.988
0.4	0.019	0.074	0.158	0.264	0.380	0.659	0.852	0.950	0.987
0.5	0.018	0.071	0.153	0.256	0.370	0.647	0.843	0.945	0.985
0.6	0.017	0.068	0.147	0.246	0.357	0.631	0.831	0.938	0.982
0.7	0.017	0.065	0.140	0.235	0.342	0.612	0.816	0.930	0.979
0.8	0.016	0.061	0.132	0.222	0.326	0.591	0.799	0.920	0.975
0.9	0.014	0.057	0.123	0.209	0.307	0.566	0.779	0.909	0.970
1.0	0.013	0.052	0.114	0.194	0.288	0.540	0.756	0.895	0.964
1.2	0.012	0.048	0.104	0.179	0.267	0.512	0.731	0.879	0.956
1.4	0.0097	0.038	0.085	0.148	0.225	0.451	0.674	0.841	0.937
1.6	0.0075	0.030	0.067	0.119	0.184	0.388	0.610	0.795	0.912
1.8	0.0056	0.022	0.051	0.092	0.145	0.325	0.541	0.739	0.878
2.0	0.0040	0.016	0.037	0.069	0.111	0.264	0.469	0.676	0.837
2.2	0.0027	0.011	0.026	0.050	0.082	0.209	0.396	0.606	0.786
2.4	0.0018	0.0075	0.018	0.034	0.059	0.161	0.327	0.532	0.726
2.6	0.0011	0.0048	0.012	0.023	0.040	0.120	0.262	0.456	0.659
2.8	0.0007	0.0030	0.0074	0.015	0.027	0.086	0.204	0.381	0.586
3.0	0.0004	0.0018	0.0045	0.0094	0.017	0.060	0.154	0.311	0.510
3.2	0.0002	0.0010	0.0027	0.0057	0.011	0.041	0.113	0.246	0.433
3.4	0.0001	0.0006	0.0015	0.0033	0.0065	0.027	0.080	0.189	0.358
3.6	0.0001	0.0003	0.0008	0.0018	0.0038	0.017	0.055	0.141	0.288
3.8		0.0002	0.0004	0.0010	0.0021	0.010	0.037	0.102	0.225
4.0		0.0001	0.0002	0.0005	0.0011	0.006	0.024	0.072	0.171
4.2				0.0001	0.0003	0.0019	0.0088	0.0320	0.090
4.4				0.0001	0.0001	0.0010	0.0051	0.0200	0.062
4.6					0.0001	0.0005	0.0029	0.0120	0.041
4.8						0.0003	0.0015	0.0073	0.027
5.0						0.0001	0.0008	0.0041	0.017
5.2						0.0001	0.0004	0.0023	0.0100
5.4							0.0002	0.0012	0.0058
5.6							0.0001	0.0006	0.0033
5.8								0.0003	0.0018

7.8   CIRCULAR ERROR PROBABILITIES

The joint probability density of two independent and normally distributed random variables X and Y, each of mean zero and having variances σ_x^2 and σ_y^2, is

$$f(x,y) = \frac{1}{2\pi\sigma_x\sigma_y} \exp\left[-\frac{1}{2}\left\{\left(\frac{x}{\sigma_x}\right)^2 + \left(\frac{y}{\sigma_y}\right)^2\right\}\right] \qquad (7.22)$$

The probability that a sample value of X and Y will lie within a circle with center at the origin and radius $K\sigma_x$ is given by

$$P(K,\sigma_x,\sigma_y) = \iint_R f(x,y)\, dx\, dy \qquad (7.23)$$

where R is the region $\sqrt{x^2 + y^2} < K\sigma_x$.

For various values of K and $c = \sigma_x/\sigma_y$ (for convenience we assume that $\sigma_y \le \sigma_x$) the following table gives the value of P.

Circular error probabilities

K	$c = 0.0$	0.2	0.4	0.6	0.8	1.0
0.1	0.0797	0.0242	0.0124	0.0083	0.0062	0.0050
0.2	0.1585	0.0885	0.0482	0.0327	0.0247	0.0198
0.3	0.2358	0.1739	0.1039	0.0719	0.0547	0.0440
0.4	0.3108	0.2635	0.1742	0.1238	0.0951	0.0769
0.5	0.3829	0.3482	0.2533	0.1857	0.1444	0.1175
0.6	0.4515	0.4256	0.3357	0.2548	0.2010	0.1647
0.7	0.5161	0.4961	0.4171	0.3280	0.2629	0.2173
0.8	0.5763	0.5604	0.4942	0.4026	0.3283	0.2739
0.9	0.6319	0.6191	0.5652	0.4759	0.3953	0.3330
1.0	0.6827	0.6724	0.6291	0.5461	0.4621	0.3935
1.2	0.7699	0.7630	0.7359	0.6714	0.5893	0.5132
1.4	0.8385	0.8340	0.8170	0.7721	0.7008	0.6247
1.6	0.8904	0.8875	0.8769	0.8478	0.7917	0.7220
1.8	0.9281	0.9263	0.9197	0.9019	0.8613	0.8021
2.0	0.9545	0.9534	0.9494	0.9388	0.9116	0.8647
2.2	0.9722	0.9715	0.9692	0.9631	0.9459	0.9111
2.4	0.9836	0.9832	0.9819	0.9785	0.9683	0.9439
2.6	0.9907	0.9905	0.9897	0.9879	0.9821	0.9660
2.8	0.9949	0.9948	0.9944	0.9934	0.9903	0.9802
3.0	0.9973	0.9972	0.9970	0.9965	0.9949	0.9889
3.2	0.9986	0.9986	0.9985	0.9982	0.9974	0.9940
3.4	0.9993	0.9993	0.9993	0.9991	0.9988	0.9969
3.6	0.9997	0.9997	0.9997	0.9996	0.9994	0.9985
3.8	0.9999	0.9999	0.9998	0.9998	0.9997	0.9993
4.0	0.9999	0.9999	0.9999	0.9999	0.9999	0.9997

CHAPTER 8

Estimation

Contents

A nonconstant function of a set of random variables is a *statistic*. It is a function of observable random variables, which does not contain any unknown parameters. A statistic is itself an observable random variable.

Let θ be a parameter appearing in the density function for the random variable X. Let g be a function that returns an approximate value $\widehat{\theta}$ of θ from a given sample $\{x_1, \ldots, x_n\}$. Then $\widehat{\theta} = g(x_1, x_2, \ldots, x_n)$ may be considered a single observation of the random variable $\widehat{\Theta} = g(X_1, X_2, \ldots, X_n)$. The random variable $\widehat{\Theta}$ is an *estimator* for the parameter θ.

8.1 DEFINITIONS

(1) $\widehat{\Theta}$ is an **unbiased estimator** for θ if $\mathrm{E}[\widehat{\Theta}] = \theta$.

(2) The **bias** of the estimator $\widehat{\Theta}$ is $\mathrm{B}[\widehat{\Theta}] = \mathrm{E}[\widehat{\Theta}] - \theta$.

(3) The **mean square error** of $\widehat{\Theta}$ is

$$\mathrm{MSE}[\widehat{\Theta}] = \mathrm{E}[(\widehat{\Theta} - \theta)^2] = \mathrm{Var}[\widehat{\Theta}] + \mathrm{B}[\widehat{\Theta}]^2.$$

(4) The **error of estimation** is $\epsilon = |\widehat{\Theta} - \theta|$.

(5) Let $\widehat{\Theta}_1$ and $\widehat{\Theta}_2$ be unbiased estimators for θ.

    (a) If $\mathrm{Var}[\widehat{\Theta}_1] < \mathrm{Var}[\widehat{\Theta}_2]$ then the estimator $\widehat{\Theta}_1$ is **relatively more efficient** than the estimator $\widehat{\Theta}_2$.

(b) The **efficiency of $\widehat{\Theta}_2$ relative to $\widehat{\Theta}_1$** is

$$\text{Efficiency} = \frac{\text{Var}\,[\widehat{\Theta}_1]}{\text{Var}\,[\widehat{\Theta}_2]}.$$

(6) $\widehat{\Theta}$ is a **consistent estimator** for θ if for every $\epsilon > 0$,

$$\lim_{n\to\infty} \text{Prob}\,[|\widehat{\Theta} - \theta| \le \epsilon] = 1 \quad \text{or, equivalently}$$

$$\lim_{n\to\infty} \text{Prob}\,[|\widehat{\Theta} - \theta| > \epsilon] = 0.$$

(7) $\widehat{\Theta}$ is a **sufficient estimator** for θ if for each value of $\widehat{\Theta}$ the conditional distribution of $X_1, X_2, \ldots, X_n$ given $\widehat{\Theta} = \theta_0$ is independent of θ.

(8) Let $\widehat{\Theta}$ be an estimator for the parameter θ and suppose $\widehat{\Theta}$ has sampling distribution $g(\widehat{\Theta})$. Then $\widehat{\Theta}$ is a **complete** statistic if for all θ, $\text{E}\,[h\,(\widehat{\Theta})] = 0$ implies $h\,(\widehat{\Theta}) = 0$ for all functions $h\,(\widehat{\Theta})$.

(9) An estimator $\widehat{\Theta}$ for θ is a **minimum variance unbiased estimator** (MVUE) if it is unbiased and has the smallest possible variance.

Example 8.45:   To determine if $\overline{X}^2$ is an unbiased estimator of μ^2, consider the following expected value:

$$\text{E}\left[\overline{X}^2\right] = \text{E}\left[\left(\frac{1}{n}\sum_{i=1}^{n} X_i\right)^2\right]$$

$$= \frac{1}{n^2}\text{E}\left[\sum_{i=1}^{n} X_i^2 + \sum_{i=1,j=1,i\neq j}^{n} X_i X_j\right]$$

$$= \frac{1}{n^2}\left[n(\sigma^2 + \mu^2) + (n^2 - n)\mu^2\right] \tag{8.1}$$

$$= \mu^2 + \frac{\sigma^2}{n}$$

$$> \mu^2$$

This shows $\overline{X}^2$ is a biased estimator of μ^2.

8.2   CRAMÉR–RAO INEQUALITY

Let $\{X_1, X_2, \ldots, X_n\}$ be a random sample from a population with probability density function $f(x)$. Let $\widehat{\Theta}$ be an unbiased estimator for θ. Under very general conditions it can be shown that

$$\text{Var}\,[\widehat{\Theta}] \ge \frac{1}{n \cdot \text{E}\left[\left(\frac{\partial \ln f(X)}{\partial\theta}\right)^2\right]} = \frac{1}{-n \cdot \text{E}\left[\frac{\partial^2 \ln f(X)}{\partial\theta^2}\right]}. \tag{8.2}$$

If equality holds then $\widehat{\Theta}$ is a minimum variance unbiased estimator (MVUE) for θ.

Example 8.46: The probability density function for a normal random variable with unknown mean θ and known variance σ^2 is $f(x;\theta) = \dfrac{1}{\sqrt{2\pi}\sigma} \exp\left[-\dfrac{(x-\theta)^2}{2\sigma^2}\right]$. Use the Cramér–Rao inequality to show the minimum variance of any unbiased estimator, $\hat{\Theta}$, for θ is at least σ^2/n.

Solution:

(S1) $\dfrac{\partial}{\partial \theta} \ln f(X;\theta) = \dfrac{(X-\theta)}{\sigma^2}$

(S2) $\left(\dfrac{\partial \ln f(X;\theta)}{\partial \theta}\right)^2 = \dfrac{(X-\theta)^2}{\sigma^4}$

(S3) $\mathrm{E}\left[\dfrac{(X-\theta)^2}{\sigma^4}\right] = \displaystyle\int_{-\infty}^{\infty} \dfrac{(x-\theta)^2}{\sigma^4} \left(\dfrac{1}{\sqrt{2\pi}\sigma} e^{-(x-\theta)^2/2\sigma^2}\right) dx = \dfrac{1}{\sigma^2}$

(S4) $\mathrm{Var}\,[\hat{\Theta}] \geq \dfrac{1}{n\frac{1}{\sigma^2}} = \dfrac{\sigma^2}{n}$

8.3 THEOREMS

(1) $\hat{\Theta}$ is a consistent estimator for θ if

   (a) $\hat{\Theta}$ is unbiased, and

   (b) $\displaystyle\lim_{n\to\infty} \mathrm{Var}\,[\hat{\Theta}] = 0$.

(2) $\hat{\Theta}$ is a sufficient estimator for the parameter θ if the joint distribution of $\{X_1, X_2, \ldots, X_n\}$ can be factored into

$$f(x_1, x_2, \ldots, x_n; \theta) = g(\hat{\Theta}, \theta) \cdot h(x_1, x_2, \ldots, x_n) \qquad (8.3)$$

where $g(\hat{\Theta}, \theta)$ depends only on the estimate $\hat{\theta}$ and the parameter θ, and $h(x_1, x_2, \ldots, x_n)$ does not depend on the parameter θ.

(3) Unbiased estimators:

   (a) An unbiased estimator may not exist.

   (b) An unbiased estimator is not unique.

   (c) An unbiased estimator may be meaningless.

   (d) An unbiased estimator is not necessarily consistent.

(4) Consistent estimators:

   (a) A consistent estimator is not unique.

   (b) A consistent estimator may be meaningless.

   (c) A consistent estimator is not necessarily unbiased.

(5) Maximum likelihood estimators (MLE):

   (a) A MLE need not be consistent.

   (b) A MLE may not be unbiased.

   (c) A MLE is not unique.

   (d) If a single sufficient statistic T exists for the parameter θ, the MLE
       of θ must be a function of T.

   (e) Let $\widehat{\Theta}$ be a MLE of θ. If $\tau(\cdot)$ is a function with a single-valued
       inverse, then a MLE of $\tau(\theta)$ is $\tau(\widehat{\Theta})$.

(6) Method of moments (MOM) estimators:

   (a) MOM estimators are not uniquely defined.

   (b) MOM estimators may not be functions of sufficient or complete
       statistics.

(7) A single sufficient estimator may not exist.

8.4   THE METHOD OF MOMENTS

The moment estimators are the solutions to the systems of equations

$$\mu'_r = \mathrm{E}\,[X^r] = \frac{1}{n}\sum_{i=1}^{n} x_i^r = m'_r, \quad r = 1, 2, \ldots, k$$

where k is the number of parameters.

Example 8.47: Suppose $X_1, X_2, \ldots, X_n$ is a random sample from a population having
a gamma distribution with parameters α and β. Use the method of moments to obtain
estimators for the parameters α and β.

Solution:

(S1) The system of equations to solve: $\mu'_1 = m'_1$; $\mu'_2 = m'_2$

(S2) $\mu'_1 = \mathrm{E}\,[X] = \alpha\beta$; $\mu'_2 = \mathrm{E}\,[X^2] = \alpha(\alpha+1)\beta^2$

(S3) Solve $\alpha\beta = m'_1$ and $\alpha(\alpha+1)\beta^2 = m'_2$ for α and β.

(S4) $\widehat{\alpha} = \dfrac{(m'_1)^2}{m'_2 - (m'_1)^2}$; $\widehat{\beta} = \dfrac{m'_2 - (m'_1)^2}{m'_1}$

(S5) Given $m'_1 = \overline{x} = \dfrac{1}{n}\sum_{i=1}^{n} x_i$ and $m'_2 = \dfrac{1}{n}\sum_{i=1}^{n} x_i^2$, then

$$\widehat{\alpha} = \frac{n\overline{x}^2}{\sum_{i=1}^{n}(x_i - \overline{x})^2} \quad \text{and} \quad \widehat{\beta} = \frac{\sum_{i=1}^{n}(x_i - \overline{x})^2}{n\overline{x}}$$

8.5   THE LIKELIHOOD FUNCTION

Let $x_1, x_2, \ldots, x_n$ be the values of a random sample from a population characterized by the parameters $\theta_1, \theta_2, \ldots, \theta_r$. The **likelihood function** of the
sample is

(1) the joint probability mass function evaluated at $(x_1, x_2, \ldots, x_n)$ if $(X_1, X_2, \ldots, X_n)$ are discrete,

$$\mathrm{L}(\theta_1, \theta_2, \ldots, \theta_r) = p(x_1, x_2, \ldots, x_n; \theta_1, \theta_2, \ldots, \theta_r) \qquad (8.4)$$

(2) the joint probability density function evaluated at $(x_1, x_2, \ldots, x_n)$ if $(X_1, X_2, \ldots, X_n)$ are continuous.

$$L(\theta_1, \theta_2, \ldots, \theta_r) = f(x_1, x_2, \ldots, x_n; \theta_1, \theta_2, \ldots, \theta_r) \qquad (8.5)$$

8.6 THE METHOD OF MAXIMUM LIKELIHOOD

The maximum likelihood estimators (MLEs) are those values of the parameters that maximize the likelihood function of the sample: $L(\theta_1, \ldots, \theta_r)$. In practice, it is often easier to maximize $\ln L(\theta_1, \ldots, \theta_r)$. This is equivalent to maximizing the likelihood function, $L(\theta_1, \ldots, \theta_r)$, since $\ln L(\theta_1, \ldots, \theta_r)$ is a monotonic function of $L(\theta_1, \ldots, \theta_r)$.

Example 8.48: Suppose $X_1, X_2, \ldots, X_n$ is a random sample from a population having a Poisson distribution with parameter λ. Find the maximum likelihood estimator for the parameter λ.

Solution:

(S1) The probability mass function for a Poisson random variable is
$$f(x; \lambda) = \frac{e^{-\lambda} \lambda^x}{x!}$$

(S2) We compute
$$L(\theta) = \left(\frac{e^{-\lambda} \lambda^{x_1}}{x_1!} \right) \left(\frac{e^{-\lambda} \lambda^{x_2}}{x_2!} \right) \cdots \left(\frac{e^{-\lambda} \lambda^{x_n}}{x_n!} \right)$$
$$= \frac{e^{-n\lambda} \lambda^{x_1 + x_2 + \cdots + x_n}}{x_1! x_2! \cdots x_n!} \qquad (8.6)$$
$$\ln L(\theta) = -n\lambda + (x_1 + x_2 + \cdots + x_n) \ln \lambda + \ln (x_1! x_2! \cdots x_n!)$$

(S3) $\dfrac{\partial \ln L(\lambda)}{\partial \lambda} = -n + \dfrac{x_1 + x_2 + \cdots + x_n}{\lambda} = 0$

(S4) Solving for λ: $\widehat{\lambda} = \dfrac{x_1 + x_2 + \cdots + x_n}{n} = \overline{x}$ is the MLE for λ.

8.7 INVARIANCE PROPERTY OF MAXIMUM LIKELIHOOD ESTIMATORS

Let $\widehat{\Theta}_1, \widehat{\Theta}_2, \ldots, \widehat{\Theta}_r$ be the maximum likelihood estimators for $\theta_1, \theta_2, \ldots, \theta_r$ and let $h(\theta_1, \theta_2, \ldots, \theta_r)$ be a function of $\theta_1, \theta_2, \ldots, \theta_r$. The maximum likelihood estimator of the parameter $h(\theta_1, \theta_2, \ldots, \theta_r)$ is $\widehat{h}(\theta_1, \theta_2, \ldots, \theta_r) = h(\widehat{\Theta}_1, \widehat{\Theta}_2, \ldots, \widehat{\Theta}_r)$.

8.8 DIFFERENT ESTIMATORS

Assume $\{x_1, x_2, \ldots, x_n\}$ is a set of observations. Let UMV stand for *uniformly minimum variance unbiased* and let MLE stand for *maximum likelihood estimator*.

(1) Normal distribution: $N(\mu, \sigma^2)$

    (a) When σ is known:

        1. $\sum x_i$ is necessary, sufficient, and complete.

    2. Point estimate for μ: $\widehat{\mu} = \overline{x} = \dfrac{1}{n}\sum x_i$ is UMV, MLE.

(b) When μ is known:

    1. $\sum(x_i - \mu)^2$ is necessary, sufficient, and complete.

    2. Point estimate for σ^2: $\widehat{\sigma^2} = \dfrac{\sum(x_i - \mu)^2}{n}$ is UMV, MLE.

(c) When μ and σ are unknown:

    1. $\{\sum x_i, \sum(x_i - \overline{x})^2\}$ are necessary, sufficient, and complete.

    2. Point estimate for μ: $\widehat{\mu} = \overline{x} = \dfrac{1}{n}\sum x_i$ is UMV, MLE.

    3. Point estimate for σ^2: $\widehat{\sigma^2} = \dfrac{\sum(x_i - \overline{x})^2}{n}$ is MLE.

    4. Point estimate for σ^2: $\widehat{\sigma^2} = \dfrac{\sum(x_i - \overline{x})^2}{n - 1}$ is UMV.

    5. Point estimate for σ: $\widehat{\sigma} = \dfrac{\Gamma\left[(n-1)/2\right]}{\sqrt{2}\Gamma(n/2)}\sqrt{\dfrac{\sum(x_i - \overline{x})^2}{n-1}}$ is UMV.

(2) Poisson distribution with parameter λ:

  (a) $\sum x_i$ is necessary, sufficient, and complete.

  (b) Point estimate for λ: $\widehat{\lambda} = \dfrac{1}{n}\sum x_i$ is UMV, MLE.

(3) Uniform distribution on an interval:

  (a) Interval is $[0, \theta]$

    1. $\max(x_i)$ is necessary, sufficient, and complete.

    2. Point estimate for θ: $\widehat{\Theta} = \max(x_i)$ is MLE.

    3. Point estimate for θ: $\widehat{\Theta} = \dfrac{n+1}{n}\max(x_i)$ is UMV.

  (b) Interval is $[\alpha, \beta]$

    1. $\{\min(x_i), \max(x_i)\}$ are necessary, sufficient, and complete.

    2. Point estimate for α: $\widehat{\alpha} = \dfrac{n\min(x_i) - \max(x_i)}{n - 1}$ is UMV.

    3. Point estimate for α: $\widehat{\alpha} = \min(x_i)$ is MLE.

    4. Point estimate for $\frac{\alpha+\beta}{2}$: $\dfrac{\widehat{\alpha + \beta}}{2} = \dfrac{\min(x_i) + \max(x_i)}{2}$ is UMV.

  (c) Interval is $\left[\theta - \frac{1}{2}, \theta + \frac{1}{2}\right]$

    1. $\{\min(x_i), \max(x_i)\}$ are necessary and sufficient.

    2. Point estimate for θ: $\widehat{\Theta} = \dfrac{\min(x_i) + \max(x_i)}{2}$ is MLE.

8.9 ESTIMATORS FOR MEAN AND STANDARD DEVIATION IN SMALL SAMPLES

In all cases below, the variance of an estimate must be multiplied by the the true variance of the sample, σ^2.

Different estimators for the mean

	Median		Midrange		Average of best two			$\frac{1}{n-2}\left(\sum_{i=2}^{n-1} x_i\right)$	
n	Var	Eff	Var	Eff	Statistic	Var	Eff	Var	Eff
2	.500	1.000	.500	1.000	$\frac{1}{2}(x_1 + x_2)$	.500	1.000		
3	.449	.743	.362	.920	$\frac{1}{2}(x_1 + x_3)$	.362	.920	.449	.743
4	.298	.838	.298	.838	$\frac{1}{2}(x_2 + x_3)$	.298	.838	.298	.838
5	.287	.697	.261	.767	$\frac{1}{2}(x_2 + x_4)$	.231	.867	.227	.881
10	.138	.723	.186	.539	$\frac{1}{2}(x_3 + x_8)$	.119	.840	.105	.949
15	.102	.656	.158	.422	$\frac{1}{2}(x_4 + x_{12})$	.081	.825	.069	.969
20	.073	.681	.143	.350	$\frac{1}{2}(x_6 + x_{15})$	.061	.824	.051	.978
∞	$\frac{1.57}{n}$	.637		.000	$\frac{1}{2}(P_{25} + P_{75})$	$\frac{1.24}{n}$	.808		1.000

Estimating standard deviation σ from the sample range w

n	Estimator	Variance	Efficiency
2	.886w	.571	1.000
3	.591w	.275	.992
4	.486w	.183	.975
5	.430w	.138	.955
6	.395w	.112	.933
7	.370w	.095	.911
8	.351w	.083	.890
9	.337w	.074	.869
10	.325w	.067	.850
15	.288w	.047	.766
20	.268w	.038	.700

Best linear estimate of the standard deviation σ

n	Estimator	Efficiency
2	$.8862(x_2 - x_1)$	1.000
3	$.5908(x_3 - x_1)$	.992
4	$.4539(x_4 - x_1) + .1102(x_3 - x_2)$	.989
5	$.3724(x_5 - x_1) + .1352(x_3 - x_2)$	.988
6	$.3175(x_6 - x_1) + .1386(x_5 - x_2) + .0432(x_4 - x_3)$	.988
7	$.2778(x_7 - x_1) + .1351(x_6 - x_2) + .0625(x_5 - x_3)$	.989

8.10 ESTIMATORS FOR MEAN AND STANDARD DEVIATION IN LARGE SAMPLES

Percentile estimates may be used to estimate both the mean and the standard deviation.

Estimators for the mean

Number of terms	Estimator using percentiles	Efficiency
1	P_{50}	.64
2	$\frac{1}{2}\left(P_{25} + P_{75}\right)$	.81
3	$\frac{1}{3}\left(P_{17} + P_{50} + P_{83}\right)$	.88
4	$\frac{1}{4}\left(P_{12.5} + P_{37.5} + P_{62.5} + P_{87.5}\right)$	.91
5	$\frac{1}{5}\left(P_{10} + P_{30} + P_{50} + P_{70} + P_{90}\right)$	.93

Estimators for the standard deviation

Number of terms	Estimator using percentiles	Efficiency
2	$.3388\left(P_{93} - P_{7}\right)$	.65
4	$.1714\left(P_{97} + P_{85} - P_{15} - P_{3}\right)$	.80
6	$.1180\left(P_{98} + P_{91} + P_{80} - P_{20} - P_{9} - P_{2}\right)$	.87

CHAPTER 9

Confidence Intervals

Contents

9.1 DEFINITIONS

A simple point estimate $\widehat{\theta}$ of a parameter θ serves as a best guess for the value of θ, but conveys no sense of confidence in the estimate. A confidence interval I, based on $\widehat{\theta}$, is used to make statements about θ when the sample size, the underlying distribution of θ, and the **confidence coefficient** $1 - \alpha$ are known. We make statements of the form:

$$\text{The probability that } \theta \text{ is in a specified interval is } 1 - \alpha. \qquad (9.1)$$

To construct a confidence interval for a parameter θ, the confidence coefficient must be specified. The usual procedure is to specify the confidence coefficient $1 - \alpha$ and then determine the confidence interval. A typical value is $\alpha = 0.05$ (also written as $\alpha = 5\%$); so that $1 - \alpha = 0.95$ (or 95% *confidence*).

The confidence interval may be denoted $(\theta_{low}, \theta_{high})$. There are many ways to specify θ_{low} and θ_{high}, depending on the parameter θ and the underlying distribution. The bounds on the confidence interval are usually defined to satisfy

The probability that $\theta < \theta_{low}$ is $\alpha/2$: $\text{Prob}\,[\theta < \theta_{low}] = \alpha/2$

$$(9.2)$$

The probability that $\theta > \theta_{high}$ is $\alpha/2$: $\text{Prob}\,[\theta > \theta_{high}] = \alpha/2$

so that $\text{Prob}\,[\theta_{low} \leq \theta \leq \theta_{high}] = 1 - \alpha$.

It is also possible to construct one-sided confidence intervals. For these,

(1) $\theta_{low} = -\infty$ and $\text{Prob}\,[\theta > \theta_{high}] = \alpha$ or $\text{Prob}\,[\theta \leq \theta_{high}] = 1 - \alpha$ (one-sided, lower-tailed confidence interval), or

(2) $\theta_{high} = \infty$ and $\text{Prob}\,[\theta < \theta_{low}] = \alpha$ or $\text{Prob}\,[\theta \geq \theta_{low}] = 1 - \alpha$ (one-sided, upper-tailed confidence interval).

Notes:

(1) When the sample size is at least 5% of the total population, a finite population correction factor may be used to modify a confidence interval. See section (9.7).

(2) If the test statistic is significant in an ANOVA, confidence intervals may then be used to determine which pairs of means differ.

9.2 COMMON CRITICAL VALUES

The formulas for common confidence intervals usually involve critical values from the normal distribution, the t distribution, or the chi–square distribution; see Tables 9.3 and 9.4. More extensive critical value tables for the normal distribution are given on page 175, for the t distribution on page 156, for the chi–square distribution on page 6.4.4, and for the F distribution on page 131.

9.3 SAMPLE SIZE CALCULATIONS

In order to construct a confidence interval of specified width, a priori parameter estimates and a bound on the error of estimation may be used to determine the necessary sample size. For a $100(1 - \alpha)\%$ confidence interval, let E = error of estimation (half the width of the confidence interval). Table 9.2 presents some common sample size calculations.

Example 9.49:   A researcher would like to estimate the probability of a success, p, in a binomial experiment. How large a sample is necessary in order to estimate

Distribution	α				
	0.10	0.05	0.01	0.001	0.0001
t distribution					
$t_{\alpha/2,10}$	1.8125	2.2281	3.1693	4.5869	6.2111
$t_{\alpha/2,100}$	1.6602	1.9840	2.6259	3.3905	4.0533
$t_{\alpha/2,1000}$	1.6464	1.9623	2.5808	3.3003	3.9063
Normal distribution					
$z_{\alpha/2}$	1.6449	1.9600	2.5758	3.2905	3.8906
χ^2 distribution					
$\chi^2_{1-\alpha/2,10}$	3.9403	3.2470	2.1559	1.2650	0.7660
$\chi^2_{\alpha/2,10}$	18.3070	20.4832	25.1882	31.4198	37.3107
$\chi^2_{1-\alpha/2,100}$	77.9295	74.2219	67.3276	59.8957	54.1129
$\chi^2_{\alpha/2,100}$	124.3421	129.5612	140.1695	153.1670	164.6591
$\chi^2_{1-\alpha/2,1000}$	927.5944	914.2572	888.5635	859.3615	835.3493
$\chi^2_{\alpha/2,1000}$	1074.6790	1089.5310	1118.9480	1153.7380	1183.4920
	0.90	0.95	0.99	0.999	0.9999
	$1 - \alpha$				

Table 9.1: Common critical values used with confidence intervals

Parameter	Estimate	Sample size	
μ	$\overline{x}$	$n = \left(\dfrac{z_{\alpha/2} \cdot \sigma}{E}\right)^2$	(1)
p	$\widehat{p}$	$n = \dfrac{(z_{\alpha/2})^2 \cdot pq}{E^2}$	(2)
$\mu_2 - \mu_2$	$\overline{x}_1 - \overline{x}_2$	$n_1 = n_2 = \dfrac{(z_{\alpha/2})^2(\sigma_1^2 + \sigma_2^2)}{E^2}$	(3)
$p_1 - p_2$	$\widehat{p}_1 - \widehat{p}_2$	$n_1 = n_2 = \dfrac{(z_{\alpha/2})^2(p_1 q_1 + p_2 q_2)}{E^2}$	(4)

Table 9.2: Common sample size calculations

this proportion to within .05 with 99% confidence, i.e., find a value of n such that $\text{Prob}\,[|\widehat{p} - p| \leq 0.05] \geq 0.99$.

Solution:

(S1) Since no a priori estimate of p is available, use $p = .5$. The bound on the error of estimation is $E = .05$ and $1 - \alpha = .99$.

(S2) From Table 9.2, $n = \dfrac{z_{.005}^2 \cdot pq}{E^2} = \dfrac{(2.5758)(.5)(.5)}{.05^2} = 663.47$

(S3) This formula produces a conservative value for the necessary sample size (since no a priori estimate of p is known). A sample size of at least 664 should be used.

9.4   SUMMARY OF COMMON CONFIDENCE INTERVALS

Table 9.3 presents a summary of common confidence intervals for one sample, Table 9.4 is for two samples. For each population parameter, the assumptions and formula for a $100(1 - \alpha)\%$ confidence interval are given.

Parameter	Assumptions (reference)	$100(1 - \alpha)\%$ Confidence interval	
μ	n large, σ^2 known, or normality, σ^2 known (§9.5.1)	$\bar{x} \pm z_{\alpha/2} \cdot \dfrac{\sigma}{\sqrt{n}}$	(1)
μ	normality, σ^2 unknown (§9.5.2)	$\bar{x} \pm t_{\alpha/2,n-1} \cdot \dfrac{s}{\sqrt{n}}$	(2)
σ^2	normality (§9.5.3)	$\left(\dfrac{(n-1)s^2}{\chi_{\alpha/2,n-1}^2} , \dfrac{(n-1)s^2}{\chi_{1-\alpha/2,n-1}^2} \right)$	(3)
p	binomial experiment, n large (§9.5.4)	$\hat{p} \pm z_{\alpha/2} \cdot \sqrt{\dfrac{\hat{p}(1 - \hat{p})}{n}}$	(4)

Table 9.3: Summary of common confidence intervals: one sample

9.5   CONFIDENCE INTERVALS: ONE SAMPLE

Let $x_1, x_2, \ldots, x_n$ be a random sample of size n.

9.5.1   Confidence interval for mean of normal population, known variance

Find a $100(1 - \alpha)\%$ confidence interval for the mean μ of a normal population with known variance σ^2, or
Find a $100(1 - \alpha)\%$ confidence interval for the mean μ of a population with known variance σ^2 where n is large.

(a) Compute the sample mean $\bar{x}$.

(b) Determine the critical value $z_{\alpha/2}$ such that $\Phi(z_{\alpha/2}) = 1 - \alpha/2$, where $\Phi(z)$ is the standard normal cumulative distribution function. That is, $z_{\alpha/2}$ is defined so that $\text{Prob}\left[Z \geq z_{\alpha/2}\right] = \alpha/2$.

(c) Compute the constant $k = \sigma z_{\alpha/2}/\sqrt{n}$. (Table 9.5 on page 200 has common values of $z_{\alpha/2}/\sqrt{n}$.)

(d) A $100(1 - \alpha)\%$ confidence interval for μ is given by $(\bar{x} - k, \bar{x} + k)$.

Parameter	Assumptions (reference)	$100(1-\alpha)\%$ Confidence interval	
$\mu_1 - \mu_2$	normality, independence, σ_1^2, σ_2^2 known or n_1, n_2 large, independence, σ_1^2, σ_2^2 known, (§9.6.1)	$(\bar{x}_1 - \bar{x}_2) \pm z_{\alpha/2} \cdot \sqrt{\dfrac{\sigma_1^2}{n_1} + \dfrac{\sigma_2^2}{n_2}}$	(1)
$\mu_1 - \mu_2$	normality, independence, $\sigma_1^2 = \sigma_2^2$ unknown (§9.6.2)	$(\bar{x}_1 - \bar{x}_2) \pm$ $t_{\frac{\alpha}{2}, n_1+n_2-2} \cdot s_p \sqrt{\dfrac{1}{n_1} + \dfrac{1}{n_2}}$ $s_p^2 = \dfrac{(n_1-1)s_1^2 + (n_2-1)s_2^2}{n_1 + n_2 - 2}$	(2)
$\mu_1 - \mu_2$	normality, independence, $\sigma_1^2 \neq \sigma_2^2$ unknown (§9.6.3)	$(\bar{x}_1 - \bar{x}_2) \pm t_{\alpha/2,\nu} \cdot \sqrt{\dfrac{s_1^2}{n_1} + \dfrac{s_2^2}{n_2}}$ $\nu \approx \dfrac{\left(\dfrac{s_1^2}{n_1} + \dfrac{s_2^2}{n_2}\right)^2}{\dfrac{(s_1^2/n_1)^2}{n_1-1} + \dfrac{(s_2^2/n_2)^2}{n_2-1}}$	(3)
$\mu_1 - \mu_2$	normality, n pairs, dependence (§9.6.4)	$\bar{d} \pm t_{\alpha/2,n-1} \cdot \dfrac{s_d}{\sqrt{n}}$	(4)
σ_1^2/σ_2^2	normality, independence (§9.6.5)	$\left(\dfrac{s_1^2}{s_2^2} \cdot \dfrac{1}{F_{\frac{\alpha}{2},n_1-1,n_2-1}}, \right.$ $\left. \dfrac{s_1^2}{s_2^2} \cdot \dfrac{1}{F_{1-\frac{\alpha}{2},n_1-1,n_2-1}}\right)$	(5)
$p_1 - p_2$	binomial experiments, n_1, n_2 large, independence (§9.6.6)	$(\hat{p}_1 - \hat{p}_2) \pm$ $z_{\alpha/2} \cdot \sqrt{\dfrac{\hat{p}_1(1-\hat{p}_1)}{n_1} + \dfrac{\hat{p}_2(1-\hat{p}_2)}{n_2}}$	(6)

Table 9.4: Summary of common confidence intervals: two samples

9.5.2 Confidence interval for mean of normal population, unknown variance

Find a $100(1-\alpha)\%$ confidence interval for the mean μ of a normal population with unknown variance σ^2.

(a) Compute the sample mean $\bar{x}$ and the sample standard deviation s.

(b) Determine the critical value $t_{\alpha/2,n-1}$ such that $F(t_{\alpha/2,n-1}) = 1 - \alpha/2$, where $F(t)$ is the cumulative distribution function for a t distribution with $n-1$ degrees of freedom. That is, $t_{\alpha/2,n-1}$ is defined so that $\text{Prob}\left[T \geq t_{\alpha/2,n-1}\right] = \alpha/2$.

n	$z_{\alpha/2}/\sqrt{n}$ when $\alpha = 0.05$	$z_{\alpha/2}/\sqrt{n}$ when $\alpha = 0.01$	n	$z_{\alpha/2}/\sqrt{n}$ when $\alpha = 0.05$	$z_{\alpha/2}/\sqrt{n}$ when $\alpha = 0.01$
2	8.99	45.0	15	0.554	0.769
3	2.48	5.73	20	0.468	0.640
4	1.59	2.92	25	0.413	0.559
5	1.24	2.06	30	0.373	0.503
6	1.05	1.65	40	0.320	0.428
7	0.925	1.40	50	0.284	0.379
8	0.836	1.24	100	0.198	0.263
9	0.769	1.12	200	0.139	0.184
10	0.715	1.03	500	0.088	0.116

Table 9.5: Common values of $z_{\alpha/2}/\sqrt{n}$

(c) Compute the constant $k = t_{\alpha/2,n-1}\, s/\sqrt{n}$.

(d) A $100(1-\alpha)\%$ confidence interval for μ is given by $(\bar{x} - k, \bar{x} + k)$.

Example 9.50: A software company conducted a survey on the size of a typical word processing file. For $n = 23$ randomly selected files, $\bar{x} = 4822$ kb and $s = 127$. Find a 95% confidence interval for the true mean size of word processing files.

Solution:

(S1) The underlying population, the size of word processing files, is assumed to be normal. The confidence interval for μ is based on a t distribution.

(S2) $1 - \alpha = .95$; $\alpha = .05$; $\alpha/2 = .025$; $t_{\alpha/2,n-1} = t_{.025,22} = 2.0739$

(S3) $k = (2.0739)(127)/\sqrt{23} = 54.92$

(S4) A 99% confidence interval for μ: $(\bar{x} - k, \bar{x} + k) = (4767.08, 4876.92)$

9.5.3 Confidence interval for variance of normal population

Find a $100(1-\alpha)\%$ confidence interval for the variance σ^2 of a normal population.

(a) Compute the sample variance s^2.

(b) Determine the critical values $\chi^2_{\alpha/2,n-1}$ and $\chi^2_{1-\alpha/2,n-1}$ such that
$$\text{Prob}\left[\chi^2 \geq \chi^2_{\alpha/2,n-1}\right] = \text{Prob}\left[\chi^2 \leq \chi^2_{1-\alpha/2,n-1}\right] = \alpha/2.$$

(c) Compute the constants $k_1 = \dfrac{(n-1)s^2}{\chi^2_{\alpha/2,n-1}}$ and $k_2 = \dfrac{(n-1)s^2}{\chi^2_{1-\alpha/2,n-1}}$.

(d) A $100(1-\alpha)\%$ confidence interval for σ^2 is given by (k_1, k_2).

(e) A $100(1-\alpha)\%$ confidence interval for σ is given by $(\sqrt{k_1}, \sqrt{k_2})$.

9.5.4 Confidence interval for the probability of a success in a binomial experiment

Find a $100(1 - \alpha)\%$ confidence interval for the probability of a success p in a binomial experiment where n is large.

(a) Compute the sample proportion of successes $\widehat{p}$.

(b) Determine the critical value $z_{\alpha/2}$ such that $\text{Prob}\left[Z \geq z_{\alpha/2}\right] = \alpha/2$.

(c) Compute the constant $k = z_{\alpha/2}\sqrt{\dfrac{\widehat{p}(1 - \widehat{p})}{n}}$.

(d) A $100(1 - \alpha)\%$ confidence interval for p is given by $(\widehat{p} - k, \widehat{p} + k)$.

9.5.5 Confidence interval for percentiles

Find an approximate $100(1 - \alpha)\%$ confidence interval for the pth percentile, ξ_p, where n is large.

(1) Compute the order statistics $\{x_{(1)}, x_{(2)}, \ldots, x_{(n)}\}$.

(2) Determine the critical value $z_{\alpha/2}$ such that $\text{Prob}\left[Z \geq z_{\alpha/2}\right] = \alpha/2$.

(3) Compute the constants $k_1 = \left\lfloor np - z_{\alpha/2}\sqrt{np(1 - p)} \right\rfloor$ and
$k_2 = \left\lceil np + z_{\alpha/2}\sqrt{np(1 - p)} \right\rceil$.

(4) A $100(1 - \alpha)\%$ confidence interval for ξ_p is given by $(x_{(k_1)}, x_{(k_2)})$.

9.5.6 Confidence interval for medians

Find an approximate $100(1 - \alpha)\%$ confidence interval for the median $\tilde{\mu}$ where n is large (based on the Wilcoxon one-sample statistic).

(1) Compute the order statistics $\{w_{(1)}, w_{(2)}, \ldots, w_{(N)}\}$ of the $N = \binom{n}{2} = \frac{n(n-1)}{2}$ averages $(x_i + x_j)/2$, for $1 \leq i < j \leq n$.

(2) Determine the critical value $z_{\alpha/2}$ such that $\text{Prob}\left[Z \geq z_{\alpha/2}\right] = \alpha/2$.

(3) Compute the constants $k_1 = \left\lfloor \dfrac{N}{2} - \dfrac{z_{\alpha/2}N}{\sqrt{3n}} \right\rfloor$ and
$k_2 = \left\lceil \dfrac{N}{2} + \dfrac{z_{\alpha/2}N}{\sqrt{3n}} \right\rceil$.

(4) A $100(1 - \alpha)\%$ confidence interval for $\tilde{\mu}$ is given by $(w_{(k_1)}, w_{(k_2)})$. (See Table 9.6.)

9.5.6.1 Table of confidence interval for medians

If the n observations $\{x_1, x_2, \ldots, x_n\}$ are arranged in ascending order $\{x_{(1)}, x_{(2)}, \ldots, x_{(n)}\}$, a $100(1 - \alpha)\%$ confidence interval for the median of the population can be found. Table 9.7 lists values of l and u such that the probability the median is between $x_{(l)}$ and $x_{(u)}$ is $(1 - \alpha)$.

| | $\alpha = .05$ | | $\alpha = .01$ | |
n	k_1	k_2	k_1	k_2
7	1	20		
8	2	26		
9	4	32		
10	6	39	1	44
11	8	47	2	53
12	11	55	4	62
13	14	64	6	72
14	17	74	9	82
15	21	84	12	93
16	26	94	15	105
17	30	106	18	118
18	35	118	22	131
19	41	130	27	144
20	46	144	31	159

Table 9.6: Confidence interval for median (see section 9.5.6)

9.5.7   Confidence interval for parameter in a Poisson distribution

The probability distribution function for a Poisson random variable is given by (see page 103)

$$f(x; \lambda) = \frac{e^{-\lambda}\lambda^x}{x!} \quad \text{for } x = 0, 1, 2, \ldots \tag{9.3}$$

For any value of x' and $\alpha < 1$, lower and upper values of λ (λ_{lower} and λ_{upper}) may be determined such that $\lambda_{\text{lower}} < \lambda_{\text{upper}}$ and

$$\sum_{x=0}^{x'} \frac{e^{-\lambda_{\text{lower}}}\lambda_{\text{lower}}^x}{x!} = \frac{\alpha}{2} \quad \text{and} \quad \sum_{x=x'}^{\infty} \frac{e^{-\lambda_{\text{upper}}}\lambda_{\text{upper}}^x}{x!} = \frac{\alpha}{2} \tag{9.4}$$

Table 9.8 lists λ_{lower} and λ_{upper} for $\alpha = 0.01$ and $\alpha = 0.05$. For $x' > 50$, λ_{upper} and λ_{lower} may be approximated by

$$\lambda_{\text{upper}} = \frac{\chi^2_{1-\alpha,n}}{2} \text{ where } 1 - F(\chi^2_{1-\alpha,n}) = \alpha, \text{ and } n = 2(x'+1)$$

$$\lambda_{\text{lower}} = \frac{\chi^2_{\alpha,n}}{2} \text{ where } F(\chi^2_{\alpha,n}) = \alpha, \text{ and } n = 2x' \tag{9.5}$$

where $F(\chi^2)$ is the cumulative distribution function for a chi–square random variable with n degrees of freedom.

Example 9.51: In a Poisson process, 5 outcomes are observed during a specified time interval. Find a 95% and a 99% confidence interval for the parameter λ in this Poisson process.

n	l	u	actual $\alpha \leq 0.05$	l	u	actual $\alpha \leq 0.01$
6	1	6	0.031			
7	1	7	0.016			
8	1	8	0.008	1	8	0.008
9	2	8	0.039	1	9	0.004
10	2	9	0.021	1	10	0.002
11	2	10	0.012	1	11	0.001
12	3	10	0.039	2	11	0.006
13	3	11	0.022	2	12	0.003
14	3	12	0.013	2	13	0.002
15	4	12	0.035	3	13	0.007
16	4	13	0.021	3	14	0.004
17	5	13	0.049	3	15	0.002
18	5	14	0.031	4	15	0.008
19	5	15	0.019	4	16	0.004
20	6	15	0.041	4	17	0.003
21	6	16	0.027	5	17	0.007
22	6	17	0.017	5	18	0.004
23	7	17	0.035	5	19	0.003
24	7	18	0.023	6	19	0.007
25	8	18	0.043	6	20	0.004
26	8	19	0.029	7	20	0.009
27	8	20	0.019	7	21	0.006
28	9	20	0.036	7	22	0.004
29	9	21	0.024	8	22	0.008
30	10	21	0.043	8	23	0.005
35	12	24	0.041	10	26	0.006
40	14	27	0.038	12	29	0.006
50	18	33	0.033	16	35	0.007
60	22	39	0.027	20	41	0.006
70	27	44	0.041	24	47	0.006
75	29	47	0.037	26	50	0.005
80	31	50	0.033	29	52	0.010
90	36	55	0.045	33	58	0.008
100	40	61	0.035	37	64	0.007
110	45	66	0.045	42	69	0.010
120	49	72	0.035	46	75	0.008

Table 9.7: Confidence intervals for medians

Solution:

(S1) Using Table 9.8 with an observed count of 5 and a 95% significance level ($\alpha = 0.05$), the confidence interval bounds are $\lambda_{lower} = 1.6$ and $\lambda_{upper} = 11.7$.

(S2) Hence, the probability is .95 that the interval $(1.6, 11.7)$ contains the true value of λ.

Observed count	Significance level				Observed count	Significance level			
	$\alpha = 0.01$		$\alpha = 0.05$			$\alpha = 0.01$		$\alpha = 0.05$	
	λ_{lower}	λ_{higher}	λ_{lower}	λ_{higher}		λ_{lower}	λ_{higher}	λ_{lower}	λ_{higher}
0	0.0	5.3	0.0	3.7	26	14.7	42.3	17.0	38.1
1	0.0	7.4	0.0	5.6	27	15.5	43.5	17.8	39.3
2	0.1	9.3	0.2	7.2	28	16.2	44.7	18.6	40.5
3	0.3	11.0	0.6	8.8	29	17.0	46.0	19.4	41.6
4	0.7	12.6	1.1	10.2	30	17.8	47.2	20.2	42.8
5	1.1	14.1	1.6	11.7	31	18.5	48.4	21.1	44.0
6	1.5	15.7	2.2	13.1	32	19.3	49.7	21.9	45.2
7	2.0	17.1	2.8	14.4	33	20.1	50.9	22.7	46.3
8	2.6	18.6	3.5	15.8	34	20.9	52.1	23.5	47.5
9	3.1	20.0	4.1	17.1	35	21.6	53.3	24.4	48.7
10	3.7	21.4	4.8	18.4	36	22.4	54.5	25.2	49.8
11	4.3	22.8	5.5	19.7	37	23.2	55.7	26.1	51.0
12	4.9	24.1	6.2	21.0	38	24.0	57.0	26.9	52.2
13	5.6	25.5	6.9	22.2	39	24.8	58.2	27.7	53.3
14	6.2	26.8	7.7	23.5	40	25.6	59.4	28.6	54.5
15	6.9	28.2	8.4	24.7	41	26.4	60.6	29.4	55.6
16	7.6	29.5	9.1	26.0	42	27.2	61.8	30.3	56.8
17	8.3	30.8	9.9	27.2	43	28.0	63.0	31.1	57.9
18	8.9	32.1	10.7	28.4	44	28.8	64.1	32.0	59.1
19	9.6	33.4	11.4	29.7	45	29.6	65.3	32.8	60.2
20	10.4	34.7	12.2	30.9	46	30.4	66.5	33.7	61.4
21	11.1	35.9	13.0	32.1	47	31.2	67.7	34.5	62.5
22	11.8	37.2	13.8	33.3	48	32.0	68.9	35.4	63.6
23	12.5	38.5	14.6	34.5	49	32.8	70.1	36.3	64.8
24	13.3	39.7	15.4	35.7	50	33.7	71.3	37.1	65.9
25	14.0	41.0	16.2	36.9					

Table 9.8: Confidence limits for the parameter in a Poisson distribution

(S3) Using Table 9.8 with an observed count of 5 and a 99% significance level ($\alpha = 0.01$), the confidence interval bounds are $\lambda_{lower} = 1.1$ and $\lambda_{upper} = 14.1$.

(S4) Hence, the probability is .99 that the interval $(1.1, 14.1)$ contains the true value of λ.

9.5.8 Confidence interval for parameter in a binomial distribution

The probability distribution function of a binomial random variable is given by (see page 84)

$$f(x; n, p) = \binom{n}{x} p^x (1 - p)^{n-x} \quad \text{for } x = 0, 1, 2, \ldots, n \qquad (9.6)$$

For known n, any value of x' less than n, and $\alpha < 1$, lower and upper values of p (p_{lower} and p_{upper}) may be determined such that $p_{lower} < p_{upper}$ and

$$\sum_{x=0}^{x'} f(x; n, p_{lower}) = \frac{\alpha}{2} \quad \text{and} \quad \sum_{x=x'}^{n} f(x; n, p_{upper}) = \frac{\alpha}{2} \qquad (9.7)$$

The tables on pages 206–221 list p_{lower} and p_{upper} for $\alpha = 0.01$ and $\alpha = 0.05$.

Example 9.52: In a binomial experiment with $n = 30$, $x = 8$ successes are observed. Determine a 95% and a 99% confidence interval for the probability of a success p.

Solution:

(S1) The table on page 210 may be used to construct a 95% confidence interval ($\alpha = 0.05$). Using this Table with $n - x = 22$ and $x = 8$ the bounds on the confidence interval are $p_{lower} = 0.123$ and $p_{upper} = 0.459$.

(S2) Hence, the probability is .95 that the interval $(0.123, 0.459)$ contains the true value of p.

(S3) Using the Table on page 218 for a 99% confidence level ($\alpha = 0.01$), the bounds on the confidence interval are $p_{lower} = 0.093$ and $p_{upper} = 0.516$.

(S4) Hence, the probability is .99 that the interval $(0.093, 0.516)$ contains the true value of p.

Confidence limits of proportions (confidence coefficient .95)

x	1	2	3	4	5	6	7	8	9

x	1	2	3	4	5	6	7	8	9
0	*0.000*	*0.000*	*0.000*	*0.000*	*0.000*	*0.000*	*0.000*	*0.000*	*0.000*
	0.975	0.842	0.708	0.602	0.522	0.459	0.410	0.369	0.336
1	*0.013*	*0.008*	*0.006*	*0.005*	*0.004*	*0.004*	*0.003*	*0.003*	*0.003*
	0.987	0.906	0.806	0.716	0.641	0.579	0.527	0.482	0.445
2	*0.094*	*0.068*	*0.053*	*0.043*	*0.037*	*0.032*	*0.028*	*0.025*	*0.023*
	0.992	0.932	0.853	0.777	0.710	0.651	0.600	0.556	0.518
3	*0.194*	*0.147*	*0.118*	*0.099*	*0.085*	*0.075*	*0.067*	*0.060*	*0.055*
	0.994	0.947	0.882	0.816	0.755	0.701	0.652	0.610	0.572
4	*0.284*	*0.223*	*0.184*	*0.157*	*0.137*	*0.122*	*0.109*	*0.099*	*0.091*
	0.995	0.957	0.901	0.843	0.788	0.738	0.692	0.651	0.614
5	*0.359*	*0.290*	*0.245*	*0.212*	*0.187*	*0.168*	*0.152*	*0.139*	*0.128*
	0.996	0.963	0.915	0.863	0.813	0.766	0.723	0.684	0.649
6	*0.421*	*0.349*	*0.299*	*0.262*	*0.234*	*0.211*	*0.192*	*0.177*	*0.163*
	0.996	0.968	0.925	0.878	0.832	0.789	0.749	0.711	0.677
7	*0.473*	*0.400*	*0.348*	*0.308*	*0.277*	*0.251*	*0.230*	*0.213*	*0.198*
	0.997	0.972	0.933	0.891	0.848	0.808	0.770	0.734	0.701
8	*0.518*	*0.444*	*0.390*	*0.349*	*0.316*	*0.289*	*0.266*	*0.247*	*0.230*
	0.997	0.975	0.940	0.901	0.861	0.823	0.787	0.753	0.722
9	*0.555*	*0.482*	*0.428*	*0.386*	*0.351*	*0.323*	*0.299*	*0.278*	*0.260*
	0.997	0.977	0.945	0.909	0.872	0.837	0.802	0.770	0.740
10	*0.587*	*0.516*	*0.462*	*0.419*	*0.384*	*0.354*	*0.329*	*0.308*	*0.289*
	0.998	0.979	0.950	0.916	0.882	0.848	0.816	0.785	0.756
11	*0.615*	*0.545*	*0.492*	*0.449*	*0.413*	*0.383*	*0.357*	*0.335*	*0.315*
	0.998	0.981	0.953	0.922	0.890	0.858	0.827	0.797	0.769
12	*0.640*	*0.572*	*0.519*	*0.476*	*0.440*	*0.410*	*0.384*	*0.361*	*0.340*
	0.998	0.982	0.957	0.927	0.897	0.867	0.837	0.809	0.782
13	*0.661*	*0.595*	*0.544*	*0.501*	*0.465*	*0.435*	*0.408*	*0.384*	*0.364*
	0.998	0.983	0.960	0.932	0.903	0.874	0.846	0.819	0.793
14	*0.680*	*0.617*	*0.566*	*0.524*	*0.488*	*0.457*	*0.430*	*0.407*	*0.385*
	0.998	0.984	0.962	0.936	0.909	0.881	0.854	0.828	0.803
15	*0.698*	*0.636*	*0.586*	*0.544*	*0.509*	*0.478*	*0.451*	*0.427*	*0.406*
	0.998	0.985	0.964	0.940	0.913	0.887	0.861	0.836	0.812

Table header: Denominator minus numerator: $n - x$ (Lower limit in italic type, upper limit in roman type)

Confidence limits of proportions (confidence coefficient .95)

x	Denominator minus numerator: $n - x$ (Lower limit in italic type, upper limit in roman type)								
	1	2	3	4	5	6	7	8	9
16	*0.713*	*0.653*	*0.604*	*0.563*	*0.528*	*0.498*	*0.471*	*0.447*	*0.425*
	0.998	0.986	0.966	0.943	0.918	0.893	0.868	0.844	0.820
17	*0.727*	*0.669*	*0.621*	*0.581*	*0.546*	*0.516*	*0.489*	*0.465*	*0.443*
	0.999	0.987	0.968	0.945	0.922	0.898	0.874	0.851	0.828
18	*0.740*	*0.683*	*0.637*	*0.597*	*0.563*	*0.533*	*0.506*	*0.482*	*0.460*
	0.999	0.988	0.970	0.948	0.925	0.902	0.879	0.857	0.835
19	*0.751*	*0.696*	*0.651*	*0.612*	*0.578*	*0.549*	*0.522*	*0.498*	*0.477*
	0.999	0.988	0.971	0.951	0.929	0.906	0.884	0.862	0.841
20	*0.762*	*0.708*	*0.664*	*0.626*	*0.593*	*0.564*	*0.537*	*0.513*	*0.492*
	0.999	0.989	0.972	0.953	0.932	0.910	0.889	0.868	0.847
22	*0.780*	*0.730*	*0.688*	*0.651*	*0.619*	*0.591*	*0.565*	*0.541*	*0.520*
	0.999	0.990	0.975	0.956	0.937	0.917	0.897	0.877	0.858
24	*0.796*	*0.749*	*0.708*	*0.673*	*0.642*	*0.614*	*0.589*	*0.566*	*0.545*
	0.999	0.991	0.977	0.960	0.942	0.923	0.904	0.885	0.867
26	*0.810*	*0.765*	*0.727*	*0.693*	*0.663*	*0.636*	*0.611*	*0.588*	*0.567*
	0.999	0.991	0.978	0.962	0.945	0.928	0.910	0.893	0.875
28	*0.822*	*0.779*	*0.742*	*0.710*	*0.681*	*0.655*	*0.631*	*0.608*	*0.588*
	0.999	0.992	0.980	0.965	0.949	0.932	0.916	0.899	0.882
30	*0.833*	*0.792*	*0.757*	*0.726*	*0.697*	*0.672*	*0.648*	*0.627*	*0.607*
	0.999	0.992	0.981	0.967	0.952	0.936	0.920	0.904	0.889
35	*0.855*	*0.818*	*0.786*	*0.758*	*0.732*	*0.708*	*0.686*	*0.666*	*0.647*
	0.999	0.993	0.983	0.971	0.958	0.944	0.930	0.916	0.902
40	*0.871*	*0.838*	*0.809*	*0.783*	*0.759*	*0.737*	*0.717*	*0.698*	*0.680*
	0.999	0.994	0.985	0.975	0.963	0.951	0.938	0.925	0.912
45	*0.885*	*0.855*	*0.828*	*0.804*	*0.782*	*0.761*	*0.742*	*0.724*	*0.707*
	0.999	0.995	0.987	0.977	0.967	0.956	0.944	0.933	0.921
50	*0.896*	*0.868*	*0.843*	*0.821*	*0.800*	*0.781*	*0.763*	*0.746*	*0.730*
	0.999	0.995	0.988	0.979	0.970	0.960	0.949	0.939	0.928
60	*0.912*	*0.888*	*0.867*	*0.848*	*0.830*	*0.813*	*0.797*	*0.781*	*0.767*
	1.000	0.996	0.990	0.983	0.975	0.966	0.957	0.948	0.939
80	*0.933*	*0.915*	*0.898*	*0.883*	*0.868*	*0.854*	*0.841*	*0.829*	*0.817*
	1.000	0.997	0.992	0.987	0.981	0.974	0.967	0.960	0.953
100	*0.946*	*0.931*	*0.917*	*0.904*	*0.892*	*0.881*	*0.870*	*0.859*	*0.849*
	1.000	0.998	0.994	0.989	0.984	0.979	0.973	0.967	0.962
∞	*1.000*	*1.000*	*1.000*	*1.000*	*1.000*	*1.000*	*1.000*	*1.000*	*1.000*
	1.000	1.000	1.000	1.000	1.000	1.000	1.000	1.000	1.000

Confidence limits of proportions (confidence coefficient .95)

x	10	11	12	13	14	15	16	17	18
0	*0.000*	*0.000*	*0.000*	*0.000*	*0.000*	*0.000*	*0.000*	*0.000*	*0.000*
	0.309	0.285	0.265	0.247	0.232	0.218	0.206	0.195	0.185
1	*0.002*	*0.002*	*0.002*	*0.002*	*0.002*	*0.002*	*0.002*	*0.001*	*0.001*
	0.413	0.385	0.360	0.339	0.320	0.302	0.287	0.273	0.260
2	*0.021*	*0.019*	*0.018*	*0.017*	*0.016*	*0.015*	*0.014*	*0.013*	*0.012*
	0.484	0.455	0.428	0.405	0.383	0.364	0.347	0.331	0.317
3	*0.050*	*0.047*	*0.043*	*0.040*	*0.038*	*0.036*	*0.034*	*0.032*	*0.030*
	0.538	0.508	0.481	0.456	0.434	0.414	0.396	0.379	0.363
4	*0.084*	*0.078*	*0.073*	*0.068*	*0.064*	*0.060*	*0.057*	*0.055*	*0.052*
	0.581	0.551	0.524	0.499	0.476	0.456	0.437	0.419	0.403
5	*0.118*	*0.110*	*0.103*	*0.097*	*0.091*	*0.087*	*0.082*	*0.078*	*0.075*
	0.616	0.587	0.560	0.535	0.512	0.491	0.472	0.454	0.437
6	*0.152*	*0.142*	*0.133*	*0.126*	*0.119*	*0.113*	*0.107*	*0.102*	*0.098*
	0.646	0.617	0.590	0.565	0.543	0.522	0.502	0.484	0.467
7	*0.184*	*0.173*	*0.163*	*0.154*	*0.146*	*0.139*	*0.132*	*0.126*	*0.121*
	0.671	0.643	0.616	0.592	0.570	0.549	0.529	0.511	0.494
8	*0.215*	*0.203*	*0.191*	*0.181*	*0.172*	*0.164*	*0.156*	*0.149*	*0.143*
	0.692	0.665	0.639	0.616	0.593	0.573	0.553	0.535	0.518
9	*0.244*	*0.231*	*0.218*	*0.207*	*0.197*	*0.188*	*0.180*	*0.172*	*0.165*
	0.711	0.685	0.660	0.636	0.615	0.594	0.575	0.557	0.540
10	*0.272*	*0.257*	*0.244*	*0.232*	*0.221*	*0.211*	*0.202*	*0.194*	*0.186*
	0.728	0.702	0.678	0.655	0.634	0.613	0.594	0.576	0.559
11	*0.298*	*0.282*	*0.268*	*0.256*	*0.244*	*0.234*	*0.224*	*0.215*	*0.207*
	0.743	0.718	0.694	0.672	0.651	0.631	0.612	0.594	0.577
12	*0.322*	*0.306*	*0.291*	*0.278*	*0.266*	*0.255*	*0.245*	*0.235*	*0.227*
	0.756	0.732	0.709	0.687	0.666	0.647	0.628	0.611	0.594
13	*0.345*	*0.328*	*0.313*	*0.299*	*0.287*	*0.275*	*0.264*	*0.255*	*0.245*
	0.768	0.744	0.722	0.701	0.680	0.661	0.643	0.626	0.609
14	*0.366*	*0.349*	*0.334*	*0.320*	*0.306*	*0.294*	*0.283*	*0.273*	*0.264*
	0.779	0.756	0.734	0.713	0.694	0.675	0.657	0.640	0.623
15	*0.387*	*0.369*	*0.353*	*0.339*	*0.325*	*0.313*	*0.302*	*0.291*	*0.281*
	0.789	0.766	0.745	0.725	0.706	0.687	0.669	0.653	0.637

Denominator minus numerator: $n - x$
(Lower limit in italic type, upper limit in roman type)

Confidence limits of proportions (confidence coefficient .95)

x	10	11	12	13	14	15	16	17	18
	Denominator minus numerator: $n - x$								
	(Lower limit in italic type, upper limit in roman type)								
16	*0.406*	*0.388*	*0.372*	*0.357*	*0.343*	*0.331*	*0.319*	*0.308*	*0.298*
	0.798	0.776	0.755	0.736	0.717	0.698	0.681	0.665	0.649
17	*0.424*	*0.406*	*0.389*	*0.374*	*0.360*	*0.347*	*0.335*	*0.324*	*0.314*
	0.806	0.785	0.765	0.745	0.727	0.709	0.692	0.676	0.660
18	*0.441*	*0.423*	*0.406*	*0.391*	*0.377*	*0.363*	*0.351*	*0.340*	*0.329*
	0.814	0.793	0.773	0.755	0.736	0.719	0.702	0.686	0.671
19	*0.457*	*0.439*	*0.422*	*0.406*	*0.392*	*0.379*	*0.367*	*0.355*	*0.344*
	0.821	0.801	0.782	0.763	0.745	0.728	0.712	0.696	0.681
20	*0.472*	*0.454*	*0.437*	*0.421*	*0.407*	*0.393*	*0.381*	*0.369*	*0.358*
	0.827	0.808	0.789	0.771	0.753	0.737	0.721	0.705	0.690
22	*0.500*	*0.482*	*0.465*	*0.449*	*0.435*	*0.421*	*0.408*	*0.396*	*0.385*
	0.839	0.820	0.803	0.785	0.769	0.753	0.737	0.722	0.707
24	*0.525*	*0.507*	*0.490*	*0.475*	*0.460*	*0.446*	*0.433*	*0.421*	*0.410*
	0.849	0.831	0.814	0.798	0.782	0.766	0.751	0.737	0.723
26	*0.548*	*0.530*	*0.513*	*0.498*	*0.483*	*0.469*	*0.456*	*0.444*	*0.433*
	0.858	0.841	0.825	0.809	0.794	0.779	0.764	0.750	0.737
28	*0.569*	*0.551*	*0.535*	*0.519*	*0.504*	*0.491*	*0.478*	*0.465*	*0.454*
	0.866	0.850	0.834	0.819	0.804	0.790	0.776	0.762	0.749
30	*0.588*	*0.570*	*0.554*	*0.539*	*0.524*	*0.510*	*0.497*	*0.485*	*0.473*
	0.873	0.858	0.843	0.828	0.814	0.800	0.786	0.773	0.760
35	*0.629*	*0.612*	*0.596*	*0.582*	*0.567*	*0.554*	*0.541*	*0.529*	*0.517*
	0.888	0.874	0.861	0.847	0.834	0.821	0.809	0.797	0.785
40	*0.663*	*0.647*	*0.632*	*0.617*	*0.603*	*0.590*	*0.578*	*0.566*	*0.555*
	0.900	0.887	0.875	0.862	0.850	0.839	0.827	0.816	0.805
45	*0.691*	*0.676*	*0.661*	*0.647*	*0.634*	*0.621*	*0.609*	*0.598*	*0.586*
	0.909	0.898	0.886	0.875	0.864	0.853	0.842	0.831	0.821
50	*0.715*	*0.700*	*0.686*	*0.673*	*0.660*	*0.648*	*0.636*	*0.625*	*0.614*
	0.917	0.906	0.896	0.885	0.875	0.865	0.855	0.845	0.835
60	*0.753*	*0.740*	*0.727*	*0.715*	*0.703*	*0.692*	*0.681*	*0.670*	*0.660*
	0.929	0.920	0.911	0.902	0.893	0.883	0.875	0.866	0.857
80	*0.805*	*0.794*	*0.783*	*0.773*	*0.763*	*0.753*	*0.743*	*0.734*	*0.725*
	0.945	0.938	0.931	0.923	0.916	0.909	0.902	0.894	0.887
100	*0.839*	*0.830*	*0.820*	*0.811*	*0.803*	*0.794*	*0.786*	*0.778*	*0.770*
	0.956	0.950	0.943	0.937	0.931	0.925	0.919	0.913	0.907
∞	*1.000*	*1.000*	*1.000*	*1.000*	*1.000*	*1.000*	*1.000*	*1.000*	*1.000*
	1.000	1.000	1.000	1.000	1.000	1.000	1.000	1.000	1.000

Confidence limits of proportions (confidence coefficient .95)

x	Denominator minus numerator: $n - x$ (Lower limit in italic type, upper limit in roman type)								
	19	20	22	24	26	28	30	35	40
0	*0.000*	*0.000*	*0.000*	*0.000*	*0.000*	*0.000*	*0.000*	*0.000*	*0.000*
	0.176	0.168	0.154	0.143	0.132	0.123	0.116	0.100	0.088
1	*0.001*	*0.001*	*0.001*	*0.001*	*0.001*	*0.001*	*0.001*	*0.001*	*0.001*
	0.249	0.238	0.220	0.204	0.190	0.178	0.167	0.145	0.129
2	*0.012*	*0.011*	*0.010*	*0.009*	*0.009*	*0.008*	*0.008*	*0.007*	*0.006*
	0.304	0.292	0.270	0.251	0.235	0.221	0.208	0.182	0.162
3	*0.029*	*0.028*	*0.025*	*0.023*	*0.022*	*0.020*	*0.019*	*0.017*	*0.015*
	0.349	0.336	0.312	0.292	0.273	0.258	0.243	0.214	0.191
4	*0.049*	*0.047*	*0.044*	*0.040*	*0.038*	*0.035*	*0.033*	*0.029*	*0.025*
	0.388	0.374	0.349	0.327	0.307	0.290	0.274	0.242	0.217
5	*0.071*	*0.068*	*0.063*	*0.058*	*0.055*	*0.051*	*0.048*	*0.042*	*0.037*
	0.422	0.407	0.381	0.358	0.337	0.319	0.303	0.268	0.241
6	*0.094*	*0.090*	*0.083*	*0.077*	*0.072*	*0.068*	*0.064*	*0.056*	*0.049*
	0.451	0.436	0.409	0.386	0.364	0.345	0.328	0.292	0.263
7	*0.116*	*0.111*	*0.103*	*0.096*	*0.090*	*0.084*	*0.080*	*0.070*	*0.062*
	0.478	0.463	0.435	0.411	0.389	0.369	0.352	0.314	0.283
8	*0.138*	*0.132*	*0.123*	*0.115*	*0.107*	*0.101*	*0.096*	*0.084*	*0.075*
	0.502	0.487	0.459	0.434	0.412	0.392	0.373	0.334	0.302
9	*0.159*	*0.153*	*0.142*	*0.133*	*0.125*	*0.118*	*0.111*	*0.098*	*0.088*
	0.523	0.508	0.480	0.455	0.433	0.412	0.393	0.353	0.320
10	*0.179*	*0.173*	*0.161*	*0.151*	*0.142*	*0.134*	*0.127*	*0.112*	*0.100*
	0.543	0.528	0.500	0.475	0.452	0.431	0.412	0.371	0.337
11	*0.199*	*0.192*	*0.180*	*0.169*	*0.159*	*0.150*	*0.142*	*0.126*	*0.113*
	0.561	0.546	0.518	0.493	0.470	0.449	0.430	0.388	0.353
12	*0.218*	*0.211*	*0.197*	*0.186*	*0.175*	*0.166*	*0.157*	*0.139*	*0.125*
	0.578	0.563	0.535	0.510	0.487	0.465	0.446	0.404	0.368
13	*0.237*	*0.229*	*0.215*	*0.202*	*0.191*	*0.181*	*0.172*	*0.153*	*0.138*
	0.594	0.579	0.551	0.525	0.502	0.481	0.461	0.418	0.383
14	*0.255*	*0.247*	*0.231*	*0.218*	*0.206*	*0.196*	*0.186*	*0.166*	*0.150*
	0.608	0.593	0.565	0.540	0.517	0.496	0.476	0.433	0.397
15	*0.272*	*0.263*	*0.247*	*0.234*	*0.221*	*0.210*	*0.200*	*0.179*	*0.161*
	0.621	0.607	0.579	0.554	0.531	0.509	0.490	0.446	0.410

Confidence limits of proportions (confidence coefficient .95)

x	Denominator minus numerator: $n - x$ (Lower limit in italic type, upper limit in roman type)								
	19	20	22	24	26	28	30	35	40
16	0.288	0.279	0.263	0.249	0.236	0.224	0.214	0.191	0.173
	0.633	0.619	0.592	0.567	0.544	0.522	0.503	0.459	0.422
17	0.304	0.295	0.278	0.263	0.250	0.238	0.227	0.203	0.184
	0.645	0.631	0.604	0.579	0.556	0.535	0.515	0.471	0.434
18	0.319	0.310	0.293	0.277	0.263	0.251	0.240	0.215	0.195
	0.656	0.642	0.615	0.590	0.567	0.546	0.527	0.483	0.445
19	0.334	0.324	0.307	0.291	0.277	0.264	0.252	0.227	0.206
	0.666	0.652	0.626	0.601	0.579	0.557	0.538	0.494	0.456
20	0.348	0.338	0.320	0.304	0.289	0.276	0.264	0.238	0.217
	0.676	0.662	0.636	0.612	0.589	0.568	0.548	0.504	0.467
22	0.374	0.364	0.346	0.329	0.314	0.300	0.287	0.260	0.237
	0.693	0.680	0.654	0.631	0.608	0.587	0.568	0.524	0.487
24	0.399	0.388	0.369	0.352	0.337	0.322	0.309	0.281	0.257
	0.709	0.696	0.671	0.648	0.626	0.605	0.586	0.543	0.505
26	0.421	0.411	0.392	0.374	0.358	0.343	0.330	0.300	0.276
	0.723	0.711	0.686	0.663	0.642	0.622	0.603	0.560	0.522
28	0.443	0.432	0.413	0.395	0.378	0.363	0.350	0.319	0.294
	0.736	0.724	0.700	0.678	0.657	0.637	0.618	0.575	0.538
30	0.462	0.452	0.432	0.414	0.397	0.382	0.368	0.337	0.311
	0.748	0.736	0.713	0.691	0.670	0.650	0.632	0.590	0.553
35	0.506	0.496	0.476	0.457	0.440	0.425	0.410	0.378	0.351
	0.773	0.762	0.740	0.719	0.700	0.681	0.663	0.622	0.586
40	0.544	0.533	0.513	0.495	0.478	0.462	0.447	0.414	0.386
	0.794	0.783	0.763	0.743	0.724	0.706	0.689	0.649	0.614
45	0.576	0.565	0.546	0.528	0.511	0.495	0.480	0.447	0.418
	0.811	0.801	0.782	0.763	0.745	0.728	0.711	0.673	0.639
50	0.604	0.594	0.575	0.557	0.540	0.524	0.510	0.476	0.447
	0.825	0.816	0.798	0.780	0.763	0.747	0.731	0.694	0.660
60	0.650	0.641	0.622	0.605	0.589	0.574	0.560	0.526	0.497
	0.849	0.840	0.824	0.808	0.792	0.777	0.763	0.728	0.697
80	0.717	0.708	0.692	0.676	0.662	0.648	0.634	0.603	0.575
	0.880	0.873	0.860	0.846	0.833	0.820	0.808	0.778	0.750
100	0.762	0.754	0.740	0.726	0.712	0.700	0.687	0.658	0.632
	0.901	0.895	0.883	0.872	0.861	0.849	0.839	0.812	0.787
∞	1.000	1.000	1.000	1.000	1.000	1.000	1.000	1.000	1.000
	1.000	1.000	1.000	1.000	1.000	1.000	1.000	1.000	1.000

Confidence limits of proportions (confidence coefficient .95)

x	Denominator minus numerator: $n - x$ (Lower limit in italics, upper limit in roman)					
	45	50	60	80	100	∞
0	*0.000*	*0.000*	*0.000*	*0.000*	*0.000*	*0.000*
	0.079	0.071	0.060	0.045	0.036	0.000
1	*0.001*	*0.001*	*0.000*	*0.000*	*0.000*	*0.000*
	0.115	0.104	0.088	0.067	0.054	0.000
2	*0.005*	*0.005*	*0.004*	*0.003*	*0.002*	*0.000*
	0.145	0.132	0.112	0.085	0.069	0.000
3	*0.013*	*0.012*	*0.010*	*0.008*	*0.006*	*0.000*
	0.172	0.157	0.133	0.102	0.083	0.000
4	*0.023*	*0.021*	*0.017*	*0.013*	*0.011*	*0.000*
	0.196	0.179	0.152	0.117	0.096	0.000
5	*0.033*	*0.030*	*0.025*	*0.019*	*0.016*	*0.000*
	0.218	0.200	0.170	0.132	0.108	0.000
6	*0.044*	*0.040*	*0.034*	*0.026*	*0.021*	*0.000*
	0.239	0.219	0.187	0.146	0.119	0.000
7	*0.056*	*0.051*	*0.043*	*0.033*	*0.027*	*0.000*
	0.258	0.237	0.203	0.159	0.130	0.000
8	*0.067*	*0.061*	*0.052*	*0.040*	*0.033*	*0.000*
	0.276	0.254	0.219	0.171	0.141	0.000
9	*0.079*	*0.072*	*0.061*	*0.047*	*0.038*	*0.000*
	0.293	0.270	0.233	0.183	0.151	0.000
10	*0.091*	*0.083*	*0.071*	*0.055*	*0.044*	*0.000*
	0.309	0.285	0.247	0.195	0.161	0.000
11	*0.102*	*0.094*	*0.080*	*0.062*	*0.050*	*0.000*
	0.324	0.300	0.260	0.206	0.170	0.000
12	*0.114*	*0.104*	*0.089*	*0.069*	*0.057*	*0.000*
	0.339	0.314	0.273	0.217	0.180	0.000
13	*0.125*	*0.115*	*0.098*	*0.077*	*0.063*	*0.000*
	0.353	0.327	0.285	0.227	0.189	0.000
14	*0.136*	*0.125*	*0.107*	*0.084*	*0.069*	*0.000*
	0.366	0.340	0.297	0.237	0.197	0.000
15	*0.147*	*0.135*	*0.117*	*0.091*	*0.075*	*0.000*
	0.379	0.352	0.308	0.247	0.206	0.000

Confidence limits of proportions (confidence coefficient .95)

x	Denominator minus numerator: $n - x$ (Lower limit in italics, upper limit in roman)					
	45	50	60	80	100	∞
16	*0.158*	*0.145*	*0.125*	*0.098*	*0.081*	*0.000*
	0.391	0.364	0.319	0.257	0.214	0.000
17	*0.169*	*0.155*	*0.134*	*0.106*	*0.087*	*0.000*
	0.402	0.375	0.330	0.266	0.222	0.000
18	*0.179*	*0.165*	*0.143*	*0.113*	*0.093*	*0.000*
	0.414	0.386	0.340	0.275	0.230	0.000
19	*0.189*	*0.175*	*0.151*	*0.120*	*0.099*	*0.000*
	0.424	0.396	0.350	0.283	0.238	0.000
20	*0.199*	*0.184*	*0.160*	*0.127*	*0.105*	*0.000*
	0.435	0.406	0.359	0.292	0.246	0.000
22	*0.218*	*0.202*	*0.176*	*0.140*	*0.117*	*0.000*
	0.454	0.425	0.378	0.308	0.260	0.000
24	*0.237*	*0.220*	*0.192*	*0.154*	*0.128*	*0.000*
	0.472	0.443	0.395	0.324	0.274	0.000
26	*0.255*	*0.237*	*0.208*	*0.167*	*0.139*	*0.000*
	0.489	0.460	0.411	0.338	0.288	0.000
28	*0.272*	*0.253*	*0.223*	*0.180*	*0.151*	*0.000*
	0.505	0.476	0.426	0.352	0.300	0.000
30	*0.289*	*0.269*	*0.237*	*0.192*	*0.161*	*0.000*
	0.520	0.490	0.440	0.366	0.313	0.000
35	*0.327*	*0.306*	*0.272*	*0.222*	*0.188*	*0.000*
	0.553	0.524	0.474	0.397	0.342	0.000
40	*0.361*	*0.340*	*0.303*	*0.250*	*0.213*	*0.000*
	0.582	0.553	0.503	0.425	0.368	0.000
45	*0.393*	*0.370*	*0.332*	*0.276*	*0.236*	*0.000*
	0.607	0.579	0.529	0.451	0.392	0.000
50	*0.421*	*0.398*	*0.359*	*0.301*	*0.259*	*0.000*
	0.630	0.602	0.552	0.474	0.415	0.000
60	*0.471*	*0.448*	*0.407*	*0.345*	*0.300*	*0.000*
	0.668	0.641	0.593	0.515	0.455	0.000
80	*0.549*	*0.526*	*0.485*	*0.420*	*0.371*	*0.000*
	0.724	0.699	0.655	0.580	0.520	0.000
100	*0.608*	*0.585*	*0.545*	*0.480*	*0.429*	*0.000*
	0.764	0.741	0.700	0.629	0.571	0.000
∞	*1.000*	*1.000*	*1.000*	*1.000*	*1.000*	*1.000*
	1.000	1.000	1.000	1.000	1.000	1.000

Confidence limits of proportions (confidence coefficient .99)

x	1	2	3	4	5	6	7	8	9

x	Denominator minus numerator: $n - x$ (Lower limit in italic type, upper limit in roman type)								
	1	2	3	4	5	6	7	8	9
0	*0.000*	*0.000*	*0.000*	*0.000*	*0.000*	*0.000*	*0.000*	*0.000*	*0.000*
	0.995	0.929	0.829	0.734	0.653	0.586	0.531	0.484	0.445
1	*0.003*	*0.002*	*0.001*	*0.001*	*0.001*	*0.001*	*0.001*	*0.001*	*0.001*
	0.997	0.959	0.889	0.815	0.746	0.685	0.632	0.585	0.544
2	*0.041*	*0.029*	*0.023*	*0.019*	*0.016*	*0.014*	*0.012*	*0.011*	*0.010*
	0.998	0.971	0.917	0.856	0.797	0.742	0.693	0.648	0.608
3	*0.111*	*0.083*	*0.066*	*0.055*	*0.047*	*0.042*	*0.037*	*0.033*	*0.030*
	0.999	0.977	0.934	0.882	0.830	0.781	0.735	0.693	0.655
4	*0.185*	*0.144*	*0.118*	*0.100*	*0.087*	*0.077*	*0.069*	*0.062*	*0.057*
	0.999	0.981	0.945	0.900	0.854	0.809	0.767	0.727	0.691
5	*0.254*	*0.203*	*0.170*	*0.146*	*0.128*	*0.114*	*0.103*	*0.094*	*0.087*
	0.999	0.984	0.953	0.913	0.872	0.831	0.791	0.755	0.720
6	*0.315*	*0.258*	*0.219*	*0.191*	*0.169*	*0.152*	*0.138*	*0.127*	*0.117*
	0.999	0.986	0.958	0.923	0.886	0.848	0.811	0.777	0.744
7	*0.368*	*0.307*	*0.265*	*0.233*	*0.209*	*0.189*	*0.172*	*0.159*	*0.147*
	0.999	0.988	0.963	0.931	0.897	0.862	0.828	0.795	0.764
8	*0.415*	*0.352*	*0.307*	*0.273*	*0.245*	*0.223*	*0.205*	*0.190*	*0.176*
	0.999	0.989	0.967	0.938	0.906	0.873	0.841	0.810	0.781
9	*0.456*	*0.392*	*0.345*	*0.309*	*0.280*	*0.256*	*0.236*	*0.219*	*0.205*
	0.999	0.990	0.970	0.943	0.913	0.883	0.853	0.824	0.795
10	*0.491*	*0.427*	*0.379*	*0.342*	*0.312*	*0.287*	*0.266*	*0.247*	*0.232*
	1.000	0.991	0.972	0.947	0.920	0.891	0.863	0.835	0.808
11	*0.523*	*0.459*	*0.411*	*0.373*	*0.341*	*0.315*	*0.293*	*0.274*	*0.257*
	1.000	0.992	0.974	0.951	0.925	0.899	0.872	0.845	0.819
12	*0.551*	*0.488*	*0.440*	*0.401*	*0.369*	*0.342*	*0.319*	*0.299*	*0.282*
	1.000	0.992	0.976	0.955	0.930	0.905	0.879	0.854	0.829
13	*0.576*	*0.514*	*0.466*	*0.427*	*0.394*	*0.367*	*0.343*	*0.323*	*0.305*
	1.000	0.993	0.978	0.957	0.935	0.910	0.886	0.862	0.838
14	*0.598*	*0.537*	*0.490*	*0.451*	*0.418*	*0.390*	*0.366*	*0.345*	*0.326*
	1.000	0.993	0.979	0.960	0.938	0.915	0.892	0.869	0.846
15	*0.619*	*0.559*	*0.512*	*0.473*	*0.440*	*0.412*	*0.388*	*0.366*	*0.347*
	1.000	0.994	0.980	0.962	0.942	0.920	0.898	0.875	0.854

Confidence limits of proportions (confidence coefficient .99)

x	\multicolumn{9}{c}{Denominator minus numerator: $n - x$ (Lower limit in italic type, upper limit in roman type)}
	1	2	3	4	5	6	7	8	9
16	*0.637*	*0.578*	*0.532*	*0.493*	*0.461*	*0.433*	*0.408*	*0.386*	*0.367*
	1.000	0.994	0.981	0.964	0.945	0.924	0.903	0.881	0.860
17	*0.654*	*0.596*	*0.550*	*0.512*	*0.480*	*0.452*	*0.427*	*0.405*	*0.385*
	1.000	0.994	0.982	0.966	0.947	0.927	0.907	0.887	0.866
18	*0.669*	*0.613*	*0.568*	*0.530*	*0.498*	*0.470*	*0.445*	*0.422*	*0.403*
	1.000	0.995	0.983	0.968	0.950	0.931	0.911	0.891	0.872
19	*0.683*	*0.628*	*0.584*	*0.547*	*0.515*	*0.486*	*0.461*	*0.439*	*0.419*
	1.000	0.995	0.984	0.969	0.952	0.934	0.915	0.896	0.877
20	*0.696*	*0.642*	*0.599*	*0.562*	*0.530*	*0.502*	*0.477*	*0.455*	*0.435*
	1.000	0.995	0.985	0.971	0.954	0.936	0.918	0.900	0.881
22	*0.719*	*0.668*	*0.626*	*0.590*	*0.559*	*0.531*	*0.507*	*0.484*	*0.464*
	1.000	0.996	0.986	0.973	0.958	0.941	0.924	0.907	0.890
24	*0.738*	*0.690*	*0.649*	*0.615*	*0.584*	*0.557*	*0.533*	*0.511*	*0.491*
	1.000	0.996	0.987	0.975	0.961	0.945	0.930	0.913	0.897
26	*0.755*	*0.709*	*0.670*	*0.637*	*0.607*	*0.581*	*0.557*	*0.535*	*0.515*
	1.000	0.996	0.988	0.977	0.963	0.949	0.934	0.919	0.904
28	*0.770*	*0.726*	*0.689*	*0.656*	*0.627*	*0.602*	*0.578*	*0.557*	*0.537*
	1.000	0.997	0.989	0.978	0.966	0.952	0.938	0.924	0.909
30	*0.784*	*0.741*	*0.705*	*0.674*	*0.646*	*0.621*	*0.597*	*0.576*	*0.557*
	1.000	0.997	0.989	0.980	0.968	0.955	0.942	0.928	0.914
35	*0.811*	*0.773*	*0.740*	*0.711*	*0.685*	*0.661*	*0.639*	*0.619*	*0.600*
	1.000	0.997	0.991	0.982	0.972	0.961	0.949	0.937	0.924
40	*0.832*	*0.797*	*0.767*	*0.741*	*0.716*	*0.694*	*0.673*	*0.654*	*0.636*
	1.000	0.997	0.992	0.984	0.975	0.965	0.955	0.944	0.933
45	*0.849*	*0.817*	*0.789*	*0.765*	*0.742*	*0.721*	*0.702*	*0.683*	*0.666*
	1.000	0.998	0.993	0.986	0.978	0.969	0.959	0.949	0.939
50	*0.863*	*0.834*	*0.808*	*0.785*	*0.763*	*0.744*	*0.725*	*0.708*	*0.692*
	1.000	0.998	0.993	0.987	0.980	0.972	0.963	0.954	0.945
60	*0.884*	*0.859*	*0.836*	*0.816*	*0.797*	*0.780*	*0.763*	*0.747*	*0.733*
	1.000	0.998	0.995	0.989	0.983	0.976	0.969	0.961	0.953
80	*0.912*	*0.892*	*0.874*	*0.858*	*0.842*	*0.828*	*0.814*	*0.801*	*0.789*
	1.000	0.999	0.996	0.992	0.987	0.982	0.976	0.970	0.964
100	*0.929*	*0.912*	*0.898*	*0.884*	*0.871*	*0.859*	*0.847*	*0.836*	*0.826*
	1.000	0.999	0.997	0.993	0.990	0.985	0.981	0.976	0.971
∞	*1.000*	*1.000*	*1.000*	*1.000*	*1.000*	*1.000*	*1.000*	*1.000*	*1.000*
	1.000	1.000	1.000	1.000	1.000	1.000	1.000	1.000	1.000

Confidence limits of proportions (confidence coefficient .99)

x	Denominator minus numerator: $n - x$ (Lower limit in italic type, upper limit in roman type)								
	10	11	12	13	14	15	16	17	18
0	*0.000*	*0.000*	*0.000*	*0.000*	*0.000*	*0.000*	*0.000*	*0.000*	*0.000*
	0.411	0.382	0.357	0.335	0.315	0.298	0.282	0.268	0.255
1	*0.000*	*0.000*	*0.000*	*0.000*	*0.000*	*0.000*	*0.000*	*0.000*	*0.000*
	0.509	0.477	0.449	0.424	0.402	0.381	0.363	0.346	0.331
2	*0.009*	*0.008*	*0.008*	*0.007*	*0.007*	*0.006*	*0.006*	*0.006*	*0.005*
	0.573	0.541	0.512	0.486	0.463	0.441	0.422	0.404	0.387
3	*0.028*	*0.026*	*0.024*	*0.022*	*0.021*	*0.020*	*0.019*	*0.018*	*0.017*
	0.621	0.589	0.560	0.534	0.510	0.488	0.468	0.450	0.432
4	*0.053*	*0.049*	*0.045*	*0.043*	*0.040*	*0.038*	*0.036*	*0.034*	*0.032*
	0.658	0.627	0.599	0.573	0.549	0.527	0.507	0.488	0.470
5	*0.080*	*0.075*	*0.070*	*0.065*	*0.062*	*0.058*	*0.055*	*0.053*	*0.050*
	0.688	0.659	0.631	0.606	0.582	0.560	0.539	0.520	0.502
6	*0.109*	*0.101*	*0.095*	*0.090*	*0.085*	*0.080*	*0.076*	*0.073*	*0.069*
	0.713	0.685	0.658	0.633	0.610	0.588	0.567	0.548	0.530
7	*0.137*	*0.128*	*0.121*	*0.114*	*0.108*	*0.102*	*0.097*	*0.093*	*0.089*
	0.734	0.707	0.681	0.657	0.634	0.612	0.592	0.573	0.555
8	*0.165*	*0.155*	*0.146*	*0.138*	*0.131*	*0.125*	*0.119*	*0.113*	*0.109*
	0.753	0.726	0.701	0.677	0.655	0.634	0.614	0.595	0.578
9	*0.192*	*0.181*	*0.171*	*0.162*	*0.154*	*0.146*	*0.140*	*0.134*	*0.128*
	0.768	0.743	0.718	0.695	0.674	0.653	0.633	0.615	0.597
10	*0.218*	*0.206*	*0.195*	*0.185*	*0.176*	*0.168*	*0.161*	*0.154*	*0.148*
	0.782	0.758	0.734	0.712	0.690	0.670	0.651	0.633	0.616
11	*0.242*	*0.229*	*0.218*	*0.207*	*0.197*	*0.189*	*0.181*	*0.173*	*0.167*
	0.794	0.771	0.748	0.726	0.706	0.686	0.667	0.649	0.632
12	*0.266*	*0.252*	*0.240*	*0.228*	*0.218*	*0.209*	*0.200*	*0.192*	*0.185*
	0.805	0.782	0.760	0.739	0.719	0.700	0.681	0.664	0.647
13	*0.288*	*0.274*	*0.261*	*0.249*	*0.238*	*0.228*	*0.219*	*0.211*	*0.203*
	0.815	0.793	0.772	0.751	0.732	0.713	0.695	0.677	0.661
14	*0.310*	*0.294*	*0.281*	*0.268*	*0.257*	*0.247*	*0.237*	*0.228*	*0.220*
	0.824	0.803	0.782	0.762	0.743	0.724	0.707	0.690	0.674
15	*0.330*	*0.314*	*0.300*	*0.287*	*0.276*	*0.265*	*0.255*	*0.246*	*0.237*
	0.832	0.811	0.791	0.772	0.753	0.735	0.718	0.701	0.685

Confidence limits of proportions (confidence coefficient .99)

x	Denominator minus numerator: $n - x$ (Lower limit in italic type, upper limit in roman type)								
	10	11	12	13	14	15	16	17	18
16	*0.349*	*0.333*	*0.319*	*0.305*	*0.293*	*0.282*	*0.272*	*0.262*	*0.253*
	0.839	0.819	0.800	0.781	0.763	0.745	0.728	0.712	0.696
17	*0.367*	*0.351*	*0.336*	*0.323*	*0.310*	*0.299*	*0.288*	*0.278*	*0.269*
	0.846	0.827	0.808	0.789	0.772	0.754	0.738	0.722	0.706
18	*0.384*	*0.368*	*0.353*	*0.339*	*0.326*	*0.315*	*0.304*	*0.294*	*0.284*
	0.852	0.833	0.815	0.797	0.780	0.763	0.747	0.731	0.716
19	*0.401*	*0.384*	*0.369*	*0.355*	*0.342*	*0.330*	*0.319*	*0.308*	*0.299*
	0.858	0.840	0.822	0.804	0.787	0.771	0.755	0.740	0.725
20	*0.417*	*0.400*	*0.384*	*0.370*	*0.357*	*0.345*	*0.333*	*0.323*	*0.313*
	0.863	0.845	0.828	0.811	0.794	0.778	0.763	0.748	0.733
22	*0.446*	*0.429*	*0.413*	*0.399*	*0.385*	*0.372*	*0.361*	*0.350*	*0.340*
	0.873	0.856	0.839	0.823	0.807	0.792	0.777	0.763	0.748
24	*0.472*	*0.455*	*0.439*	*0.425*	*0.411*	*0.398*	*0.386*	*0.375*	*0.364*
	0.881	0.865	0.849	0.834	0.819	0.804	0.789	0.776	0.762
26	*0.496*	*0.479*	*0.463*	*0.449*	*0.435*	*0.422*	*0.410*	*0.398*	*0.388*
	0.888	0.873	0.858	0.843	0.829	0.815	0.801	0.787	0.774
28	*0.518*	*0.502*	*0.486*	*0.471*	*0.457*	*0.444*	*0.432*	*0.420*	*0.409*
	0.894	0.880	0.866	0.852	0.838	0.824	0.811	0.798	0.785
30	*0.539*	*0.522*	*0.506*	*0.492*	*0.478*	*0.465*	*0.452*	*0.441*	*0.430*
	0.900	0.886	0.873	0.859	0.846	0.833	0.820	0.807	0.795
35	*0.583*	*0.567*	*0.551*	*0.537*	*0.523*	*0.510*	*0.498*	*0.486*	*0.475*
	0.912	0.900	0.887	0.875	0.863	0.851	0.839	0.828	0.816
40	*0.620*	*0.604*	*0.589*	*0.575*	*0.561*	*0.549*	*0.536*	*0.525*	*0.514*
	0.921	0.910	0.899	0.888	0.877	0.866	0.855	0.844	0.833
45	*0.650*	*0.635*	*0.621*	*0.607*	*0.594*	*0.582*	*0.570*	*0.558*	*0.548*
	0.929	0.919	0.908	0.898	0.888	0.878	0.867	0.857	0.848
50	*0.676*	*0.662*	*0.648*	*0.635*	*0.622*	*0.610*	*0.599*	*0.587*	*0.577*
	0.935	0.926	0.916	0.906	0.897	0.888	0.878	0.869	0.860
60	*0.719*	*0.705*	*0.692*	*0.680*	*0.668*	*0.657*	*0.646*	*0.636*	*0.626*
	0.945	0.937	0.928	0.920	0.912	0.903	0.895	0.887	0.879
80	*0.777*	*0.766*	*0.755*	*0.744*	*0.734*	*0.724*	*0.714*	*0.705*	*0.696*
	0.957	0.951	0.944	0.938	0.931	0.924	0.918	0.911	0.905
100	*0.815*	*0.806*	*0.796*	*0.787*	*0.778*	*0.769*	*0.761*	*0.752*	*0.744*
	0.965	0.960	0.955	0.949	0.944	0.938	0.933	0.927	0.921
∞	*1.000*	*1.000*	*1.000*	*1.000*	*1.000*	*1.000*	*1.000*	*1.000*	*1.000*
	1.000	1.000	1.000	1.000	1.000	1.000	1.000	1.000	1.000

Confidence limits of proportions (confidence coefficient .99)

x	19	20	22	24	26	28	30	35	40
	Denominator minus numerator: $n - x$ (Lower limit in italic type, upper limit in roman type)								
0	*0.000*	*0.000*	*0.000*	*0.000*	*0.000*	*0.000*	*0.000*	*0.000*	*0.000*
	0.243	0.233	0.214	0.198	0.184	0.172	0.162	0.140	0.124
1	*0.000*	*0.000*	*0.000*	*0.000*	*0.000*	*0.000*	*0.000*	*0.000*	*0.000*
	0.317	0.304	0.281	0.262	0.245	0.230	0.216	0.189	0.168
2	*0.005*	*0.005*	*0.004*	*0.004*	*0.004*	*0.003*	*0.003*	*0.003*	*0.003*
	0.372	0.358	0.332	0.310	0.291	0.274	0.259	0.227	0.203
3	*0.016*	*0.015*	*0.014*	*0.013*	*0.012*	*0.011*	*0.011*	*0.009*	*0.008*
	0.416	0.401	0.374	0.351	0.330	0.311	0.295	0.260	0.233
4	*0.031*	*0.029*	*0.027*	*0.025*	*0.023*	*0.022*	*0.020*	*0.018*	*0.016*
	0.453	0.438	0.410	0.385	0.363	0.344	0.326	0.289	0.259
5	*0.048*	*0.046*	*0.042*	*0.039*	*0.037*	*0.034*	*0.032*	*0.028*	*0.025*
	0.485	0.470	0.441	0.416	0.393	0.373	0.354	0.315	0.284
6	*0.066*	*0.064*	*0.059*	*0.055*	*0.051*	*0.048*	*0.045*	*0.039*	*0.035*
	0.514	0.498	0.469	0.443	0.419	0.398	0.379	0.339	0.306
7	*0.085*	*0.082*	*0.076*	*0.070*	*0.066*	*0.062*	*0.058*	*0.051*	*0.045*
	0.539	0.523	0.493	0.467	0.443	0.422	0.403	0.361	0.327
8	*0.104*	*0.100*	*0.093*	*0.087*	*0.081*	*0.076*	*0.072*	*0.063*	*0.056*
	0.561	0.545	0.516	0.489	0.465	0.443	0.424	0.381	0.346
9	*0.123*	*0.119*	*0.110*	*0.103*	*0.096*	*0.091*	*0.086*	*0.076*	*0.067*
	0.581	0.565	0.536	0.509	0.485	0.463	0.443	0.400	0.364
10	*0.142*	*0.137*	*0.127*	*0.119*	*0.112*	*0.106*	*0.100*	*0.088*	*0.079*
	0.599	0.583	0.554	0.528	0.504	0.482	0.461	0.417	0.380
11	*0.160*	*0.155*	*0.144*	*0.135*	*0.127*	*0.120*	*0.114*	*0.100*	*0.090*
	0.616	0.600	0.571	0.545	0.521	0.498	0.478	0.433	0.396
12	*0.178*	*0.172*	*0.161*	*0.151*	*0.142*	*0.134*	*0.127*	*0.113*	*0.101*
	0.631	0.616	0.587	0.561	0.537	0.514	0.494	0.449	0.411
13	*0.196*	*0.189*	*0.177*	*0.166*	*0.157*	*0.148*	*0.141*	*0.125*	*0.112*
	0.645	0.630	0.601	0.575	0.551	0.529	0.508	0.463	0.425
14	*0.213*	*0.206*	*0.193*	*0.181*	*0.171*	*0.162*	*0.154*	*0.137*	*0.123*
	0.658	0.643	0.615	0.589	0.565	0.543	0.522	0.477	0.439
15	*0.229*	*0.222*	*0.208*	*0.196*	*0.185*	*0.176*	*0.167*	*0.149*	*0.134*
	0.670	0.655	0.628	0.602	0.578	0.556	0.535	0.490	0.451

Confidence limits of proportions (confidence coefficient .99)

x	19	20	22	24	26	28	30	35	40
	Denominator minus numerator: $n - x$								
	(Lower limit in italic type, upper limit in roman type)								
16	*0.245*	*0.237*	*0.223*	*0.211*	*0.199*	*0.189*	*0.180*	*0.161*	*0.145*
	0.681	0.667	0.639	0.614	0.590	0.568	0.548	0.502	0.464
17	*0.260*	*0.252*	*0.237*	*0.224*	*0.213*	*0.202*	*0.193*	*0.172*	*0.156*
	0.692	0.677	0.650	0.625	0.602	0.580	0.559	0.514	0.475
18	*0.275*	*0.267*	*0.252*	*0.238*	*0.226*	*0.215*	*0.205*	*0.184*	*0.167*
	0.701	0.687	0.660	0.636	0.612	0.591	0.570	0.525	0.486
19	*0.289*	*0.281*	*0.265*	*0.251*	*0.239*	*0.227*	*0.217*	*0.195*	*0.177*
	0.711	0.697	0.670	0.646	0.623	0.601	0.581	0.536	0.497
20	*0.303*	*0.295*	*0.279*	*0.264*	*0.251*	*0.239*	*0.229*	*0.206*	*0.187*
	0.719	0.705	0.679	0.655	0.632	0.611	0.591	0.546	0.507
22	*0.330*	*0.321*	*0.304*	*0.289*	*0.275*	*0.263*	*0.251*	*0.227*	*0.207*
	0.735	0.721	0.696	0.672	0.650	0.629	0.609	0.565	0.526
24	*0.354*	*0.345*	*0.328*	*0.312*	*0.298*	*0.285*	*0.273*	*0.247*	*0.226*
	0.749	0.736	0.711	0.688	0.666	0.646	0.626	0.582	0.543
26	*0.377*	*0.368*	*0.350*	*0.334*	*0.319*	*0.306*	*0.293*	*0.266*	*0.244*
	0.761	0.749	0.725	0.702	0.681	0.661	0.642	0.598	0.560
28	*0.399*	*0.389*	*0.371*	*0.354*	*0.339*	*0.325*	*0.313*	*0.285*	*0.262*
	0.773	0.761	0.737	0.715	0.694	0.675	0.656	0.613	0.575
30	*0.419*	*0.409*	*0.391*	*0.374*	*0.358*	*0.344*	*0.331*	*0.303*	*0.279*
	0.783	0.771	0.749	0.727	0.707	0.687	0.669	0.626	0.589
35	*0.464*	*0.454*	*0.435*	*0.418*	*0.402*	*0.387*	*0.374*	*0.343*	*0.318*
	0.805	0.794	0.773	0.753	0.734	0.715	0.697	0.657	0.620
40	*0.503*	*0.493*	*0.474*	*0.457*	*0.440*	*0.425*	*0.411*	*0.380*	*0.353*
	0.823	0.813	0.793	0.774	0.756	0.738	0.721	0.682	0.647
45	*0.537*	*0.527*	*0.508*	*0.491*	*0.474*	*0.459*	*0.445*	*0.413*	*0.386*
	0.838	0.829	0.810	0.792	0.775	0.758	0.742	0.704	0.670
50	*0.567*	*0.557*	*0.538*	*0.521*	*0.505*	*0.489*	*0.475*	*0.443*	*0.415*
	0.851	0.842	0.824	0.807	0.791	0.775	0.759	0.723	0.690
60	*0.616*	*0.607*	*0.589*	*0.572*	*0.556*	*0.541*	*0.527*	*0.495*	*0.466*
	0.871	0.863	0.847	0.832	0.817	0.802	0.788	0.755	0.724
80	*0.687*	*0.679*	*0.662*	*0.647*	*0.632*	*0.618*	*0.605*	*0.574*	*0.547*
	0.898	0.892	0.879	0.866	0.853	0.841	0.829	0.800	0.773
100	*0.736*	*0.729*	*0.714*	*0.700*	*0.686*	*0.674*	*0.661*	*0.632*	*0.606*
	0.916	0.910	0.899	0.888	0.878	0.867	0.857	0.832	0.807
∞	*1.000*	*1.000*	*1.000*	*1.000*	*1.000*	*1.000*	*1.000*	*1.000*	*1.000*
	1.000	1.000	1.000	1.000	1.000	1.000	1.000	1.000	1.000

Confidence limits of proportions (confidence coefficient .99)

x	Denominator minus numerator: $n - x$ (Lower limit in italics, upper limit in roman)					
	45	50	60	80	100	∞
0	*0.000*	*0.000*	*0.000*	*0.000*	*0.000*	*0.000*
	0.111	0.101	0.085	0.064	0.052	0.000
1	*0.000*	*0.000*	*0.000*	*0.000*	*0.000*	*0.000*
	0.151	0.137	0.116	0.088	0.071	0.000
2	*0.002*	*0.002*	*0.002*	*0.001*	*0.001*	*0.000*
	0.183	0.166	0.141	0.108	0.088	0.000
3	*0.007*	*0.007*	*0.005*	*0.004*	*0.003*	*0.000*
	0.211	0.192	0.164	0.126	0.102	0.000
4	*0.014*	*0.013*	*0.011*	*0.008*	*0.007*	*0.000*
	0.235	0.215	0.184	0.142	0.116	0.000
5	*0.022*	*0.020*	*0.017*	*0.013*	*0.010*	*0.000*
	0.258	0.237	0.203	0.158	0.129	0.000
6	*0.031*	*0.028*	*0.024*	*0.018*	*0.015*	*0.000*
	0.279	0.256	0.220	0.172	0.141	0.000
7	*0.041*	*0.037*	*0.031*	*0.024*	*0.019*	*0.000*
	0.298	0.275	0.237	0.186	0.153	0.000
8	*0.051*	*0.046*	*0.039*	*0.030*	*0.024*	*0.000*
	0.317	0.292	0.253	0.199	0.164	0.000
9	*0.061*	*0.055*	*0.047*	*0.036*	*0.029*	*0.000*
	0.334	0.308	0.267	0.211	0.174	0.000
10	*0.071*	*0.065*	*0.055*	*0.043*	*0.035*	*0.000*
	0.350	0.324	0.281	0.223	0.185	0.000
11	*0.081*	*0.074*	*0.063*	*0.049*	*0.040*	*0.000*
	0.365	0.338	0.295	0.234	0.194	0.000
12	*0.092*	*0.084*	*0.072*	*0.056*	*0.045*	*0.000*
	0.379	0.352	0.308	0.245	0.204	0.000
13	*0.102*	*0.094*	*0.080*	*0.062*	*0.051*	*0.000*
	0.393	0.365	0.320	0.256	0.213	0.000
14	*0.112*	*0.103*	*0.088*	*0.069*	*0.056*	*0.000*
	0.406	0.378	0.332	0.266	0.222	0.000
15	*0.122*	*0.112*	*0.097*	*0.076*	*0.062*	*0.000*
	0.418	0.390	0.343	0.276	0.231	0.000
16	*0.133*	*0.122*	*0.105*	*0.082*	*0.067*	*0.000*
	0.430	0.401	0.354	0.286	0.239	0.000
17	*0.143*	*0.131*	*0.113*	*0.089*	*0.073*	*0.000*
	0.442	0.413	0.364	0.295	0.248	0.000

Confidence limits of proportions (confidence coefficient .99)

x	45	50	60	80	100	∞
	Denominator minus numerator: $n - x$					
	(Lower limit in italics, upper limit in roman)					
18	*0.152*	*0.140*	*0.121*	*0.095*	*0.079*	*0.000*
	0.452	0.423	0.374	0.304	0.256	0.000
19	*0.162*	*0.149*	*0.129*	*0.102*	*0.084*	*0.000*
	0.463	0.433	0.384	0.313	0.264	0.000
20	*0.171*	*0.158*	*0.137*	*0.108*	*0.090*	*0.000*
	0.473	0.443	0.393	0.321	0.271	0.000
22	*0.190*	*0.176*	*0.153*	*0.121*	*0.101*	*0.000*
	0.492	0.462	0.411	0.338	0.286	0.000
24	*0.208*	*0.193*	*0.168*	*0.134*	*0.112*	*0.000*
	0.509	0.479	0.428	0.353	0.300	0.000
26	*0.225*	*0.209*	*0.183*	*0.147*	*0.122*	*0.000*
	0.526	0.495	0.444	0.368	0.314	0.000
28	*0.242*	*0.225*	*0.198*	*0.159*	*0.133*	*0.000*
	0.541	0.511	0.459	0.382	0.326	0.000
30	*0.258*	*0.241*	*0.212*	*0.171*	*0.143*	*0.000*
	0.555	0.525	0.473	0.395	0.339	0.000
35	*0.296*	*0.277*	*0.245*	*0.200*	*0.168*	*0.000*
	0.587	0.557	0.505	0.426	0.368	0.000
40	*0.330*	*0.310*	*0.276*	*0.227*	*0.193*	*0.000*
	0.614	0.585	0.534	0.453	0.394	0.000
45	*0.362*	*0.341*	*0.305*	*0.253*	*0.216*	*0.000*
	0.638	0.609	0.559	0.478	0.418	0.000
50	*0.391*	*0.369*	*0.332*	*0.277*	*0.238*	*0.000*
	0.659	0.631	0.581	0.501	0.440	0.000
60	*0.441*	*0.419*	*0.380*	*0.321*	*0.278*	*0.000*
	0.695	0.668	0.620	0.541	0.479	0.000
80	*0.522*	*0.499*	*0.459*	*0.396*	*0.349*	*0.000*
	0.747	0.723	0.679	0.604	0.543	0.000
100	*0.582*	*0.560*	*0.521*	*0.457*	*0.407*	*0.000*
	0.784	0.762	0.722	0.651	0.593	0.000
∞	*1.000*	*1.000*	*1.000*	*1.000*	*1.000*	*1.000*
	1.000	1.000	1.000	1.000	1.000	1.000

9.6 CONFIDENCE INTERVALS: TWO SAMPLES

Let $x_1, x_2, \ldots, x_{n_1}$ be a random sample of size n_1 from population 1 and $y_1, y_2, \ldots, y_{n_2}$ a random sample of size n_2 from population 2.

9.6.1 Confidence interval for difference in means, known variances

Find a $100(1 - \alpha)\%$ confidence interval for the difference in means $\mu_1 - \mu_2$ if the populations are normal, σ_1^2 and σ_2^2 are known, and the samples are independent, or
Find a $100(1 - \alpha)\%$ confidence interval for the difference in means $\mu_1 - \mu_2$ if n_1 and n_2 are large, σ_1^2 and σ_2^2 are known, and the samples are independent.

(a) Compute the sample means $\bar{x}_1$ and $\bar{x}_2$.

(b) Determine the critical value $z_{\alpha/2}$ such that $\text{Prob}\left[Z \geq z_{\alpha/2}\right] = \alpha/2$.

(c) Compute the constant $k = z_{\alpha/2}\sqrt{\dfrac{\sigma_1^2}{n_1} + \dfrac{\sigma_2^2}{n_2}}$.

(d) A $100(1 - \alpha)\%$ confidence interval for $\mu_1 - \mu_2$ is given by $((\bar{x}_1 - \bar{x}_2) - k, (\bar{x}_1 - \bar{x}_2) + k)$.

9.6.2 Confidence interval for difference in means, equal unknown variances

Find a $100(1 - \alpha)\%$ confidence interval for the difference in means $\mu_1 - \mu_2$ if the populations are normal, the samples are independent, and the variances are unknown but assumed equal ($\sigma_1^2 = \sigma_2^2 = \sigma^2$).

(a) Compute the sample means $\bar{x}_1$ and $\bar{x}_2$, the sample variances s_1^2 and s_2^2, and the pooled estimate of the common variance σ^2

$$s_p^2 = \frac{(n_1 - 1)s_1^2 + (n_2 - 1)s_2^2}{n_1 + n_2 - 2}. \tag{9.8}$$

(b) Determine the critical value $t_{\alpha/2, n_1+n_2-2}$ such that
$\text{Prob}\left[T \geq t_{\alpha/2, n_1+n_2-2}\right] = \alpha/2$.

(c) Compute the constant $k = t_{\alpha/2, n_1+n_2-2} \cdot s_p \sqrt{\dfrac{1}{n_1} + \dfrac{1}{n_2}}$,

(d) A $100(1 - \alpha)\%$ confidence interval for $\mu_1 - \mu_2$ is given by $((\bar{x}_1 - \bar{x}_2) - k, (\bar{x}_1 - \bar{x}_2) + k)$.

9.6.3 Confidence interval for difference in means, unequal unknown variances

Find an approximate $100(1 - \alpha)\%$ confidence interval for the difference in means $\mu_1 - \mu_2$ if the populations are normal, the samples are independent, and the variances are unknown and unequal.

(a) Compute the sample means $\bar{x}_1$ and $\bar{x}_2$, the sample variances s_1^2 and s_2^2, and the approximate degrees of freedom

$$\nu = \frac{\left(\frac{s_1^2}{n_1} + \frac{s_2^2}{n_2}\right)^2}{\frac{(s_1^2/n_1)^2}{n_1-1} + \frac{(s_2^2/n_2)^2}{n_2-1}}. \tag{9.9}$$

Round ν to the nearest integer.

(b) Determine the critical value $t_{\alpha/2,\nu}$ such that $\text{Prob}\left[T \geq t_{\alpha/2,\nu}\right] = \alpha/2$.

(c) Compute the constant $k = t_{\alpha/2,\nu} \cdot \sqrt{\dfrac{s_1^2}{n_1} + \dfrac{s_2^2}{n_2}}$.

(d) An approximate $100(1-\alpha)\%$ confidence interval for $\mu_1 - \mu_2$ is given by $((\bar{x}_1 - \bar{x}_2) - k, (\bar{x}_1 - \bar{x}_2) + k)$.

9.6.4 Confidence interval for difference in means, paired observations

Find a $100(1-\alpha)\%$ confidence interval for the difference in means $\mu_1 - \mu_2$ if the populations are normal and the observations are paired (dependent).

(a) Compute the paired differences $x_1 - y_1, x_2 - y_2, \ldots, x_n - y_n$, the sample mean for the differences $\bar{d}$, and the sample variance for the differences s_d^2.

(b) Determine the critical value $t_{\alpha/2,n-1}$ such that $\text{Prob}\left[T \geq t_{\alpha/2,n-1}\right] = \alpha/2$.

(c) Compute the constant $k = t_{\alpha/2,n-1}s_d/\sqrt{n}$.

(d) A $100(1-\alpha)\%$ confidence interval for $\mu_1 - \mu_2$ is given by $(\bar{d} - k, \bar{d} + k)$.

9.6.5 Ratio of variances

Find a $100(1-\alpha)\%$ confidence interval for the ratio of variances σ_1^2/σ_2^2 if the populations are normal and the samples are independent.

(a) Compute the sample variances s_1^2 and s_2^2.

(b) Determine the critical values $F_{\alpha/2,n_1-1,n_2-1}$ and $F_{1-\alpha/2,n_1-2,n_2-1}$ such that $\text{Prob}\left[F \geq F_{\alpha/2,n_1-1,n_2-1}\right] = 1 - \alpha/2$ and $\text{Prob}\left[F \leq F_{1-\alpha/2,n_1-2,n_2-1}\right] = \alpha/2$.

(c) Compute the constants $k_1 = 1/F_{\alpha/2,n_1-1,n_2-1}$ and $k_2 = 1/F_{1-\alpha/2,n_1-1,n_2-1}$.

(d) A $100(1-\alpha)\%$ confidence interval for $\frac{\sigma_1^2}{\sigma_2^2}$ is $\left(\frac{s_1^2}{s_2^2}k_1, \frac{s_1^2}{s_2^2}k_2\right)$.

9.6.6 Difference in success probabilities

Find a $100(1-\alpha)\%$ confidence interval for the difference in success probabilities $p_1 - p_2$ if the samples are from a binomial experiment, the sample sizes are large, and the samples are independent.

(a) Compute the proportion of success for each sample $\hat{p}_1$ and $\hat{p}_2$.

(b) Determine the critical value $z_{\alpha/2}$ such that $\text{Prob}\left[Z \geq z_{\alpha/2}\right] = \alpha/2$.

(c) Compute the constant $k = z_{\alpha/2}\sqrt{\dfrac{\hat{p}_1(1-\hat{p}_1)}{n_1} + \dfrac{\hat{p}_2(1-\hat{p}_2)}{n_2}}$.

(d) A $100(1-\alpha)\%$ confidence interval for $\hat{p}_1 - \hat{p}_2$ is given by
$((\hat{p}_1 - \hat{p}_2) - k, (\hat{p}_1 - \hat{p}_2) + k)$.

Example 9.53: A researcher would like to compare the quality of incoming students at a public college and a private college. One measure of the strength of a class is the proportion of students who took an Advanced Placement test in High School. Random samples were selected at each institution. For the public college, $\hat{p}_1 = 190/500 = .38$ and for the private college, $\hat{p}_2 = 215/500 = .43$. Find a 95% confidence interval for the difference in proportions of students who took an AP test in high school.

Solution:

(S1) The samples are assumed to be from binomial experiments and independent, and the samples are large.

(S2) $1 - \alpha = .95$; $\alpha = .05$; $\alpha/2 = .025$; $z_{\alpha/2} = z_{.025} = 1.96$

(S3) $k = (1.96)\sqrt{\dfrac{(.38)(.62)}{500} + \dfrac{(.43)(.57)}{500}} = .0608$

(S4) A 95% confidence interval for $p_1 - p_2$: $((\hat{p}_1 - \hat{p}_2) - k, (\hat{p}_1 - \hat{p}_2) + k) = (-.1108, .0108)$

9.6.7 Difference in medians

The following technique, based on the Mann–Whitney–Wilcoxon procedure, may be used to find an approximate $100(1 - \alpha)\%$ confidence interval for the difference in medians, $\tilde{\mu}_1 - \tilde{\mu}_2$. Assume the sample sizes are large and the samples are independent.

(1) Compute the order statistics $\{w_{(1)}, w_{(2)}, \ldots, w_{(N)}\}$ for the $N = n_1 n_2$ differences $x_i - y_j$, for $1 \leq i \leq n_1$ and $1 \leq j \leq n_2$.

(2) Determine the critical value $z_{\alpha/2}$ such that $\text{Prob}\left[Z \geq z_{\alpha/2}\right] = \alpha/2$.

(3) Compute the constants
$$k_1 = \left\lceil \frac{n_1 n_2}{2} + 0.5 - z_{\alpha/2}\sqrt{\frac{n_1 n_2(n_1 + n_2 + 1)}{12}} \right\rceil \text{ and}$$

$$k_2 = \left\lceil \frac{n_1 n_2}{2} + 0.5 + z_{\alpha/2}\sqrt{\frac{n_1 n_2(n_1 + n_2 + 1)}{12}} \right\rceil.$$

(4) An approximate $100(1 - \alpha)\%$ confidence interval for $\tilde{\mu}_1 - \tilde{\mu}_2$ is given by $(w_{(k_1)}, w_{(k_2)})$.

9.7   FINITE POPULATION CORRECTION FACTOR

Suppose a sample of size n is taken without replacement from a (finite) population of size N. If n is large or a significant portion of the population then, intuitively, a point estimate based on this sample should be more accurate than if the population were infinite. In such cases, therefore, the standard deviation of the sample mean and the standard deviation of the sample proportion are corrected (multiplied) by the **finite population correction factor**:

$$\sqrt{\frac{N-n}{N-1}}. \tag{9.10}$$

When constructing a confidence interval, the *critical distance* is multiplied by this function of n and N to yield a more accurate interval estimate. If the sample size is *less* than 5% of the total population, the finite population correction factor is usually not applied.

Confidence intervals constructed using the finite population correction factor:

(1) Suppose a random sample of size n is taken from a population of size N. If the population is assumed normal, the endpoints for a $100(1-\alpha)\%$ confidence interval for the population mean μ are

$$\bar{x} \pm z_{\alpha/2} \cdot \frac{s}{\sqrt{n}} \cdot \sqrt{\frac{N-n}{N-1}}. \tag{9.11}$$

(2) In a binomial experiment, suppose a random sample of size n is taken from a population of size N. The endpoints for a $100(1-\alpha)\%$ confidence interval for the population proportion p are

$$\hat{p} \pm z_{\alpha/2} \cdot \sqrt{\frac{\hat{p}(1-\hat{p})}{n}} \cdot \sqrt{\frac{N-n}{N-1}}. \tag{9.12}$$

CHAPTER 10

Hypothesis Testing

Contents

10.1 INTRODUCTION

A hypothesis test is a formal procedure used to investigate a claim about one or more population parameters. Using the information in the sample the claim is either rejected or not rejected.

There are four parts to every hypothesis test:

(1) The **null hypothesis**, H_0, is a claim about the value of one or more population parameters; assumed to be true.

(2) The **alternative**, or (**research**), **hypothesis**, H_a, is an opposing statement; believed to be true if the null hypothesis is false.

(3) The **test statistic**, TS, is a quantity computed from the sample and used to decide whether or not to reject the null hypothesis.

(4) The **rejection region**, RR, is a set or interval of numbers selected in such a way that if the value of the test statistic lies in the rejection region the null hypothesis is rejected. One or more **critical values** separate the rejection region from the remaining values of the test statistic.

227

Decision

		Do not reject H_0	Reject H_0
Nature	H_0 True	Correct decision	Type I error: α
	H_0 False	Type II error: β	Correct decision

Table 10.1: Hypothesis test errors

There are two error probabilities associated with hypothesis testing; they are illustrated in Table 10.1 and described below.

(1) A **type I error** occurs if the null hypothesis is rejected when it is really true. The probability of a type I error is usually denoted by α, so that Prob [type I error] $= \alpha$. Common values of α include 0.05, 0.01, and 0.001.

(2) A **type II error** occurs if the null hypothesis is accepted when it is really false. The probability of a type II error depends upon the true value of the population parameter(s) and is usually denoted by β (or $\beta(\theta)$), so that Prob [type II error] $= \beta$. The **power** of the hypothesis test is $1 - \alpha$.

Note:

(1) α is the **significance level** of the hypothesis test. The test statistic is **significant** if it lies in the rejection region.

(2) The values α and β are inversely related, that is, when α increases then β decreases, and conversely.

(3) To decrease both α and β, increase the sample size.

The **p-value** is the smallest value of α (the smallest significance level) that would result in rejecting the null hypothesis. A p-value for a hypothesis test is often reported rather than whether or not the value of the test statistic lies in the rejection region.

10.1.1   Tables

Tables 10.2 and 10.3 contain hypothesis tests for one and two samples. The small numbers on the right-hand side of each table are for referencing these tests.

Example 10.54:   A breakfast cereal manufacturer claims each box is filled with 24 ounces of cereal. To check this claim, a consumer group randomly selected 17 boxes and carefully weighed the contents. The summary statistics: $\bar{x} = 23.55$ and $s = 1.5$. Is there any evidence to suggest the cereal boxes are underfilled? Use $\alpha = .05$.

Solution:

(S1) This is a question about a population mean μ. The distribution of cereal box weights is assumed normal and the population variance is unknown. A one-sample

Null hypothesis, assumptions	Alternative hypotheses	Test statistic	Rejection regions	
$\mu = \mu_0$, n large, σ^2 known, or normality, σ^2 known	$\mu > \mu_0$	$Z = \dfrac{\overline{X} - \mu_0}{\sigma/\sqrt{n}}$	$Z \geq z_\alpha$	(1)
	$\mu < \mu_0$		$Z \leq -z_\alpha$	(2)
	$\mu \neq \mu_0$		$|Z| \geq z_{\alpha/2}$	(3)
$\mu = \mu_0$, normality, σ^2 unknown	$\mu > \mu_0$	$T = \dfrac{\overline{X} - \mu_0}{S/\sqrt{n}}$	$T \geq t_{\alpha,n-1}$	(4)
	$\mu < \mu_0$		$T \leq -t_{\alpha,n-1}$	(5)
	$\mu \neq \mu_0$		$|T| \geq t_{\alpha/2,n-1}$	(6)
$\sigma^2 = \sigma_0^2$, normality	$\sigma^2 > \sigma_0^2$	$\chi^2 = \dfrac{(n-1)S^2}{\sigma_0^2}$	$\chi^2 \geq \chi^2_{\alpha,n-1}$	(7)
	$\sigma^2 < \sigma_0^2$		$\chi^2 \leq \chi^2_{1-\alpha,n-1}$	(8)
	$\sigma^2 \neq \sigma_0^2$		$\chi^2 \leq \chi^2_{1-\alpha/2,n-1}$, or $\chi^2 \geq \chi^2_{\alpha/2,n-1}$	(9)
$p = p_0$, binomial experiment, n large	$p > p_0$	$Z = \dfrac{\hat{p} - p_0}{\sqrt{p_0(1-p_0)/n}}$	$Z \geq z_\alpha$	(10)
	$p < p_0$		$Z \leq -z_\alpha$	(11)
	$p \neq p_0$		$|Z| \geq z_{\alpha/2}$	(12)

Table 10.2: Hypothesis tests: one sample

t test is appropriate (Table 10.2, number (5)).

(S2) The four parts to the hypothesis test are:

H_0: $\mu = 24 = \mu_0$

H_a: $\mu < 24$

TS: $T = \dfrac{\overline{X} - \mu_0}{S/\sqrt{n}}$

RR: $T \leq -t_{\alpha,n-1} = -t_{.05,16} = -1.7459$

(S3) $T = \dfrac{23.55 - 24}{1.5/\sqrt{17}} = -1.2369$

(S4) Conclusion: The value of the test statistic does not lie in the rejection region (equivalently, $p = .1170 > .05$). There is no evidence to suggest the population mean is less than 24 ounces.

Example 10.55: A newspaper article claimed the proportion of local residents in favor of a property tax increase to fund new educational programs is .45. A school board member selected 192 random residents and found 65 were in favor of the tax increase. Is there any evidence to suggest the proportion reported in the newspaper article is wrong? Use $\alpha = 0.1$.

Solution:

(S1) This is a question about a population proportion p. A binomial experiment is assumed and n is large. A one-sample test based on a Z statistic is appropriate (Table 10.2, number (12)).

(S2) The four parts to the hypothesis test are:

Assumptions				
Null hypothesis	Alternative hypotheses	Test statistic	Rejection regions	
n_1, n_2 large, independence, σ_1^2, σ_2^2 known, or normality, independence, σ_1^2, σ_2^2 known				
$\mu_1 - \mu_2 = \Delta_0$	$\mu_1 - \mu_2 > \Delta_0$	$Z = \dfrac{(\overline{X}_1 - \overline{X}_2) - \Delta_0}{\sqrt{\dfrac{\sigma_1^2}{n_1} + \dfrac{\sigma_2^2}{n_2}}}$	$Z \geq z_\alpha$	(1)
	$\mu_1 - \mu_2 < \Delta_0$		$Z \leq -z_\alpha$	(2)
	$\mu_1 - \mu_2 \neq \Delta_0$		$|Z| \geq z_{\alpha/2}$	(3)
normality, independence, $\sigma_1^2 = \sigma_2^2$ unknown				
$\mu_1 - \mu_2 = \Delta_0$	$\mu_1 - \mu_2 > \Delta_0$	$T = \dfrac{(\overline{X}_1 - \overline{X}_2) - \Delta_0}{S_p\sqrt{\dfrac{1}{n_1} + \dfrac{1}{n_2}}}$	$T \geq t_{\alpha, n_1 + n_2 - 2}$	(4)
	$\mu_1 - \mu_2 < \Delta_0$		$T \leq -t_{\alpha, n_1 + n_2 - 2}$	(5)
	$\mu_1 - \mu_2 \neq \Delta_0$		$|T| \geq t_{\frac{\alpha}{2}, n_1 + n_2 - 2}$	(6)
		$S_p = \dfrac{(n_1 - 1)S_1^2 + (n_2 - 1)S_2^2}{n_1 + n_2 - 2}$		
normality, independence, σ_1^2, σ_2^2 unknown, $\sigma_1^2 \neq \sigma_2^2$				
$\mu_1 - \mu_2 = \Delta_0$	$\mu_1 - \mu_2 > \Delta_0$	$T' = \dfrac{(\overline{X}_1 - \overline{X}_2) - \Delta_0}{\sqrt{\dfrac{S_1^2}{n_1} + \dfrac{S_2^2}{n_2}}}$	$T' \geq t_{\alpha, \nu}$	(7)
	$\mu_1 - \mu_2 < \Delta_0$		$T' \leq -t_{\alpha, \nu}$	(8)
	$\mu_1 - \mu_2 \neq \Delta_0$		$|T'| \geq t_{\alpha/2, \nu}$	(9)
		$\nu \approx \dfrac{\left(\dfrac{s_1^2}{n_1} + \dfrac{s_2^2}{n_2}\right)^2}{\dfrac{(s_1^2/n_1)^2}{n_1 - 1} + \dfrac{(s_2^2/n_2)^2}{n_2 - 1}}$		
normality, n pairs, dependence				
$\mu_D = \Delta_0$	$\mu_D > \Delta_0$	$T = \dfrac{\overline{D} - \Delta_0}{S_D/\sqrt{n}}$	$T \geq t_{\alpha, n-1}$	(10)
	$\mu_D < \Delta_0$		$T \leq -t_{\alpha, n-1}$	(11)
	$\mu_D \neq \Delta_0$		$|T| \geq t_{\alpha/2, n-1}$	(12)
normality, independence				
$\sigma_1^2 = \sigma_2^2$	$\sigma_1^2 > \sigma_2^2$	$F = S_1^2/S_2^2$	$F \geq F_{\alpha, n_1 - 1, n_2 - 1}$	(13)
	$\sigma_1^2 < \sigma_2^2$		$F \leq F_{1-\alpha, n_1 - 1, n_2 - 1}$	(14)
	$\sigma_1^2 \neq \sigma_2^2$		$F \leq F_{1-\frac{\alpha}{2}, n_1 - 1, n_2 - 1}$ or $F \geq F_{\frac{\alpha}{2}, n_1 - 1, n_2 - 1}$	(15)
binomial experiments, n_1, n_2 large, independence				
$p_1 = p_2 = 0$	$p_1 - p_2 > 0$	$Z = \dfrac{\widehat{p}_1 - \widehat{p}_2}{\sqrt{\widehat{p}\widehat{q}(1/n_1 + 1/n_2)}}$	$Z \geq z_\alpha$	(16)
	$p_1 - p_2 < 0$		$Z \leq -z_\alpha$	(17)
	$p_1 - p_2 \neq 0$	$\widehat{p} = \dfrac{X_1 + X_2}{n_1 + n_2}, \quad \widehat{q} = 1 - \widehat{p}$	$|Z| \geq z_{\alpha/2}$	(18)
binomial experiments, n_1, n_2 large, independence				
$p_1 - p_2 = \Delta_0$	$p_1 - p_2 > \Delta_0$	$Z = \dfrac{(\widehat{p}_1 - \widehat{p}_2) - \Delta_0}{\sqrt{\dfrac{\widehat{p}_1(1-\widehat{p}_1)}{n_1} + \dfrac{\widehat{p}_2(1-\widehat{p}_2)}{n_2}}}$	$Z \geq z_\alpha$	(19)
	$p_1 - p_2 < \Delta_0$		$Z \leq -z_\alpha$	(20)
	$p_1 - p_2 \neq \Delta_0$		$|Z| \geq z_{\alpha/2}$	(21)

Table 10.3: Hypothesis tests: two samples

H_0: $p = .45 = p_0$

H_a: $p \neq .45$

TS: $Z = \dfrac{\hat{p} - p_0}{\sqrt{p_0(1 - p_0)/n}}$

RR: $|Z| \geq z_{\alpha/2} = z_{.005} = 2.5758$

(S3) $\hat{p} = \dfrac{65}{192} = .3385$; $Z = \dfrac{.3385 - .45}{\sqrt{(.3385)(.6615)/192}} = -3.1044$

(S4) Conclusion: The value of the test statistic lies in the rejection region (equivalently, $p = .0019 < .005$). There is evidence to suggest the true proportion of residents in favor of the property tax increase is different from .45.

Example 10.56:   An automobile parts seller claims a new product when attached to an engine's air filter will significantly improve gas mileage. To test this claim, a consumer group randomly selected 10 cars and drivers. The miles per gallon for each automobile was recorded without the product and then using the new product. The summary statistics for the differences (before − after) were: $\bar{d} = -1.2$ and $s_D = 3.5$. Is there any evidence to suggest the new product improves gas mileage? Use $\alpha = .01$.

Solution:

(S1) This is a question about a difference in population means, μ_D. The data are assumed to be from a normal distribution and the observations are dependent. A paired t test is appropriate (Table 10.3, number (5)).

(S2) The four parts to the hypothesis test are:

H_0: $\mu_D = 0 = \Delta_0$

H_a: $\mu_D < 0$

TS: $T = \dfrac{\bar{D} - \Delta_0}{S_d/\sqrt{n}}$

RR: $T \leq -t_{\alpha, n-1} = -t_{.01, 9} = -2.8214$

(S3) $T = \dfrac{-1.2 - 0}{3.5/\sqrt{10}} = -1.0842$

(S4) Conclusion: The value of the test statistic does not lie in the rejection region (equivalently, $p = .1532 > .01$). There is no evidence to suggest the new product improves gas mileage.

10.2   THE NEYMAN–PEARSON LEMMA

Given the null hypothesis H_0: $\theta = \theta_0$ versus the alternative hypothesis H_a: $\theta = \theta_a$, let $L(\theta)$ be the likelihood function evaluated at θ. For a given α, the test that maximizes the power at θ_a has a rejection region determined by

$$\frac{L(\theta_0)}{L(\theta_a)} < k \tag{10.1}$$

This statistical test is the most powerful test of H_0 versus H_a.

10.3   LIKELIHOOD RATIO TESTS

Given the null hypothesis H_0: $\boldsymbol{\theta} \in \Omega_0$ versus the alternative hypothesis H_a: $\boldsymbol{\theta} \in \Omega_a$ with $\Omega_0 \cap \Omega_a = \phi$ and $\Omega = \Omega_0 \cup \Omega_a$. Let $L(\widehat{\Omega}_0)$ be the likelihood function with all unknown parameters replaced by their maximum likelihood estimators subject to the constraint $\boldsymbol{\theta} \in \Omega_0$, and let $L(\widehat{\Omega})$ be defined similarly so that $\boldsymbol{\theta} \in \Omega$. Define

$$\lambda = \frac{L(\widehat{\Omega}_0)}{L(\widehat{\Omega})}. \qquad (10.2)$$

A likelihood ratio test of H_0 versus H_a uses λ as a test statistic and has a rejection region given by $\lambda \leq k$ (for $0 < k < 1$).

Under very general conditions and for large n, $-2\ln\lambda$ has approximately a chi–square distribution with degrees of freedom equal to the number of parameters or functions of parameters assigned specific values under H_0.

10.4   GOODNESS OF FIT TEST

Let n_i be the number of observations falling into the ith category (for $i = 1, 2, \ldots, k$) and let $n = n_1 + n_2 + \cdots + n_k$.

H_0: $p_1 = p_{10}, \ p_2 = p_{20}, \ldots, \ p_k = p_{k0}$

H_a: $p_i \neq p_{i0}$ for at least one i

TS: $\chi^2 = \displaystyle\sum_{i=1}^{k} \frac{(\text{observed} - \text{estimated expected})^2}{\text{estimated expected}} = \sum_{i=1}^{k} \frac{(n_i - np_{i0})^2}{np_{i0}}$

Under the null hypothesis χ^2 has approximately a chi–square distribution with $k - 1$ degrees of freedom. The approximation is satisfactory if $n\,p_{i0} \geq 5$ for all i.

RR: $\chi^2 \geq \chi^2_{\alpha, k-1}$

Example 10.57:   The bookstore at a large university stocks four brands of graphing calculators. Recent sales figures indicated 55% of all graphing calculator sales were Texas Instruments (TI), 25% were Hewlett Packard (HP), 15% were Casio, and 5% were Sharp. This semester 200 graphing calculators were sold according to the table given below. Is there any evidence to suggest the sales proportions have changed? Use $\alpha = .05$.

Calculator Sales

TI	HP	Casio	Sharp
120	47	21	12

Solution:

(S1) There are $k = 4$ categories (of calculators) with unequal expected frequencies. The bookstore would like to determine if sales are consistent with previous results. This problem involves a goodness of fit test based on a chi–square distribution.

(S2) The four parts to the hypothesis test are:

H_0: $p_1 = .55$, $p_2 = .25$, $p_3 = .15$, $p_4 = .05$.

H_a: $p_i \neq p_{i0}$ for at least one i

TS: $\chi^2 = \displaystyle\sum_{i=1}^{4} \frac{(\text{observed} - \text{estimated expected})^2}{\text{estimated expected}} = \sum_{i=1}^{4} \frac{(n_i - np_{i0})^2}{np_{i0}}$

RR: $\chi^2 \geq \chi^2_{\alpha, k-1} = \chi^2_{.05, 3} = 7.8147$

(S3) $\chi^2 = \dfrac{(120 - 110)^2}{110} + \dfrac{(50 - 47)^2}{50} + \dfrac{(30 - 21)^2}{30} + \dfrac{10 - 12)^2}{10} = 4.1891$

(S4) Conclusion: The value of the test statistic does not lie in the rejection region (equivalently, $p = .2418 > .05$). There is no evidence to suggest the proportions of graphing calculator sales have changed.

If $k = 2$, this test is equivalent to a one proportion Z test, Table 10.2, number (3). This result follows from section 6.18.3 (page 149): If Z is a standard normal random variable, then Z^2 has a chi–square distribution with 1 degree of freedom.

10.5 CONTINGENCY TABLES

The general $I \times J$ contingency table has the form:

	Treatment 1	Treatment 1	...	Treatment J	Totals
Sample 1	n_{11}	n_{12}	...	n_{1J}	$n_{1.}$
Sample 2	n_{21}	n_{22}	...	n_{2J}	$n_{2.}$
$\vdots$	$\vdots$	$\vdots$	$\ddots$	$\vdots$	$\vdots$
Sample I	n_{I1}	n_{I2}	...	n_{IJ}	$n_{I.}$
Totals	$n_{.1}$	$n_{.2}$	...	$n_{.J}$	n

where $n_{k.} = \sum_{j=1}^{J} n_{kj}$ and $n_{.k} = \sum_{i=1}^{I} n_{ik}$. If complete independence is assumed, then the probability of any specific configuration, given the row and column totals $\{n_{.k}, n_{k.}\}$, is

$$\text{Prob}\,[n_{11}, \ldots, n_{IJ} \mid n_{1.}, \ldots, n_{.J}] = \frac{(\Pi_i^I n_{i.}!)(\Pi_j^J n_{.j}!)}{n!\ \Pi_i^I \Pi_j^J n_{ij}!} \quad (10.3)$$

Let a contingency table contain I rows and J columns, let n_{ij} be the count in the $(i, j)^{\text{th}}$ cell, and let $\hat{\epsilon}_{ij}$ be the estimated expected count in that cell. The test statistic is

$$\chi^2 = \sum_{\text{all cells}} \frac{(\text{observed} - \text{estimated expected})^2}{\text{estimated expected}} = \sum_{i=1}^{I} \sum_{j=1}^{J} \frac{(n_{ij} - \hat{\epsilon}_{ij})^2}{\hat{\epsilon}_{ij}} \quad (10.4)$$

where

$$\hat{\epsilon}_{ij} = \frac{(i^{\text{th}} \text{ row total})(j^{\text{th}} \text{ column total})}{\text{grand total}} = \frac{n_{i.}n_{.j}}{n} \qquad (10.5)$$

Under the null hypothesis χ^2 has approximately a chi–square distribution with $(I-1)(J-1)$ degrees of freedom. The approximation is satisfactory if $\hat{\epsilon}_{ij} \geq 5$ for all i and j.

Example 10.58: Recent reports indicate meals served during flights are rated similar regardless of airline. A survey given to randomly selected passengers asked each to rate the quality of in-flight meals. The results are given in the table below.

	A	B	C	D
Poor	42	35	22	23
Acceptable	50	75	33	28
Good	10	17	21	18

Airline

Is there any evidence to suggest the quality of meals differs by airline? Use $\alpha = .01$.

Solution:

(S1) The contingency table has $I = 3$ rows and $J = 4$ columns. To determine if the meal ratings differ by airline, a contingency table analysis is appropriate. The test statistic is based on a chi–square distribution.

(S2) The four parts to the hypothesis test are:

H_0: Airline and meal ratings are independent

H_a: Airline and meal ratings are dependent

TS: $\displaystyle \chi^2 = \sum_{i=1}^{3}\sum_{j=1}^{4}\frac{(n_{ij}-\hat{\epsilon}_{ij})^2}{\hat{\epsilon}_{ij}}$

RR: $\chi^2 \geq \chi^2_{.01,6} = 18.5476$

(S3) $\displaystyle \chi^2 = \frac{(42-33.27)^2}{33.27} + \frac{(35-41.43)^2}{41.43} + \frac{(22-24.79)^2}{24.79} + \frac{(23-22.51)^2}{22.51}$

$\displaystyle + \frac{(50-50.73)^2}{50.73} + \frac{(75-63.16)^2}{63.16} + \frac{(33-37.80)^2}{37.80} + \frac{(28-34.32)^2}{34.32}$

$\displaystyle + \frac{(10-18.00)^2}{18.00} + \frac{(17-22.41)^2}{22.41} + \frac{(21-13.41)^2}{13.41} + \frac{(18-12.18)^2}{12.18}$

$= 19.553$

(S4) The value of the test statistic lies in the rejection region (equivalently, $p = .003 <$ .01). There is evidence to suggest the meal rating proportions differ by airline.

10.6   BARTLETT'S TEST

Let there be k independent samples with n_i (for $i = 1, 2, \ldots, k$) observations in each sample, $N = n_1 + n_2 + \cdots + n_k$, and let S_i^2 be the ith sample variance.

H_0: $\sigma_1^2 = \sigma_2^2 = \cdots = \sigma_k^2$

H_a: the variances are not all equal

TS: $B = \dfrac{\left[(S_1^2)^{n_1-1}(S_2^2)^{n_2-1}\cdots(S_k^2)^{n_k-1}\right]^{1/(N-k)}}{S_p^2}$

$S_p^2 = \dfrac{\sum\limits_{i=1}^{k}(n_i - 1)S_i^2}{N - k}$

RR: $B \leq b_{\alpha,k,n}$ $\quad (n_1 = n_2 = \cdots = n_k = n)$

$B \leq b_{\alpha,k,n_1,n_2,\ldots,n_k}$ $\quad$ (when sample sizes are unequal)

where $b_{\alpha,k,n_1,n_2,\ldots,n_k} \approx \dfrac{n_1 b_{\alpha,k,n_1} + n_2 b_{\alpha,k,n_2} + \cdots + n_k b_{\alpha,k,n_k}}{N}$

Here $b_{\alpha,k,n}$ is a critical value for Bartlett's test with α being the significance level, k is the number of populations, and n is the sample size from each population. A table of values is in section 10.6.2.

10.6.1 Approximate test procedure

Let $\nu_i = n_i - 1$

TS: $\chi^2 = M/C$ $\quad$ where

$$M = \left(\sum_{i=1}^{k}\nu_i\right)\ln \overline{S}^2 - \sum_{i=1}^{k}\nu_i \ln S_i^2, \quad \overline{S}^2 = \sum_{i=1}^{k}\nu_i S_i^2 \left/ \sum_{i=1}^{k}\nu_i\right.$$

$$C = 1 + \frac{1}{3(k-1)}\left[\sum_{i=1}^{k}\frac{1}{\nu_i} - \left(1 \left/ \sum_{i=1}^{k}\nu_i\right.\right)\right]$$

Under the null hypothesis χ^2 has approximately a chi–square distribution with $k - 1$ degrees of freedom.

RR: $\chi^2 \geq \chi^2_{\alpha,k-1}$

10.6.2 Tables for Bartlett's test

These tables contain critical values, $b_{\alpha,k,n}$, for Bartlett's test where α is the significance level, k is the number of populations, and n is the sample size from each population. These tables are from D. D. Dyer and J. P. Keating, "On the Determination of Critical Values for Bartlett's Test", *JASA*, Volume 75, 1980, pages 313–319. Reprinted with permission from the *Journal of American Statistical Association*. Copyright 1980 by the American Statistical Association. All rights reserved.

Critical values for Bartlett's test, $b_{\alpha,k,n}$

$\alpha = .05$ k

n	2	3	4	5	6	7	8	9	10
3	.3123	.3058	.3173	.3299	*	*	*	*	*
4	.4780	.4699	.4803	.4921	.5028	.5122	.5204	.5277	.5341
5	.5845	.5762	.5850	.5952	.6045	.6126	.6197	.6260	.6315
6	.6563	.6483	.6559	.6646	.6727	.6798	.6860	.6914	.6961
7	.7075	.7000	.7065	.7142	.7213	.7275	.7329	.7376	.7418
8	.7456	.7387	.7444	.7512	.7574	.7629	.7677	.7719	.7757
9	.7751	.7686	.7737	.7798	.7854	.7903	.7946	.7984	.8017
10	.7984	.7924	.7970	.8025	.8076	.8121	.8160	.8194	.8224
11	.8175	.8118	.8160	.8210	.8257	.8298	.8333	.8365	.8392
12	.8332	.8280	.8317	.8364	.8407	.8444	.8477	.8506	.8531
13	.8465	.8415	.8450	.8493	.8533	.8568	.8598	.8625	.8648
14	.8578	.8532	.8564	.8604	.8641	.8673	.8701	.8726	.8748
15	.8676	.8632	.8662	.8699	.8734	.8764	.8790	.8814	.8834
16	.8761	.8719	.8747	.8782	.8815	.8843	.8868	.8890	.8909
17	.8836	.8796	.8823	.8856	.8886	.8913	.8936	.8957	.8975
18	.8902	.8865	.8890	.8921	.8949	.8975	.8997	.9016	.9033
19	.8961	.8926	.8949	.8979	.9006	.9030	.9051	.9069	.9086
20	.9015	.8980	.9003	.9031	.9057	.9080	.9100	.9117	.9132
21	.9063	.9030	.9051	.9078	.9103	.9124	.9143	.9160	.9175
22	.9106	.9075	.9095	.9120	.9144	.9165	.9183	.9199	.9213
23	.9146	.9116	.9135	.9159	.9182	.9202	.9219	.9235	.9248
24	.9182	.9153	.9172	.9195	.9217	.9236	.9253	.9267	.9280
25	.9216	.9187	.9205	.9228	.9249	.9267	.9283	.9297	.9309
26	.9246	.9219	.9236	.9258	.9278	.9296	.9311	.9325	.9336
27	.9275	.9249	.9265	.9286	.9305	.9322	.9337	.9350	.9361
28	.9301	.9276	.9292	.9312	.9330	.9347	.9361	.9374	.9385
29	.9326	.9301	.9316	.9336	.9354	.9370	.9383	.9396	.9406
30	.9348	.9325	.9340	.9358	.9376	.9391	.9404	.9416	.9426
40	.9513	.9495	.9506	.9520	.9533	.9545	.9555	.9564	.9572
50	.9612	.9597	.9606	.9617	.9628	.9637	.9645	.9652	.9658
60	.9677	.9665	.9672	.9681	.9690	.9698	.9705	.9710	.9716
80	.9758	.9749	.9754	.9761	.9768	.9774	.9779	.9783	.9787
100	.9807	.9799	.9804	.9809	.9815	.9819	.9823	.9827	.9830

Critical values for Bartlett's test, $b_{\alpha,k,n}$

$\alpha = .01$ $\hspace{4cm}$ k

n	2	3	4	5	6	7	8	9	10
3	.1411	.1672	*	*	*	*	*	*	*
4	.2843	.3165	.3475	.3729	.3937	.4110	*	*	*
5	.3984	.4304	.4607	.4850	.5046	.5207	.5343	.5458	.5558
6	.4850	.5149	.5430	.5653	.5832	.5978	.6100	.6204	.6293
7	.5512	.5787	.6045	.6248	.6410	.6542	.6652	.6744	.6824
8	.6031	.6282	.6518	.6704	.6851	.6970	.7069	.7153	.7225
9	.6445	.6676	.6892	.7062	.7197	.7305	.7395	.7471	.7536
10	.6783	.6996	.7195	.7352	.7475	.7575	.7657	.7726	.7786
11	.7063	.7260	.7445	.7590	.7703	.7795	.7871	.7935	.7990
12	.7299	.7483	.7654	.7789	.7894	.7980	.8050	.8109	.8160
13	.7501	.7672	.7832	.7958	.8056	.8135	.8201	.8256	.8303
14	.7674	.7835	.7985	.8103	.8195	.8269	.8330	.8382	.8426
15	.7825	.7977	.8118	.8229	.8315	.8385	.8443	.8491	.8532
16	.7958	.8101	.8235	.8339	.8421	.8486	.8541	.8586	.8625
17	.8076	.8211	.8338	.8436	.8514	.8576	.8627	.8670	.8707
18	.8181	.8309	.8429	.8523	.8596	.8655	.8704	.8745	.8780
19	.8275	.8397	.8512	.8601	.8670	.8727	.8773	.8811	.8845
20	.8360	.8476	.8586	.8671	.8737	.8791	.8835	.8871	.8903
21	.8437	.8548	.8653	.8734	.8797	.8848	.8890	.8926	.8956
22	.8507	.8614	.8714	.8791	.8852	.8901	.8941	.8975	.9004
23	.8571	.8673	.8769	.8844	.8902	.8949	.8988	.9020	.9047
24	.8630	.8728	.8820	.8892	.8948	.8993	.9030	.9061	.9087
25	.8684	.8779	.8867	.8936	.8990	.9034	.9069	.9099	.9124
26	.8734	.8825	.8911	.8977	.9029	.9071	.9105	.9134	.9158
27	.8781	.8869	.8951	.9015	.9065	.9105	.9138	.9166	.9190
28	.8824	.8909	.8988	.9050	.9099	.9138	.9169	.9196	.9219
29	.8864	.8946	.9023	.9083	.9130	.9167	.9198	.9224	.9246
30	.8902	.8981	.9056	.9114	.9159	.9195	.9225	.9250	.9271
40	.9175	.9235	.9291	.9335	.9370	.9397	.9420	.9439	.9455
50	.9339	.9387	.9433	.9468	.9496	.9518	.9536	.9551	.9564
60	.9449	.9489	.9527	.9557	.9580	.9599	.9614	.9626	.9637
80	.9586	.9617	.9646	.9668	.9685	.9699	.9711	.9720	.9728
100	.9669	.9693	.9716	.9734	.9748	.9759	.9769	.9776	.9783

10.7   COCHRAN'S TEST

Let there be k independent samples with n observations in each sample, and let S_i^2 be the ith sample variance (for $i = 1, 2, \ldots, k$).

H_0: $\sigma_1^2 = \sigma_2^2 = \cdots = \sigma_k^2$

H_a: the variances are not all equal

TS: $G = \text{largest } S_i^2 \left/ \displaystyle\sum_{i=1}^{k} S_i^2 \right.$

RR: $G \geq g_{\alpha,k,n}$

Here $g_{\alpha,k,n}$ is a critical value for Cochran's test with α being the significance level, k is the number of populations, and n is the sample size from each population. A table of values is in section 10.7.1.

10.7.1   Tables for Cochran's test

These tables contain critical values, $g_{\alpha,k,n}$, for Cochran's test where α is the significance level, k is the number of independent estimates of variance, each of which is based on n degrees of freedom. These tables are from C. Eisenhart, M. W. Hastay, and W. A. Wallis, *Techniques of Statistical Analysis*, McGraw-Hill Book Company, 1947, Tables 15.1 and 15.2 (pages 390-391). Reprinted courtesy of *The McGraw-Hill Companies*.

Critical values for Cochran's test, $g_{\alpha,k,n}$

$\alpha = .05$

k	n=1	2	3	4	5	6	7	8	9	10	16	36	144	∞
2	.9985	.9750	.9392	.9057	.8772	.8534	.8332	.8159	.8010	.7880	.7341	.6602	.5813	.5000
3	.9669	.8709	.7977	.7457	.7071	.6771	.6530	.6333	.6167	.6025	.5466	.4748	.4031	.3333
4	.9065	.7679	.6841	.6287	.5895	.5598	.5365	.5175	.5017	.4884	.4366	.3720	.3093	.2500
5	.8412	.6838	.5981	.5441	.5065	.4783	.4564	.4387	.4241	.4118	.3645	.3066	.2513	.2000
6	.7808	.6161	.5321	.4803	.4447	.4184	.3980	.3817	.3682	.3568	.3135	.2612	.2119	.1667
7	.7271	.5612	.4800	.4307	.3974	.3726	.3535	.3384	.3259	.3154	.2756	.2278	.1833	.1429
8	.6798	.5157	.4377	.3910	.3595	.3362	.3185	.3043	.2926	.2829	.2462	.2022	.1616	.1250
9	.6385	.4775	.4027	.3584	.3286	.3067	.2901	.2768	.2659	.2568	.2226	.1820	.1446	.1111
10	.6020	.4450	.3733	.3311	.3029	.2823	.2666	.2541	.2439	.2353	.2032	.1655	.1308	.1000
12	.5410	.3924	.3264	.2880	.2624	.2439	.2299	.2187	.2098	.2020	.1737	.1403	.1100	.0833
15	.4709	.3346	.2758	.2419	.2195	.2034	.1911	.1815	.1736	.1671	.1429	.1144	.0889	.0667
20	.3894	.2705	.2205	.1921	.1735	.1602	.1501	.1422	.1357	.1303	.1108	.0879	.0675	.0500
24	.3434	.2354	.1907	.1656	.1493	.1374	.1286	.1216	.1160	.1113	.0942	.0743	.0567	.0417
30	.2929	.1980	.1593	.1377	.1237	.1137	.1061	.1002	.0958	.0921	.0771	.0604	.0457	.0333
40	.2370	.1576	.1259	.1082	.0968	.0887	.0827	.0780	.0745	.0713	.0595	.0462	.0347	.0250
60	.1737	.1131	.0895	.0765	.0682	.0623	.0583	.0552	.0520	.0497	.0411	.0316	.0234	.0167
120	.0998	.0632	.0495	.0419	.0371	.0337	.0312	.0292	.0279	.0266	.0218	.0165	.0120	.0083
∞	0	0	0	0	0	0	0	0	0	0	0	0	0	0

Critical values for Cochran's test, $g_{\alpha,k,n}$

$\alpha = .05$

k	1	2	3	4	5	6	7	8	9	10	16	36	144	∞
2	.9999	.9950	.9794	.9586	.9373	.9172	.8988	.8823	.8674	.8539	.7949	.7067	.6062	.5000
3	.9933	.9423	.8831	.8335	.7933	.7606	.7335	.7107	.6912	.6743	.6059	.5153	.4230	.3333
4	.9676	.8643	.7814	.7212	.6761	.6410	.6129	.5897	.5702	.5536	.4884	.4057	.3251	.2500
5	.9279	.7885	.6957	.6329	.5875	.5531	.5259	.5037	.4854	.4697	.4094	.3351	.2644	.2000
6	.8828	.7218	.6258	.5635	.5195	.4866	.4608	.4401	.4229	.4084	.3529	.2858	.2229	.1667
7	.8376	.6644	.5685	.5080	.4659	.4347	.4105	.3911	.3751	.3616	.3105	.2494	.1929	.1429
8	.7945	.6152	.5209	.4627	.4226	.3932	.3704	.3522	.3373	.3248	.2779	.2214	.1700	.1250
9	.7544	.5727	.4810	.4251	.3870	.3592	.3378	.3207	.3067	.2950	.2514	.1992	.1521	.1111
10	.7175	.5358	.4469	.3934	.3572	.3308	.3106	.2945	.2813	.2704	.2297	.1811	.1376	.1000
12	.6528	.4751	.3919	.3428	.3099	.2861	.2680	.2535	.2419	.2320	.1961	.1535	.1157	.0833
15	.5747	.4069	.3317	.2882	.2593	.2386	.2228	.2104	.2002	.1918	.1612	.1251	.0934	.0667
20	.4799	.3297	.2654	.2288	.2048	.1877	.1748	.1646	.1567	.1501	.1248	.0960	.0709	.0500
24	.4247	.2871	.2295	.1970	.1759	.1608	.1495	.1406	.1338	.1283	.1060	.0810	.0595	.0417
30	.3632	.2412	.1913	.1635	.1454	.1327	.1232	.1157	.1100	.1054	.0867	.0658	.0480	.0333
40	.2940	.1915	.1508	.1281	.1135	.1033	.0957	.0898	.0853	.0816	.0668	.0503	.0363	.0250
60	.2151	.1371	.1069	.0902	.0796	.0722	.0668	.0625	.0594	.0567	.0461	.0344	.0245	.0167
120	.1225	.0759	.0585	.0489	.0429	.0387	.0357	.0334	.0316	.0302	.0242	.0178	.0125	.0083
∞	0	0	0	0	0	0	0	0	0	0	0	0	0	0

10.8 NUMBER OF OBSERVATIONS REQUIRED FOR THE COMPARISON OF A POPULATION VARIANCE WITH A STANDARD VALUE USING THE CHI–SQUARE TEST

Suppose $x_1, x_2, \ldots, x_{n+1}$ is a random sample from a population with variance $sigma_1^2$. The sample variance, s_1^2 has n degrees of freedom, and may be used to test the hypothesis that $\sigma_1^2 = \sigma_0^2$. Let R be the ratio of the variances σ_0^2 and σ_1^2. The table below shows the value of the ratio R for which a chi-square test, with significance level α, will not be able to detect the difference in the variances with probability β. Note that when R is far from one few samples will be required to distinguish σ_0^2 from σ_1^2, while for R near one large samples will be required.

Example 10.59: Testing for an increase in variance. Let $\alpha = 0.05$, $\beta = 0.01$, and $R = 4$. Using the table below with these values the value $R = 4$ occurs between the rows corresponding to $n = 15$ and $n = 20$. Using rough, linear, interpolation, the table indicates that the estimate of variance should be based on 19 degrees of freedom.

Example 10.60: Testing for a decrease in variance. Let $\alpha = 0.05$, $\beta = 0.01$, and $R = 0.33$. Using the table below with $\alpha' = \beta = 0.01$, $\beta' = \alpha = 0.05$ and $R' = 1/R = 3$, the value $R = 3$ occurs between the rows corresponding to $n = 24$ and $n = 30$. Using rough, linear, interpolation, the table indicates that the estimate of variance should be based on 26 degrees of freedom.

Values of R given n, α, and β

n	$\alpha = 0.01$				$\alpha = 0.05$			
	$\beta = 0.01$	$\beta = 0.05$	$\beta = 0.1$	$\beta = 0.5$	$\beta = 0.01$	$\beta = 0.05$	$\beta = 0.1$	$\beta = 0.5$
1	42236.852	1687.350	420.176	14.584	24454.206	976.938	243.272	8.444
2	458.211	89.781	43.709	6.644	298.073	58.404	28.433	4.322
3	98.796	32.244	19.414	4.795	68.054	22.211	13.373	3.303
4	44.686	18.681	12.483	3.955	31.933	13.349	8.920	2.827
5	27.217	13.170	9.369	3.467	19.972	9.665	6.875	2.544
6	19.278	10.280	7.627	3.144	14.438	7.699	5.713	2.354
7	14.911	8.524	6.521	2.911	11.353	6.490	4.965	2.217
8	12.202	7.352	5.757	2.736	9.418	5.675	4.444	2.112
9	10.377	6.516	5.198	2.597	8.103	5.088	4.059	2.028
10	9.072	5.890	4.770	2.484	7.156	4.646	3.763	1.960
15	5.847	4.211	3.578	2.133	4.780	3.442	2.925	1.743
20	4.548	3.462	3.019	1.943	3.803	2.895	2.524	1.624
25	3.845	3.033	2.690	1.821	3.267	2.577	2.286	1.547
30	3.403	2.752	2.471	1.735	2.927	2.367	2.125	1.492
40	2.874	2.403	2.192	1.619	2.516	2.103	1.919	1.418
50	2.564	2.191	2.021	1.544	2.272	1.942	1.791	1.368
75	2.150	1.898	1.779	1.431	1.945	1.716	1.609	1.294
100	1.938	1.743	1.649	1.367	1.775	1.596	1.510	1.252
150	1.715	1.575	1.506	1.297	1.594	1.464	1.400	1.206
∞	1.000	1.000	1.000	1.000	1.000	1.000	1.000	1.000

10.9 CRITICAL VALUES FOR TESTING OUTLIERS

Tests for outliers may be based on the largest deviation $\max\limits_{i=1,2,\ldots} (x_i - \bar{x})$ of the observations from their mean (which has to be normalized by the standard deviation or an estimate of the standard deviation). An alternative technique is to look at ratios of approximations to the range.

(a) To determine if the smallest element in a sample, $x_{(1)}$, is an outlier compute

$$r_{10} = \frac{x_{(2)} - x_{(1)}}{x_{(n)} - x_{(1)}} \tag{10.6}$$

Equivalently, to determine if the largest element in a sample, $x_{(n)}$, is an outlier compute

$$r_{10} = \frac{x_{(n)} - x_{(n-1)}}{x_{(n)} - x_{(1)}} \tag{10.7}$$

(b) To determine if the smallest element in a sample, $x_{(1)}$, is an outlier, and the value $x_{(n)}$ is not to be used, then compute

$$r_{11} = \frac{x_{(2)} - x_{(1)}}{x_{(n-1)} - x_{(1)}} \tag{10.8}$$

Equivalently, to determine if the largest element in a sample, $x_{(n)}$, is an outlier, without using the value $x_{(1)}$, compute

$$r_{11} = \frac{x_{(n)} - x_{(n-1)}}{x_{(n)} - x_{(2)}} \tag{10.9}$$

(c) To determine if the smallest element in a sample, $x_{(1)}$, is an outlier, and the value $x_{(2)}$ is not to be used, then compute

$$r_{20} = \frac{x_{(3)} - x_{(1)}}{x_{(n)} - x_{(1)}} \tag{10.10}$$

Equivalently, to determine if the largest element in a sample, $x_{(n)}$, is an outlier, without using the value $x_{(n-1)}$, compute

$$r_{20} = \frac{x_{(n)} - x_{(n-2)}}{x_{(n)} - x_{(2)}} \tag{10.11}$$

The following tables contain critical values for r_{10}, r_{11}, and r_{20}. See W. J. Dixon, *Annals of Mathematical Statistics*, **22**, 1951, pages 68–78.

Percentage values for r_{10}   ($\mathrm{Prob}\,[r_{10} > R] = \alpha$)

n	$\alpha = .005$	.01	.02	.05	.10	.50	.90	.95
3	.994	.988	.976	.941	.886	.500	.114	.059
4	.926	.889	.846	.745	.679	.324	.065	.033
5	.821	.780	.729	.642	.557	.250	.048	.023
6	.740	.698	.644	.560	.482	.210	.038	.018
7	.680	.637	.586	.507	.434	.184	.032	.016
8	.634	.590	.543	.468	.399	.166	.029	.014
9	.598	.555	.510	.437	.370	.152	.026	.013
10	.568	.527	.483	.412	.349	.142	.025	.012
15	.475	.438	.399	.338	.285	.111	.019	.010
20	.425	.391	.356	.300	.252	.096	.017	.008
25	.393	.362	.329	.277	.230	.088	.015	.008
30	.372	.341	.309	.260	.215	.082	.014	.007

Percentage values for r_{11}   ($\mathrm{Prob}\,[r_{11} > R] = \alpha$)

n	$\alpha = .005$	.01	.02	.05	.10	.50	.90	.95
4	.995	.991	.981	.955	.910	.554	.131	.069
5	.937	.916	.876	.807	.728	.369	.078	.039
6	.839	.805	.763	.689	.609	.288	.056	.028
7	.782	.740	.689	.610	.530	.241	.045	.022
8	.725	.683	.631	.554	.479	.210	.037	.019
9	.677	.635	.587	.512	.441	.189	.033	.016
10	.639	.597	.551	.477	.409	.173	.030	.014
15	.522	.486	.445	.381	.323	.129	.023	.011
20	.464	.430	.392	.334	.282	.110	.019	.010
25	.426	.394	.359	.394	.255	.098	.017	.009
30	.399	.369	.336	.283	.236	.090	.016	.008

Percentage values for r_{20}   ($\mathrm{Prob}\,[r_{20} > R] = \alpha$)

n	$\alpha = .005$	.01	.02	.05	.10	.50	.90	.95
4	.996	.992	.987	.967	.935	.676	.321	.235
5	.950	.929	.901	.845	.782	.500	.218	.155
6	.865	.836	.800	.736	.670	.411	.172	.126
7	.814	.778	.732	.661	.596	.355	.144	.099
8	.746	.719	.670	.607	.545	.317	.125	.085
9	.700	.667	.627	.565	.505	.288	.114	.077
10	.664	.632	.592	.531	.474	.268	.104	.070
15	.554	.522	.486	.430	.382	.209	.079	.052
20	.494	.464	.430	.372	.333	.179	.067	.046
25	.456	.428	.395	.343	.304	.161	.060	.041
30	.428	.402	.372	.322	.285	.149	.056	.039

10.10   TEST OF SIGNIFICANCE IN 2×2 CONTINGENCY TABLES

A 2×2 contingency table (see section 10.5) is a special case that occurs often. Suppose n elements are simultaneously classified as having either property 1 or 2 and as having property I or II. The 2×2 contingency table may be written as:

	I	II	Totals
1	a	$A - a$	A
2	b	$B - b$	B
Totals	r	$n - r$	n

If the marginal totals are fixed, the probability of a given configuration may be written as

$$f(a \mid r, A, B) = \frac{\binom{A}{a}\binom{B}{b}}{\binom{n}{r}} = \frac{A! \, B! \, r! \, (n - r)!}{n! \, a! \, b! \, (A - a)! \, (B - b)!} \tag{10.12}$$

The following tables are designed to be used in conducting a hypothesis test concerning the difference between observed and expected frequencies in a 2×2 contingency table. For given values of a, A, and B, table entries show the largest value of b (in bold type, with $b < a$) for which there is a significant difference (between observed and expected frequencies, or equivalently, between a/A and b/B). Critical values of b (probability levels) are presented for $\alpha = .05, .025, .01,$ and $.005$. The tables also satisfy the following conditions:

(1) Categories 1 and 2 are determined so that $A \geq B$.

(2) $\dfrac{a}{A} \geq \dfrac{b}{B}$ or, $aB \geq bA$.

(3) If b is less than or equal to the integer in bold type, then a/A is significantly greater than b/B (for a one tailed–test) at the probability level (α) indicated by the column heading. For a two-tailed test the significance level is 2α.

(4) A dash in the body of the table indicates no 2×2 table may show a significant effect at that probability level and combination of a, A, and B.

(5) For a given r, the probability b is less than the integer in bold type is shown in small type following an entry.

Note that as A and B get large, this test may be approximated by a two-sample Z test of proportions.

Example 10.61: In order to compare the probability of a success in two populations, the following 2 × 2 contingency table was obtained.

	Success	Failure	Totals
Sample from population 1	7	2	9
Sample from population 2	3	3	6
Totals	10	5	15

Is there any evidence to suggest the two population proportions are different? Use $\alpha = .05$.

Solution:

(S1) In this 2 × 2 contingency table, $a = 7$, $A = 9$, and $B = 6$. For $\alpha = .05$ the table entry is **1.035**.

(S2) The critical value for b is 1. If $b \leq 1$ then the null hypothesis H_0: $p_1 = p_2$ is rejected.

(S3) Conclusion: The value of the test statistic does not lie in the rejection region, $b = 3$. There is no evidence to suggest the population proportions are different.

(S4) Note there are six 2 × 2 tables with the same marginal totals as the table in this example (that is, $A = 9$, $B = 6$, and $r = 10$):

9	0		8	1		7	2		6	3		5	4		4	5
1	5		2	4		3	3		4	2		5	1		6	0

Assuming independence, the probability of obtaining each of these six tables (using equation (10.12), rounded) is {.002, .045, .24, .42, .25, .042}. That is, the first configuration is the least likely, and the fourth configuration is the most likely.

Contingency tables: 2×2

		a	Probability			
			0.05	0.025	0.01	0.005
$A=3$	$B=3$	3	0.050	–	–	–
$A=4$	$B=4$	4	0.014	0.014	–	–
	$B=3$	4	0.029	–	–	–
$A=5$	$B=5$	5	1.024	1.024	0.004	0.004
		4	0.024	0.024	–	–
	$B=4$	5	1.048	0.008	0.008	–
		4	0.040	–	–	–
	$B=3$	5	0.018	0.018	–	–
	$B=2$	5	0.048	–	–	–
$A=6$	$B=6$	6	2.030	1.008	1.008	0.001
		5	1.039	0.008	0.008	–
		4	0.030	–	–	–
	$B=5$	6	1.015	1.015	0.002	0.002
		5	0.013	0.013	–	–
		4	0.045	–	–	–
	$B=4$	6	1.033	0.005	0.005	0.005
		5	0.024	0.024	–	–
	$B=3$	6	0.012	0.012	–	–
		5	0.048	–	–	–
	$B=2$	6	0.036	–	–	–
$A=7$	$B=7$	7	3.035	2.010	1.002	1.002
		6	2.049	1.014	0.002	0.002
		5	1.049	0.010	–	–
		4	0.035	–	–	–
	$B=6$	7	2.021	2.021	1.005	1.005
		6	1.024	1.024	0.004	0.004
		5	0.016	0.016	–	–
		4	0.049	–	–	–
	$B=5$	7	2.045	1.010	0.001	0.001
		6	1.044	0.008	0.008	–
		5	0.027	–	–	–
	$B=4$	7	1.024	1.024	0.003	0.003
		6	0.015	0.015	–	–
		5	0.045	–	–	–
	$B=3$	7	0.008	0.008	0.008	–
		6	0.033	–	–	–
	$B=2$	7	0.028	–	–	–
$A=8$	$B=8$	8	4.038	3.013	2.003	2.003
		7	2.020	2.020	1.005	1.005
		6	1.020	1.020	0.003	0.003
		5	0.013	0.013	–	–
		4	0.038	–	–	–
	$B=7$	8	3.026	2.007	2.007	1.001
		7	2.034	1.009	1.009	0.001
		6	1.030	0.006	0.006	–
		5	0.019	0.019	–	–
	$B=6$	8	2.015	2.015	1.003	1.003
		7	1.016	1.016	0.002	0.002
		6	1.049	0.009	0.009	–
		5	0.028	–	–	–
	$B=5$	8	2.035	1.007	1.007	0.001
		7	1.031	0.005	0.005	0.005

		a	Probability			
			0.05	0.025	0.01	0.005
$A=8$	$B=5$	6	0.016	0.016	–	–
		5	0.044	–	–	–
	$B=4$	8	1.018	1.018	0.002	0.002
		7	0.010	0.010	–	–
		6	0.030	–	–	–
	$B=3$	8	0.006	0.006	0.006	–
		7	0.024	0.024	–	–
	$B=2$	8	0.022	0.022	–	–
$A=9$	$B=9$	9	5.041	4.015	3.005	3.005
		8	3.024	3.024	2.007	1.002
		7	2.027	1.007	1.007	0.001
		6	1.024	1.024	0.005	0.005
		5	0.015	0.015	–	–
		4	0.041	–	–	–
	$B=8$	9	4.029	3.009	3.009	2.002
		8	3.041	2.013	1.003	1.003
		7	2.041	1.012	0.002	0.002
		6	1.035	0.007	0.007	–
		5	0.020	0.020	–	–
	$B=7$	9	3.019	3.019	2.005	2.005
		8	2.024	2.024	1.006	0.001
		7	1.020	1.020	0.003	0.003
		6	0.010	0.010	–	–
		5	0.029	–	–	–
	$B=6$	9	3.044	2.011	1.002	1.002
		8	2.045	1.011	0.001	0.001
		7	1.034	0.006	0.006	–
		6	0.017	0.017	–	–
		5	0.042	–	–	–
	$B=5$	9	2.027	1.005	1.005	1.005
		8	1.022	1.022	0.003	0.003
		7	0.010	0.010	–	–
		6	0.028	–	–	–
	$B=4$	9	1.014	1.014	0.001	0.001
		8	0.007	0.007	0.007	–
		7	0.021	0.021	–	–
		6	0.049	–	–	–
	$B=3$	9	1.045	0.005	0.005	0.005
		8	0.018	0.018	–	–
		7	0.045	–	–	–
	$B=2$	9	0.018	0.018	–	–
$A=10$	$B=10$	10	6.043	5.016	4.005	3.002
		9	4.027	3.010	3.010	2.003
		8	3.032	2.011	1.003	1.003
		7	2.032	1.010	1.010	0.002
		6	1.027	0.005	0.005	–
		5	0.016	0.016	–	–
		4	0.043	–	–	–
	$B=9$	10	5.033	4.011	3.003	3.003
		9	4.046	3.017	2.005	2.005
		8	2.018	2.018	1.004	1.004
		7	2.047	1.014	0.002	0.002
		6	1.038	0.008	0.008	–
		5	0.022	0.022	–	–

Contingency tables: 2×2

	a	0.05	0.025	0.01	0.005
$A=10$ $B=8$	10	4.023	4.023	3.007	2.002
	9	3.030	2.009	2.009	1.002
	8	2.029	1.007	1.007	0.001
	7	1.022	1.022	0.004	0.004
	6	0.011	0.011	–	–
	5	0.029	–	–	–
$B=7$	10	3.015	3.015	2.003	2.003
	9	2.017	2.017	1.004	1.004
	8	2.049	1.013	0.002	0.002
	7	1.035	0.006	0.006	–
	6	0.017	0.017	–	–
	5	0.041	–	–	–
$B=6$	10	3.036	2.008	2.008	1.001
	9	2.034	1.007	1.007	0.001
	8	1.024	1.024	0.003	0.003
	7	0.010	0.010	–	–
	6	0.026	–	–	–
$B=5$	10	2.022	2.022	1.004	1.004
	9	1.017	1.017	0.002	0.002
	8	1.045	0.007	0.007	–
	7	0.019	0.019	–	–
	6	0.042	–	–	–
$B=4$	10	1.011	1.011	0.001	0.001
	9	1.040	0.005	0.005	0.005
	8	0.015	0.015	–	–
	7	0.035	–	–	–
$B=3$	10	1.038	0.003	0.003	0.003
	9	0.014	0.014	–	–
	8	0.035	–	–	–
$B=2$	10	0.015	0.015	–	–
	9	0.045	–	–	–
$A=11$ $B=11$	11	7.045	6.018	5.006	4.002
	10	5.030	4.011	3.004	3.004
	9	4.036	3.014	2.004	2.004
	8	3.039	2.014	1.004	1.004
	7	2.036	1.011	0.002	0.002
	6	1.030	0.006	0.006	–
	5	0.018	0.018	–	–
	4	0.045	–	–	–
$B=10$	11	6.035	5.012	4.004	4.004
	10	4.020	4.020	3.006	2.002
	9	3.022	3.022	2.007	1.002
	8	2.021	2.021	1.006	0.001
	7	1.016	1.016	0.003	0.003
	6	1.040	0.009	0.009	–
	5	0.023	0.023	–	–
$B=9$	11	5.026	4.008	4.008	3.002
	10	4.036	3.012	2.003	2.003
	9	3.037	2.012	1.003	1.003
	8	2.032	1.009	1.009	0.001
	7	1.024	1.024	0.004	0.004
	6	0.012	0.012	–	–
	5	0.030	–	–	–
$B=8$	11	4.018	4.018	3.005	3.005
	10	3.023	3.023	2.006	1.001

	a	0.05	0.025	0.01	0.005
$A=11$ $B=8$	9	2.020	2.020	1.005	1.005
	8	1.014	1.014	0.002	0.002
	7	1.035	0.007	0.007	–
	6	0.017	0.017	–	–
	5	0.040	–	–	–
$B=7$	11	4.043	3.011	2.002	2.002
	10	3.045	2.012	1.002	1.002
	9	2.036	1.009	1.009	0.001
	8	1.024	1.024	0.004	0.004
	7	0.010	0.010	–	–
	6	0.025	0.025	–	–
$B=6$	11	3.029	2.006	2.006	1.001
	10	2.027	1.005	1.005	0.001
	9	1.017	1.017	0.002	0.002
	8	1.041	0.007	0.007	–
	7	0.017	0.017	–	–
	6	0.037	–	–	–
$B=5$	11	2.018	2.018	1.003	1.003
	10	1.013	1.013	0.001	0.001
	9	1.034	0.005	0.005	0.005
	8	0.013	0.013	–	–
	7	0.029	–	–	–
$B=4$	11	1.009	1.009	1.009	0.001
	10	1.032	0.004	0.004	0.004
	9	0.011	0.011	–	–
	8	0.026	–	–	–
$B=3$	11	1.033	0.003	0.003	0.003
	10	0.011	0.011	–	–
	9	0.027	–	–	–
$B=2$	11	0.013	0.013	–	–
	10	0.038	–	–	–
$A=12$ $B=12$	12	8.047	7.019	6.007	5.002
	11	6.032	5.013	4.005	4.005
	10	5.040	4.017	3.006	2.002
	9	4.044	3.018	2.006	1.001
	8	3.044	2.017	1.005	1.005
	7	2.040	1.013	0.002	0.002
	6	1.032	0.007	0.007	–
	5	0.019	0.019	–	–
	4	0.047	–	–	–
$B=11$	12	7.037	6.014	5.005	5.005
	11	5.023	5.023	4.008	3.002
	10	4.027	3.010	3.010	2.003
	9	3.027	2.009	2.009	1.002
	8	2.024	2.024	1.007	0.001
	7	1.018	1.018	0.003	0.003
	6	1.041	0.009	0.009	–
	5	0.024	0.024	–	–
$B=10$	12	6.029	5.010	5.010	4.003
	11	5.041	4.015	3.005	3.005
	10	4.043	3.016	2.005	2.005
	9	3.041	2.014	1.003	1.003
	8	2.034	1.010	1.010	0.002
	7	1.025	1.025	0.005	0.005
	6	0.012	0.012	–	–

Contingency tables: 2×2

		a	Probability			
			0.05	0.025	0.01	0.005
A = 12	B = 10	5	0.030	–	–	–
	B = 9	12	5.021	5.021	4.006	3.002
		11	4.028	3.009	3.009	2.002
		10	3.027	2.008	2.008	1.002
		9	2.022	2.022	1.006	0.001
		8	1.015	1.015	0.002	0.002
		7	1.035	0.007	0.007	–
		6	0.017	0.017	–	–
		5	0.039	–	–	–
	B = 8	12	5.049	4.014	3.004	3.004
		11	3.017	3.017	2.004	2.004
		10	3.048	2.015	1.003	1.003
		9	2.037	1.010	1.010	0.001
		8	1.024	1.024	0.004	0.004
		7	0.010	0.010	–	–
		6	0.024	0.024	–	–
	B = 7	12	4.036	3.009	3.009	2.002
		11	3.036	2.009	2.009	1.002
		10	2.028	1.006	1.006	0.001
		9	1.017	1.017	0.002	0.002
		8	1.038	0.007	0.007	–
		7	0.016	0.016	–	–
		6	0.034	–	–	–
	B = 6	12	3.025	3.025	2.005	2.005
		11	2.021	2.021	1.004	1.004
		10	1.012	1.012	0.002	0.002
		9	1.030	0.005	0.005	0.005
		8	0.011	0.011	–	–
		7	0.025	0.025	–	–
		6	0.050	–	–	–
	B = 5	12	2.015	2.015	1.002	1.002
		11	1.010	1.010	1.010	0.001
		10	1.027	0.003	0.003	0.003
		9	0.009	0.009	0.009	–
		8	0.020	0.020	–	–
		7	0.041	–	–	–
	B = 4	12	2.050	1.007	1.007	0.001
		11	1.026	0.003	0.003	0.003
		10	0.008	0.008	0.008	–
		9	0.019	0.019	–	–
		8	0.038	–	–	–
	B = 3	12	1.029	0.002	0.002	0.002
		11	0.009	0.009	0.009	–
		10	0.022	0.022	–	–
		9	0.044	–	–	–
	B = 2	12	0.011	0.011	–	–
		11	0.033	–	–	–
A = 13	B = 13	13	9.048	8.020	7.007	6.003
		12	7.034	6.014	5.005	4.002
		11	6.043	5.019	4.007	3.002
		10	5.048	4.021	3.008	2.002
		9	4.049	3.021	2.007	1.002
		8	3.048	2.019	1.005	0.001
		7	2.043	1.014	0.003	0.003
		6	1.034	0.007	0.007	–

		a	Probability			
			0.05	0.025	0.01	0.005
A = 13	B = 13	5	0.020	0.020	–	–
		4	0.048	–	–	–
	B = 12	13	8.039	7.015	6.005	5.002
		12	6.025	6.025	5.010	4.003
		11	5.030	4.012	3.004	3.004
		10	4.032	3.012	2.004	2.004
		9	3.030	2.011	1.003	1.003
		8	2.026	1.008	1.008	0.001
		7	1.019	1.019	0.004	0.004
		6	1.043	0.010	0.010	–
		5	0.024	0.024	–	–
	B = 11	13	7.031	6.011	5.003	5.003
		12	6.045	5.017	4.006	3.002
		11	5.049	4.020	3.007	2.002
		10	4.048	3.019	2.006	1.001
		9	3.044	2.016	1.004	1.004
		8	2.036	1.011	0.002	0.002
		7	1.026	0.005	0.005	0.005
		6	0.013	0.013	–	–
		5	0.030	–	–	–
	B = 10	13	6.024	6.024	5.007	4.002
		12	5.032	4.011	3.003	3.003
		11	4.033	3.011	2.003	2.003
		10	3.030	2.010	2.010	1.002
		9	2.024	2.024	1.006	0.001
		8	1.016	1.016	0.003	0.003
		7	1.035	0.007	0.007	–
		6	0.017	0.017	–	–
		5	0.038	–	–	–
	B = 9	13	5.017	5.017	4.005	4.005
		12	4.022	4.022	3.006	2.001
		11	3.020	3.020	2.006	1.001
		10	3.048	2.016	1.004	1.004
		9	2.036	1.010	1.010	0.001
		8	1.023	1.023	0.004	0.004
		7	1.048	0.010	–	–
		6	0.023	0.023	–	–
		5	0.049	–	–	–
	B = 8	13	5.042	4.012	3.003	3.003
		12	4.045	3.013	2.003	2.003
		11	3.038	2.011	1.002	1.002
		10	2.027	1.006	1.006	0.001
		9	1.016	1.016	0.002	0.002
		8	1.035	0.006	0.006	–
		7	0.015	0.015	–	–
		6	0.032	–	–	–
	B = 7	13	4.031	3.007	3.007	2.001
		12	3.029	2.007	2.007	1.001
		11	2.021	2.021	1.004	1.004
		10	2.048	1.012	0.002	0.002
		9	1.027	0.004	0.004	0.004
		8	0.010	0.010	–	–
		7	0.022	0.022	–	–
		6	0.044	–	–	–
	B = 6	13	3.021	3.021	2.004	2.004

Contingency tables: 2×2

	a	Probability			
		0.05	0.025	0.01	0.005
$A = 13\ B = 6$	12	2.017	2.017	1.003	1.003
	11	2.043	1.009	1.009	0.001
	10	1.023	1.023	0.003	0.003
	9	1.046	0.008	0.008	—
	8	0.017	0.017	—	—
	7	0.034	—	—	—
$B = 5$	13	2.012	2.012	1.002	1.002
	12	2.042	1.008	1.008	0.001
	11	1.021	1.021	0.002	0.002
	10	1.045	0.007	0.007	—
	9	0.015	0.015	—	—
	8	0.029	—	—	—
$B = 4$	13	2.044	1.006	1.006	0.000
	12	1.022	1.022	0.002	0.002
	11	0.006	0.006	0.006	—
	10	0.015	0.015	—	—
	9	0.029	—	—	—
$B = 3$	13	1.025	1.025	0.002	0.002
	12	0.007	0.007	0.007	—
	11	0.018	0.018	—	—
	10	0.036	—	—	—
$B = 2$	13	0.010	0.010	0.010	—
	12	0.029	—	—	—
$A = 14\ B = 14$	14	10.049	9.020	8.008	7.003
	13	8.036	7.015	6.006	5.002
	12	7.045	6.021	5.008	4.003
	11	5.024	5.024	4.010	3.003
	10	4.025	4.025	3.010	2.003
	9	3.024	3.024	2.008	1.002
	8	2.021	2.021	1.006	0.001
	7	2.045	1.015	0.003	0.003
	6	1.036	0.008	0.008	—
	5	0.020	0.020	—	—
	4	0.049	—	—	—
$B = 13$	14	9.041	8.016	7.006	6.002
	13	7.027	6.011	5.004	5.004
	12	6.033	5.014	4.005	4.005
	11	5.036	4.015	3.005	2.001
	10	4.036	3.014	2.004	2.004
	9	3.033	2.012	1.003	1.003
	8	2.028	1.008	1.008	0.001
	7	1.020	1.020	0.004	0.004
	6	1.044	0.010	—	—
	5	0.025	0.025	—	—
$B = 12$	14	8.033	7.012	6.004	6.004
	13	7.048	6.020	5.007	4.002
	12	5.023	5.023	4.008	3.003
	11	4.023	4.023	3.008	2.002
	10	3.021	3.021	2.007	1.002
	9	3.046	2.017	1.005	1.005
	8	2.037	1.012	0.002	0.002
	7	1.026	0.005	0.005	—
	6	0.013	0.013	—	—
	5	0.030	—	—	—
$B = 11$	14	7.026	6.009	6.009	5.003

	a	Probability			
		0.05	0.025	0.01	0.005
$A = 14\ B = 11$	13	6.037	5.013	4.004	4.004
	12	5.039	4.015	3.005	3.005
	11	4.037	3.013	2.004	2.004
	10	3.032	2.011	1.002	1.002
	9	2.025	2.025	1.007	0.001
	8	1.016	1.016	0.003	0.003
	7	1.035	0.007	0.007	—
	6	0.017	0.017	—	—
	5	0.038	—	—	—
$B = 10$	14	6.020	6.020	5.006	4.002
	13	5.026	4.008	4.008	3.002
	12	4.026	3.008	3.008	2.002
	11	3.022	3.022	2.007	1.001
	10	3.048	2.017	1.004	1.004
	9	2.036	1.010	0.002	0.002
	8	1.023	1.023	0.004	0.004
	7	1.047	0.010	0.010	—
	6	0.022	0.022	—	—
	5	0.047	—	—	—
$B = 9$	14	6.047	5.014	4.004	4.004
	13	4.017	4.017	3.005	3.005
	12	4.047	3.016	2.004	2.004
	11	3.037	2.011	1.002	1.002
	10	2.027	1.007	1.007	0.001
	9	1.016	1.016	0.002	0.002
	8	1.033	0.006	0.006	—
	7	0.014	0.014	—	—
	6	0.030	—	—	—
$B = 8$	14	5.036	4.010	4.010	3.002
	13	4.037	3.011	2.002	2.002
	12	3.030	2.008	2.008	1.001
	11	2.020	2.020	1.005	1.005
	10	2.043	1.011	0.002	0.002
	9	1.025	1.025	0.004	0.004
	8	1.048	0.009	0.009	—
	7	0.020	0.020	—	—
	6	0.040	—	—	—
$B = 7$	14	4.026	3.006	3.006	2.001
	13	3.024	3.024	2.005	1.001
	12	2.016	2.016	1.003	1.003
	11	2.038	1.009	1.009	0.001
	10	1.020	1.020	0.003	0.003
	9	1.040	0.007	0.007	—
	8	0.015	0.015	—	—
	7	0.030	—	—	—
$B = 6$	14	3.018	3.018	2.003	2.003
	13	2.014	2.014	1.002	1.002
	12	2.035	1.007	1.007	0.001
	11	1.017	1.017	0.002	0.002
	10	1.036	0.005	0.005	—
	9	0.012	0.012	—	—
	8	0.024	0.024	—	—
	7	0.044	—	—	—
$B = 5$	14	2.010	2.010	1.001	1.001
	13	2.036	1.006	1.006	0.001

Contingency tables: 2×2

	a	Probability 0.05	0.025	0.01	0.005
$A = 14$ $B = 5$	12	1.017	1.017	0.002	0.002
	11	1.036	0.005	0.005	0.005
	10	0.011	0.011	–	–
	9	0.022	0.022	–	–
	8	0.040	–	–	–
$B = 4$	14	2.039	1.005	1.005	1.005
	13	1.018	1.018	0.002	0.002
	12	1.042	0.005	0.005	0.005
	11	0.011	0.011	–	–
	10	0.023	0.023	–	–
	9	0.041	–	–	–
$B = 3$	14	1.022	1.022	0.001	0.001
	13	0.006	0.006	0.006	–
	12	0.015	0.015	–	–
	11	0.029	–	–	–
$B = 2$	14	0.008	0.008	0.008	–
	13	0.025	0.025	–	–
	12	0.050	–	–	–
$A = 15$ $B = 15$	15	11.050	10.021	9.008	8.003
	14	9.037	8.016	7.007	6.002
	13	8.047	7.022	6.010	5.004
	12	6.026	5.011	4.004	4.004
	11	5.028	4.012	3.004	3.004
	10	4.028	3.011	2.004	2.004
	9	3.026	2.010	2.010	1.002
	8	2.022	2.022	1.007	0.001
	7	2.047	1.016	0.003	0.003
	6	1.037	0.008	0.008	–
	5	0.021	0.021	–	–
	4	0.050	–	–	–
$B = 14$	15	10.042	9.017	8.006	7.002
	14	8.029	7.012	6.004	6.004
	13	7.036	6.016	5.006	4.002
	12	6.039	5.018	4.007	3.002
	11	5.040	4.018	3.006	2.002
	10	4.039	3.016	2.005	1.001
	9	3.035	2.013	1.003	1.003
	8	2.029	1.009	1.009	0.001
	7	1.021	1.021	0.004	0.004
	6	1.045	0.011	–	–
	5	0.025	–	–	–
$B = 13$	15	9.035	8.013	7.005	7.005
	14	7.022	7.022	6.008	5.003
	13	6.026	5.010	4.003	4.003
	12	5.027	4.011	3.003	3.003
	11	4.026	3.010	3.010	2.003
	10	3.023	3.023	2.008	1.002
	9	3.047	2.018	1.005	1.005
	8	2.038	1.012	0.002	0.002
	7	1.027	0.005	0.005	–
	6	0.013	0.013	–	–
	5	0.031	–	–	–
$B = 12$	15	8.028	7.010	7.010	6.003
	14	7.040	6.016	5.005	4.002
	13	6.044	5.018	4.006	3.002

	a	Probability 0.05	0.025	0.01	0.005
$A = 15$ $B = 12$	12	5.043	4.017	3.006	2.001
	11	4.039	3.015	2.004	2.004
	10	3.033	2.011	1.003	1.003
	9	2.025	1.007	1.007	0.001
	8	1.016	1.016	0.003	0.003
	7	1.035	0.007	0.007	–
	6	0.017	0.017	–	–
	5	0.037	–	–	–
$B = 11$	15	7.022	7.022	6.007	5.002
	14	6.030	5.011	4.003	4.003
	13	5.031	4.011	3.003	3.003
	12	4.028	3.010	3.010	2.003
	11	3.023	3.023	2.007	1.002
	10	3.048	2.017	1.004	1.004
	9	2.036	1.010	0.002	0.002
	8	1.023	1.023	0.004	0.004
	7	1.045	0.010	0.010	–
	6	0.022	0.022	–	–
	5	0.046	–	–	–
$B = 10$	15	6.017	6.017	5.005	5.005
	14	5.021	5.021	4.007	3.002
	13	4.020	4.020	3.006	2.001
	12	4.047	3.017	2.005	2.005
	11	3.037	2.012	1.003	1.003
	10	2.026	1.007	1.007	0.001
	9	1.015	1.015	0.002	0.002
	8	1.031	0.006	0.006	–
	7	0.013	0.013	–	–
	6	0.028	–	–	–
$B = 9$	15	6.042	5.012	4.003	4.003
	14	5.044	4.014	3.004	3.004
	13	4.038	3.012	2.003	2.003
	12	3.029	2.008	2.008	1.002
	11	2.020	2.020	1.005	1.005
	10	2.040	1.011	0.002	0.002
	9	1.023	1.023	0.004	0.004
	8	1.044	0.009	0.009	–
	7	0.019	0.019	–	–
	6	0.037	–	–	–
$B = 8$	15	5.032	4.008	4.008	3.002
	14	4.031	3.008	3.008	2.002
	13	3.024	3.024	2.006	1.001
	12	2.016	2.016	1.003	1.003
	11	2.033	1.008	1.008	0.001
	10	1.018	1.018	0.003	0.003
	9	1.035	0.006	0.006	–
	8	0.013	0.013	–	–
	7	0.026	–	–	–
	6	0.050	–	–	–
$B = 7$	15	4.023	4.023	3.005	3.005
	14	3.020	3.020	2.004	2.004
	13	3.049	2.013	1.002	1.002
	12	2.030	1.006	1.006	0.001
	11	1.015	1.015	0.002	0.002
	10	1.030	0.005	0.005	0.005

Contingency tables: 2 × 2

		a	Probability			
			0.05	0.025	0.01	0.005
A = 15	B = 7	9	0.010	0.010	–	–
		8	0.020	0.020	–	–
		7	0.038	–	–	–
	B = 6	15	3.015	3.015	2.003	2.003
		14	2.011	2.011	1.002	1.002
		13	2.029	1.005	1.005	0.001
		12	1.013	1.013	0.002	0.002
		11	1.028	0.004	0.004	0.004
		10	0.009	0.009	0.009	–
		9	0.017	0.017	–	–
		8	0.032	–	–	–
	B = 5	15	2.009	2.009	2.009	1.001
		14	2.031	1.005	1.005	1.005
		13	1.014	1.014	0.001	0.001
		12	1.029	0.004	0.004	0.004
		11	0.008	0.008	0.008	–
		10	0.016	0.016	–	–
		9	0.030	–	–	–
	B = 4	15	2.035	1.004	1.004	1.004
		14	1.015	1.015	0.001	0.001
		13	1.036	0.004	0.004	0.004
		12	0.009	0.009	0.009	–
		11	0.018	0.018	–	–
		10	0.033	–	–	–
	B = 3	15	1.020	1.020	0.001	0.001
		14	0.005	0.005	0.005	0.005
		13	0.012	0.012	–	–
		12	0.025	0.025	–	–
		11	0.043	–	–	–
	B = 2	15	0.007	0.007	0.007	–
		14	0.022	0.022	–	–
		13	0.044	–	–	–
A = 16	B = 16	16	11.022	11.022	10.009	9.003
		15	10.038	9.017	8.007	7.003
		14	9.049	8.024	6.004	6.004
		13	7.028	6.013	5.005	4.002
		12	6.031	5.014	4.006	3.002
		11	5.032	4.014	3.005	2.002
		10	4.031	3.013	2.004	2.004
		9	3.028	2.011	1.003	1.003
		8	2.024	2.024	1.007	0.001
		7	2.049	1.017	0.003	0.003
		6	1.038	0.009	0.009	–
		5	0.022	0.022	–	–
	B = 15	16	11.043	10.018	9.007	8.002
		15	9.030	8.013	7.005	6.002
		14	8.038	7.017	6.007	5.003
		13	7.043	6.020	5.008	4.003
		12	6.044	5.021	4.008	3.003
		11	5.044	4.020	3.007	2.002
		10	4.041	3.018	2.006	1.001
		9	3.037	2.014	1.004	1.004
		8	2.030	1.010	1.010	0.002
		7	1.022	1.022	0.004	0.004
		6	1.046	0.011	–	–

		a	Probability			
			0.05	0.025	0.01	0.005
A = 16	B = 15	5	0.026	–	–	–
	B = 14	16	10.037	9.014	8.005	7.002
		15	8.024	8.024	7.009	6.003
		14	7.029	6.012	5.004	5.004
		13	6.031	5.013	4.005	4.005
		12	5.030	4.013	3.004	3.004
		11	4.028	3.011	2.003	2.003
		10	3.024	3.024	2.008	1.002
		9	3.048	2.019	1.005	0.001
		8	2.039	1.013	0.002	0.002
		7	1.027	0.006	0.006	–
		6	0.013	0.013	–	–
		5	0.031	–	–	–
	B = 13	16	9.030	8.011	7.004	7.004
		15	8.043	7.018	6.006	5.002
		14	7.048	6.021	5.008	4.002
		13	6.048	5.021	4.008	3.002
		12	5.045	4.019	3.007	2.002
		11	4.040	3.016	2.005	1.001
		10	3.034	2.012	1.003	1.003
		9	2.026	1.007	1.007	0.0q1
		8	1.017	1.017	0.003	0.003
		7	1.035	0.007	0.007	–
		6	0.017	0.017	–	–
		5	0.037	–	–	–
	B = 12	16	8.024	8.024	7.008	6.002
		15	7.034	6.012	5.004	5.004
		14	6.036	5.014	4.005	4.005
		13	5.034	4.013	3.004	3.004
		12	4.030	3.011	2.003	2.003
		11	3.024	3.024	2.008	1.002
		10	3.047	2.017	1.004	1.004
		9	2.035	1.010	0.002	0.002
		8	1.022	1.022	0.004	0.004
		7	1.044	0.010	0.010	–
		6	0.021	0.021	–	–
		5	0.044	–	–	–
	B = 11	16	7.019	7.019	6.006	5.002
		15	6.025	6.025	5.008	4.002
		14	5.025	5.025	4.008	3.002
		13	4.022	4.022	3.007	2.002
		12	4.046	3.017	2.005	2.005
		11	3.036	2.012	1.003	1.003
		10	2.025	1.007	1.007	0.001
		9	2.048	1.015	0.002	0.002
		8	1.030	0.006	0.006	–
		7	0.013	0.013	–	–
		6	0.027	–	–	–
	B = 10	16	7.046	6.014	5.004	5.004
		15	5.018	5.018	4.005	3.001
		14	5.046	4.016	3.005	3.005
		13	4.038	3.013	2.003	2.003
		12	3.028	2.008	2.008	1.002
		11	2.019	2.019	1.005	1.005
		10	2.037	1.010	0.002	0.002

Contingency tables: 2×2

	a	Probability					a	Probability			
		0.05	0.025	0.01	0.005			0.05	0.025	0.01	0.005
$A=16\ B=10$	9	1.022	1.022	0.004	0.004	$A=16\ B=4$	13	0.007	0.007	0.007	—
	8	1.041	0.008	0.008	—		12	0.014	0.014	—	—
	7	0.017	0.017	—	—		11	0.026	—	—	—
	6	0.035	—	—	—		10	0.043	—	—	—
$B=9$	16	6.037	5.010	5.010	4.002	$B=3$	16	1.018	1.018	0.001	0.001
	15	5.038	4.011	3.003	3.003		15	1.050	0.004	0.004	0.004
	14	4.031	3.009	3.009	2.002		14	0.010	0.010	—	—
	13	3.023	3.023	2.006	1.001		13	0.021	0.021	—	—
	12	3.047	2.015	1.003	1.003		12	0.036	—	—	—
	11	2.030	1.008	1.008	0.001	$B=2$	16	0.007	0.007	0.007	—
	10	1.016	1.016	0.002	0.002		15	0.020	0.020	—	—
	9	1.031	0.006	0.006	—		14	0.039	—	—	—
	8	0.012	0.012	—	—	$A=17\ B=17$	17	12.022	12.022	11.009	10.004
	7	0.024	0.024	—	—		16	11.039	10.018	9.008	8.003
	6	0.045	—	—	—		15	9.025	8.012	7.005	7.005
$B=8$	16	5.028	4.007	4.007	3.001		14	8.030	7.014	6.006	5.002
	15	4.026	3.007	3.007	2.001		13	7.033	6.016	5.007	4.002
	14	3.019	3.019	2.005	2.005		12	6.035	5.016	4.007	3.002
	13	3.043	2.012	1.002	1.002		11	5.035	4.016	3.006	2.002
	12	2.026	1.006	1.006	0.001		10	4.033	3.014	2.005	2.005
	11	2.049	1.013	0.002	0.002		9	3.030	2.012	1.003	1.003
	10	1.026	0.004	0.004	0.004		8	2.025	1.008	1.008	0.001
	9	1.047	0.009	0.009	—		7	1.018	1.018	0.004	0.004
	8	0.017	0.017	—	—		6	1.039	0.009	0.009	—
	7	0.033	—	—	—		5	0.022	0.022	—	—
$B=7$	16	4.020	4.020	3.004	3.004	$B=16$	17	12.044	11.018	10.007	9.003
	15	3.017	3.017	2.003	2.003		16	10.032	9.014	8.006	7.002
	14	3.042	2.010	1.002	1.002		15	9.040	8.019	7.008	6.003
	13	2.024	2.024	1.005	1.005		14	8.045	7.022	6.010	5.004
	12	2.047	1.011	0.001	0.001		13	7.048	6.023	4.004	4.004
	11	1.023	1.023	0.003	0.003		12	6.048	5.023	4.010	3.003
	10	1.041	0.007	0.007	—		11	5.046	4.022	3.008	2.003
	9	0.014	0.014	—	—		10	4.043	3.019	2.007	1.002
	8	0.026	—	—	—		9	3.038	2.015	1.004	1.004
	7	0.047	—	—	—		8	2.032	1.010	0.002	0.002
$B=6$	16	3.013	3.013	2.002	2.002		7	1.022	1.022	0.005	0.005
	15	3.044	2.009	2.009	1.001		6	1.046	0.011	—	—
	14	2.024	2.024	1.004	1.004		5	0.026	—	—	—
	13	2.049	1.011	0.001	0.001	$B=15$	17	11.038	10.015	9.006	8.002
	12	1.022	1.022	0.003	0.003		16	9.025	8.010	7.004	7.004
	11	1.041	0.006	0.006	—		15	8.031	7.014	6.005	5.002
	10	0.012	0.012	—	—		14	7.034	6.015	5.006	4.002
	9	0.023	0.023	—	—		13	6.034	5.015	4.006	3.002
	8	0.040	—	—	—		12	5.033	4.014	3.005	3.005
$B=5$	16	3.048	2.008	2.008	1.001		11	4.030	3.012	2.004	2.004
	15	2.027	1.004	1.004	1.004		10	3.025	2.009	2.009	1.002
	14	1.011	1.011	0.001	0.001		9	3.049	2.020	1.006	0.001
	13	1.024	1.024	0.003	0.003		8	2.040	1.013	0.002	0.002
	12	1.045	0.006	0.006	—		7	1.028	0.006	0.006	—
	11	0.012	0.012	—	—		6	0.014	0.014	—	—
	10	0.023	0.023	—	—		5	0.031	—	—	—
	9	0.039	—	—	—	$B=14$	17	10.032	9.012	8.004	8.004
$B=4$	16	2.032	1.004	1.004	1.004		16	9.046	8.019	7.007	6.003
	15	1.013	1.013	0.001	0.001		15	7.023	7.023	6.009	5.003
	14	1.031	0.003	0.003	0.003		14	6.024	6.024	5.010	4.003

Contingency tables: 2×2

	a	0.05	0.025	0.01	0.005
$A = 17\ B = 14$	13	5.023	5.023	4.009	3.003
	12	5.047	4.021	3.007	2.002
	11	4.041	3.017	2.005	1.001
	10	3.034	2.013	1.003	1.003
	9	2.026	1.008	1.008	0.001
	8	2.050	1.017	0.003	0.003
	7	1.035	0.007	0.007	–
	6	0.017	0.017	–	–
	5	0.036	–	–	–
$B = 13$	17	9.026	8.009	8.009	7.003
	16	8.037	7.014	6.005	6.005
	15	7.040	6.016	5.006	4.002
	14	6.039	5.016	4.006	3.002
	13	5.035	4.014	3.005	3.005
	12	4.030	3.011	2.003	2.003
	11	3.024	3.024	2.008	1.002
	10	3.046	2.018	1.005	1.005
	9	2.035	1.011	0.002	0.002
	8	1.022	1.022	0.004	0.004
	7	1.043	0.010	0.010	–
	6	0.021	0.021	–	–
	5	0.043	–	–	–
$B = 12$	17	8.021	8.021	7.007	6.002
	16	7.028	6.010	5.003	5.003
	15	6.029	5.011	4.003	4.003
	14	5.027	4.010	4.010	3.003
	13	4.023	4.023	3.008	2.002
	12	4.045	3.018	2.005	1.001
	11	3.035	2.012	1.003	1.003
	10	2.025	2.025	1.007	0.001
	9	2.046	1.015	0.002	0.002
	8	1.029	0.006	0.006	–
	7	0.012	0.012	–	–
	6	0.026	–	–	–
$B = 11$	17	7.016	7.016	6.005	6.005
	16	6.021	6.021	5.007	4.002
	15	5.020	5.020	4.006	3.002
	14	5.045	4.017	3.005	2.001
	13	4.037	3.013	2.003	2.003
	12	3.027	2.008	2.008	1.002
	11	2.018	2.018	1.004	1.004
	10	2.035	1.010	1.010	0.001
	9	1.020	1.020	0.004	0.004
	8	1.039	0.008	0.008	–
	7	0.016	0.016	–	–
	6	0.033	–	–	–
$B = 10$	17	7.041	6.012	5.003	5.003
	16	6.044	5.014	4.004	4.004
	15	5.039	4.013	3.003	3.003
	14	4.030	3.010	3.010	2.002
	13	3.022	3.022	2.006	1.001
	12	3.043	2.014	1.003	1.003
	11	2.028	1.007	1.007	0.001
	10	1.015	1.015	0.002	0.002
	9	1.029	0.005	0.005	–

	a	0.05	0.025	0.01	0.005
$A = 17\ B = 10$	8	0.011	0.011	–	–
	7	0.022	0.022	–	–
	6	0.042	–	–	–
$B = 9$	17	6.032	5.008	5.008	4.002
	16	5.033	4.009	4.009	3.002
	15	4.026	3.007	3.007	2.002
	14	3.018	3.018	2.005	2.005
	13	3.038	2.011	1.002	1.002
	12	2.023	2.023	1.005	0.001
	11	2.043	1.012	0.002	0.002
	10	1.023	1.023	0.004	0.004
	9	1.041	0.008	0.008	–
	8	0.016	0.016	–	–
	7	0.030	–	–	–
$B = 8$	17	5.024	5.024	4.006	3.001
	16	4.022	4.022	3.005	2.001
	15	3.016	3.016	2.004	2.004
	14	3.035	2.009	2.009	1.002
	13	2.020	2.020	1.004	1.004
	12	2.039	1.010	1.010	0.001
	11	1.019	1.019	0.003	0.003
	10	1.035	0.006	0.006	–
	9	0.012	0.012	–	–
	8	0.022	0.022	–	–
	7	0.040	–	–	–
$B = 7$	17	4.017	4.017	3.003	3.003
	16	3.014	3.014	2.003	2.003
	15	3.035	2.008	2.008	1.001
	14	2.019	2.019	1.004	1.004
	13	2.038	1.008	1.008	0.001
	12	1.017	1.017	0.002	0.002
	11	1.032	0.005	0.005	0.005
	10	0.010	0.010	0.010	–
	9	0.019	0.019	–	–
	8	0.033	–	–	–
$B = 6$	17	3.011	3.011	2.002	2.002
	16	3.038	2.008	2.008	1.001
	15	2.020	2.020	1.003	1.003
	14	2.042	1.008	1.008	0.001
	13	1.017	1.017	0.002	0.002
	12	1.032	0.005	0.005	0.005
	11	0.009	0.009	0.009	–
	10	0.017	0.017	–	–
	9	0.030	–	–	–
	8	0.050	–	–	–
$B = 5$	17	3.043	2.006	2.006	1.001
	16	2.023	2.023	1.003	1.003
	15	1.009	1.009	1.009	0.001
	14	1.020	1.020	0.002	0.002
	13	1.037	0.005	0.005	0.005
	12	0.010	0.010	0.010	–
	11	0.018	0.018	–	–
	10	0.030	–	–	–
	9	0.049	–	–	–
$B = 4$	17	2.029	1.003	1.003	1.003

Contingency tables: 2×2

a		Probability			
		0.05	0.025	0.01	0.005
$A = 17\ B = 4$	16	1.011	1.011	0.001	0.001
	15	1.027	0.003	0.003	0.003
	14	0.006	0.006	0.006	−
	13	0.012	0.012	−	−
	12	0.021	0.021	−	−
	11	0.035	−	−	−
$B = 3$	17	1.016	1.016	0.001	0.001
	16	1.045	0.004	0.004	0.004
	15	0.009	0.009	0.009	−
	14	0.018	0.018	−	−
	13	0.031	−	−	−
	12	0.049	−	−	−
$B = 2$	17	0.006	0.006	0.006	−
	16	0.018	0.018	−	−
	15	0.035	−	−	−
$A = 18\ B = 18$	18	13.023	13.023	12.010	11.004
	17	12.040	11.019	10.008	9.003
	16	10.026	9.012	8.005	7.002
	15	9.032	8.015	7.007	6.003
	14	8.035	7.017	6.008	5.003
	13	7.037	6.019	5.008	4.003
	12	6.038	5.019	4.008	3.003
	11	5.037	4.017	3.007	2.002
	10	4.035	3.015	2.005	1.001
	9	3.032	2.012	1.003	1.003
	8	2.026	1.008	1.008	0.001
	7	1.019	1.019	0.004	0.004
	6	1.040	0.010	0.010	−
	5	0.023	0.023	−	−
$B = 17$	18	13.045	12.019	11.008	10.003
	17	11.033	10.015	9.006	8.002
	16	10.042	9.020	8.009	7.004
	15	9.048	8.024	6.004	6.004
	14	7.026	6.012	5.005	5.005
	13	6.026	5.012	4.004	4.004
	12	5.025	4.011	3.004	3.004
	11	5.049	4.023	3.009	2.003
	10	4.045	3.020	2.007	1.002
	9	3.040	2.016	1.005	1.005
	8	2.032	1.011	0.002	0.002
	7	1.023	1.023	0.005	0.005
	6	1.047	0.011	−	−
	5	0.026	−	−	−
$B = 16$	18	12.039	11.016	10.006	9.002
	17	10.027	9.011	8.004	8.004
	16	9.033	8.015	7.006	6.002
	15	8.037	7.017	6.007	5.003
	14	7.038	6.018	5.007	4.003
	13	6.037	5.017	4.007	3.002
	12	5.035	4.015	3.006	2.002
	11	4.031	3.013	2.004	2.004
	10	3.026	2.010	2.010	1.002
	9	3.050	2.020	1.006	0.001
	8	2.040	1.013	0.002	0.002
	7	1.028	0.006	0.006	−

a		Probability			
		0.05	0.025	0.01	0.005
$A = 18\ B = 16$	6	0.014	0.014	−	−
	5	0.031	−	−	−
$B = 15$	18	11.033	10.013	9.005	9.005
	17	10.049	9.021	8.008	7.003
	16	8.026	7.011	6.004	6.004
	15	7.027	6.012	5.004	5.004
	14	6.027	5.011	4.004	4.004
	13	5.025	5.025	3.003	3.003
	12	5.048	4.022	3.008	2.002
	11	4.042	3.018	2.006	1.001
	10	3.035	2.013	1.003	1.003
	9	2.026	1.008	1.008	0.001
	8	2.050	1.017	0.003	0.003
	7	1.034	0.007	0.007	−
	6	0.017	0.017	−	−
	5	0.036	−	−	−
$B = 14$	18	10.028	9.010	9.010	8.003
	17	9.040	8.016	7.006	6.002
	16	8.044	7.019	6.007	5.002
	15	7.043	6.019	5.007	4.002
	14	6.041	5.018	4.006	3.002
	13	5.036	4.015	3.005	2.001
	12	4.031	3.012	2.004	2.004
	11	3.025	3.025	2.008	1.002
	10	3.046	2.018	1.005	1.005
	9	2.034	1.011	0.002	0.002
	8	1.022	1.022	0.004	0.004
	7	1.042	0.009	0.009	−
	6	0.020	0.020	−	−
	5	0.043	−	−	−
$B = 13$	18	9.023	9.023	8.008	7.002
	17	8.031	7.012	6.004	6.004
	16	7.033	6.013	5.004	5.004
	15	6.032	5.012	4.004	4.004
	14	5.028	4.011	3.003	3.003
	13	4.023	4.023	3.008	2.002
	12	4.044	3.018	2.005	1.001
	11	3.034	2.012	1.003	1.003
	10	2.024	2.024	1.007	0.001
	9	2.045	1.014	0.002	0.002
	8	1.028	0.006	0.006	−
	7	0.012	0.012	−	−
	6	0.025	−	−	−
$B = 12$	18	8.018	8.018	7.006	6.002
	17	7.024	7.024	6.008	5.002
	16	6.024	6.024	5.008	4.003
	15	5.022	5.022	4.007	3.002
	14	5.044	4.018	3.006	2.001
	13	4.035	3.013	2.004	2.004
	12	3.026	2.008	2.008	1.002
	11	3.048	2.018	1.004	1.004
	10	2.033	1.010	1.010	0.001
	9	1.019	1.019	0.003	0.003
	8	1.037	0.007	0.007	−
	7	0.016	0.016	−	−

Contingency tables: 2×2

	a	Probability					a	Probability			
		0.05	0.025	0.01	0.005			0.05	0.025	0.01	0.005
$A=18$ $B=12$	6	0.031	–	–	–	$A=18$ $B=7$	14	2.031	1.007	1.007	0.001
$B=11$	18	8.045	7.014	6.004	6.004		13	1.013	1.013	0.002	0.002
	17	6.018	6.018	5.005	4.001		12	1.025	1.025	0.004	0.004
	16	6.045	5.016	4.005	3.001		11	1.043	0.007	0.007	–
	15	5.038	4.013	3.004	3.004		10	0.013	0.013	–	–
	14	4.029	3.010	3.010	2.002		9	0.024	0.024	–	–
	13	3.021	3.021	2.006	1.001		8	0.040	–	–	–
	12	3.039	2.013	1.003	1.003	$B=6$	18	3.010	3.010	3.010	2.001
	11	2.026	1.007	1.007	0.001		17	3.034	2.006	2.006	1.001
	10	2.046	1.014	0.002	0.002		16	2.017	2.017	1.003	1.003
	9	1.027	0.005	0.005	0.005		15	2.035	1.007	1.007	0.001
	8	1.048	0.010	–	–		14	1.014	1.014	0.002	0.002
	7	0.020	0.020	–	–		13	1.026	0.003	0.003	0.003
	6	0.039	–	–	–		12	1.045	0.007	0.007	–
$B=10$	18	7.037	6.010	5.003	5.003		11	0.013	0.013	–	–
	17	6.038	5.012	4.003	4.003		10	0.022	0.022	–	–
	16	5.033	4.010	3.003	3.003		9	0.037	–	–	–
	15	4.025	4.025	3.007	2.002	$B=5$	18	3.040	2.006	2.006	1.001
	14	4.049	3.017	2.005	2.005		17	2.020	2.020	1.003	1.003
	13	3.034	2.010	1.002	1.002		16	2.045	1.008	1.008	0.001
	12	2.021	2.021	1.005	1.005		15	1.017	1.017	0.002	0.002
	11	2.038	1.010	0.001	0.001		14	1.031	0.004	0.004	0.004
	10	1.020	1.020	0.003	0.003		13	0.007	0.007	0.007	–
	9	1.037	0.007	0.007	–		12	0.014	0.014	–	–
	8	0.014	0.014	–	–		11	0.024	0.024	–	–
	7	0.027	–	–	–		10	0.038	–	–	–
	6	0.049	–	–	–	$B=4$	18	2.026	1.003	1.003	1.003
$B=9$	18	6.029	5.007	5.007	4.002		17	1.010	1.010	1.010	0.001
	17	5.028	4.008	4.008	3.002		16	1.023	1.023	0.002	0.002
	16	4.022	4.022	3.006	2.001		15	1.044	0.005	0.005	0.005
	15	4.046	3.015	2.003	2.003		14	0.010	0.010	0.010	–
	14	3.030	2.008	2.008	1.002		13	0.017	0.017	–	–
	13	2.018	2.018	1.004	1.004		12	0.029	–	–	–
	12	2.033	1.008	1.008	0.001		11	0.045	–	–	–
	11	1.016	1.016	0.002	0.002	$B=3$	18	1.014	1.014	0.001	0.001
	10	1.030	0.005	0.005	–		17	1.041	0.003	0.003	0.003
	9	0.010	0.010	–	–		16	0.008	0.008	0.008	–
	8	0.020	0.020	–	–		15	0.015	0.015	–	–
	7	0.036	–	–	–		14	0.026	–	–	–
$B=8$	18	5.022	5.022	4.005	4.005		13	0.042	–	–	–
	17	4.019	4.019	3.004	3.004	$B=2$	18	0.005	0.005	0.005	–
	16	4.047	3.013	2.003	2.003		17	0.016	0.016	–	–
	15	3.029	2.007	2.007	1.001		16	0.032	–	–	–
	14	2.016	2.016	1.003	1.003	$A=19$ $B=19$	19	14.023	14.023	13.010	12.004
	13	2.031	1.007	1.007	0.001		18	13.041	12.020	11.009	10.004
	12	1.014	1.014	0.002	0.002		17	11.027	10.013	9.006	8.002
	11	1.026	0.004	0.004	0.004		16	10.033	9.017	8.008	7.003
	10	1.045	0.008	0.008	–		15	9.037	8.019	7.009	6.004
	9	0.016	0.016	–	–		14	8.040	7.020	6.009	5.004
	8	0.028	–	–	–		13	7.041	6.021	5.009	4.004
	7	0.048	–	–	–		12	6.041	5.020	4.009	3.003
$B=7$	18	4.015	4.015	3.003	3.003		11	5.040	4.019	3.008	2.002
	17	4.050	3.012	2.002	2.002		10	4.037	3.017	2.006	1.001
	16	3.030	2.007	2.007	1.001		9	3.033	2.013	1.004	1.004
	15	2.016	2.016	1.003	1.003		8	2.027	1.009	1.009	0.002

Contingency tables: 2×2

		a	\multicolumn{4}{c}{Probability}			a	\multicolumn{4}{c}{Probability}
			0.05	0.025	0.01	0.005				0.05	0.025	0.01	0.005
$A=19$	$B=19$	7	1.020	1.020	0.004	0.004	$A=19$	$B=15$	12	4.031	3.012	2.004	2.004
		6	1.041	0.010	0.010	–			11	3.025	3.025	2.009	1.002
		5	0.023	0.023	–	–			10	3.045	2.018	1.005	1.005
	$B=18$	19	14.046	13.020	12.008	11.003			9	2.034	1.011	0.002	0.002
		18	12.034	11.016	10.007	9.003			8	1.022	1.022	0.004	0.004
		17	11.044	10.021	9.010	8.004			7	1.042	0.009	0.009	–
		16	10.050	8.012	7.005	6.002			6	0.020	0.020	–	–
		15	8.028	7.013	6.006	5.002			5	0.042	–	–	–
		14	7.029	6.014	5.006	4.002		$B=14$	19	10.024	10.024	9.008	8.003
		13	6.029	5.013	4.005	3.002			18	9.034	8.013	7.005	7.005
		12	5.027	4.012	3.004	3.004			17	8.037	7.015	6.006	5.002
		11	4.025	4.025	2.003	2.003			16	7.036	6.015	5.005	4.002
		10	4.046	3.021	2.008	1.002			15	6.033	5.014	4.005	4.005
		9	3.041	2.017	1.005	1.005			14	5.028	4.011	3.004	3.004
		8	2.033	1.011	0.002	0.002			13	4.023	4.023	3.008	2.002
		7	1.023	1.023	0.005	0.005			12	4.043	3.018	2.006	1.001
		6	1.047	0.012	–	–			11	3.034	2.012	1.003	1.003
		5	0.027	–	–	–			10	2.024	2.024	1.007	0.001
	$B=17$	19	13.040	12.016	11.006	10.002			9	2.043	1.014	0.002	0.002
		18	11.028	10.012	9.005	9.005			8	1.027	0.005	0.005	–
		17	10.035	9.016	8.007	7.003			7	0.012	0.012	–	–
		16	9.039	8.019	7.008	6.003			6	0.024	0.024	–	–
		15	8.041	7.020	6.009	5.003			5	0.049	–	–	–
		14	7.041	6.020	5.008	4.003		$B=13$	19	9.020	9.020	8.006	7.002
		13	6.039	5.019	4.008	3.003			18	8.027	7.010	7.010	6.003
		12	5.036	4.016	3.006	2.002			17	7.028	6.010	5.003	5.003
		11	4.032	3.014	2.004	2.004			16	6.026	5.010	5.010	4.003
		10	3.027	2.010	1.003	1.003			15	5.022	5.022	4.008	3.002
		9	2.021	2.021	1.006	0.001			14	5.043	4.018	3.006	2.002
		8	2.040	1.014	0.002	0.002			13	4.034	3.013	2.004	2.004
		7	1.028	0.006	0.006	–			12	3.025	2.008	2.008	1.002
		6	0.014	0.014	–	–			11	3.046	2.017	1.004	1.004
		5	0.031	–	–	–			10	2.032	1.009	1.009	0.001
	$B=16$	19	12.035	11.013	10.005	10.005			9	1.019	1.019	0.003	0.003
		18	10.023	10.023	9.009	8.003			8	1.035	0.007	0.007	–
		17	9.028	8.012	7.005	7.005			7	0.015	0.015	–	–
		16	8.030	7.013	6.005	5.002			6	0.030	–	–	–
		15	7.030	6.013	5.005	4.002		$B=12$	19	9.049	8.016	7.005	7.005
		14	6.029	5.013	4.005	4.005			18	7.020	7.020	6.007	5.002
		13	5.026	4.011	3.004	3.004			17	6.020	6.020	5.007	4.002
		12	5.049	4.023	3.009	2.003			16	6.044	5.017	4.006	3.002
		11	4.042	3.018	2.006	1.001			15	5.036	4.014	3.004	3.004
		10	3.035	2.013	1.004	1.004			14	4.028	3.010	3.010	2.003
		9	2.027	1.008	1.008	0.001			13	3.020	3.020	2.006	1.001
		8	2.049	1.017	0.003	0.003			12	3.037	2.013	1.003	1.003
		7	1.034	0.007	0.007	–			11	2.024	2.024	1.006	0.001
		6	0.017	0.017	–	–			10	2.043	1.013	0.002	0.002
		5	0.036	–	–	–			9	1.025	1.025	0.005	0.005
	$B=15$	19	11.029	10.011	9.004	9.004			8	1.045	0.010	0.010	–
		18	10.042	9.018	8.007	7.002			7	0.019	0.019	–	–
		17	9.047	8.021	7.008	6.003			6	0.037	–	–	–
		16	8.048	7.022	6.009	5.003		$B=11$	19	8.041	7.012	6.003	6.003
		15	7.045	6.021	5.008	4.003			18	7.044	6.015	5.004	5.004
		14	6.042	5.019	4.007	3.002			17	6.039	5.013	4.004	4.004
		13	5.037	4.016	3.006	2.002			16	5.031	4.011	3.003	3.003

Contingency tables: 2×2

	a	Probability					a	Probability			
		0.05	0.025	0.01	0.005			0.05	0.025	0.01	0.005
$A=19\ B=11$	15	4.023	4.023	3.007	2.002	$A=19\ B=7$	12	1.034	0.005	0.005	–
	14	4.044	3.016	2.004	2.004		11	0.010	0.010	0.010	–
	13	3.031	2.010	2.010	1.002		10	0.017	0.017	–	–
	12	2.019	2.019	1.005	1.005		9	0.030	–	–	–
	11	2.035	1.010	1.010	0.001		8	0.048	–	–	–
	10	1.019	1.019	0.003	0.003	$B=6$	19	4.050	3.009	3.009	2.001
	9	1.034	0.006	0.006	–		18	3.030	2.005	2.005	1.001
	8	0.013	0.013	–	–		17	2.014	2.014	1.002	1.002
	7	0.025	0.025	–	–		16	2.030	1.005	1.005	0.000
	6	0.046	–	–	–		15	1.011	1.011	0.001	0.001
$B=10$	19	7.033	6.009	6.009	5.002		14	1.021	1.021	0.003	0.003
	18	6.034	5.010	4.003	4.003		13	1.037	0.005	0.005	–
	17	5.028	4.008	4.008	3.002		12	0.010	0.010	0.010	–
	16	4.020	4.020	3.006	2.001		11	0.017	0.017	–	–
	15	4.041	3.013	2.003	2.003		10	0.028	–	–	–
	14	3.027	2.008	2.008	1.001		9	0.045	–	–	–
	13	3.048	2.016	1.003	1.003	$B=5$	19	3.036	2.005	2.005	2.005
	12	2.029	1.007	1.007	0.001		18	2.018	2.018	1.002	1.002
	11	1.015	1.015	0.002	0.002		17	2.040	1.006	1.006	0.000
	10	1.027	0.005	0.005	0.005		16	1.014	1.014	0.001	0.001
	9	1.046	0.009	0.009	–		15	1.026	0.003	0.003	0.003
	8	0.018	0.018	–	–		14	1.044	0.006	0.006	–
	7	0.032	–	–	–		13	0.011	0.011	–	–
$B=9$	19	6.026	5.006	5.006	4.001		12	0.019	0.019	–	–
	18	5.024	5.024	4.006	3.001		11	0.030	–	–	–
	17	4.018	4.018	3.005	3.005		10	0.047	–	–	–
	16	4.039	3.012	2.003	2.003	$B=4$	19	2.024	2.024	1.002	1.002
	15	3.025	3.025	2.006	1.001		18	1.009	1.009	1.009	0.001
	14	3.045	2.014	1.003	1.003		17	1.020	1.020	0.002	0.002
	13	2.026	1.006	1.006	0.001		16	1.038	0.004	0.004	0.004
	12	2.045	1.012	0.002	0.002		15	0.008	0.008	0.008	–
	11	1.022	1.022	0.004	0.004		14	0.014	0.014	–	–
	10	1.039	0.007	0.007	–		13	0.024	0.024	–	–
	9	0.013	0.013	–	–		12	0.037	–	–	–
	8	0.024	0.024	–	–	$B=3$	19	1.013	1.013	0.001	0.001
	7	0.043	–	–	–		18	1.037	0.003	0.003	0.003
$B=8$	19	5.019	5.019	4.004	4.004		17	0.006	0.006	0.006	–
	18	4.016	4.016	3.004	3.004		16	0.013	0.013	–	–
	17	4.040	3.011	2.002	2.002		15	0.023	0.023	–	–
	16	3.024	3.024	2.006	1.001		14	0.036	–	–	–
	15	3.046	2.013	1.002	1.002	$B=2$	19	0.005	0.005	0.005	0.005
	14	2.025	2.025	1.005	0.001		18	0.014	0.014	–	–
	13	2.044	1.011	0.001	0.001		17	0.029	–	–	–
	12	1.020	1.020	0.003	0.003		16	0.048	–	–	–
	11	1.035	0.006	0.006	–	$A=20\ B=20$	20	15.024	15.024	13.004	13.004
	10	0.011	0.011	–	–		19	14.042	13.020	12.009	11.004
	9	0.020	0.020	–	–		18	12.028	11.014	10.006	9.003
	8	0.034	–	–	–		17	11.034	10.018	9.008	8.004
$B=7$	19	4.013	4.013	3.002	3.002		16	10.039	9.020	8.010	7.004
	18	4.044	3.010	2.002	2.002		15	9.041	8.022	6.005	6.005
	17	3.026	2.005	2.005	1.001		14	8.043	7.023	5.005	5.005
	16	2.013	2.013	1.002	1.002		13	7.044	6.023	4.004	4.004
	15	2.026	1.005	1.005	0.001		12	6.043	5.022	4.010	3.004
	14	2.046	1.011	0.001	0.001		11	5.041	4.020	3.008	2.003
	13	1.020	1.020	0.003	0.003		10	4.039	3.018	2.006	1.002

Contingency tables: 2×2

$A = 20$ $B = 20$	a	0.05	0.025	0.01	0.005
	9	3.034	2.014	1.004	1.004
	8	2.028	1.009	1.009	0.002
	7	1.020	1.020	0.004	0.004
	6	1.042	0.010	−	−
	5	0.024	0.024	−	−
$B = 19$	20	15.047	14.020	13.008	12.003
	19	13.035	12.016	11.007	10.003
	18	12.045	11.023	9.004	9.004
	17	10.027	9.013	8.006	7.002
	16	9.030	8.015	7.006	6.003
	15	8.031	7.015	6.007	5.003
	14	7.031	6.015	5.007	4.002
	13	6.031	5.014	4.006	3.002
	12	5.029	4.013	3.005	3.005
	11	4.026	3.011	2.004	2.004
	10	4.047	3.022	2.008	1.002
	9	3.042	2.017	1.005	0.001
	8	2.034	1.011	0.002	0.002
	7	1.024	1.024	0.005	−
	6	1.048	0.012	−	−
	5	0.027	−	−	−
$B = 18$	20	14.041	13.017	12.007	11.003
	19	12.029	11.013	10.005	9.002
	18	11.037	10.018	9.008	8.003
	17	10.041	9.020	8.009	7.004
	16	9.044	8.022	7.010	6.004
	15	8.044	7.022	5.004	5.004
	14	7.043	6.021	5.009	4.004
	13	6.041	5.020	4.008	3.003
	12	5.038	4.017	3.007	2.002
	11	4.033	3.014	2.005	2.005
	10	3.028	2.010	1.003	1.003
	9	2.021	2.021	1.006	0.001
	8	2.041	1.014	0.003	0.003
	7	1.029	0.006	0.006	−
	6	0.014	0.014	−	−
	5	0.031	−	−	−
$B = 17$	20	13.036	12.014	11.005	10.002
	19	11.024	11.024	9.004	9.004
	18	10.030	9.013	8.005	7.002
	17	9.032	8.015	7.006	6.002
	16	8.033	7.015	6.006	5.002
	15	7.032	6.015	5.006	4.002
	14	6.030	5.014	4.005	3.002
	13	5.027	4.012	3.004	3.004
	12	5.049	4.023	3.009	2.003
	11	4.043	3.019	2.006	1.002
	10	3.035	2.014	1.004	1.004
	9	2.027	1.008	1.008	0.001
	8	2.049	1.017	0.003	0.003
	7	1.034	0.008	0.008	−
	6	0.017	0.017	−	−
	5	0.036	−	−	−
$B = 16$	20	12.031	11.012	10.004	10.004
	19	11.045	10.019	9.008	8.003

$A = 20$ $B = 16$	a	0.05	0.025	0.01	0.005
	18	9.023	9.023	8.010	7.004
	17	8.024	8.024	6.004	6.004
	16	8.050	7.024	5.004	5.004
	15	7.047	6.022	5.009	4.003
	14	6.042	5.020	4.008	3.003
	13	5.037	4.016	3.006	2.002
	12	4.031	3.013	2.004	2.004
	11	3.025	3.025	2.009	1.002
	10	3.045	2.018	1.005	1.005
	9	2.034	1.011	0.002	0.002
	8	1.021	1.021	0.004	0.004
	7	1.041	0.009	0.009	−
	6	0.020	0.020	−	−
	5	0.041	−	−	−
$B = 15$	20	11.026	10.009	10.009	9.003
	19	10.037	9.015	8.005	7.002
	18	9.040	8.017	7.007	6.002
	17	8.040	7.018	6.007	5.002
	16	7.037	6.016	5.006	4.002
	15	6.033	5.014	4.005	3.002
	14	5.029	4.012	3.004	3.004
	13	4.023	4.023	3.009	2.003
	12	4.042	3.018	2.006	1.001
	11	3.033	2.012	1.003	1.003
	10	2.023	2.023	1.007	0.001
	9	2.042	1.014	0.002	0.002
	8	1.027	0.005	0.005	−
	7	1.049	0.012	−	−
	6	0.024	0.024	−	−
	5	0.048	−	−	−
$B = 14$	20	10.022	10.022	9.007	8.002
	19	9.030	8.011	7.004	7.004
	18	8.031	7.012	6.004	6.004
	17	7.030	6.012	5.004	5.004
	16	6.027	5.010	4.003	4.003
	15	5.022	5.022	4.008	3.003
	14	5.042	4.018	3.006	2.002
	13	4.033	3.013	2.004	2.004
	12	3.025	3.025	2.008	1.002
	11	3.044	2.016	1.004	1.004
	10	2.031	1.009	1.009	0.001
	9	1.018	1.018	0.003	0.003
	8	1.034	0.007	0.007	−
	7	0.014	0.014	−	−
	6	0.029	−	−	−
$B = 13$	20	9.017	9.017	8.005	7.002
	19	8.023	8.023	7.008	6.002
	18	7.023	7.023	6.008	5.003
	17	6.021	6.021	5.008	4.002
	16	6.043	5.018	4.006	3.002
	15	5.035	4.014	3.004	3.004
	14	4.027	3.010	3.010	2.003
	13	4.048	3.019	2.006	1.001
	12	3.035	2.012	1.003	1.003
	11	2.023	2.023	1.006	0.001

Contingency tables: 2×2

	a	Probability					a	Probability			
		0.05	0.025	0.01	0.005			0.05	0.025	0.01	0.005
$A=20$ $B=13$	10	2.041	1.012	0.002	0.002	$A=20$ $B=9$	13	2.036	1.009	1.009	0.001
	9	1.024	1.024	0.004	0.004		12	1.017	1.017	0.002	0.002
	8	1.042	0.009	0.009	–		11	1.029	0.005	0.005	0.005
	7	0.018	0.018	–	–		10	1.048	0.009	0.009	–
	6	0.035	–	–	–		9	0.017	0.017	–	–
$B=12$	20	9.044	8.014	7.004	7.004		8	0.029	–	–	–
	19	8.049	7.017	6.005	5.002		7	0.050	–	–	–
	18	7.045	6.017	5.005	4.001	$B=8$	20	5.017	5.017	4.003	4.003
	17	6.038	5.014	4.004	4.004		19	4.014	4.014	3.003	3.003
	16	5.030	4.011	3.003	3.003		18	4.035	3.009	3.009	2.002
	15	4.022	4.022	3.007	2.002		17	3.021	3.021	2.005	2.005
	14	4.041	3.015	2.004	2.004		16	3.039	2.010	1.002	1.002
	13	3.028	2.009	2.009	1.002		15	2.020	2.020	1.004	1.004
	12	3.049	2.018	1.004	1.004		14	2.036	1.008	1.008	0.001
	11	2.032	1.009	1.009	0.001		13	1.015	1.015	0.002	0.002
	10	1.017	1.017	0.003	0.003		12	1.027	0.004	0.004	0.004
	9	1.031	0.006	0.006	–		11	1.044	0.008	0.008	–
	8	0.012	0.012	–	–		10	0.014	0.014	–	–
	7	0.023	0.023	–	–		9	0.024	0.024	–	–
	6	0.043	–	–	–		8	0.041	–	–	–
$B=11$	20	8.037	7.010	6.003	6.003	$B=7$	20	4.012	4.012	3.002	3.002
	19	7.039	6.013	5.004	5.004		19	4.040	3.009	3.009	2.001
	18	6.033	5.011	4.003	4.003		18	3.022	3.022	2.004	2.004
	17	5.026	4.008	4.008	3.002		17	3.045	2.011	1.002	1.002
	16	4.019	4.019	3.006	2.001		16	2.022	2.022	1.004	1.004
	15	4.036	3.012	2.003	2.003		15	2.039	1.008	1.008	0.001
	14	3.024	3.024	2.007	1.001		14	1.016	1.016	0.002	0.002
	13	3.043	2.014	1.003	1.003		13	1.027	0.004	0.004	0.004
	12	2.026	1.007	1.007	0.001		12	1.044	0.007	0.007	–
	11	2.045	1.013	0.002	0.002		11	0.013	0.013	–	–
	10	1.024	1.024	0.004	0.004		10	0.022	0.022	–	–
	9	1.042	0.008	0.008	–		9	0.036	–	–	–
	8	0.016	0.016	–	–	$B=6$	20	4.046	3.008	3.008	2.001
	7	0.029	–	–	–		19	3.027	2.005	2.005	2.005
$B=10$	20	7.030	6.008	6.008	5.002		18	2.012	2.012	1.002	1.002
	19	6.029	5.008	5.008	4.002		17	2.026	1.004	1.004	1.004
	18	5.024	5.024	4.007	3.002		16	2.047	1.009	1.009	0.001
	17	5.049	4.017	3.005	3.005		15	1.018	1.018	0.002	0.002
	16	4.034	3.011	2.003	2.003		14	1.030	0.004	0.004	0.004
	15	3.022	3.022	2.006	1.001		13	1.048	0.007	0.007	–
	14	3.039	2.012	1.002	1.002		12	0.013	0.013	–	–
	13	2.022	2.022	1.005	0.001		11	0.022	0.022	–	–
	12	2.039	1.011	0.001	0.001		10	0.035	–	–	–
	11	1.019	1.019	0.003	0.003	$B=5$	20	3.033	2.004	2.004	2.004
	10	1.034	0.006	0.006	–		19	2.016	2.016	1.002	1.002
	9	0.012	0.012	–	–		18	2.036	1.005	1.005	0.000
	8	0.022	0.022	–	–		17	1.012	1.012	0.001	0.001
	7	0.038	–	–	–		16	1.022	1.022	0.002	0.002
$B=9$	20	6.023	6.023	5.005	4.001		15	1.038	0.005	0.005	0.005
	19	5.021	5.021	4.005	3.001		14	0.009	0.009	0.009	–
	18	4.015	4.015	3.004	3.004		13	0.015	0.015	–	–
	17	4.033	3.010	3.010	2.002		12	0.024	0.024	–	–
	16	3.020	3.020	2.005	1.001		11	0.038	–	–	–
	15	3.038	2.011	1.002	1.002	$B=4$	20	2.022	2.022	1.002	1.002
	14	2.021	2.021	1.004	1.004		19	1.008	1.008	1.008	0.000

10.11   DETERMINING VALUES IN BERNOULLI TRIALS

Suppose the probability of heads for a *biased* coin is either (H_0) $p = \alpha$ or (H_1) $p = \beta$ (with $\alpha < \beta$). Assume the values α and β are known and the purpose of an experiment is to determine the true value of p. Toss the coin repeatedly and let the random variable Y be the number of tosses until the rth head appears. Let $1 - \theta$ be the probability the identification is correct under either hypothesis and define N by

$$1 - \theta \approx \text{Prob}\,[Y \leq N \mid p = \beta]$$
$$\approx \text{Prob}\,[Y > N \mid p = \alpha] \tag{10.13}$$

This hypothesis test has significance level θ and power $1 - \theta$.

The random variable Y has a negative binomial distribution with mean r/p and variance $r(1 - p)/p^2$ but may be approximated by a normal distribution. If ξ is defined by $\Phi(\xi) = \theta$, then

$$r \approx \left[\frac{\xi}{\beta - \alpha} \left\{ \alpha\sqrt{1 - \beta} + \beta\sqrt{1 - \alpha} \right\} \right]^2$$
$$N \approx \frac{r - \xi\sqrt{r(1 - \alpha)}}{\alpha} \tag{10.14}$$

Example 10.62: If $\theta = 0.05$, equation (10.14) yields the values below. In this table, $\text{E}\,[T]$ is the expected number of tosses required to reach a decision.

α	β	r	N	$\text{E}\,[T]$
0.1	0.2	21	139	122
0.3	0.4	87	247	232
0.5	0.6	148	268	257
0.7	0.8	153	203	197
0.1	0.6	4	10	8
0.4	0.9	7	9	8

See G. J. Manas and D. H. Meyer, "On a problem of coin identification," *SIAM Review*, **31**, Number 1, March 1989, SIAM, Philadelphia, pages 114–117.

CHAPTER 11

Regression Analysis

Contents

11.1 SIMPLE LINEAR REGRESSION

Let $(x_1, y_1), (x_2, y_2), \ldots, (x_n, y_n)$ be n pairs of observations such that y_i is an observed value of the random variable Y_i. Assume there exist constants β_0 and β_1 such that

$$Y_i = \beta_0 + \beta_1 x_i + \epsilon_i \tag{11.1}$$

where $\epsilon_1, \epsilon_2, \ldots, \epsilon_n$ are independent, normal random variables having mean 0 and variance σ^2.

Assumptions	
In terms of ϵ_i's	In terms of Y_i's
ϵ_i's are normally distributed	Y_i's are normally distributed
$E[\epsilon_i] = 0$	$E[Y_i] = \beta_0 + \beta_1 x_i$
$\text{Var}[\epsilon_i] = \sigma^2$	$\text{Var}[Y_i] = \sigma^2$
$\text{Cov}[\epsilon_i, \epsilon_j] = 0, \ i \neq j$	$\text{Cov}[Y_i, Y_j] = 0, \ i \neq j$

Principle of least squares: The sum of squared deviations about the true regression line is

$$S(\beta_0, \beta_1) = \sum_{i=1}^{n} [y_i - (\beta_0 + \beta_1 x_i)]^2. \tag{11.2}$$

The point estimates of β_0 and β_1, denoted by $\widehat{\beta}_0$ and $\widehat{\beta}_1$, are those values that minimize $S(\beta_0, \beta_1)$. The estimates $\widehat{\beta}_0$ and $\widehat{\beta}_1$ are called the **least squares estimates**. The estimated regression line or least squares line is $\widehat{y} = \widehat{\beta}_0 + \widehat{\beta}_1 x$.

The **normal equations** for $\widehat{\beta}_0$ and $\widehat{\beta}_1$ are

$$\left(\sum_{i=1}^{n} y_i \right) = \widehat{\beta}_0 \, n + \widehat{\beta}_1 \left(\sum_{i=1}^{n} x_i \right)$$

$$\left(\sum_{i=1}^{n} x_i y_i \right) = \widehat{\beta}_0 \left(\sum_{i=1}^{n} x_i \right) + \widehat{\beta}_1 \left(\sum_{i=1}^{n} x_i^2 \right) \tag{11.3}$$

Notation:

$$S_{xx} = \sum_{i=1}^{n} (x_i - \overline{x})^2 = \sum_{i=1}^{n} x_i^2 - \frac{1}{n} \left(\sum_{i=1}^{n} x_i \right)^2$$

$$S_{yy} = \sum_{i=1}^{n} (y_i - \overline{y})^2 = \sum_{i=1}^{n} y_i^2 - \frac{1}{n} \left(\sum_{i=1}^{n} y_i \right)^2$$

$$S_{xy} = \sum_{i=1}^{n} (x_i - \overline{x})(y_i - \overline{y})^2 = \sum_{i=1}^{n} x_i y_i - \frac{1}{n} \left(\sum_{i=1}^{n} x_i \right) \left(\sum_{i=1}^{n} y_i \right)$$

11.1.1 Least squares estimates

$$\widehat{\beta}_1 = \frac{S_{xy}}{S_{xx}} = \frac{n\left(\sum\limits_{i=1}^{n} x_i y_i\right) - \left(\sum\limits_{i=1}^{n} x_i\right)\left(\sum\limits_{i=1}^{n} y_i\right)}{n\left(\sum\limits_{i=1}^{n} x_i^2\right) - \left(\sum\limits_{i=1}^{n} x_i\right)^2} \tag{11.4}$$

$$\widehat{\beta}_0 = \frac{\sum\limits_{i=1}^{n} y_i - \widehat{\beta}_1 \sum\limits_{i=1}^{n} x_i}{n} = \overline{y} - \widehat{\beta}_1 \overline{x}$$

The ith **predicted (fitted) value**: $\widehat{y}_i = \widehat{\beta}_0 + \widehat{\beta}_1 x_i$ (for $i = 1, 2, \ldots, n$).
The ith **residual**: $e_i = y_i - \widehat{y}_i$ (for $i = 1, 2, \ldots, n$).
Properties:

(1) $\mathrm{E}\left[\widehat{\beta}_1\right] = \beta_1$, $\quad \mathrm{Var}\left[\widehat{\beta}_1\right] = \dfrac{\sigma^2}{\sum\limits_{i=1}^{n}(x_i - \overline{x})^2} = \dfrac{\sigma^2}{S_{xx}}$

(2) $\mathrm{E}\left[\widehat{\beta}_0\right] = \beta_0$, $\quad \mathrm{Var}\left[\widehat{\beta}_0\right] = \dfrac{\sigma^2 \sum\limits_{i=1}^{n} x_i}{n \sum\limits_{i=1}^{n}(x_i - \overline{x})^2} = \dfrac{\sigma^2 \sum\limits_{i=1}^{n} x_i}{n S_{xx}} = \dfrac{\sigma^2 \overline{x}}{S_{xx}}$

(3) $\widehat{\beta}_0$ and $\widehat{\beta}_1$ are normally distributed.

11.1.2 Sum of squares

$$\underbrace{\sum_{i=1}^{n}(y_i - \overline{y})^2}_{\text{SST}} = \underbrace{\sum_{i=1}^{n}(\widehat{y}_i - \overline{y})^2}_{\text{SSR}} + \underbrace{\sum_{i=1}^{n}(y_i - \widehat{y}_i)^2}_{\text{SSE}}$$

SST = total sum of squares = S_{yy}
SSR = sum of squares due to regression = $\widehat{\beta}_2 S_{xy}$
SSE = sum of squares due to error

$$= \sum_{i=1}^{n}[y_i - (\widehat{\beta}_0 + \widehat{\beta}_1 x_i)]^2 = \sum_{i=1}^{n} y_i^2 - \widehat{\beta}_0 \sum_{i=1}^{n} y_i - \widehat{\beta}_1 \sum_{i=1}^{n} x_i y_i$$

$$= S_{yy} - 2\widehat{\beta}_1 S_{xy} + \widehat{\beta}_1^2 S_{xx} = S_{yy} - \widehat{\beta}_1^2 S_{xx} = S_{yy} - \widehat{\beta}_1 S_{xy}$$

$$\widehat{\sigma}^2 = s^2 = \frac{\text{SSE}}{n-2}, \quad \mathrm{E}\left[S^2\right] = \sigma^2$$

Sample coefficient of determination: $r^2 = \dfrac{\text{SSR}}{\text{SST}} = 1 - \dfrac{\text{SSE}}{\text{SST}}$

11.1.3   Inferences concerning the regression coefficients

The parameter $\widehat{\beta}_1$

(1) $T = \dfrac{\widehat{\beta}_1 - \beta_1}{S/\sqrt{S_{xx}}} = \dfrac{\widehat{\beta}_1 - \beta_1}{S_{\widehat{\beta}_1}}$

has a t distribution with $n - 2$ degrees of freedom, where
$S_{\widehat{\beta}_1} = S/\sqrt{S_{xx}}$ is an estimate for the standard deviation of $\widehat{\beta}_1$.

(2) A $100(1 - \alpha)\%$ confidence interval for β_1 has as endpoints

$\widehat{\beta}_1 \pm t_{\alpha/2, n-2} \cdot s_{\widehat{\beta}_1}$

(3) Hypothesis test:

Null hypothesis	Alternative hypotheses	Test statistic	Rejection regions	
$\beta_1 = \beta_{10}$	$\beta_1 > \beta_{10}$		$T \geq t_{\alpha, n-2}$	(1)
	$\beta_1 < \beta_{10}$	$T = \dfrac{\widehat{\beta}_1 - \beta_{10}}{S_{\widehat{\beta}_1}}$	$T \leq -t_{\alpha, n-2}$	(2)
	$\beta_1 \neq \beta_{10}$		$|T| \geq t_{\alpha/2, n-2}$	(3)

The parameter $\widehat{\beta}_0$

(1) $T = \dfrac{\widehat{\beta}_0 - \beta_0}{S\sqrt{\sum\limits_{i=1}^{n} x_i^2 / nS_{xx}}} = \dfrac{\widehat{\beta}_0 - \beta_0}{S_{\widehat{\beta}_0}}$

has a t distribution with $n - 2$ degrees of freedom, where
$S_{\widehat{\beta}_0}$ denotes the estimate for the standard deviation of $\widehat{\beta}_0$.

(2) A $100(1 - \alpha)\%$ confidence interval for β_1 has as endpoints

$\widehat{\beta}_1 \pm t_{\alpha/2, n-2} \cdot s_{\widehat{\beta}_0}$

(3) Hypothesis test:

Null hypothesis	Alternative hypotheses	Test statistic	Rejection regions	
$\beta_0 = \beta_{00}$	$\beta_0 > \beta_{00}$		$T \geq t_{\alpha, n-2}$	(1)
	$\beta_0 < \beta_{00}$	$T = \dfrac{\widehat{\beta}_0 - \beta_{00}}{S_{\widehat{\beta}_0}}$	$T \leq -t_{\alpha, n-2}$	(2)
	$\beta_0 \neq \beta_{00}$		$|T| \geq t_{\alpha/2, n-2}$	(3)

11.1.4   The mean response

The mean response of Y given $x = x_0$ is $\mu_{Y|x_0} = \beta_0 + \beta_1 x_0$. The random variable $\widehat{Y}_0 = \widehat{\beta}_0 + \widehat{\beta}_1 x_0$ is used to estimate $\mu_{Y|x_0}$.

(1) $E[\widehat{Y}_0] = \beta_0 + \beta_1 x_0$

(2) $\mathrm{Var}[\widehat{Y}_0] = \sigma^2 \left[\dfrac{1}{n} + \dfrac{(x_0 - \bar{x})^2}{S_{xx}}\right]$

(3) $\widehat{Y}_0$ has a normal distribution.

(4) $T = \dfrac{\widehat{Y}_0 - \mu_{Y|x_0}}{S\sqrt{(1/n) + [(x_0 - \bar{x})^2/S_{xx}]}} = \dfrac{\widehat{Y}_0 - \mu_{Y|x_0}}{S_{\widehat{Y}_0}}$

has a t distribution with $n - 2$ degrees of freedom, where $S_{\widehat{Y}_0}$ denotes the estimate for the standard deviation of $\widehat{Y}_0$.

(5) A $100(1 - \alpha)\%$ confidence interval for $\mu_{Y|x_0}$ has as endpoints

$\widehat{y}_0 \pm t_{\alpha/2, n-2} \cdot s_{\widehat{Y}_0}.$

(6) Hypothesis test:

Null hypothesis	Alternative hypotheses	Test statistic	Rejection regions	
$\beta_0 + \beta_1 x_0 = y_0 = \mu_0$	$y_0 > \mu_0$		$T \geq t_{\alpha, n-2}$	(1)
	$y_0 < \mu_0$	$T = \dfrac{\widehat{Y}_0 - \mu_0}{S_{\widehat{Y}_0}}$	$T \leq -t_{\alpha, n-2}$	(2)
	$y_0 \neq \mu_0$		$|T| \geq t_{\alpha/2, n-2}$	(3)

11.1.5 Prediction interval

A prediction interval for a value y_0 of the random variable $Y_0 = \beta_0 + \beta_1 x_0 + \epsilon_0$ is obtained by considering the random variable $\widehat{Y}_0 - Y_0$.

(1) $E[\widehat{Y}_0 - Y_0] = 0$

(2) $\mathrm{Var}[\widehat{Y}_0 - Y_0] = \sigma^2 \left[1 + \dfrac{1}{n} + \dfrac{(x_0 - \bar{x})^2}{S_{xx}}\right]$

(3) $\widehat{Y}_0 - Y_0$ has a normal distribution.

(4) $T = \dfrac{\widehat{Y}_0 - Y_0}{S\sqrt{1 + (1/n) + [(x_0 - \bar{x})^2/S_{xx}]}} = \dfrac{\widehat{Y}_0 - Y_0}{S_{\widehat{Y}_0 - Y_0}}$

has a t distribution with $n - 2$ degrees of freedom.

(5) A $100(1 - \alpha)\%$ prediction interval for y_0 has as endpoints

$\widehat{y}_0 \pm t_{\alpha/2, n-2} \cdot s_{\widehat{Y}_0 - Y_0}$

11.1.6   Analysis of variance table

Source of variation	Sum of squares	Degrees of freedom	Mean square	Computed F
Regression	SSR	1	$MSR = \dfrac{SSR}{1}$	MSR/MSE
Error	SSE	$n - 2$	$MSE = \dfrac{SSE}{n - 2}$	
Total	SST	$n - 1$		

Hypothesis test of significant regression:

Null hypothesis	Alternative hypothesis	Test statistic	Rejection region
$\beta_1 = 0$	$\beta_1 \neq 0$	$F = MSR/MSE$	$F \geq F_{\alpha,1,n-2}$

11.1.7   Test for linearity of regression

Suppose there are k distinct values of x, $\{x_1, x_2, \ldots, x_k\}$, n_i observations for x_i, and $n = n_1 + n_2 + \cdots + n_k$.

Definitions:

(1) $y_{ij} =$ the jth observation on the random variable Y_i.

(2) $T_i = \displaystyle\sum_{j=1}^{n_i} y_{ij}, \quad \overline{y}_{i.} = T_i/n_i$

(3) SSPE = sum of squares due to pure error

$$= \sum_{i=1}^{k}\sum_{j=1}^{n_i}(y_{ij} - \overline{y}_{i.})^2 = \sum_{i=1}^{k}\sum_{j=1}^{n_i} y_{ij}^2 - \sum_{i=1}^{k}\frac{T_i^2}{n_i}$$

(4) SSLF = Sum of squares due to lack of fit = SSE $-$ SSPE

Hypothesis test:

Null hypothesis	Alternative hypothesis	Test statistic	Rejection region
Linear regression	Lack of fit	$F = \dfrac{\text{SSLF}/(k-2)}{\text{SSPE}/(n-k)}$	$F \geq F_{\alpha,k-2,n-k}$

11.1.8   Sample correlation coefficient

The sample correlation coefficient is a measure of linear association and is defined by

$$r = \widehat{\beta}_1 \sqrt{\frac{S_{xx}}{S_{yy}}} = \frac{S_{xy}}{\sqrt{S_{xx}S_{yy}}}. \tag{11.5}$$

Hypothesis tests:

Null hypothesis	Alternative hypothesis	Test statistic	Rejection region	
$\rho = 0$	$\rho > 0$		$T \geq t_{\alpha,n-2}$	(1)
	$\rho < 0$	$T = \dfrac{R\sqrt{n-2}}{\sqrt{1-R^2}} = \dfrac{\widehat{\beta}_1}{S_{\widehat{\beta}_1}}$	$T \leq -t_{\alpha,n-2}$	(2)
	$\rho \neq 0$		$|T| \geq t_{\alpha/2,n-2}$	(3)

If X and Y have a bivariate normal distribution:

$\rho = \rho_0$	$\rho > \rho_0$	$Z = \dfrac{\sqrt{n-3}}{2}$	$Z \geq z_\alpha$	(4)
	$\rho < \rho_0$		$Z \leq -z_\alpha$	(5)
	$\rho \neq \rho_0$	$\times \ln\left[\dfrac{(1+R)(1-\rho_0)}{(1-R)(1+\rho_0)}\right]$	$|Z| \geq z_{\alpha/2}$	(6)

11.1.9 Example

Example 11.63: A recent study at a manufacturing facility examined the relationship between the noon temperature (Fahrenheit) inside the plant and the number of defective items produced during the day shift. The data are given in the following table.

Noon temperature (x)	Number defective (y)	Noon temperature (x)	Number defective (y)
68	27	74	48
78	52	65	33
71	39	72	45
69	22	73	51
66	21	67	29
75	66	77	65

(1) Find the regression equation using temperature as the independent variable and construct the anova table.

(2) Test for a significant regression. Use $\alpha = .05$.

(3) Find a 95% confidence interval for the mean number of defective items produced when the temperature is 68° F.

(4) Find a 99% prediction interval for a temperature of 75° F.

Solution:

(S1) $\widehat{\beta}_1 = \dfrac{S_{xy}}{S_{xx}} = \dfrac{640.5}{204.25} = 3.1359$

$\widehat{\beta}_0 = \overline{y} - \widehat{\beta}_1\overline{x} = 41.5 - (3.1359)(71.25) = -181.93$

Regression line: $\widehat{y} = -181.93 + 3.1359x$

(S2)

Source of variation	Sum of squares	Degrees of freedom	Mean square	Computed F
Regression	2008.5	1	2008.5	31.16
Error	644.5	10	64.4	
Total	2653.0	11		

(S3) Hypothesis test of significant regression:

H_0: $\beta_1 = 0$

H_a: $\beta_1 \neq 0$

TS: $F = \text{MSR}/\text{MSE}$

RR: $F \geq F_{.05,1,10} = 4.96$

Conclusion: The value of the test statistic ($F = 31.16$) lies in the rejection region. There is evidence to suggest a significant regression. Note: this test is equivalent to the t test in section 11.1.3 with $\beta_{10} = 0$.

(S4) A 95% confidence interval for $\mu_{Y|68}$:

$$\widehat{y}_0 \pm t_{.025,10} \cdot s_{\widehat{Y}_0} = 31.31 \pm (2.228)(2.9502) = (24.74, 37.88)$$

(S5) A 99% prediction interval for $y_0 = -181.93 + (3.1359)(75) = 53.26$

$$\widehat{y}_0 \pm t_{.005,10} \cdot s_{\widehat{Y}_0 - Y_0} = 53.26 \pm (3.169)(8.6172) = (25.95, 80.56)$$

11.2 MULTIPLE LINEAR REGRESSION

Let there be n observations of the form $(x_{1i}, x_{2i}, \ldots, x_{ki}, y_i)$ such that y_i is an observed value of the random variable Y_i. Assume there exist constants $\beta_0, \beta_1, \ldots, \beta_k$ such that

$$Y_i = \beta_0 + \beta_1 x_{1i} + \cdots + \beta_k x_{ki} + \epsilon_i \qquad (11.6)$$

where $\epsilon_1, \epsilon_2, \ldots, \epsilon_n$ are independent, normal random variables having mean 0 and variance σ^2.

Assumptions	
In terms of ϵ_i's	In terms of Y_i's
ϵ_i's are normally distributed	Y_i's are normally distributed
$\text{E}[\epsilon_i] = 0$	$\text{E}[Y_i] = \beta_0 + \beta_1 x_{1i} + \cdots + \beta_k x_{ki}$
$\text{Var}[\epsilon_i] = \sigma^2$	$\text{Var}[Y_i] = \sigma^2$
$\text{Cov}[\epsilon_i, \epsilon_j] = 0, \; i \neq j$	$\text{Cov}[Y_i, Y_j] = 0, \; i \neq j$

Notation:

Let $\mathbf{Y}$ be the random vector of responses, $\mathbf{y}$ be the vector of observed responses, $\boldsymbol{\beta}$ be the vector of regression coefficients, $\boldsymbol{\epsilon}$ be the vector of random errors, and let $\mathbf{X}$ be the design matrix:

$$\mathbf{Y} = \begin{bmatrix} Y_1 \\ Y_2 \\ \vdots \\ Y_n \end{bmatrix} \quad \mathbf{y} = \begin{bmatrix} y_1 \\ y_2 \\ \vdots \\ y_n \end{bmatrix} \quad \boldsymbol{\beta} = \begin{bmatrix} \beta_0 \\ \beta_1 \\ \vdots \\ \beta_k \end{bmatrix} \quad \boldsymbol{\epsilon} = \begin{bmatrix} \epsilon_1 \\ \epsilon_2 \\ \vdots \\ \epsilon_n \end{bmatrix} \quad \mathbf{X} = \begin{bmatrix} 1 & x_{11} & x_{21} & \cdots & x_{k1} \\ 1 & x_{12} & x_{22} & \cdots & x_{k2} \\ \vdots & \vdots & \vdots & \ddots & \vdots \\ 1 & x_{1n} & x_{2n} & \cdots & x_{kn} \end{bmatrix}$$

$$(11.7)$$

The model can now be written as $\mathbf{Y} = \mathbf{X}\boldsymbol{\beta} + \boldsymbol{\epsilon}$ where $\boldsymbol{\epsilon} \sim N_n(\mathbf{0}, \sigma^2 \mathbf{I}_n)$ or equivalently $\mathbf{Y} \sim N_n(\mathbf{X}\boldsymbol{\beta}, \sigma^2 \mathbf{I}_n)$.

Principle of least squares: The sum of squared deviations about the true regression line is

$$S(\boldsymbol{\beta}) = \sum_{i=1}^{n}[y_i - (\beta_0 + \beta_1 x_{1i} + \cdots + \beta_k x_{ki})]^2 = \|\mathbf{y} - \mathbf{X}\boldsymbol{\beta}\|^2. \qquad (11.8)$$

The vector $\widehat{\boldsymbol{\beta}} = [\widehat{\beta}_0, \widehat{\beta}_1, \ldots, \widehat{\beta}_k]^T$ that minimizes $S(\boldsymbol{\beta})$ is the vector of least squares estimates. The estimated regression line or least squares line is $y = \widehat{\beta}_0 + \widehat{\beta}_1 x_1 + \cdots + \widehat{\beta}_k x_k$.

The normal equations may be written as $(\mathbf{X}^T \mathbf{X})\widehat{\boldsymbol{\beta}} = \mathbf{X}^T \mathbf{y}$.

11.2.1 Least squares estimates

If the matrix $\mathbf{X}^T \mathbf{X}$ is non–singular, then $\widehat{\boldsymbol{\beta}} = (\mathbf{X}^T \mathbf{X})^{-1} \mathbf{X}^T \mathbf{y}$.

The ith predicted (fitted) value: $\widehat{y}_i = \widehat{\beta}_0 + \widehat{\beta}_1 x_{1i} + \cdots + \widehat{\beta}_k x_{ki}$ (for $i = 1, 2, \ldots, n$), $\widehat{\mathbf{y}} = \mathbf{X}\widehat{\boldsymbol{\beta}}$.

The ith residual: $e_i = y_i - \widehat{y}_i$, $i = 1, 2, \ldots, n$, $\mathbf{e} = \mathbf{y} - \widehat{\mathbf{y}}$.

Properties: For $i = 0, 1, 2, \ldots, k$ and $j = 0, 1, 2, \ldots, k$:

(1) $E[\widehat{\beta}_i] = \beta_i$.

(2) $\text{Var}[\widehat{\beta}_i] = c_{ii}\sigma^2$, where c_{ij} is the value in the ith row and j^{th} column of the matrix $(\mathbf{X}^T \mathbf{X})^{-1}$.

(3) $\widehat{\beta}_i$ is normally distributed.

(4) $\text{Cov}[\widehat{\beta}_i, \widehat{\beta}_j] = c_{ij}\sigma^2$, $i \neq j$.

11.2.2 Sum of squares

$$\underbrace{\sum_{i=1}^{n}(y_i - \overline{y})^2}_{\text{SST}} = \underbrace{\sum_{i=1}^{n}(\widehat{y}_i - \overline{y})^2}_{\text{SSR}} + \underbrace{\sum_{i=1}^{n}(y_i - \widehat{y}_i)^2}_{\text{SSE}}$$

SST = total sum of squares = $\|\mathbf{y} - \overline{y}\mathbf{1}\|^2 = \mathbf{y}^T\mathbf{y} - n\overline{y}^2$

SSR = sum of squares due to regression = $\|\mathbf{X}\widehat{\boldsymbol{\beta}} - \overline{y}\mathbf{1}\|^2 = \widehat{\boldsymbol{\beta}}^T\mathbf{X}^T\mathbf{y} - n\overline{y}^2$

SSE = sum of squares due to error = $\|\mathbf{y} - \mathbf{X}\widehat{\boldsymbol{\beta}}\|^2 = \mathbf{y}^T\mathbf{y} - \widehat{\boldsymbol{\beta}}\mathbf{X}^T\mathbf{y}$

where $\mathbf{1} = [1, 1, \ldots, 1]^T$ is a column vector of all 1's.

(1) $\widehat{\sigma}^2 = s^2 = \dfrac{\text{SSE}}{n - k - 1}$, $\quad E[S^2] = \sigma^2$

(2) $\dfrac{(n - k - 1)S^2}{\sigma^2}$ has a chi–square distribution with $n - k - 1$ degrees of freedom, and S^2 and $\widehat{\beta}_i$ are independent.

(3) **The coefficient of multiple determination:**

$$R^2 = \frac{\text{SSR}}{\text{SST}} = 1 - \frac{\text{SSE}}{\text{SST}}$$

(4) **Adjusted coefficient of multiple determination:**

$$R_a^2 = 1 - \left(\frac{n-1}{n-k-1}\right)\frac{\text{SSE}}{\text{SST}} = 1 - (1-R^2)\left(\frac{n-1}{n-k-1}\right)$$

11.2.3   Inferences concerning the regression coefficients

(1) $T = \dfrac{\widehat{\beta}_i - \beta_i}{S\sqrt{c_{ii}}}$ has a t distribution with $n-k-1$ degrees of freedom.

(2) A $100(1-\alpha)\%$ confidence interval for β_i has as endpoints

$$\widehat{\beta}_i \pm t_{\alpha/2,n-k-1} \cdot s\sqrt{c_{ii}}$$

(3) Hypothesis test for β_i:

Null hypothesis	Alternative hypotheses	Test statistic	Rejection regions	
$\beta_i = \beta_{i0}$	$\beta_i > \beta_{i0}$	$T = \dfrac{\widehat{\beta}_i - \beta_{i0}}{S\sqrt{c_{ii}}}$	$T \geq t_{\alpha,n-k-1}$	(1)
	$\beta_i < \beta_{i0}$		$T \leq -t_{\alpha,n-k-1}$	(2)
	$\beta_i \neq \beta_{i0}$		$\lvert T \rvert \geq t_{\alpha/2,n-k-1}$	(3)

11.2.4   The mean response

The mean response of Y given $\mathbf{x} = \mathbf{x}_0 = [1, x_{10}, x_{20}, \ldots, x_{k0}]^{\text{T}}$ is $\mu_{Y|x_{10},x_{20},\ldots,x_{k0}} = \beta_0 + \beta_1 x_{10} + \cdots + \beta_k x_{k0}$. The random variable $\widehat{Y}_0 = \widehat{\boldsymbol{\beta}}^{\text{T}}\mathbf{x}_0 = \widehat{\beta}_0 + \widehat{\beta}_1 x_{10} + \cdots + \widehat{\beta}_k x_{k0}$ is used to estimate $\mu_{Y|x_{10},x_{20},\ldots,x_{k0}}$.

(1) $\mathrm{E}\left[\widehat{Y}_0\right] = \beta_0 + \beta_1 x_{10} + \cdots + \beta_k x_{k0}$

(2) $\mathrm{Var}\left[\widehat{Y}_0\right] = \sigma^2 \mathbf{x}_0^{\text{T}}(\mathbf{X}^{\text{T}}\mathbf{X})^{-1}\mathbf{x}_0$

(3) $\widehat{Y}_0$ has a normal distribution.

(4) $T = \dfrac{\widehat{Y}_0 - \mu_{Y|x_{10},x_{20},\ldots,x_{k0}}}{S\sqrt{\mathbf{x}_0^{\text{T}}(\mathbf{X}^{\text{T}}\mathbf{X})^{-1}\mathbf{x}_0}}$

   has a t distribution with $n - k - 1$ degrees of freedom.

(5) A $100(1-\alpha)\%$ confidence interval for $\mu_{Y|x_{10},x_{20},\ldots,x_{k0}}$ has as endpoints

$$\widehat{y}_0 \pm t_{\alpha/2,n-k-1} \cdot s\sqrt{\mathbf{x}_0^{\text{T}}(\mathbf{X}^{\text{T}}\mathbf{X})^{-1}\mathbf{x}_0}. \text{ ''}$$

(6) Hypothesis test:

Null hypothesis	Alternative hypotheses	Test statistic	Rejection regions	
$\beta_0 + \beta_1 x_{10} + \cdots + \beta_k x_{k0} = y_0 = \mu_0$				
	$y_0 > \mu_0$	$T = \dfrac{\widehat{Y}_0 - \mu_0}{S\sqrt{\mathbf{x}_0^{\mathrm{T}}(\mathbf{X}^{\mathrm{T}}\mathbf{X})^{-1}\mathbf{x}_0}}$	$T \geq t_{\alpha,n-k-1}$	(1)
	$y_0 < \mu_0$		$T \leq -t_{\alpha,n-k-1}$	(2)
	$y_0 \neq \mu_0$		$\lvert T \rvert \geq t_{\alpha/2,n-k-1}$	(3)

11.2.5 Prediction interval

A prediction interval for a value y_0 of the random variable $Y_0 = \beta_0 + \beta_1 x_{10} + \cdots + \beta_k x_{k0} + \epsilon_0$ is obtained by considering the random variable $\widehat{Y}_0 - Y_0$.

(1) $E[\widehat{Y}_0 - Y_0] = 0$

(2) $\mathrm{Var}[\widehat{Y}_0 - Y_0] = \sigma^2 \left[1 + \mathbf{x}_0^{\mathrm{T}}(\mathbf{X}^{\mathrm{T}}\mathbf{X})^{-1}\mathbf{x}_0\right]$

(3) $\widehat{Y}_0 - Y_0$ has a normal distribution.

(4) $T = \dfrac{\widehat{Y}_0 - Y_0}{S\sqrt{1 + \mathbf{x}_0^{\mathrm{T}}(\mathbf{X}^{\mathrm{T}}\mathbf{X})^{-1}\mathbf{x}_0}}$

has a t distribution with $n - k - 1$ degrees of freedom.

(5) A $100(1 - \alpha)\%$ prediction interval for y_0 has as endpoints

$$\widehat{y}_0 \pm t_{\alpha/2,n-2} \cdot s\sqrt{1 + \mathbf{x}_0^{\mathrm{T}}(\mathbf{X}^{\mathrm{T}}\mathbf{X})^{-1}\mathbf{X}_0}$$

11.2.6 Analysis of variance table

Source of variation	Sum of squares	Degrees of freedom	Mean square	Computed F
Regression	SSR	k	$MSR = \dfrac{SSR}{k}$	MSR/MSE
Error	SSE	$n - k - 1$	$MSE = \dfrac{SSE}{n - k - 1}$	
Total	SST	$n - 1$		

Hypothesis test of significant regression:

$H_0: \beta_1 = \beta_2 = \cdots = \beta_k = 0$

$H_a: \beta_i \neq 0$ for some i

TS: $F = MSR/MSE$

RR: $F \geq F_{\alpha,k,n-k-1}$

11.2.7   Sequential sum of squares

Define

$$\mathbf{g} = \begin{bmatrix} g_0 \\ g_1 \\ \vdots \\ g_k \end{bmatrix} = \mathbf{X}^T\mathbf{y} = \begin{bmatrix} \sum_{i=1}^{n} y_i \\ \sum_{i=1}^{n} x_{1i}y_i \\ \vdots \\ \sum_{i=1}^{n} x_{ki}y_i \end{bmatrix} \tag{11.9}$$

$$\mathrm{SSR} = \sum_{j=0}^{k} \widehat{\beta}_j g_j - n\bar{y}^2$$

$$\mathrm{SS}(\beta_1, \beta_2, \ldots, \beta_r) = \text{the sum of squares due to } \beta_1, \beta_2, \ldots, \beta_r$$

$$= \sum_{j=1}^{r} \widehat{\beta}_j g_j - n\bar{y}^2$$

$$\mathrm{SS}(\beta_1) = \text{the regression sum of squares due to } x_1$$

$$= \sum_{j=0}^{1} \widehat{\beta}_j g_j - n\bar{y}^2$$

$$\mathrm{SS}(\beta_2|\beta_1) = \text{the regression sum of squares due to } x_2 \text{ given}$$
$$x_1 \text{ is in the model}$$

$$= \mathrm{SS}(\beta_1, \beta_2) - \mathrm{SS}(\beta_1) = \widehat{\beta}_2 g_2$$

$$\mathrm{SS}(\beta_3|\beta_1, \beta_2) = \text{the regression sum of squares due to } x_3 \text{ given}$$
$$x_1, x_2 \text{ are in the model}$$

$$= \mathrm{SS}(\beta_1, \beta_2, \beta_3) - \mathrm{SS}(\beta_1, \beta_2) = \widehat{\beta}_3 g_3$$

$$\vdots$$

$$\mathrm{SS}(\beta_r|\beta_1, \ldots, \beta_{r-1}) = \text{the regression sum of squares due to } x_r \text{ given}$$
$$x_1, \ldots, x_{r-1} \text{ are in the model}$$

$$= \mathrm{SS}(\beta_1, \ldots, \beta_r) - \mathrm{SS}(\beta_1, \ldots, \beta_{r-1}) = \widehat{\beta}_r g_r$$

11.2.8   Partial F test

Definitions:

(1) Full model:
$$y_i = \beta_0 + \beta_1 x_{1i} + \cdots + \beta_r x_{ri} + \beta_{r+1} x_{(r+1)i} + \cdots + \beta_k x_{ki} + \epsilon_i$$

(2) SSE(F) = sum of squares due to error in the full model.
$$= \sum_{i=1}^{n} (y_i - \widehat{y}_i)^2 \quad \text{where}$$

$$\widehat{y}_i = \hat{\beta}_0 + \hat{\beta}_1 x_{1i} + \cdots + \hat{\beta}_r x_{ri} + \hat{\beta}_{r+1} x_{(r+1)i} + \cdots + \hat{\beta}_k x_{ki}$$

(3) Reduced model: $y_i = \beta_0 + \beta_1 x_{1i} + \cdots + \beta_r x_{ri} + \epsilon_i$

(4) SSE(R) = sum of squares due to error in the reduced model.

$$= \sum_{i=1}^{n} (y_i - \widehat{y}_i)^2 \quad \text{where} \quad \widehat{y}_i = \hat{\beta}_0 + \hat{\beta}_1 x_{1i} + \cdots + \hat{\beta}_r x_{ri}$$

(5) $SS(\beta_{r+1}, \ldots, |\beta_1, \ldots, \beta_r) =$ regression sum of squares due to $x_{r+1}, \ldots, x_k$ given $x_1, \ldots, x_r$ are in the model. It is given by:

$$SS(\beta_1, \ldots, \beta_r, \beta_{r+1}, \ldots, \beta_k) - SS(\beta_1, \ldots, \beta_r) = \sum_{j=r+1}^{k} \hat{\beta}_j g_j$$

Hypothesis test:

H_0: $\beta_{r+1} = \beta_{r+2} = \cdots = \beta_k = 0$

H_a: $\beta_m \neq 0$ for some $m = r+1, r+2, \ldots, k$

TS: $F = \dfrac{[SSE(R) - SSE(F)]/(k-r)}{SSE(F)/(n-k-1)}$

$\quad = \dfrac{SS(\beta_{r+1}, \ldots, \beta_k | \beta_1, \ldots, \beta_r)/(k-r)}{SSE(R)/(n-k-1)}$

RR: $F \geq F_{\alpha, k-r, n-k-1}$

11.2.9 Residual analysis

Let h_{ii} be the diagonal entries of the **HAT matrix** defined by $\mathbf{H} = \mathbf{X}(\mathbf{X}^{\mathrm{T}}\mathbf{X})^{-1}\mathbf{X}^{\mathrm{T}}$.

Standardized residuals: $\dfrac{e_i}{\sqrt{\text{MSE}}} = \dfrac{e_i}{s}, \quad i = 1, 2, \ldots, n$

Studentized residuals: $e_i^* = \dfrac{e_i}{s\sqrt{1 - h_{ii}}}, \quad i = 1, 2, \ldots, n$

Deleted studentized residuals:

$$d_i^* = e_i \sqrt{\dfrac{n-k-2}{s^2(1 - h_{ii}) - e_i^2}}, \quad i = 1, 2, \ldots, n$$

Cook's distance: $D_i = \dfrac{e_i^2}{(k+1)s^2} \left[\dfrac{h_{ii}}{(1 - h_{ii})^2}\right], \quad i = 1, 2, \ldots, n$

Press residuals: $\delta_i = y_i - \widehat{y}_{i,-i} = \dfrac{e_i}{1 - h_{ii}}, \quad i = 1, 2, \ldots, n$

where $\widehat{y}_{i,-i}$ is the ith predicted value by the model without using the ith observation in calculating the regression coefficients.

Prediction sum of squares $= \text{PRESS} = \sum_{i=1}^{n} \delta_i^2$

$\sum_{i=1}^{n} |\delta_i|$: is used for cross validation, it is less sensitive to large press residuals.

11.2.10   Example

Example 11.64:   A university foundation office recently investigated factors that might contribute to alumni donations. Fifteen years were randomly selected and the total donations (in millions of dollars), United States savings rate, the unemployment rate, and the number of games won by the school basketball team are given in the table below.

Donations (y)	Savings rate (%) (x_1)	Unemployment rate (%) (x_2)	Games won
27.80	15.1	4.7	26
17.41	11.3	5.5	14
18.51	13.6	5.6	24
30.09	16.1	4.9	15
34.22	17.8	5.1	17
20.28	13.3	5.6	18
26.98	15.5	4.6	20
27.32	16.7	4.7	9
18.74	13.8	5.8	18
35.52	17.4	4.5	17
15.52	10.9	6.1	19
17.75	12.5	5.7	16
22.94	12.9	5.0	21
41.47	18.8	4.2	19
22.95	16.1	5.4	18

(1) Find the regression equation using donation as the dependent variable and construct the anova table.
(2) Test for a significant regression. Use $\alpha = .05$.
(3) Is there any evidence to suggest the number of games won by the school basketball team affects alumni donations. Use $\alpha = .10$.

Solution:

(S1) Regression line: $\widehat{y} = 23.89 - 5.54x_1 + 1.95x_2 + .057x_3$

(S2)
Source of variation	Sum of squares	Degrees of freedom	Mean square	Computed F
Regression	755.64	3	251.88	38.87
Error	71.28	11	6.48	
Total	826.92	14		

(S3) Hypothesis test of significant regression:

H_0: $\beta_1 = \beta_2 = \beta_3 = 0$

H_a: $\beta_i \neq 0$ for some i

TS: $F = \text{MSR}/\text{MSE}$

RR: $F \geq F_{.05,3,11} = 3.58$

Conclusion: The value of the test statistic ($F = 38.87$) lies in the rejection region. There is evidence to suggest a significant regression.

(S4) Hypothesis test for β_3:

H_0: $\beta_3 = 0$

H_a: $\beta_3 \neq 0$

TS: $T = \dfrac{\widehat{\beta}_3 - 0}{S\sqrt{c_{ii}}}$

RR: $|T| \geq t_{.05,11} = 1.796$

$t = .057/.1719 = .33$

Conclusion: The value of the test statistic does not lie in the rejection region. There is no evidence to suggest the number of games won by the basketball team affects alumni donations.

11.3 ORTHOGONAL POLYNOMIALS

Polynomial regression models may contain several independent variables and each independent variable may appear in the model to various powers. Let $(x_1, y_1), (x_2, y_2), \ldots, (x_n, y_n)$ be n pairs of observations such that y_i is an observed value of the random variable Y_i. Assume there exist constants $\beta_0, \beta_1, \ldots, \beta_p$ such that

$$Y_i = \beta_0 + \beta_1 x_i + \beta_2 x_i^2 + \cdots + \beta_p x_i^p + \epsilon_i \tag{11.10}$$

where $\epsilon_1, \epsilon_2, \ldots, \epsilon_n$ are independent, normal random variables having mean 0 and variance σ^2.

In order to determine the *best* polynomial model, the regression coefficients $\{\beta_j\}$ must be recalculated for various values of p. If the values $\{x_i\}$ are evenly spaced, then orthogonal polynomials are often used to determine the best model. This technique presents certain computational advantages and quickly isolates significant effects.

The orthogonal polynomial model is

$$Y_i = \alpha_0 \xi_0(x) + \alpha_1 \xi_1(x) + \cdots + \alpha_p \xi_p(x) + \epsilon_i \tag{11.11}$$

where $\xi_i(x)$ are orthogonal polynomials in x of degree i. The orthogonal polynomials have the property

$$\sum_{i=1}^{n} \xi_h(x_i) \cdot \xi_j(x_i) = 0, \text{ when } h \neq j \text{ and } x_i = x_0 + i\delta \tag{11.12}$$

The least square estimator for α_j is given by

$$\widehat{\alpha}_j = \frac{\sum\limits_{i=1}^{n} y_i \xi_j(x_i)}{\sum\limits_{i=1}^{n} [\xi_j(x_i)]^2} \tag{11.13}$$

The estimator $\widehat{\alpha}_j$ (for $j = 1, 2, \ldots, n-1$) is a normal random variable with mean α_j if $j \le p$ and mean 0 if $j > p$, and variance $\sigma^2 \bigg/ \sum\limits_{i=1}^{n} [\xi_j(x_i)]^2$. An estimate of σ^2 is given by

$$s^2 = \frac{\sum\limits_{i=1}^{n} y_i^2 - \sum\limits_{j=0}^{p} (\widehat{\alpha}_j)^2 \left(\sum\limits_{i=1}^{n} [\xi_j(x_i)]^2 \right)}{n - p - 1}. \tag{11.14}$$

Each ratio $\dfrac{(\alpha_j - \widehat{\alpha}_j) \sqrt{\sum\limits_{i} [\xi_j(x_i)]^2}}{s}$ (with $\widehat{\alpha}_j = 0$ for $j > p$) has a t distribution.

The contribution of each term may be tested; if it is not significant then the term may be discarded without recalculating the previously obtained coefficients (i.e., the tests for significance effects are isolated).

In order to tabulate the values of the orthogonal polynomials for repeated use, the values of x_i are assumed to be one unit apart, and $\xi'_j(x)$ is defined to be a multiple, λ_j, of $\xi_j(x)$ so that $\xi_j(x)$ has a leading coefficient of unity. This adjustment makes all the tabulated values $\xi'_j(x_i)$ integers. Thus, in particular:

$$\xi'_1(x) = \lambda_1 \xi_1(x) = \lambda_1 (x - \overline{x})$$

$$\xi'_2(x) = \lambda_2 \xi_2(x) = \lambda_2 \left[(x - \overline{x})^2 - \frac{n^2 - 1}{12} \right]$$

$$\xi'_3(x) = \lambda_3 \xi_3(x) = \lambda_3 \left[(x - \overline{x})^3 - (x - \overline{x}) \left(\frac{3n^2 - 7}{20} \right) \right] \tag{11.15}$$

$$\xi'_4(x) = \lambda_4 \xi_4(x) = \lambda_4 \left\{ (x - \overline{x})^4 \right.$$

$$\left. - (x - \overline{x})^2 \left(\frac{3n^2 - 13}{14} \right) + \frac{3(n^2 - 1)(n^2 - 9)}{560} \right\}$$

and, in general, $\xi'_r(x) = \lambda_r \xi_r$ with the recursion relation

$$\xi_0 = 1, \qquad \xi_1 = (x - \overline{x})$$

$$\xi_{r+1} = \xi_1 \xi_r - \frac{r^2 (n^2 - r^2)}{4(4r^2 - 1)} \xi_{r-1} \tag{11.16}$$

The tables below provide values of $\{\xi'_j(x_i)\}$ for various values of n and j. When $n > 10$ only half of the table is shown (the other half can be found

by symmetry). The two values at the bottom of each column are the values
$$D_j = \sum_{i=1}^{n} [\xi_j(x_i)]^2 \text{ and } \lambda_j.$$

Example 11.65: Consider $n = 11$ observations of data from a quadratic with noise: specifically the values are $\{1, 4, 9, \ldots, 121\} + 0.1\{n_1, n_2, \ldots, n_{11}\}$ where each $\{n_i\}$ is a standard normal random variable. In order to fit a polynomial regression we must assume some maximum polynomial power of interest; we presume that terms up to degree three might be of interest. The $\{\xi'_j(x_i)\}$ table for $n = 11$ (see page 279) gives the values for $x \le 0$; the values for $x > 0$ are given by symmetry. From the data and the $\{\xi'_j(x_i)\}$ table we compute

x_i	y_i	ξ'_1	ξ'_2	ξ'_3
1	1.53	-5	15	-30
2	4.27	-4	6	6
3	8.94	-3	-1	22
4	15.92	-2	-6	23
5	24.99	-1	-9	14
6	35.94	0	-10	0
7	48.77	1	-9	-14
8	64.13	2	-6	-23
9	80.80	3	-1	-22
10	99.89	4	6	-6
11	120.99	5	15	30
$D_i = \sum \xi_i^2$	$\ldots$	110	858	4290
λ_i	$\ldots$	1	1	$5/6$
$\sum y_i \xi_i$	$\ldots$	1315	869.5	-12.26

The values for $\hat{\alpha}_i$ are found by:

$$\bar{x} = 6 \qquad \bar{y} = 46.01 \qquad \hat{\alpha}_1 = \frac{1315}{110} = 11.96$$

$$\hat{\alpha}_2 = \frac{869.5}{858} = 1.013 \qquad \hat{\alpha}_3 = \frac{-12.26}{4290} = -.00029$$

(11.17)

Using the value of n and $\{\lambda_i\}$ in equation (11.15), the orthogonal polynomials are

$$\xi'_1(x) = x - 8 \qquad \xi'_2(x) = x^2 - 12x + 26$$

$$\xi'_3(x) = \frac{1}{6}(5x^3 - 90x^2 + 451x - 546)$$

(11.18)

The regression equation is $\hat{y} = \bar{y} + \hat{\alpha}_1 \xi'_1(x) + \hat{\alpha}_2 \xi'_2(x) + \hat{\alpha}_3 \xi'_3(x) = .85 - .42x + 1.06x^2 - .002x^3$. The predicted values are $\{1.49, 4.22, 9.04, \ldots, 120.92\}$.

The significance of the coefficients may be determined by computing:

Quantity	Sum of squares	Degrees freedom	Mean square	Error mean square	Computed F statistic
(1) $\dfrac{n\sum y_i^2 - (\sum y_i)^2}{n}$	16615	10			
(2) linear regression $\left(\sum y_i \xi_1'\right)^2 \Big/ \sum \xi_1^2$	15733	1	15733	881.4	17.9
(3) quadratic regression $\left(\sum y_i \xi_2'\right)^2 \Big/ \sum \xi_2^2$	881.2	1	881.2	0.15	5958
(3) cubic regression $\left(\sum y_i \xi_3'\right)^2 \Big/ \sum \xi_3^2$	0.035	1	0.035	0.11	0.31
residual sum of squares: $(1) - (2) - (3) - (4)$	0.018	7	0.0026	0.095	0.027

where the computed F statistic is given by $F = $ (mean square)/(error mean square).

Conclusion: The cubic term should probably not be included in the model. The regression equation then becomes $\hat{y} = \bar{y} + \hat{\alpha}_1 \xi_1'(x) + \hat{\alpha}_2 \xi_2'(x) + 0 \cdot \xi_3'(x) = .59 - .20x + 1.0134x^2$. The predicted values are then $\{1.40, 4.24, 9.11, \ldots, 121.00\}$.

11.3.1   Tables for orthogonal polynomials

$n = 3$ points		$n = 4$ points			$n = 5$ points				$n = 6$ points				
ξ_1'	ξ_2'	ξ_1'	ξ_2'	ξ_3'	ξ_1'	ξ_2'	ξ_3'	ξ_4'	ξ_1'	ξ_2'	ξ_3'	ξ_4'	ξ_5'
−1	1	−3	1	−1	−2	2	−1	1	−5	5	−5	1	−1
0	−2	−1	−1	3	−1	−1	2	−4	−3	−1	7	−3	5
1	1	1	−1	−3	0	−2	0	6	−1	−4	4	2	−10
D 2	6	3	1	1	1	−1	−2	−4	1	−4	−4	2	10
λ 1	3	D 20	4	20	2	2	1	1	3	−1	−7	−3	−5
		λ 2	1	10/3	D 10	14	10	70	5	5	5	1	1
					λ 1	1	5/6	35/12	D 70	84	180	28	252
									λ 2	3/2	5/3	7/12	21/10

$n = 7$ points					$n = 8$ points					$n = 9$ points				
ξ_1'	ξ_2'	ξ_3'	ξ_4'	ξ_5'	ξ_1'	ξ_2'	ξ_3'	ξ_4'	ξ_5'	ξ_1'	ξ_2'	ξ_3'	ξ_4'	ξ_5'
−3	5	−1	3	−1	−7	7	−7	7	−7	−4	28	−14	14	−4
−2	0	1	−7	4	−5	1	5	−13	23	−3	7	7	−21	11
−1	−3	1	1	−5	−3	−3	7	−3	−17	−2	−8	13	−11	−4
0	−4	0	6	0	−1	−5	3	9	−15	−1	−17	9	9	−9
1	−3	−1	1	5	1	−5	−3	9	15	0	−20	0	18	0
2	0	−1	−7	−4	3	−3	−7	−3	17	1	−17	−9	9	9
3	5	1	3	1	5	1	−5	−13	−23	2	−8	−13	−11	4
D 28	84	6	154	84	7	7	7	7	7	3	7	−7	−21	−11
λ 1	1	1/6	7/12	7/20	D 168	168	264	616	2184	4	28	14	14	4
					λ 2	1	2/3	7/12	7/10	D 60	2772	990	2002	468
										λ 1	3	5/6	7/12	3/20

$n = 10$ points

ξ'_1	ξ'_2	ξ'_3	ξ'_4	ξ'_5
-9	6	-42	18	-6
-7	2	14	-22	14
-5	-1	35	-17	-1
-3	-3	31	3	-11
-1	-4	12	18	-6
1	-4	-12	18	6
3	-3	-31	3	11
5	-1	-35	-17	1
7	2	-14	-22	-14
9	6	42	18	6
D 330	132	8580	2860	780
λ 2	$1/2$	$5/3$	$5/12$	$1/10$

$n = 11$ points

ξ'_1	ξ'_2	ξ'_3	ξ'_4	ξ'_5
-5	15	-30	6	-3
-4	6	6	-6	6
-3	-1	22	-6	1
-2	-6	23	-1	-4
-1	-9	14	4	-4
0	-10	0	6	0
D 110	858	4290	286	156
λ 1	1	$5/6$	$1/12$	$1/40$

$n = 12$ points

ξ'_1	ξ'_2	ξ'_3	ξ'_4	ξ'_5
-11	55	-33	33	-33
-9	25	3	-27	57
-7	1	21	-33	21
-5	-17	25	-13	-29
-3	-29	19	12	-44
-1	-35	7	28	-20
D 572	12012	5148	8008	15912
λ 2	3	$2/3$	$7/24$	$3/20$

$n = 13$ points

ξ'_1	ξ'_2	ξ'_3	ξ'_4	ξ'_5
-6	22	-11	99	-22
-5	11	0	-66	33
-4	2	6	-96	18
-3	-5	8	-54	-11
-2	-10	7	11	-26
-1	-13	4	64	-20
0	-14	0	84	0
D 182	2002	572	68068	6188
λ 1	1	$1/6$	$7/12$	$7/120$

$n = 14$ points

ξ'_1	ξ'_2	ξ'_3	ξ'_4	ξ'_5
-13	13	-143	143	-143
-11	7	-11	-77	187
-9	2	66	-132	132
-7	-2	98	-92	-28
-5	-5	95	-13	-139
-3	-7	67	63	-145
-1	-8	24	108	-60
D 910	728	97240	136136	235144
λ 2	$1/2$	$5/3$	$7/12$	$7/30$

$n = 15$ points

ξ'_1	ξ'_2	ξ'_3	ξ'_4	ξ'_5
-7	91	-91	1001	-1001
-6	52	-13	-429	1144
-5	19	35	-869	979
-4	-8	58	-704	44
-3	-29	61	-249	-751
-2	-44	49	251	-1000
-1	-53	27	621	-675
0	-56	0	756	0
D 280	37128	39780	6466460	10581480
λ 1	3	$5/6$	$35/12$	$21/20$

$n = 16$ points

ξ'_1	ξ'_2	ξ'_3	ξ'_4	ξ'_5
-15	35	-455	273	-143
-13	21	-91	-91	143
-11	9	143	-221	143
-9	-1	267	-201	33
-7	-9	301	-101	-77
-5	-15	265	23	-131
-3	-19	179	129	-115
-1	-21	63	189	-45
D 1360	5712	1007760	470288	201552
λ 2	1	$10/3$	$7/12$	$1/10$

$n = 17$ points

ξ'_1	ξ'_2	ξ'_3	ξ'_4	ξ'_5
-8	40	-28	52	-104
-7	25	-7	-13	91
-6	12	7	-39	104
-5	1	15	-39	39
-4	-8	18	-24	-36
-3	-15	17	-3	-83
-2	-20	13	17	-88
-1	-23	7	31	-55
0	-24	0	36	0
D 408	7752	3876	16796	100776
λ 1	1	$1/6$	$1/12$	$1/20$

$n = 18$ points

ξ'_1	ξ'_2	ξ'_3	ξ'_4	ξ'_5
-17	68	-68	68	-884
-15	44	-20	-12	676
-13	23	13	-47	871
-11	5	33	-51	429
-9	-10	42	-36	-156
-7	-22	42	-12	-588
-5	-31	35	13	-733
-3	-37	23	33	-583
-1	-40	8	44	-220
D 1938	23256	23256	28424	6953544
λ 2	$3/2$	$1/3$	$1/12$	$3/10$

ξ_1'	ξ_2'	ξ_3'	ξ_4'	ξ_5'
		$n = 19$ points		
−9	51	−204	612	−102
−8	34	−68	−68	68
−7	19	28	−388	98
−6	6	89	−453	58
−5	−5	120	−354	−3
−4	−14	126	−168	−54
−3	−21	112	42	−79
−2	−26	83	227	−74
−1	−29	44	352	−44
0	−30	0	396	0
D 570	13566	213180	2288132	89148
λ 1	1	5/6	7/12	1/40

ξ_1'	ξ_2'	ξ_3'	ξ_4'	ξ_5'
		$n = 20$ points		
−19	57	−969	1938	−1938
−17	39	−357	−102	1122
−15	23	85	−1122	1802
−13	9	377	−1402	1222
−11	−3	539	−1187	187
−9	−13	591	−687	−771
−7	−21	553	−77	−1351
−5	−27	445	503	−1441
−3	−31	287	948	−1076
−1	−33	99	1188	−396
D 2660	17556	4903140	22881320	31201800
λ 2	1	10/3	35/24	7/20

ξ_1'	ξ_2'	ξ_3'	ξ_4'	ξ_5'
		$n = 21$ points		
−10	190	−285	969	−3876
−9	133	−114	0	1938
−8	82	12	−510	3468
−7	37	98	−680	2618
−6	−2	149	−615	788
−5	−35	170	−406	−1063
−4	−62	166	−130	−2354
−3	−83	142	150	−2819
−2	−98	103	385	−2444
−1	−107	54	540	−1404
0	−110	0	594	0
D 770	201894	432630	5720330	121687020
λ 1	3	5/6	7/12	21/40

ξ_1'	ξ_2'	ξ_3'	ξ_4'	ξ_5'
		$n = 22$ points		
−21	35	−133	1197	−2261
−19	25	−57	57	969
−17	16	0	−570	1938
−15	8	40	−810	1598
−13	1	65	−775	663
−11	−5	77	−563	−363
−9	−10	78	−258	−1158
−7	−14	70	70	−1554
−5	−17	55	365	−1509
−3	−19	35	585	−1079
−1	−20	12	702	−390
D 3542	7084	96140	8748740	40562340
λ 2	1/2	1/3	7/12	7/30

ξ_1'	ξ_2'	ξ_3'	ξ_4'	ξ_5'
		$n = 23$ points		
−11	77	−77	1463	−209
−10	56	−35	133	76
−9	37	−3	−627	171
−8	20	20	−950	152
−7	5	35	−955	77
−6	−8	43	−747	−12
−5	−19	45	−417	−87
−4	−28	42	−42	−132
−3	−35	35	315	−141
−2	−40	25	605	−116
−1	−43	13	793	−65
0	−44	0	858	0
D 1012	35420	32890	13123110	340860
λ 1	1	1/6	7/12	1/60

ξ_1'	ξ_2'	ξ_3'	ξ_4'	ξ_5'
		$n = 24$ points		
−23	253	−1771	253	−4807
−21	187	−847	33	1463
−19	127	−133	−97	3743
−17	73	391	−157	3553
−15	25	745	−165	2071
−13	−17	949	−137	169
−11	−53	1023	−87	−1551
−9	−83	987	−27	−2721
−7	−107	861	33	−3171
−5	−125	665	85	−2893
−3	−137	419	123	−2005
−1	−143	143	143	−715
D 4600	394680	17760600	394680	177928920
λ 2	3	10/3	1/12	3/10

CHAPTER 12

Analysis of Variance

Contents

12.1   ONE-WAY ANOVA

Let there be k treatments, or populations, and independent random samples of size n_i (for $i = 1, 2, \ldots, k$) from each population, and let $N = n_1 + n_2 + \cdots + n_k$. Let Y_{ij} be the jth random observation in the ith treatment group. Assume each population is normally distributed with mean μ_i and common variance σ^2. In a **fixed effects**, or **model I**, **experiment** the treatment levels are predetermined. In a **random effects**, or **model II**, **experiment**, the treatment levels are selected at random.

Notation:

(1) Dot notation is used to indicate a sum over all values of the selected subscript.

(2) The random error term is ϵ_{ij} and an observed value is e_{ij}.

Fixed effects experiment:

Model: $Y_{ij} = \mu_i + \epsilon_{ij} = \mu + \alpha_i + \epsilon_{ij}$

$\qquad\qquad\quad i = 1, 2, \ldots, k, \quad j = 1, 2, \ldots, n_i$

Assumptions: The ϵ_{ij}'s are independent, normally distributed with

$$\text{mean 0 and variance } \sigma^2 \ (\epsilon_{ij} \overset{\text{ind}}{\sim} N(0, \sigma^2)), \ \sum_{i=1}^{k} \alpha_i = 0$$

Random effects experiment:

Model: $Y_{ij} = \mu_i + \epsilon_{ij} = \mu + A_i + \epsilon_{ij}$

$\qquad\qquad\quad i = 1, 2, \ldots, k, \quad j = 1, 2, \ldots, n_i$

Assumptions: $\epsilon_{ij} \overset{\text{ind}}{\sim} N(0, \sigma^2), \quad A_i \overset{\text{ind}}{\sim} N(0, \sigma_\alpha^2)$

$\qquad\qquad\quad$ The ϵ_{ij}'s are independent of the A_i's.

12.1.1   Sum of squares

$$\underbrace{\sum_{i=1}^{k} \sum_{j=1}^{n_i} (y_{ij} - \bar{y}_{..})^2}_{\text{SST}} = \underbrace{\sum_{i=1}^{k} n_i (\bar{y}_{i.} - \bar{y}_{..})^2}_{\text{SSA}} + \underbrace{\sum_{i=1}^{k} \sum_{j=1}^{n_i} (y_{ij} - \bar{y}_{i.})^2}_{\text{SSE}} \qquad (12.1)$$

Notation:

y_{ij} = observed value of Y_{ij}

$$\bar{y}_{i.} = \frac{1}{n_i} \sum_{j=1}^{n_i} y_{ij} \qquad = \text{mean of the observations in the } i^{\text{th}} \text{ sample}$$

$$\bar{y}_{..} = \frac{1}{N} \sum_{i=1}^{k} \sum_{j=1}^{n_i} y_{ij} = \text{mean of all observations}$$

$$T_{i.} = \sum_{j=1}^{n_i} y_{ij} \quad = \text{sum of the observations in the } i^{\text{th}} \text{ sample}$$

$$T_{..} = \sum_{i=1}^{k} \sum_{j=1}^{n_i} y_{ij} = \text{sum of all observations}$$

$$\text{SST} = \text{total sum of squares} = \sum_{i=1}^{k} \sum_{j=1}^{n_i} (y_{ij} - \bar{y}_{..})^2 = \sum_{i=1}^{k} \sum_{j=1}^{n_i} y_{ij}^2 - \frac{T_{..}^2}{N}$$

$$\text{SSA} = \text{sum of squares due to treatment}$$

$$= \sum_{i=1}^{k} n_i (\bar{y}_{i.} - \bar{y}_{..})^2 = \sum_{i=1}^{k} \frac{T_{i.}^2}{n_i} - \frac{T_{..}^2}{N}$$

$$\text{SSE} = \text{sum of squares due to error} = \sum_{i=1}^{k} \sum_{j=1}^{n_i} (y_{ij} - \bar{y}_{i.})^2 = \text{SST} - \text{SSA}$$

12.1.2 Properties

Mean square	Expected value	
	Fixed model	Random model
$\text{MSA} = S_A^2 = \dfrac{\text{SSA}}{k-1}$	$\sigma^2 + \dfrac{\sum_{i=1}^{k} n_i \alpha_i^2}{k-1}$	$\sigma^2 + \dfrac{1}{k-1}\left(n - \dfrac{\sum_{i=1}^{k} n_i^2}{n}\right)\sigma_\alpha^2$
$\text{MSE} = S^2 = \dfrac{\text{SSE}}{N-k}$	σ^2	σ^2

$F = S_A^2 / S^2$ has an F distribution with $k - 1$ and $N - k$ degrees of freedom.

12.1.3 Analysis of variance table

Source of variation	Sum of squares	Degrees of freedom	Mean square	Computed F
Treatments (between groups)	SSA	$k - 1$	MSA	MSA/MSE
Error (within groups)	SSE	$N - k$	MSE	
Total	SST	$N - 1$		

Hypothesis test of significant regression:

H_0: $\mu_1 = \mu_2 = \cdots = \mu_k$
    (Fixed effects model: $\alpha_1 = \alpha_2 = \cdots = \alpha_k = 0$)
    (Random effects model: $\sigma_\alpha^2 = 0$)

H_a: at least two of the means are unequal
    (Fixed effects model: $\alpha_i \neq 0$ for some i)
    (Random effects model: $\sigma_\alpha^2 \neq 0$)

TS: $F = S_A^2 / S^2$

RR: $F \geq F_{\alpha, k-1, N-k}$

12.1.4 Multiple comparison procedures

12.1.4.1 Tukey's procedure

Equal sample sizes:

Let $n = n_1 = n_2 = \cdots = n_k$ and let Q_{α, ν_1, ν_2} be a critical value of the Studentized Range distribution (see page 76). The set of intervals with endpoints

$$(\bar{y}_{i.} - \bar{y}_{j.}) \pm Q_{\alpha, k, k(n-1)} \cdot \sqrt{s^2/n} \text{ for all } i \text{ and } j, \, i \neq j \qquad (12.2)$$

is a collection of simultaneous $100(1-\alpha)\%$ confidence intervals for the differences between the true treatment means, $\mu_i - \mu_j$. Each confidence interval that does not include zero suggests $\mu_i \neq \mu_j$ at the α significance level.

Unequal sample sizes:

The set of confidence intervals with endpoints ($N = n_1 + n_2 + \cdots + n_k$)

$$(\bar{y}_{i.} - \bar{y}_{j.}) \pm \frac{1}{\sqrt{2}} Q_{\alpha, k, N-k} \cdot s \sqrt{\frac{1}{n_i} + \frac{1}{n_j}} \text{ for all } i \text{ and } j, \, i \neq j \qquad (12.3)$$

is a collection of simultaneous $100(1-\alpha)\%$ confidence intervals for the differences between the true treatment means, $\mu_i - \mu_j$.

12.1.4.2 Duncan's multiple range test

Let $n = n_1 = n_2 = \cdots = n_k$ and let r_{α, ν_1, ν_2} be a critical value for Duncan's multiple range test (see page 285). Duncan's procedure for determining significant differences between each treatment group at the joint significance level α is:

(1) Define $R_p = r_{\alpha, p, k(n-1)} \cdot \sqrt{\dfrac{s^2}{n}}$ for $p = 2, 3, \ldots, k$.

(2) List the sample means in increasing order.

(3) Compare the range of every subset of p sample means (for $p = 2, 3, \ldots, k$) in the ordered list with R_p.

(4) If the range of a p–subset is less than R_p then that subset of ordered means in not significantly different.

12.1.4.3 Duncan's multiple range test

These tables contain critical values for the least significant studentized ranges, $r_{\alpha,p,\nu}$, for Duncan's multiple range test where α is the significance level, p is the number of successive values from an ordered list of k means of equal sample sizes ($p = 2, 3, \ldots, k$), and n is the degrees of freedom for the independent estimate s^2. These tables are from L. Hunter, "Critical Values for Duncan's New Multiple Range Test", *Biometrics*, 1960, Volume 16, pages 671–685. Reprinted with permission from the *Journal of American Statistical Association*. Copyright 1960 by the American Statistical Association. All rights reserved.

Critical values for Duncan's test, $r_{\alpha,p,n}$, for $\alpha = .05$

n	$p=2$	3	4	5	6	7	8	9	10	11	12	13	14	15	16	17	18	19	20
1	17.97	17.97	17.97	17.97	17.97	17.97	17.97	17.97	17.97	17.97	17.97	17.97	17.97	17.97	17.97	17.97	17.97	17.97	17.97
2	6.085	6.085	6.085	6.085	6.085	6.085	6.085	6.085	6.085	6.085	6.085	6.085	6.085	6.085	6.085	6.085	6.085	6.085	6.085
3	4.501	4.516	4.516	4.516	4.516	4.516	4.516	4.516	4.516	4.516	4.516	4.516	4.516	4.516	4.516	4.516	4.516	4.516	4.516
4	3.927	4.013	4.033	4.033	4.033	4.033	4.033	4.033	4.033	4.033	4.033	4.033	4.033	4.033	4.033	4.033	4.033	4.033	4.033
5	3.635	3.749	3.797	3.814	3.814	3.814	3.814	3.814	3.814	3.814	3.814	3.814	3.814	3.814	3.814	3.814	3.814	3.814	3.814
6	3.461	3.587	3.649	3.680	3.694	3.697	3.697	3.697	3.697	3.697	3.697	3.697	3.697	3.697	3.697	3.697	3.697	3.697	3.697
7	3.344	3.477	3.548	3.588	3.611	3.622	3.626	3.626	3.626	3.626	3.626	3.626	3.626	3.626	3.626	3.626	3.626	3.626	3.626
8	3.261	3.399	3.475	3.521	3.549	3.566	3.575	3.579	3.579	3.579	3.579	3.579	3.579	3.579	3.579	3.579	3.579	3.579	3.579
9	3.199	3.339	3.420	3.470	3.502	3.523	3.536	3.544	3.547	3.547	3.547	3.547	3.547	3.547	3.547	3.547	3.547	3.547	3.547
10	3.151	3.293	3.376	3.430	3.465	3.489	3.505	3.516	3.522	3.525	3.526	3.526	3.526	3.526	3.526	3.526	3.526	3.526	3.526
11	3.113	3.256	3.342	3.397	3.435	3.462	3.480	3.493	3.501	3.506	3.509	3.510	3.510	3.510	3.510	3.510	3.510	3.510	3.510
12	3.082	3.225	3.313	3.370	3.410	3.439	3.459	3.474	3.484	3.491	3.496	3.498	3.499	3.499	3.499	3.499	3.499	3.499	3.499
13	3.055	3.200	3.289	3.348	3.389	3.419	3.442	3.458	3.470	3.478	3.484	3.488	3.488	3.490	3.490	3.490	3.490	3.490	3.490
14	3.033	3.178	3.268	3.329	3.372	3.403	3.426	3.444	3.457	3.467	3.474	3.479	3.482	3.482	3.484	3.485	3.485	3.485	3.485
15	3.014	3.160	3.250	3.312	3.356	3.389	3.413	3.432	3.446	3.457	3.465	3.471	3.476	3.478	3.480	3.481	3.481	3.481	3.481
16	2.998	3.144	3.235	3.298	3.343	3.376	3.402	3.422	3.437	3.449	3.458	3.465	3.470	3.473	3.477	3.478	3.478	3.478	3.478
17	2.984	3.130	3.222	3.285	3.331	3.366	3.392	3.412	3.429	3.441	3.451	3.459	3.465	3.469	3.473	3.475	3.476	3.476	3.476
18	2.971	3.118	3.210	3.274	3.321	3.356	3.383	3.405	3.421	3.435	3.445	3.454	3.460	3.465	3.470	3.472	3.474	3.474	3.474
19	2.960	3.107	3.199	3.264	3.311	3.347	3.375	3.397	3.415	3.429	3.440	3.449	3.456	3.462	3.467	3.470	3.472	3.473	3.474
20	2.950	3.097	3.190	3.255	3.303	3.339	3.368	3.391	3.409	3.424	3.436	3.445	3.453	3.459	3.464	3.467	3.470	3.472	3.473
24	2.919	3.066	3.160	3.226	3.276	3.315	3.345	3.370	3.390	3.406	3.420	3.432	3.441	3.449	3.456	3.461	3.465	3.469	3.471
30	2.888	3.035	3.131	3.199	3.250	3.290	3.322	3.349	3.371	3.389	3.405	3.418	3.430	3.439	3.447	3.454	3.460	3.466	3.470
40	2.858	3.006	3.102	3.171	3.224	3.266	3.300	3.328	3.352	3.373	3.390	3.405	3.418	3.429	3.439	3.448	3.456	3.463	3.469
60	2.829	2.976	3.073	3.143	3.198	3.241	3.277	3.307	3.333	3.355	3.374	3.391	3.406	3.419	3.431	3.442	3.451	3.460	3.467
120	2.800	2.947	3.045	3.116	3.172	3.217	3.254	3.287	3.314	3.337	3.359	3.377	3.394	3.409	3.423	3.435	3.446	3.457	3.466
∞	2.772	2.918	3.017	3.089	3.146	3.193	3.232	3.265	3.294	3.320	3.343	3.363	3.382	3.399	3.414	3.428	3.442	3.454	3.466

Critical values for Duncan's test, $r_{\alpha,p,n}$, for $\alpha = .01$

n	p = 2	3	4	5	6	7	8	9	10	11	12	13	14	15	16	17	18	19	20
1	90.03	90.03	90.03	90.03	90.03	90.03	90.03	90.03	90.03	90.03	90.03	90.03	90.03	90.03	90.03	90.03	90.03	90.03	90.03
2	14.04	14.04	14.04	14.04	14.04	14.04	14.04	14.04	14.04	14.04	14.04	14.04	14.04	14.04	14.04	14.04	14.04	14.04	14.04
3	8.261	8.321	8.321	8.321	8.321	8.321	8.321	8.321	8.321	8.321	8.321	8.321	8.321	8.321	8.321	8.321	8.321	8.321	8.321
4	6.512	6.677	6.740	6.756	6.756	6.756	6.756	6.756	6.756	6.756	6.756	6.756	6.756	6.756	6.756	6.756	6.756	6.756	6.756
5	5.702	5.893	5.989	6.040	6.065	6.074	6.074	6.074	6.074	6.074	6.074	6.074	6.074	6.074	6.074	6.074	6.074	6.074	6.074
6	5.243	5.439	5.549	5.614	5.655	5.680	5.694	5.701	5.703	5.703	5.703	5.703	5.703	5.703	5.703	5.703	5.703	5.703	5.703
7	4.949	5.145	5.260	5.334	5.383	5.416	5.439	5.454	5.464	5.470	5.472	5.472	5.472	5.472	5.472	5.472	5.472	5.472	5.472
8	4.746	4.939	5.057	5.135	5.189	5.227	5.256	5.276	5.291	5.302	5.309	5.314	5.316	5.317	5.317	5.317	5.317	5.317	5.317
9	4.596	4.787	4.906	4.986	5.043	5.086	5.118	5.142	5.160	5.174	5.185	5.193	5.199	5.203	5.205	5.206	5.206	5.206	5.206
10	4.482	4.671	4.790	4.871	4.931	4.975	5.010	5.037	5.058	5.074	5.088	5.098	5.106	5.112	5.117	5.120	5.122	5.124	5.124
11	4.392	4.579	4.697	4.780	4.841	4.887	4.924	4.952	4.975	4.994	5.009	5.021	5.031	5.039	5.045	5.050	5.054	5.057	5.059
12	4.320	4.504	4.622	4.706	4.767	4.815	4.852	4.883	4.907	4.927	4.944	4.958	4.969	4.978	4.986	4.993	4.998	5.002	5.006
13	4.260	4.442	4.560	4.644	4.706	4.755	4.793	4.824	4.850	4.872	4.889	4.904	4.917	4.928	4.937	4.944	4.950	4.956	4.960
14	4.210	4.391	4.508	4.591	4.654	4.704	4.743	4.775	4.802	4.824	4.843	4.859	4.872	4.884	4.894	4.902	4.910	4.916	4.921
15	4.168	4.347	4.463	4.547	4.610	4.660	4.700	4.733	4.760	4.783	4.803	4.820	4.834	4.846	4.857	4.866	4.874	4.881	4.887
16	4.131	4.309	4.425	4.509	4.572	4.622	4.663	4.696	4.724	4.748	4.768	4.786	4.800	4.813	4.825	4.835	4.844	4.851	4.858
17	4.099	4.275	4.391	4.475	4.539	4.589	4.630	4.664	4.693	4.717	4.738	4.756	4.771	4.785	4.797	4.807	4.816	4.824	4.832
18	4.071	4.246	4.362	4.445	4.509	4.560	4.601	4.635	4.664	4.689	4.711	4.729	4.745	4.759	4.772	4.783	4.792	4.801	4.808
19	4.046	4.220	4.335	4.419	4.483	4.534	4.575	4.610	4.639	4.665	4.686	4.705	4.722	4.736	4.749	4.761	4.771	4.780	4.788
20	4.024	4.197	4.312	4.395	4.459	4.510	4.552	4.587	4.617	4.642	4.664	4.684	4.701	4.716	4.729	4.741	4.751	4.761	4.769
24	3.956	4.126	4.239	4.322	4.386	4.437	4.480	4.516	4.546	4.573	4.596	4.616	4.634	4.651	4.665	4.678	4.690	4.700	4.710
30	3.889	4.056	4.168	4.250	4.314	4.366	4.409	4.445	4.477	4.504	4.528	4.550	4.569	4.586	4.601	4.615	4.628	4.640	4.650
40	3.825	3.988	4.098	4.180	4.244	4.296	4.339	4.376	4.408	4.436	4.461	4.483	4.503	4.521	4.537	4.553	4.566	4.579	4.591
60	3.762	3.922	4.031	4.111	4.174	4.226	4.270	4.307	4.340	4.368	4.394	4.417	4.438	4.456	4.474	4.490	4.504	4.518	4.530
120	3.702	3.858	3.965	4.044	4.107	4.158	4.202	4.239	4.272	4.301	4.327	4.351	4.372	4.392	4.410	4.426	4.442	4.456	4.469
∞	3.643	3.796	3.900	3.978	4.040	4.091	4.135	4.172	4.205	4.235	4.261	4.285	4.307	4.327	4.345	4.363	4.379	4.394	4.408

Critical values for Duncan's test, $r_{\alpha,p,n}$, for $\alpha = .001$

n	p = 2	3	4	5	6	7	8	9	10	11	12	13	14	15	16	17	18	19	20
1	900.3	900.3	900.3	900.3	900.3	900.3	900.3	900.3	900.3	900.3	900.3	900.3	900.3	900.3	900.3	900.3	900.3	900.3	900.3
2	44.69	44.69	44.69	44.69	44.69	44.69	44.69	44.69	44.69	44.69	44.69	44.69	44.69	44.69	44.69	44.69	44.69	44.69	44.69
3	18.28	18.45	18.45	18.45	18.45	18.45	18.45	18.45	18.45	18.45	18.45	18.45	18.45	18.45	18.45	18.45	18.45	18.45	18.45
4	12.18	12.52	12.67	12.73	12.75	12.75	12.75	12.75	12.75	12.75	12.75	12.75	12.75	12.75	12.75	12.75	12.75	12.75	12.75
5	9.714	10.05	10.24	10.35	10.42	10.46	10.48	10.49	10.49	10.49	10.49	10.49	10.49	10.49	10.49	10.49	10.49	10.49	10.49
6	8.427	8.743	8.932	9.055	9.139	9.198	9.241	9.272	9.294	9.309	9.319	9.325	9.328	9.329	9.329	9.329	9.329	9.329	9.329
7	7.648	7.943	8.127	8.252	8.342	8.409	8.460	8.500	8.530	8.555	8.574	8.589	8.600	8.609	8.616	8.621	8.624	8.626	8.627
8	7.130	7.407	7.584	7.708	7.799	7.869	7.924	7.968	8.004	8.033	8.057	8.078	8.094	8.108	8.119	8.129	8.137	8.143	8.149
9	6.762	7.024	7.195	7.316	7.407	7.478	7.535	7.582	7.619	7.652	7.679	7.702	7.722	7.739	7.753	7.766	7.777	7.786	7.794
10	6.487	6.738	6.902	7.021	7.111	7.182	7.240	7.287	7.327	7.361	7.390	7.415	7.437	7.456	7.472	7.487	7.500	7.511	7.522
11	6.275	6.516	6.676	6.791	6.880	6.950	7.008	7.056	7.097	7.132	7.162	7.188	7.211	7.231	7.250	7.266	7.280	7.293	7.304
12	6.106	6.340	6.494	6.607	6.695	6.765	6.822	6.870	6.911	6.947	6.978	7.005	7.029	7.050	7.069	7.086	7.102	7.116	7.128
13	5.970	6.195	6.346	6.457	6.543	6.612	6.670	6.718	6.759	6.795	6.826	6.854	6.878	6.900	6.920	6.937	6.954	6.968	6.982
14	5.856	6.075	6.223	6.332	6.416	6.485	6.542	6.590	6.631	6.667	6.699	6.727	6.752	6.774	6.794	6.812	6.829	6.844	6.858
15	5.760	5.974	6.119	6.225	6.309	6.377	6.433	6.481	6.522	6.558	6.590	6.619	6.644	6.666	6.687	6.706	6.723	6.739	6.753
16	5.678	5.888	6.030	6.135	6.217	6.284	6.340	6.388	6.429	6.465	6.497	6.525	6.551	6.574	6.595	6.614	6.631	6.647	6.661
17	5.608	5.813	5.953	6.056	6.138	6.204	6.260	6.307	6.348	6.384	6.416	6.444	6.470	6.493	6.514	6.533	6.551	6.567	6.582
18	5.546	5.748	5.886	5.988	6.068	6.134	6.189	6.236	6.277	6.313	6.345	6.373	6.399	6.422	6.443	6.462	6.480	6.497	6.512
19	5.492	5.691	5.826	5.927	6.007	6.072	6.127	6.174	6.214	6.250	6.281	6.310	6.336	6.359	6.380	6.400	6.418	6.434	6.450
20	5.444	5.640	5.774	5.873	5.952	6.017	6.071	6.117	6.158	6.193	6.225	6.254	6.279	6.303	6.324	6.344	6.362	6.379	6.394
24	5.297	5.484	5.612	5.708	5.784	5.846	5.899	5.945	5.984	6.020	6.051	6.079	6.105	6.129	6.150	6.170	6.188	6.205	6.221
30	5.156	5.335	5.457	5.549	5.622	5.682	5.734	5.778	5.817	5.851	5.882	5.910	5.935	5.958	5.980	6.000	6.018	6.036	6.051
40	5.022	5.191	5.308	5.396	5.466	5.524	5.574	5.617	5.654	5.688	5.718	5.745	5.770	5.793	5.814	5.834	5.852	5.869	5.885
60	4.894	5.055	5.166	5.249	5.317	5.372	5.420	5.461	5.498	5.530	5.559	5.586	5.610	5.632	5.653	5.672	5.690	5.707	5.723
120	4.771	4.924	5.029	5.109	5.173	5.226	5.271	5.311	5.346	5.377	5.405	5.431	5.454	5.476	5.496	5.515	5.532	5.549	5.565
∞	4.654	4.798	4.898	4.974	5.034	5.085	5.128	5.166	5.199	5.229	5.256	5.280	5.303	5.324	5.343	5.361	5.378	5.394	5.409

12.1.4.4 Dunnett's procedure

Let $n = n_0 = n_1 = n_2 = \cdots = n_k$ where treatment 0 is the control group and let d_{α,ν_1,ν_2} be a critical value for Dunnett's procedure (see page 289). Dunnett's procedure for determining significant differences between each treatment and the control at the joint significance level α is given in the following table. For $i = 1, 2, \ldots, k$:

Null hypothesis	Alternative hypotheses	Test statistic	Rejection regions	
$\mu_0 = \mu_i$	$\mu_0 > \mu_i$	$D_i = \dfrac{\overline{y}_{i.} - \overline{y}_{0.}}{\sqrt{2S^2/n}}$	$D_i \geq d_{\alpha,k,k(n-1)}$	(1)
	$\mu_0 < \mu_i$		$D_i \leq -d_{\alpha,k,k(n-1)}$	(2)
	$\mu_0 \neq \mu_i$		$\lvert D_i\rvert \geq d_{\alpha,k,k(n-1)}$	(3)

12.1.4.5 Tables for Dunnett's procedure

This table contains critical values $d_{\alpha/2,k,\nu}$ and $d_{\alpha,k,n}$ for simultaneous comparisons of each treatment group with a control group; α is the significance level, k is the number of treatment groups, and n is the degrees of freedom of the independent estimate s^2. These tables are from C. W. Dunnett, "A Multiple Comparison Procedure for Comparing Several Treatments with a Control", *JASA*, Volume 50, 1955, pages 1096–1121. Reprinted with permission from the *Journal of American Statistical Association*. Copyright 1980 by the American Statistical Association. All rights reserved.

Values of $d_{\alpha/2,k,n}$ for two–sided comparisons ($\alpha = .05$)

n	$k=1$	2	3	4	5	6	7	8	9
5	2.57	3.03	3.39	3.66	3.88	4.06	4.22	4.36	4.49
6	2.45	2.86	3.18	3.41	3.60	3.75	3.88	4.00	4.11
7	2.36	2.75	3.04	3.24	3.41	3.54	3.66	3.76	3.86
8	2.31	2.67	2.94	3.13	3.28	3.40	3.51	3.60	3.68
9	2.26	2.61	2.86	3.04	3.18	3.29	3.39	3.48	3.55
10	2.23	2.57	2.81	2.97	3.11	3.21	3.31	3.39	3.46
11	2.20	2.53	2.76	2.92	3.05	3.15	3.24	3.31	3.38
12	2.18	2.50	2.72	2.88	3.00	3.10	3.18	3.25	3.32
13	2.16	2.48	2.69	2.84	2.96	3.06	3.14	3.21	3.27
14	2.14	2.46	2.67	2.81	2.93	3.02	3.10	3.17	3.23
15	2.13	2.44	2.64	2.79	2.90	2.99	3.07	3.13	3.19
16	2.12	2.42	2.63	2.77	2.88	2.96	3.04	3.10	3.16
17	2.11	2.41	2.61	2.75	2.85	2.94	3.01	3.08	3.13
18	2.10	2.40	2.59	2.73	2.84	2.92	2.99	3.05	3.11
19	2.09	2.39	2.58	2.72	2.82	2.90	2.97	3.04	3.09
20	2.09	2.38	2.57	2.70	2.81	2.89	2.96	3.02	3.07
24	2.06	2.35	2.53	2.66	2.76	2.84	2.91	2.96	3.01
30	2.04	2.32	2.50	2.62	2.72	2.79	2.86	2.91	2.96
40	2.02	2.29	2.47	2.58	2.67	2.75	2.81	2.86	2.90
60	2.00	2.27	2.43	2.55	2.63	2.70	2.76	2.81	2.85
120	1.98	2.24	2.40	2.51	2.59	2.66	2.71	2.76	2.80
∞	1.96	2.21	2.37	2.47	2.55	2.62	2.67	2.71	2.75

Values of $d_{\alpha/2,k,n}$ for two–sided comparisons ($\alpha = .01$)

n	$k=1$	2	3	4	5	6	7	8	9
5	4.03	4.63	5.09	5.44	5.73	5.97	6.18	6.36	6.53
6	3.71	4.22	4.60	4.88	5.11	5.30	5.47	5.61	5.74
7	3.50	3.95	4.28	4.52	4.71	4.87	5.01	5.13	5.24
8	3.36	3.77	4.06	4.27	4.44	4.58	4.70	4.81	4.90
9	3.25	3.63	3.90	4.09	4.24	4.37	4.48	4.57	4.65
10	3.17	3.53	3.78	3.95	4.10	4.21	4.31	4.40	4.47
11	3.11	3.45	3.68	3.85	3.98	4.09	4.18	4.26	4.33
12	3.05	3.39	3.61	3.76	3.89	3.99	4.08	4.15	4.22
13	3.01	3.33	3.54	3.69	3.81	3.91	3.99	4.06	4.13
14	2.98	3.29	3.49	3.64	3.75	3.84	3.92	3.99	4.05
15	2.95	3.25	3.45	3.59	3.70	3.79	3.86	3.93	3.99
16	2.92	3.22	3.41	3.55	3.65	3.74	3.82	3.88	3.93
17	2.90	3.19	3.38	3.51	3.62	3.70	3.77	3.83	3.89
18	2.88	3.17	3.35	3.48	3.58	3.67	3.74	3.80	3.85
19	2.86	3.15	3.33	3.46	3.55	3.64	3.70	3.76	3.81
20	2.85	3.13	3.31	3.43	3.53	3.61	3.67	3.73	3.78
24	2.80	3.07	3.24	3.36	3.45	3.52	3.58	3.64	3.69
30	2.75	3.01	3.17	3.28	3.37	3.44	3.50	3.55	3.59
40	2.70	2.95	3.10	3.21	3.29	3.36	3.41	3.46	3.50
60	2.66	2.90	3.04	3.14	3.22	3.28	3.33	3.38	3.42
120	2.62	2.84	2.98	3.08	3.15	3.21	3.25	3.30	3.33
∞	2.58	2.79	2.92	3.01	3.08	3.14	3.18	3.22	3.25

Values of $d_{\alpha/2,k,n}$ for one–sided comparisons ($\alpha = .05$)

n	$k=1$	2	3	4	5	6	7	8	9
5	2.02	2.44	2.68	2.85	2.98	3.08	3.16	3.24	3.30
6	1.94	2.34	2.56	2.71	2.83	2.92	3.00	3.07	3.12
7	1.89	2.27	2.48	2.62	2.73	2.82	2.89	2.95	3.01
8	1.86	2.22	2.42	2.55	2.66	2.74	2.81	2.87	2.92
9	1.83	2.18	2.37	2.50	2.60	2.68	2.75	2.81	2.86
10	1.81	2.15	2.34	2.47	2.56	2.64	2.70	2.76	2.81
11	1.80	2.13	2.31	2.44	2.53	2.60	2.67	2.72	2.77
12	1.78	2.11	2.29	2.41	2.50	2.58	2.64	2.69	2.74
13	1.77	2.09	2.27	2.39	2.48	2.55	2.61	2.66	2.71
14	1.76	2.08	2.25	2.37	2.46	2.53	2.59	2.64	2.69
15	1.75	2.07	2.24	2.36	2.44	2.51	2.57	2.62	2.67
16	1.75	2.06	2.23	2.34	2.43	2.50	2.56	2.61	2.65
17	1.74	2.05	2.22	2.33	2.42	2.49	2.54	2.59	2.64
18	1.73	2.04	2.21	2.32	2.41	2.48	2.53	2.58	2.62
19	1.73	2.03	2.20	2.31	2.40	2.47	2.52	2.57	2.61
20	1.72	2.03	2.19	2.30	2.39	2.46	2.51	2.56	2.60
24	1.71	2.01	2.17	2.28	2.36	2.43	2.48	2.53	2.57
30	1.70	1.99	2.15	2.25	2.33	2.40	2.45	2.50	2.54
40	1.68	1.97	2.13	2.23	2.31	2.37	2.42	2.47	2.51
60	1.67	1.95	2.10	2.21	2.28	2.35	2.39	2.44	2.48
120	1.66	1.93	2.08	2.18	2.26	2.32	2.37	2.41	2.45
∞	1.64	1.92	2.06	2.16	2.23	2.29	2.34	2.38	2.42

Values of $d_{\alpha/2,k,n}$ for one–sided comparisons ($\alpha = .01$)

n	$k=1$	2	3	4	5	6	7	8	9
5	3.37	3.90	4.21	4.43	4.60	4.73	4.85	4.94	5.03
6	3.14	3.61	3.88	4.07	4.21	4.33	4.43	4.51	4.59
7	3.00	3.42	3.66	3.83	3.96	4.07	4.15	4.23	4.30
8	2.90	3.29	3.51	3.67	3.79	3.88	3.96	4.03	4.09
9	2.82	3.19	3.40	3.55	3.66	3.75	3.82	3.89	3.94
10	2.76	3.11	3.31	3.45	3.56	3.64	3.71	3.78	3.83
11	2.72	3.06	3.25	3.38	3.48	3.56	3.63	3.69	3.74
12	2.68	3.01	3.19	3.32	3.42	3.50	3.56	3.62	3.67
13	2.65	2.97	3.15	3.27	3.37	3.44	3.51	3.56	3.61
14	2.62	2.94	3.11	3.23	3.32	3.40	3.46	3.51	3.56
15	2.60	2.91	3.08	3.20	3.29	3.36	3.42	3.47	3.52
16	2.58	2.88	3.05	3.17	3.26	3.33	3.39	3.44	3.48
17	2.57	2.86	3.03	3.14	3.23	3.30	3.36	3.41	3.45
18	2.55	2.84	3.01	3.12	3.21	3.27	3.33	3.38	3.42
19	2.54	2.83	2.99	3.10	3.18	3.25	3.31	3.36	3.40
20	2.53	2.81	2.97	3.08	3.17	3.23	3.29	3.34	3.38
24	2.49	2.77	2.92	3.03	3.11	3.17	3.22	3.27	3.31
30	2.46	2.72	2.87	2.97	3.05	3.11	3.16	3.21	3.24
40	2.42	2.68	2.82	2.92	2.99	3.05	3.10	3.14	3.18
60	2.39	2.64	2.78	2.87	2.94	3.00	3.04	3.08	3.12
120	2.36	2.60	2.73	2.82	2.89	2.94	2.99	3.03	3.06
∞	2.33	2.56	2.68	2.77	2.84	2.89	2.93	2.97	3.00

12.1.5   Contrasts

A **contrast** L is a linear combination of the means μ_i such that the coefficients c_i sum to zero:

$$L = \sum_{i=1}^{k} c_i \mu_i \quad \text{where} \quad \sum_{i=1}^{k} c_i = 0. \tag{12.4}$$

Let $\widehat{L} = \sum_{i=1}^{k} c_i \bar{y}_{i.}$, then

(1) $\widehat{L}$ has a normal distribution, $\mathrm{E}\left[\widehat{L}\right] = \sum_{i=1}^{k} c_i \mu_i$, $\mathrm{Var}\left[\widehat{L}\right] = \sigma^2 \sum_{i=1}^{k} \frac{c_i^2}{n_i}$.

(2) A $100(1-\alpha)\%$ confidence interval for L has as endpoints

$$\widehat{l} \pm t_{\alpha/2,N-k} \cdot s \sqrt{\sum_{i=1}^{k} c_i^2/n_i}. \tag{12.5}$$

(3) Single degree of freedom test: i^{th} sample

$$H_0: \sum_{i=1}^{k} c_i \mu_i = c$$

$$H_a: \sum_{i=1}^{k} c_i \mu_i > c, \quad \sum_{i=1}^{k} c_i \mu_i < c, \quad \sum_{i=1}^{k} c_i \mu_i \neq c$$

TS: $T = \dfrac{L - c}{s\sqrt{\sum\limits_{i=1}^{k} c_i^2 / n_i}}$ or $F = T^2 = \dfrac{(L - c)^2}{s^2 \sum\limits_{i=1}^{k} c_i^2 / n_i}$

RR: $T \geq t_{\alpha, N-k}, \quad T \leq -t_{\alpha, N-k}, \quad |T| \geq t_{\alpha/2, N-k}$

  or $F \geq F_{\alpha, 1, N-k}$

(4) The set of confidence intervals with endpoints

$$\hat{l} \pm \sqrt{(k - 1)F_{\alpha, k-1, N-k}} \cdot s\sqrt{\sum_{i=1}^{k} c_i^2 / n_i} \tag{12.6}$$

is the collection of simultaneous $100(1 - \alpha)\%$ confidence intervals for all possible contrasts.

(5) Let $n = n_i$, $i = 1, 2, \ldots, k$, then the contrast sum of squares, SSL, is given by

$$\text{SSL} = \dfrac{\left(\sum\limits_{i=1}^{k} c_i T_{i.} \right)^2}{n \sum\limits_{i=1}^{k} c_i^2}. \tag{12.7}$$

(6) Two contrasts $L_1 = \sum\limits_{i=1}^{k} b_i \mu_i$ and $L_2 = \sum\limits_{i=1}^{k} c_i \mu_i$ are orthogonal if

$$\sum_{i=1}^{k} \frac{b_i c_i}{n_i} = 0.$$

12.1.6 Example

Example 12.66: A telephone company recently surveyed the length of long distance calls originating in four different parts of the country. The length of each randomly selected call (in minutes) is given in the table below.

North:	11.0	9.5	10.3	8.7	10.6	7.9
	11.1	10.7	8.4	10.8		
South:	13.6	12.2	12.5	17.5	9.7	16.4
	13.0	15.6	12.1	10.2	14.6	16.2
	13.6	17.2	13.1	13.3	11.9	12.0
Midwest:	15.0	14.2	11.9	14.5	12.7	17.1
	10.6	13.8	16.8	11.4	15.7	11.0
	13.4	10.9	11.1	11.0	10.3	10.4
West:	12.0	12.7	13.0	13.3	11.6	11.4
	12.9	15.2	14.9	11.4	11.3	

Is there any evidence to suggest the mean lengths of long distance calls from these four parts of the country are different? Use $\alpha = .05$.

Solution:

(S1) There are $k = 4$ treatments and each sample is assumed to be independent and randomly selected. Each population is assumed to be normally distributed with a mean of μ_i (for $i = 1, 2, 3, 4$) and common variance σ^2.

(S2) Summary statistics:

$$T_{1.} = 99, \quad T_{2.} = 244.7, \quad T_{3.} = 231.8, \quad T_{4.} = 139.7, \quad T_{..} = 715.2$$

(S3) Sum of squares:

$$SST = \sum_{i=1}^{4} \sum_{j=1}^{n_i} y_{ij}^2 - \frac{T_{..}^2}{N} = 9269.7 - 715.2^2/57 = 295.82$$

$$SSA = \sum_{i=1}^{4} \frac{T_{i.}^2}{n_i} - \frac{T_{..}^2}{N} = \left(\frac{99^2}{10} + \frac{244.7^2}{18} + \frac{231.8^2}{18} + \frac{139.7^2}{11} \right) - 715.2^2/57$$

$$= 9065.92 - 8973.88 = 92.04$$

$$SSE = SST - SSA = 295.82 - 92.04 = 203.78$$

(S4) The analysis of variance table:

Source of variation	Sum of squares	Degrees of freedom	Mean square	Computed F
Treatments	92.04	3	30.68	7.98
Error	203.78	53	3.84	
Total	295.82	56		

(S5) Hypothesis test for significant regression (see section 12.1.3):

$H_0: \mu_1 = \mu_2 = \mu_3 = \mu_4 = 0$

H_a: at least two of the means are unequal

TS: $F = S_A^2/S^2$

RR: $F \geq F_{.05,3,53} = 2.78$

Conclusion: The value of the test statistic lies in the rejection region. There is evidence to suggest at least two of the mean lengths are different.

12.2   TWO-WAY ANOVA

12.2.1   One observation per cell

12.2.1.1   Models and assumptions

Let Y_{ij} be the random observation in the ith row and the jth column for $i = 1, 2, \ldots, r$ and $j = 1, 2, \ldots, c$.

Fixed effects experiment:

Model: $Y_{ij} = \mu + \alpha_i + \beta_j + \epsilon_{ij}$

Assumptions: $\epsilon_{ij} \overset{\text{ind}}{\sim} N(0, \sigma^2), \quad \sum_{i=1}^{r} \alpha_i = \sum_{j=1}^{c} \beta_j = 0$

Random effects experiment:

Model: $Y_{ij} = \mu + A_i + B_j + \epsilon_{ij}$

Assumptions: $\epsilon_{ij} \overset{\text{ind}}{\sim} N(0, \sigma^2), \quad A_i \overset{\text{ind}}{\sim} N(0, \sigma_\alpha^2), \quad B_j \overset{\text{ind}}{\sim} N(0, \sigma_\beta^2)$

The ϵ_{ij}'s, A_i's, and B_j's are independent.

Mixed Effects Experiment:

Model: $Y_{ij} = \mu + A_i + \beta_j + \epsilon_{ij}$

Assumptions: $\epsilon_{ij} \overset{\text{ind}}{\sim} N(0, \sigma^2), \quad A_i \overset{\text{ind}}{\sim} N(0, \sigma_\alpha^2), \quad \sum_{j=1}^{c} \beta_j = 0$

The ϵ_{ij}'s and A_i's are independent.

12.2.1.2   Sum of squares

Dots in the subscript of $\overline{y}$ and T indicate the mean and sum of y_{ij}, respectively, over the appropriate subscript(s).

$$\text{SST} = \sum_{i=1}^{r} \sum_{j=1}^{c} (y_{ij} - \overline{y}_{..})^2 = \sum_{i=1}^{r} \sum_{j=1}^{c} y_{ij}^2 - \frac{T_{..}^2}{rc}$$

$$\text{SSR} = c \sum_{i=1}^{r} (\overline{y}_{i.} - \overline{y}_{..})^2 = \frac{\sum_{i=1}^{r} T_{i.}^2}{c} - \frac{T_{..}^2}{rc}$$

$$\text{SSC} = r \sum_{j=1}^{c} (\overline{y}_{.j} - \overline{y}_{..})^2 = \frac{\sum_{j=1}^{c} T_{.j}^2}{r} - \frac{T_{..}^2}{rc}$$

$$\text{SSE} = \sum_{i=1}^{r} \sum_{j=1}^{c} (y_{ij} - \overline{y}_{i.} - \overline{y}_{.j} + \overline{y}_{..})^2 = \text{SST} - \text{SSR} - \text{SSC}$$

12.2.1.3 Mean squares and properties

$$\text{MSR} = \frac{\text{SSR}}{r-1} \qquad = S_R^2 = \text{mean square due to rows}$$

$$\text{MSC} = \frac{\text{SSC}}{c-1} \qquad = S_C^2 = \text{mean square due to columns}$$

$$\text{MSE} = \frac{\text{SSE}}{(r-1)(c-1)} = S^2 = \text{mean square due to error}$$

Mean Square	Expected value		
	Fixed model	Random model	Mixed model
MSR	$\sigma^2 + c\left(\dfrac{\sum\limits_{i=1}^{r} \alpha_i^2}{r-1}\right)$	$\sigma^2 + c\sigma_\alpha^2$	$\sigma^2 + c\sigma_\alpha^2$
MSC	$\sigma^2 + r\left(\dfrac{\sum\limits_{j=1}^{c} \beta_i^2}{c-1}\right)$	$\sigma^2 + r\sigma_\beta^2$	$\sigma^2 + r\left(\dfrac{\sum\limits_{j=1}^{c} \beta_i^2}{c-1}\right)$
MSE	σ^2	σ^2	σ^2

(1) $F = S_R^2/S^2$ has an F distribution with $r-1$ and $(r-1)(c-1)$ degrees of freedom.

(2) $F = S_C^2/S^2$ has an F distribution with $c-1$ and $(r-1)(c-1)$ degrees of freedom.

12.2.2 Analysis of variance table

Source of variation	Sum of squares	Degrees of freedom	Mean square	Computed F
Rows	SSR	$r-1$	MSR	MSR/MSE
Columns	SSC	$c-1$	MSC	MSC/MSE
Error	SSE	$(r-1)(c-1)$	MSE	
Total	SST	$rc-1$		

Hypothesis tests:

(1) Test for significant row effect

H_0: There is no effect due to rows

(Fixed effects model: $\alpha_1 = \alpha_2 = \cdots = \alpha_r = 0$)

(Random effects model: $\sigma_\alpha^2 = 0$)

(Mixed effects model: $\sigma_\alpha^2 = 0$)

H_a: There is an effect due to rows
  (Fixed effects model: $\alpha_i \neq 0$ for some i)
  (Random effects model: $\sigma_\alpha^2 \neq 0$)
  (Mixed effects model: $\sigma_\alpha^2 \neq 0$)

TS: $F = S_{\mathrm{R}}^2/S^2$

RR: $F \geq F_{\alpha,r-1,(r-1)(c-1)}$

(2) Test for significant column effect

H_0: There is no effect due to columns
  (Fixed effects model: $\beta_1 = \beta_2 = \cdots = \beta_c = 0$)
  (Random effects model: $\sigma_\beta^2 = 0$)
  (Mixed effects model: $\beta_1 = \beta_2 = \cdots = \beta_c = 0$)

H_a: There is an effect due to columns
  (Fixed effects model: $\beta_j \neq 0$ for some j)
  (Random effects model: $\sigma_\beta^2 \neq 0$)
  (Mixed effects model: $\beta_j \neq 0$ for some j)

TS: $F = S_{\mathrm{C}}^2/S^2$

RR: $F \geq F_{\alpha,c-1,(r-1)(c-1)}$

12.2.3 Nested classifications with equal samples

12.2.3.1 Models and assumptions

Let Y_{ijk} be the kth random observation for the ith level of factor A and the jth level of factor B. There are n observations for each factor combination: $i = 1, 2, \ldots, a$, $j = 1, 2, \ldots, b$, and $k = 1, 2, \ldots, n$.

Fixed effects experiment:

Model: $\qquad Y_{ijk} = \mu + \alpha_i + \beta_{j(i)} + \epsilon_{k(ij)}$

Assumptions: $\epsilon_{k(ij)} \overset{\mathrm{ind}}{\sim} N(0, \sigma^2)$, $\displaystyle\sum_{i=1}^{a} \alpha_i = 0$, $\displaystyle\sum_{j=1}^{b} \beta_{j(i)} = 0$ for all i

Random effects experiment:

Model: $\qquad Y_{ijk} = \mu + A_i + B_{j(i)} + \epsilon_{k(ij)}$

Assumptions: $\epsilon_{k(ij)} \overset{\mathrm{ind}}{\sim} N(0, \sigma^2)$, $A_i \overset{\mathrm{ind}}{\sim} N(0, \sigma_\alpha^2)$, $B_{j(i)} \overset{\mathrm{ind}}{\sim} N(0, \sigma_\beta^2)$

  The $\epsilon_{k(ij)}$'s, A_i's, and $B_{j(i)}$'s are independent.

Mixed effects experiment (α):

Model: $\qquad Y_{ijk} = \mu + \alpha_i + B_{j(i)} + \epsilon_{k(ij)}$

Assumptions: $\epsilon_{k(ij)} \overset{\text{ind}}{\sim} N(0, \sigma^2), \quad \sum_{i=1}^{a} \alpha_i = 0, \quad B_{j(i)} \overset{\text{ind}}{\sim} N(0, \sigma_\beta^2)$

The $\epsilon_{k(ij)}$'s and $B_{j(i)}$'s are independent.

Mixed effects experiment (β):

Model: $\qquad Y_{ijk} = \mu + A_i + \beta_{j(i)} + \epsilon_{k(ij)}$

Assumptions: $\epsilon_{k(ij)} \overset{\text{ind}}{\sim} N(0, \sigma^2), \quad A_i \overset{\text{ind}}{\sim} N(0, \sigma_\alpha^2), \quad \sum_{j=1}^{b} \beta_{j(i)} = 0 \text{ for all } i$

The $\epsilon_{k(ij)}$'s and A_i's are independent.

12.2.3.2 Sum of squares

Dots in the subscript of $\bar{y}$ and T indicate the mean and sum of y_{ijk}, respectively, over the appropriate subscript(s).

$$\text{SST} = \sum_{i=1}^{a} \sum_{j=1}^{b} \sum_{k=1}^{n} (y_{ijk} - \bar{y}...)^2 = \sum_{i=1}^{a} \sum_{j=1}^{b} \sum_{k=1}^{n} y_{ijk}^2 - \frac{T^2...}{abn}$$

$$\text{SSA} = \sum_{i=1}^{a} n_{i.} (\bar{y}_{i..} - \bar{y}...)^2 = \frac{\sum_{i=1}^{a} T_{i..}^2}{n_{i.}} - \frac{T^2...}{abn}$$

$$\text{SSB(A)} = \sum_{i=1}^{a} \sum_{j=1}^{b} n_{ij} (\bar{y}_{ij.} - \bar{y}_{i..})^2 = \sum_{i=1}^{a} \sum_{j=1}^{b} \frac{T_{ij.}^2}{n_{ij}} - \sum_{i=1}^{a} \frac{T_{i..}^2}{n_{i.}}$$

$$\text{SSE} = \sum_{i=1}^{a} \sum_{j=1}^{b} \sum_{k=1}^{n} (y_{ijk} - \bar{y}_{ij.})^2 = \text{SST} - \text{SSA} - \text{SSB(A)}$$

12.2.3.3 Mean squares and properties

$$\text{MSA} = \frac{\text{SSA}}{a-1} \qquad = S_A^2 \qquad = \text{mean square due to factor A}$$

$$\text{MSB(A)} = \frac{\text{SSB(A)}}{a(b-1)} \qquad = S_{B(A)}^2 = \text{mean square due to factor B}$$

$$\text{MSE} = \frac{\text{SSE}}{ab(n-1)} \qquad = S^2 \qquad = \text{mean square due to error}$$

Mean square	Expected value	
	Fixed model Mixed model (α)	Random model Mixed model (β)
MSA	$\sigma^2 + bn\dfrac{\sum\limits_{i=1}^{a}\alpha_i^2}{a-1}$	$\sigma^2 + bn\sigma_\alpha^2 + n\sigma_\beta^2$
	$\sigma^2 + n\sigma_\beta^2 + bn\left(\dfrac{\sum\limits_{i=1}^{a}\alpha_i^2}{a-1}\right)$	$\sigma^2 + bn\sigma_\alpha^2$
MSB(A)	$\sigma^2 + n\dfrac{\sum\limits_{i=1}^{a}\sum\limits_{j=1}^{b}\beta_{j(i)}^2}{a(b-1)}$	$\sigma^2 + n\sigma_\beta^2$
	$\sigma^2 + n\sigma_\beta^2$	$\sigma^2 + n\dfrac{\sum\limits_{i=1}^{a}\sum\limits_{j=1}^{b}\beta_{j(i)}^2}{a(b-1)}$
MSE	σ^2	σ^2
	σ^2	σ^2

(1) $F = S_A^2/S^2$ has an F distribution with $a-1$ and $ab(n-1)$ degrees of freedom.

(2) $F = S_{B(A)}^2/S^2$ has an F distribution with $a(b-1)$ and $ab(n-1)$ degrees of freedom.

12.2.3.4 Analysis of variance table

Source of variation	Sum of squares	Degrees of freedom	Mean square	Computed F
Between main groups	SSA	$a-1$	MSA	$\dfrac{\text{MSA}}{\text{MSE}}$
Subgroups within main groups	SSB(A)	$a(b-1)$	MSB(A)	$\dfrac{\text{MSB(A)}}{\text{MSE}}$
Error	SSE	$ab(n-1)$	MSE	
Total	SST	$abn-1$		

Hypothesis tests:

(1) Test for significant factor A main effect

H_0: There is no effect due to factor A
(Fixed effects model: $\alpha_1 = \alpha_2 = \cdots = \alpha_a = 0$)
(Random effects model: $\sigma_\alpha^2 = 0$)
(Mixed effects model (α): $\alpha_1 = \alpha_2 = \cdots = \alpha_a = 0$)
(Mixed effects model (β): $\sigma_\alpha^2 = 0$)

H_a: There is an effect due to factor A
(Fixed effects model: $\alpha_i \neq 0$ for some i)
(Random effects model: $\sigma_\alpha^2 \neq 0$)
(Mixed effects model (α): $\alpha_i \neq 0$ for some i)
(Mixed effects model (β): $\sigma_\alpha^2 \neq 0$)

TS: $F = S_A^2 / S^2$

RR: $F \geq F_{\alpha, a-1, ab(n-1)}$

(2) Test for significant factor B specific effect
H_0: There is no effect due to factor B
(Fixed effects model: all $\beta_{j(i)} = 0$)
(Random effects model: $\sigma_\beta^2 = 0$)
(Mixed effects model (α): $\sigma_\beta^2 = 0$)
(Mixed effects model (β): all $\beta_{j(i)} = 0$)

H_a: There is an effect due to factor B
(Fixed effects model: not all $\beta_{j(i)} = 0$)
(Random effects model: $\sigma_\beta^2 \neq 0$)
(Mixed effects model (α): $\sigma_\beta^2 \neq 0$)
(Mixed effects model (β): not all $\beta_{j(i)} = 0$)

TS: $F = S_{B(A)}^2 / S^2$

RR: $F \geq F_{\alpha, a(b-1), ab(n-1)}$

12.2.4 Nested classifications with unequal samples

12.2.4.1 Models and assumptions

Let Y_{ijk} be the kth random observation for the ith level of factor A and the jth level of factor B. There are n_{ij} observations for each factor combination:

$$i = 1, 2, \ldots, a, \; j = 1, 2, \ldots, m_i, \; k = 1, 2, \ldots, n_{ij}, \text{ and } \sum_{i=1}^{a} \sum_{j=1}^{m_i} n_{ij} = n.$$

Fixed effects experiment:

Model: $Y_{ijk} = \mu + \alpha_i + \beta_{j(i)} + \epsilon_{k(ij)}$

Assumptions: $\epsilon_{k(ij)} \overset{\text{ind}}{\sim} N(0, \sigma^2)$, $\displaystyle\sum_{i=1}^{a} \alpha_i = 0$, $\displaystyle\sum_{j=1}^{m_i} \beta_{j(i)} = 0$ for all i

Random effects experiment:

Model: $Y_{ijk} = \mu + A_i + B_{j(i)} + \epsilon_{k(ij)}$

Assumptions: $\epsilon_{k(ij)} \overset{\text{ind}}{\sim} N(0, \sigma^2)$, $A_i \overset{\text{ind}}{\sim} N(0, \sigma_\alpha^2)$, $B_{j(i)} \overset{\text{ind}}{\sim} N(0, \sigma_\beta^2)$

The $\epsilon_{k(ij)}$'s, A_i's, and $B_{j(i)}$'s are independent.

Mixed effects experiment (α):

Model: $Y_{ijk} = \mu + \alpha_i + B_{j(i)} + \epsilon_{k(ij)}$

Assumptions: $\epsilon_{k(ij)} \overset{\text{ind}}{\sim} N(0, \sigma^2)$, $\displaystyle\sum_{i=1}^{a} \alpha_i = 0$, $B_{j(i)} \overset{\text{ind}}{\sim} N(0, \sigma_\beta^2)$

The $\epsilon_{k(ij)}$'s and $B_{j(i)}$'s are independent.

Mixed effects experiment (β):

Model: $Y_{ijk} = \mu + A_i + \beta_{j(i)} + \epsilon_{k(ij)}$

Assumptions: $\epsilon_{k(ij)} \overset{\text{ind}}{\sim} N(0, \sigma^2)$, $A_i \overset{\text{ind}}{\sim} N(0, \sigma_\alpha^2)$, $\displaystyle\sum_{j=1}^{m_i} \beta_{j(i)} = 0$ for all i

The $\epsilon_{k(ij)}$'s and A_i's are independent.

12.2.4.2   Sum of squares

Dots in the subscript of $\bar{y}$ and T indicate the mean and sum of y_{ijk}, respectively, over the appropriate subscript(s).

$$\text{SST} = \sum_{i=1}^{a}\sum_{j=1}^{m_i}\sum_{k=1}^{n_{ij}}(y_{ijk} - \bar{y}_{...})^2 = \sum_{i=1}^{a}\sum_{j=1}^{m_i}\sum_{k=1}^{n_{ij}} y_{ijk}^2 - \frac{T_{...}^2}{n}$$

$$\text{SSA} = \sum_{i=1}^{a} n_{i.}(\bar{y}_{i..} - \bar{y}_{...})^2 = \frac{\displaystyle\sum_{i=1}^{a} T_{i..}^2}{n_{i.}} - \frac{T_{...}^2}{n}$$

$$\text{SSB(A)} = \sum_{i=1}^{a}\sum_{j=1}^{m_i} n_{ij}(\bar{y}_{ij.} - \bar{y}_{i..})^2 = \sum_{i=1}^{a}\sum_{j=1}^{m_i} \frac{T_{ij.}^2}{n_{ij}} - \sum_{i=1}^{a} \frac{T_{i..}^2}{n_{i.}}$$

$$\text{SSE} = \sum_{i=1}^{a}\sum_{j=1}^{m_i}\sum_{k=1}^{n_{ij}}(y_{ijk} - \bar{y}_{ij.})^2 = \text{SST} - \text{SSA} - \text{SSB(A)}$$

12.2.4.3 Mean squares and properties

$$\text{MSA} = \frac{\text{SSA}}{a-1} = S_A^2 = \text{mean square due to factor A}$$

$$\text{MSB(A)} = \frac{\text{SSB(A)}}{\sum\limits_{i=1}^{a} m_i} = S_{B(A)}^2 = \text{mean square due to factor B}$$

$$\text{MSE} = \frac{\text{SSE}}{n - \sum\limits_{i=1}^{a} m_i} = S^2 = \text{mean square due to error}$$

Mean	Expected value	
square	Fixed model	Random model
	Mixed model (α)	Mixed model (β)
MSA	$\sigma^2 + \dfrac{\sum\limits_{i=1}^{a} m_i \alpha_i^2}{a-1}$	$\sigma^2 + c_1 \sigma_\alpha^2 + c_2 \sigma_\beta^2$
	$\sigma^2 + c_2 \sigma_\beta^2 + \dfrac{\sum\limits_{i=1}^{a} m_i \alpha_i^2}{a-1}$	$\sigma^2 + c_1 \sigma_\alpha^2$
MSB(A)	$\sigma^2 + \dfrac{\sum\limits_{i=1}^{a} \sum\limits_{j=1}^{m_i} n_{ij} \beta_{j(i)}^2}{\sum\limits_{i=1}^{a} m_i - a}$	$\sigma^2 + c_3 \sigma_\beta^2$
	$\sigma^2 + c_3 \sigma_\beta^2$	$\sigma^2 + \dfrac{\sum\limits_{i=1}^{a} \sum\limits_{j=1}^{m_i} n_{ij} \beta_{j(i)}^2}{\sum\limits_{i=1}^{a} m_i - a}$
MSE	σ^2	σ^2
	σ^2	σ^2

where

$$c_1 = \frac{\sum\limits_{i=1}^{a} \dfrac{\sum\limits_{j=1}^{m_i} n_{ij}^2}{m_i} - \dfrac{\sum\limits_{i=1}^{a} \sum\limits_{j=1}^{m_i} n_{ij}^2}{n}}{a-1}, \quad c_2 = \frac{n - \dfrac{\sum\limits_{i=1}^{a} m_i^2}{n}}{a-1}, \quad c_3 = \frac{n - \sum\limits_{i=1}^{a} \dfrac{\sum\limits_{j=1}^{m_i} n_{ij}^2}{m_i}}{\sum\limits_{i=1}^{a} m_i - k}.$$

(1) $F = S_A^2 / S^2$ has an F distribution with $a-1$ and $n - \sum\limits_{i=1}^{a} m_i$ degrees of freedom.

(2) $F = S^2_{B(A)}/S^2$ has an F distribution with $\sum_{i=1}^{a} m_i - a$ and $n - \sum_{i=1}^{a} m_i$ degrees of freedom.

12.2.4.4 Analysis of variance table

Source of variation	Sum of squares	Degrees of freedom	Mean square	Computed F
Between main groups	SSA	$a - 1$	MSA	$\dfrac{\text{MSA}}{\text{MSE}}$
Subgroups within main groups	SSB(A)	$\sum_{i=1}^{a} m_i - a$	MSB(A)	$\dfrac{\text{MSB(A)}}{\text{MSE}}$
Error	SSE	$n - \sum_{i=1}^{a} m_i$	MSE	
Total	SST	$n - 1$		

Hypothesis tests:

(1) Test for significant factor A main effect

H_0: There is no effect due to factor A

(Fixed effects model: $\alpha_1 = \alpha_2 = \cdots = \alpha_a = 0$)

(Random effects model: $\sigma^2_\alpha = 0$)

(Mixed effects model (α): $\alpha_1 = \alpha_2 = \cdots = \alpha_a = 0$)

(Mixed effects model (β): $\sigma^2_\alpha = 0$)

H_a: There is an effect due to factor A

(Fixed effects model: $\alpha_i \neq 0$ for some i)

(Random effects model: $\sigma^2_\alpha \neq 0$)

(Mixed effects model (α): $\alpha_i \neq 0$ for some i)

(Mixed effects model (β): $\sigma^2_\alpha \neq 0$)

TS: $F = S^2_A/S^2$

RR: $F \geq F_{\alpha, a-1, n-\sum_{i=1}^{a} m_i}$

(2) Test for significant factor B specific effect

H_0: There is no effect due to factor B
(Fixed effects model: all $\beta_{j(i)} = 0$)
(Random effects model: $\sigma_\beta^2 = 0$)
(Mixed effects model (α): $\sigma_\beta^2 = 0$)
(Mixed effects model (β): all $\beta_{j(i)} = 0$)

H_a: There is an effect due to factor B
(Fixed effects model: not all $\beta_{j(i)} = 0$)
(Random effects model: $\sigma_\beta^2 \neq 0$)
(Mixed effects model (α): $\sigma_\beta^2 \neq 0$)
(Mixed effects model (β): not all $\beta_{j(i)} = 0$)

TS: $F = S_{B(A)}^2 / S^2$

RR: $F \geq F_{\alpha, \sum_{i=1}^{a} m_i - a, n - \sum_{i=1}^{a} m_i}$

12.2.5 Two-factor experiments

12.2.5.1 Models and assumptions

Let Y_{ijk} be the kth random observation for the ith level of factor A and the jth level of factor B. There are n observations for each factor combination: $i = 1, 2, \ldots, a$, $j = 1, 2, \ldots, b$, and $k = 1, 2, \ldots, n$.

Fixed effects experiment:

Model: $Y_{ijk} = \mu + \alpha_i + \beta_j + (\alpha\beta)_{ij} + \epsilon_{ijk}$

Assumptions: $\epsilon_{ijk} \overset{\text{ind}}{\sim} N(0, \sigma^2)$, $\displaystyle\sum_{i=1}^{a} \alpha_i = 0$, $\displaystyle\sum_{j=1}^{b} \beta_j = 0$

$$\sum_{i=1}^{a} (\alpha\beta)_{ij} = \sum_{j=1}^{b} (\alpha\beta)_{ij} = 0$$

Random effects experiment:

Model: $Y_{ijk} = \mu + A_i + B_j + (AB)_{ij} + \epsilon_{ijk}$

Assumptions: $\epsilon_{ijk} \overset{\text{ind}}{\sim} N(0, \sigma^2)$, $A_i \overset{\text{ind}}{\sim} N(0, \sigma_\alpha^2)$, $B_j \overset{\text{ind}}{\sim} N(0, \sigma_\beta^2)$

$(AB)_{ij} \overset{\text{ind}}{\sim} N(0, \sigma_{\alpha\beta}^2)$

The ϵ_{ijk}'s, A_i's, B_j's, and $(AB)_{ij}$'s are independent.

Mixed effects experiment (α):

Model: $Y_{ijk} = \mu + \alpha_i + B_j + (\alpha B)_{ij} + \epsilon_{ijk}$

Assumptions: $\epsilon_{ijk} \overset{\text{ind}}{\sim} N(0, \sigma^2)$, $\displaystyle\sum_{i=1}^{a} \alpha_i = 0$, $\displaystyle\sum_{i=1}^{a} (\alpha B)_{ij} = 0$ for all j

$B_j \overset{\text{ind}}{\sim} N(0, \sigma_\beta^2)$, $(\alpha B)_{ij} \overset{\text{ind}}{\sim} N(0, \frac{a-1}{a}\sigma_{\alpha\beta}^2)$

The ϵ_{ijk}'s, B_j's, and $(\alpha B)_{ij}$'s are independent.

12.2.5.2 *Sum of squares*

Dots in the subscript of $\bar{y}$ and T indicate the mean and sum of y_{ijk}, respectively, over the appropriate subscript(s).

$$\text{SST} = \sum_{i=1}^{a}\sum_{j=1}^{b}\sum_{k=1}^{n}(y_{ijk} - \bar{y}_{...})^2 = \sum_{i=1}^{a}\sum_{j=1}^{b}\sum_{k=1}^{n} y_{ijk}^2 - \frac{T_{...}^2}{abn}$$

$$\text{SSA} = bn\sum_{i=1}^{a}(\bar{y}_{i..} - \bar{y}_{...})^2 = \frac{\sum\limits_{i=1}^{a} T_{i..}^2}{bn} - \frac{T_{...}^2}{abn}$$

$$\text{SSB} = an\sum_{j=1}^{b}(\bar{y}_{.j.} - \bar{y}_{...})^2 = \frac{\sum\limits_{j=1}^{b} T_{.j.}^2}{an} - \frac{T_{...}^2}{abn}$$

$$\text{SS(AB)} = n\sum_{i=1}^{a}\sum_{j=1}^{b}(\bar{y}_{ij.} - \bar{y}_{i..} - \bar{y}_{.j.} + \bar{y}_{...})^2$$

$$= \frac{\sum\limits_{i=1}^{a}\sum\limits_{j=1}^{b} T_{ij.}^2}{n} - \frac{\sum\limits_{i=1}^{a} T_{i..}^2}{bn} - \frac{\sum\limits_{j=1}^{b} T_{.j.}^2}{an} + \frac{T_{...}^2}{abn}$$

$$\text{SSE} = \sum_{i=1}^{a}\sum_{j=1}^{b}\sum_{k=1}^{n}(y_{ijk} - \bar{y}_{ij.})^2 = \text{SST} - \text{SSA} - \text{SSB} - \text{SS(AB)}$$

12.2.5.3 *Mean squares and properties*

$$\text{MSA} = \frac{\text{SSA}}{a-1} \qquad = S_A^2 \quad = \text{mean square due to factor A}$$

$$\text{MSB} = \frac{\text{SSB}}{b-1} \qquad = S_B^2 \quad = \text{mean square due to factor B}$$

$$\text{MS(AB)} = \frac{\text{SS(AB)}}{(a-1)(b-1)} = S_{AB}^2 = \text{mean square due to interaction}$$

$$\text{MSE} = \frac{\text{SSE}}{ab(n-1)} \qquad = S^2 \quad = \text{mean square due to error}$$

Mean square		
Expected value		
Fixed model	Random model	Mixed model (α)
MSA		
$\sigma^2 + nb\dfrac{\sum\limits_{i=1}^{a} \alpha_i^2}{a-1}$	$\sigma^2 + nb\sigma_\alpha^2 + \sigma_{\alpha\beta}^2$	$\sigma^2 + nb\dfrac{\sum\limits_{i=1}^{a} \alpha_i^2}{a-1} + n\sigma_{\alpha\beta}^2$
MSB		
$\sigma^2 + na\dfrac{\sum\limits_{j=1}^{b} \beta_j^2}{b-1}$	$\sigma^2 + na\sigma_\beta^2 + n\sigma_{\alpha\beta}^2$	$\sigma^2 + na\sigma_\beta^2$
MS(AB)		
$\sigma^2 + n\dfrac{\sum\limits_{i=1}^{a}\sum\limits_{j=1}^{b} (\alpha\beta)_{ij}^2}{(a-1)(b-1)}$	$\sigma^2 + n\sigma_{\alpha\beta}^2$	$\sigma^2 + n\sigma_{\alpha\beta}^2$
MSE		
σ^2	σ^2	σ^2

(1) $F = S_A^2/S^2$ has an F distribution with $a-1$ and $ab(n-1)$ degrees of freedom.

(2) $F = S_B^2/S^2$ has an F distribution with $b-1$ and $ab(n-1)$ degrees of freedom.

(3) $F = S_{AB}^2/S^2$ has an F distribution with $(a-1)(b-1)$ and $ab(n-1)$ degrees of freedom.

12.2.5.4 Analysis of variance table

Source of variation	Sum of squares	Degrees of freedom	Mean square	Computed F
Factor A	SSA	$a-1$	MSA	$\dfrac{\text{MSA}}{\text{MSE}}$
Factor B	SSB	$b-1$	MSB	$\dfrac{\text{MSB}}{\text{MSE}}$
Interaction AB	SS(AB)	$(a-1)(b-1)$	MS(AB)	$\dfrac{\text{MS(AB)}}{\text{MSE}}$
Error	SSE	$ab(n-1)$	MSE	
Total	SST	$abn-1$		

Hypothesis tests:

(1) Test for significant factor A main effect

H_0: There is no effect due to factor A
  (Fixed effects model: $\alpha_1 = \alpha_2 = \cdots = \alpha_a = 0$)
  (Random effects model: $\sigma_\alpha^2 = 0$)
  (Mixed effects model (α): $\alpha_1 = \alpha_2 = \cdots = \alpha_a = 0$)

H_a: There is an effect due to factor A
  (Fixed effects model: $\alpha_i \neq 0$ for some i)
  (Random effects model: $\sigma_\alpha^2 \neq 0$)
  (Mixed effects model (α): $\alpha_i \neq 0$ for some i)

TS: $F = S_A^2 / S^2$

RR: $F \geq F_{\alpha, a-1, ab(n-1)}$

(2) Test for significant factor B main effect

H_0: There is no effect due to factor B
  (Fixed effects model: $\beta_1 = \beta_2 = \cdots = \beta_b = 0$)
  (Random effects model: $\sigma_\beta^2 = 0$)
  (Mixed effects model (α): $\sigma_\beta^2 = 0$)

H_a: There is an effect due to factor B
  (Fixed effects model: $\beta_j \neq 0$ for some j)
  (Random effects model: $\sigma_\beta^2 \neq 0$)
  (Mixed effects model (α): $\sigma_\beta^2 \neq 0$)

TS: $F = S_B^2 / S^2$

RR: $F \geq F_{\alpha, b-1, ab(n-1)}$

(3) Test for significant AB interaction effect

H_0: There is no effect due to interaction
  (Fixed effects model: $(\alpha\beta)_{11} = (\alpha\beta)_{12} = \cdots = (\alpha\beta)_{ab} = 0$)
  (Random effects model: $\sigma_{\alpha\beta}^2 = 0$
  (Mixed effects model (α): $\sigma_{\alpha\beta}^2 = 0$

H_a: There is an effect due to interaction
  (Fixed effects model: $(\alpha\beta)_{ij} \neq 0$ for some ij)
  (Random effects model: $\sigma_{\alpha\beta}^2 \neq 0$
  (Mixed effects model (α): $\sigma_{\alpha\beta}^2 \neq 0$

TS: $F = S_{AB}^2 / S^2$

RR: $F \geq F_{\alpha, (a-1)(b-1), ab(n-1)}$

12.2.6   Example

Example 12.67: An electrical engineer believes the brand of battery and the style of music played most often may have an effect on the lifetime of batteries in a portable CD player. Random samples were selected and the lifetime of each battery (in hours) was recorded. The data are given in the table below.

		Battery Brand			
		A	B	C	D
Style	Easy listening	61.1 58.3	60.3 68.8	58.0 59.9	55.7 48.9
		61.3 69.2	69.5 61.9	66.4 60.3	64.1 60.5
	Country	62.2 55.9	64.0 53.4	64.5 59.2	61.8 59.0
		63.1 67.0	64.3 64.8	66.0 56.4	54.1 50.5
	Rock	64.2 57.0	58.7 61.1	62.8 63.0	58.7 57.3
		59.1 48.7	62.3 70.8	65.1 64.1	60.9 63.0

Is there any evidence to suggest a difference in battery life due to brand or music style?

Solution:

(S1) A fixed effects experiment is assumed. There are $i = 3$ styles of music, $j = 4$ battery brands, and $k = 4$ observations for each factor combination.

(S2) The analysis of variance table:

Source of variation	Sum of squares	Degrees of freedom	Mean square	Computed F
Style	10.2	2	5.1	0.22
Brand	199.7	3	66.6	2.91
Interaction	125.8	6	21.0	0.92
Error	822.5	36	22.8	
Total	1158.1	47		

(S3) Test for significant interaction effect:

H_0: There is no effect due to interaction.

H_a: There is an effect due to interaction.

TS: $F = S_{AB}^2/S^2$

RR: $F \geq F_{.05,6,36} = 2.36$

Conclusion: The value of the test statistic ($F = 0.92$) does not lie in the rejection region. There is no evidence to suggest an interaction effect. Tests for main effects may be analyzed as though there were no interaction.

(S4) Test for significant effect due to style:

H_0: There is no effect due to style.

H_a: There is an effect due to style.

TS: $F = S_A^2/S^2$

RR: $F \geq F_{.05,2,36} = 3.26$

Conclusion: The value of the test statistic ($F = 0.22$) does not lie in the rejection region. There is no evidence to suggest a difference in battery life due to style of music.

(S5) Test for significant effect due to brand:

H_0: There is no effect due to brand.

H_a: There is an effect due to brand.

TS: $F = S_B^2 / S^2$

RR: $F \geq F_{.05, 3, 36} = 2.86$

Conclusion: The value of the test statistic ($F = 2.91$) lies in the rejection region. There is some evidence to suggest a difference in battery life due to brand.

12.3   THREE-FACTOR EXPERIMENTS

12.3.1   Models and assumptions

Let Y_{ijkl} be the lth random observation for the ith level of factor A, the jth level of factor B, and the kth level of factor C. There are n observations for each factor combination: $i = 1, 2, \ldots, a$, $j = 1, 2, \ldots, b$, and $k = 1, 2, \ldots, c$, $l = 1, 2, \ldots, n$.

Fixed effects experiment:

Model:

$$Y_{ijkl} = \mu + \alpha_i + \beta_j + \gamma_k + (\alpha\beta)_{ij} + (\alpha\gamma)_{ik} + (\beta\gamma)_{jk} + (\alpha\beta\gamma)_{ijk} + \epsilon_{ijkl}$$

Assumptions:

$$\epsilon_{ijkl} \overset{\text{ind}}{\sim} N(0, \sigma^2), \quad \sum_{i=1}^{a} \alpha_i = 0, \quad \sum_{j=1}^{b} \beta_j = 0, \quad \sum_{k=1}^{c} \gamma_k = 0,$$

$$\sum_{i=1}^{a} (\alpha\beta)_{ij} = \sum_{j=1}^{b} (\alpha\beta)_{ij} = \sum_{i=1}^{c} (\alpha\gamma)_{ik} = \sum_{k=1}^{c} (\alpha\gamma)_{ik} = \sum_{j=1}^{b} (\beta\gamma)_{jk} = \sum_{k=1}^{c} (\beta\gamma)_{jk} = 0,$$

$$\sum_{i=1}^{a} (\alpha\beta\gamma)_{ijk} = \sum_{j=1}^{b} (\alpha\beta\gamma)_{ijk} = \sum_{k=1}^{c} (\alpha\beta\gamma)_{ijk} = 0$$

Random effects experiment:

Model:

$$Y_{ijkl} = \mu + A_i + B_j + C_k + (AB)_{ij} + (AC)_{ik} + (BC)_{jk} + (ABC)_{ijk} + \epsilon_{ijkl}$$

Assumptions:

$$\epsilon_{ijkl} \overset{\text{ind}}{\sim} N(0, \sigma^2), \quad A_i \overset{\text{ind}}{\sim} N(0, \sigma_\alpha^2), \quad B_j \overset{\text{ind}}{\sim} N(0, \sigma_\beta^2), \quad C_k \overset{\text{ind}}{\sim} N(0, \sigma_\gamma^2)$$

$$(AB)_{ij} \overset{\text{ind}}{\sim} N(0, \sigma_{\alpha\beta}^2), \quad (AC)_{ik} \overset{\text{ind}}{\sim} N(0, \sigma_{\alpha\gamma}^2), \quad (BC)_{jk} \overset{\text{ind}}{\sim} N(0, \sigma_{\beta\gamma}^2),$$

$$(ABC)_{ijk} \overset{\text{ind}}{\sim} N(0, \sigma_{\alpha\beta\gamma}^2).$$

The ϵ_{ijkl}'s are independent of the other random components.

Mixed effects experiment (α):

Model:

$$Y_{ijkl} = \mu + \alpha_i + B_j + C_k + (\alpha B)_{ij} + (\alpha C)_{ik} + (BC)_{jk} + (\alpha BC)_{ijk} + \epsilon_{ijkl}$$

Assumptions:

$$\epsilon_{ijk} \stackrel{\text{ind}}{\sim} N(0, \sigma^2), \sum_{i=1}^{a} \alpha_i = \sum_{i=1}^{a} (\alpha B)_{ij} = \sum_{i=1}^{a} (\alpha C)_{ik} = \sum_{i=1}^{a} (\alpha BC)_{ijk} = 0 \,,$$

$$B_j \stackrel{\text{ind}}{\sim} N(0, \sigma_\beta^2), \quad C_k \stackrel{\text{ind}}{\sim} N(0, \sigma_\gamma^2), \quad (\alpha B)_{ij} \stackrel{\text{ind}}{\sim} N(0, \sigma_{\alpha\beta}^2)$$

$$(\alpha C)_{ik} \stackrel{\text{ind}}{\sim} N(0, \sigma_{\alpha\gamma}^2), \quad (BC)_{jk} \stackrel{\text{ind}}{\sim} N(0, \sigma_{\beta\gamma}^2), \quad (\alpha BC)_{ijk} \stackrel{\text{ind}}{\sim} N(0, \sigma_{\alpha\beta\gamma}^2)$$

The ϵ_{ijkl}'s are independent of the other random components.

Mixed effects experiment (α, β):

Model:

$$Y_{ijkl} = \mu + \alpha_i + \beta_j + C_k + (\alpha\beta)_{ij} + (\alpha C)_{ik} + (\beta C)_{jk} + (\alpha\beta C)_{ijk} + \epsilon_{ijkl}$$

Assumptions:

$$\epsilon_{ijkl} \stackrel{\text{ind}}{\sim} N(0, \sigma^2), \quad \sum_{i=1}^{a} \alpha_i = \sum_{j=1}^{b} \beta_j = \sum_{i=1}^{a} (\alpha C)_{ik} = \sum_{j=1}^{b} (\beta C)_{jk} = 0,$$

$$\sum_{i=1}^{a} (\alpha\beta)_{ij} = \sum_{j=1}^{b} (\alpha\beta)_{ij} = \sum_{i=1}^{a} (\alpha\beta C)_{ijk} = \sum_{j=1}^{b} (\alpha\beta C)_{ijk} = 0,$$

$$C_k \stackrel{\text{ind}}{\sim} N(0, \sigma_\gamma^2), \quad (\alpha C)_{ik} \stackrel{\text{ind}}{\sim} N(0, \sigma_{\alpha\gamma}^2), \quad (\beta C)_{jk} \stackrel{\text{ind}}{\sim} N(0, \sigma_{\beta\gamma}^2),$$

$$(\alpha\beta C)_{ijk} \stackrel{\text{ind}}{\sim} N(0, \sigma_{\alpha\beta\gamma}^2)$$

The ϵ_{ijkl}'s are independent of the other random components.

12.3.2 Sum of squares

Dots in the subscript of $\bar{y}$ and T indicate the mean and sum of y_{ijkl}, respectively, over the appropriate subscript(s).

$$SST = \sum_{i=1}^{a}\sum_{j=1}^{b}\sum_{k=1}^{c}\sum_{l=1}^{n}(y_{ijkl} - \overline{y}_{....})^2 = \sum_{i=1}^{a}\sum_{j=1}^{b}\sum_{k=1}^{c}\sum_{l=1}^{n} y_{ijkl}^2 - \frac{T_{....}^2}{abcn}$$

$$SSA = bcn\sum_{i=1}^{a}(\overline{y}_{i...} - \overline{y}_{....})^2 = \frac{\sum_{i=1}^{a} T_{i...}^2}{bn} - \frac{T_{....}^2}{abcn}$$

$$SSB = acn\sum_{j=1}^{b}(\overline{y}_{.j..} - \overline{y}_{....})^2 = \frac{\sum_{j=1}^{b} T_{.j..}^2}{acn} - \frac{T_{....}^2}{abcn}$$

$$SSC = abn\sum_{k=1}^{c}(\overline{y}_{..k.} - \overline{y}_{....})^2 = \frac{\sum_{k=1}^{c} T_{..k.}^2}{abn} - \frac{T_{....}^2}{abcn}$$

$$SS(AB) = cn\sum_{i=1}^{a}\sum_{j=1}^{b}(\overline{y}_{ij..} - \overline{y}_{i...} - \overline{y}_{.j..} + \overline{y}_{....})^2$$

$$= \frac{\sum_{i=1}^{a}\sum_{j=1}^{b} T_{ij..}^2}{cn} - \frac{\sum_{i=1}^{a} T_{i...}^2}{bcn} - \frac{\sum_{j=1}^{b} T_{.j..}^2}{acn} + \frac{T_{....}^2}{abcn}$$

$$SS(AC) = bn\sum_{i=1}^{a}\sum_{k=1}^{c}(\overline{y}_{i.k.} - \overline{y}_{i...} - \overline{y}_{..k.} + \overline{y}_{....})^2$$

$$= \frac{\sum_{i=1}^{a}\sum_{k=1}^{c} T_{i.k.}^2}{bn} - \frac{\sum_{i=1}^{a} T_{i...}^2}{bcn} - \frac{\sum_{k=1}^{c} T_{..k.}^2}{abn} + \frac{T_{....}^2}{abcn}$$

$$SS(BC) = an\sum_{j=1}^{b}\sum_{k=1}^{c}(\overline{y}_{.jk.} - \overline{y}_{.j..} - \overline{y}_{..k.} + \overline{y}_{....})^2$$

$$= \frac{\sum_{j=1}^{b}\sum_{k=1}^{c} T_{.jk.}^2}{an} - \frac{\sum_{j=1}^{b} T_{.j..}^2}{acn} - \frac{\sum_{k=1}^{c} T_{..k.}^2}{abn} + \frac{T_{....}^2}{abcn}$$

$$SS(ABC) = \sum_{i=1}^{a}\sum_{j=1}^{b}\sum_{k=1}^{c}(\overline{y}_{ijk.} - \overline{y}_{ij..} - \overline{y}_{i.k.} - \overline{y}_{.jk.} + \overline{y}_{i...} + \overline{y}_{.j..} + \overline{y}_{..k.} - \overline{y}_{....})^2$$

$$= \frac{\sum_{i=1}^{a}\sum_{j=1}^{b}\sum_{k=1}^{c} T_{ijk.}^2}{n} - \frac{\sum_{i=1}^{a}\sum_{j=1}^{b} T_{ij..}^2}{cn} - \frac{\sum_{i=1}^{a}\sum_{j=1}^{b} T_{i.k.}^2}{bn} - \frac{\sum_{j=1}^{b}\sum_{k=1}^{c} T_{.jk.}^2}{an}$$

$$+ \frac{\sum_{i=1}^{a} T_{i...}^2}{bcn} + \frac{\sum_{j=1}^{b} T_{.j..}^2}{acn} + \frac{\sum_{k=1}^{c} T_{..k.}^2}{abn} + \frac{T_{....}^2}{abcn}$$

$$SSE = \sum_{i=1}^{a}\sum_{j=1}^{b}\sum_{k=1}^{c}\sum_{l=1}^{n}(y_{ijkl} - \overline{y}_{ijk.})^2$$

$$= SST - SSA - SSB - SSC - SS(AB) - SS(AC) - SS(BC)$$
$$- SS(ABC)$$

12.3.3 Mean squares and properties

$$MSA = \frac{SSA}{a-1} = S_A^2 = \text{mean square due to factor A}$$

$$MSB = \frac{SSB}{b-1} = S_B^2 = \text{mean square due to factor B}$$

$$MSC = \frac{SSC}{c-1} = S_C^2 = \text{mean square due to factor C}$$

$$MS(AB) = \frac{SS(AB)}{(a-1)(b-1)}$$
$$= S_{AB}^2 = \text{mean square due to AB interaction}$$

$$MS(AC) = \frac{SS(AC)}{(a-1)(c-1)}$$
$$= S_{AC}^2 = \text{mean square due to AC interaction}$$

$$MS(BC) = \frac{SS(BC)}{(b-1)(c-1)}$$
$$= S_{BC}^2 = \text{mean square due to BC interaction}$$

$$MS(ABC) = \frac{SS(ABC)}{(a-1)(b-1)(c-1)}$$
$$= S_{ABC}^2 = \text{mean square due to ABC interaction}$$

$$MSE = \frac{SSE}{abc(n-1)} = S^2 = \text{mean square due to error}$$

Mean square	
Expected value	
Fixed model	Random model
MSA	
$\sigma^2 + bcn\dfrac{\displaystyle\sum_{i=1}^{a} \alpha_i^2}{a-1}$	$\sigma^2 + bcn\sigma_\alpha^2 + cn\sigma_{\alpha\beta}^2 + bn\sigma_{\alpha\gamma}^2 + n\sigma_{\alpha\beta\gamma}^2$
MSB	
$\sigma^2 + acn\dfrac{\displaystyle\sum_{j=1}^{b} \beta_j^2}{b-1}$	$\sigma^2 + acn\sigma_\beta^2 + cn\sigma_{\alpha\beta}^2 + an\sigma_{\beta\gamma}^2 + n\sigma_{\alpha\beta\gamma}^2$
MSC	
$\sigma^2 + abn\dfrac{\displaystyle\sum_{k=1}^{c} \gamma_k^2}{c-1}$	$\sigma^2 + abn\sigma_\gamma^2 + bn\sigma_{\alpha\gamma}^2 + an\sigma_{\beta\gamma}^2 + n\sigma_{\alpha\beta\gamma}^2$
MS(AB)	
$\sigma^2 + cn\dfrac{\displaystyle\sum_{i=1}^{a}\sum_{j=1}^{b} (\alpha\beta)_{ij}^2}{(a-1)(b-1)}$	$\sigma^2 + cn\sigma_{\alpha\beta}^2 + n\sigma_{\alpha\beta\gamma}^2$
MS(AC)	
$\sigma^2 + bn\dfrac{\displaystyle\sum_{i=1}^{a}\sum_{k=1}^{c} (\alpha\gamma)_{ik}^2}{(a-1)(c-1)}$	$\sigma^2 + bn\sigma_{\alpha\gamma}^2 + n\sigma_{\alpha\beta\gamma}^2$
MS(BC)	
$\sigma^2 + an\dfrac{\displaystyle\sum_{j=1}^{b}\sum_{k=1}^{c} (\beta\gamma)_{jk}^2}{(b-1)(c-1)}$	$\sigma^2 + an\sigma_{\beta\gamma}^2 + n\sigma_{\alpha\beta\gamma}^2$
MS(ABC)	
$\sigma^2 + n\dfrac{\displaystyle\sum_{i=1}^{a}\sum_{j=1}^{b}\sum_{k=1}^{c} (\alpha\beta\gamma)_{ijk}^2}{(a-1)(b-1)(c-1)}$	$\sigma^2 + n\sigma_{\alpha\beta\gamma}^2$
MSE	
σ^2	σ^2

Mean square	
Expected value	
Mixed model (α)	Mixed model (α, β)
MSA	
$\sigma^2 + bcn\dfrac{\sum_{i=1}^{a}\alpha_i^2}{a-1} + cn\sigma_{\alpha\beta}^2$ $+ bn\sigma_{\alpha\gamma}^2 + n\sigma_{\alpha\beta\gamma}^2$	$\sigma^2 + bcn\dfrac{\sum_{i=1}^{a}\alpha_i^2}{a-1} + bn\sigma_{\alpha\gamma}^2$
MSB	
$\sigma^2 + acn\sigma_{\beta}^2 + an\sigma_{\beta\gamma}^2$	$\sigma^2 + acn\dfrac{\sum_{j=1}^{b}\beta_j^2}{b-1} + an\sigma_{\beta\gamma}^2$
MSC	
$\sigma^2 + abn\sigma_{\gamma}^2 + an\sigma_{\beta\gamma}^2$	$\sigma^2 + abn\sigma_{\gamma}^2$
MS(AB)	
$\sigma^2 + cn\sigma_{\alpha\beta}^2 + n\sigma_{\alpha\beta\gamma}^2$	$\sigma^2 + cn\dfrac{\sum_{i=1}^{a}\sum_{j=1}^{b}(\alpha\beta)_{ij}^2}{(a-1)(b-1)} + n\sigma_{\alpha\beta\gamma}^2$
MS(AC)	
$\sigma^2 + bn\sigma_{\alpha\gamma}^2 + n\sigma_{\alpha\beta\gamma}^2$	$\sigma^2 + bn\sigma_{\alpha\gamma}^2$
MS(BC)	
$\sigma^2 + an\sigma_{\beta\gamma}^2$	$\sigma^2 + an\sigma_{\beta\gamma}^2$
MS(ABC)	
$\sigma^2 + n\sigma_{\alpha\beta\gamma}^2$	$\sigma^2 + n\sigma_{\alpha\beta\gamma}^2$
MSE	
σ^2	σ^2

The following statistics have F distributions with the stated degrees of freedom.

Statistic	Numerator df	Denominator df
S_A^2/S^2	$a-1$	$abc(n-1)$
S_B^2/S^2	$b-1$	$abc(n-1)$
S_C^2/S^2	$c-1$	$abc(n-1)$
S_{AB}^2/S^2	$(a-1)(b-1)$	$abc(n-1)$
S_{AC}^2/S^2	$(a-1)(c-1)$	$abc(n-1)$
S_{BC}^2/S^2	$(b-1)(c-1)$	$abc(n-1)$
S_{ABC}^2/S^2	$(a-1)(b-1)(c-1)$	$abc(n-1)$

12.3.4   Analysis of variance table

Source of variation	Sum of squares	Degrees of freedom	Mean square	Computed F
Factor A	SSA	$a-1$	MSA	$\dfrac{\text{MSA}}{\text{MSE}}$
Factor B	SSB	$b-1$	MSB	$\dfrac{\text{MSB}}{\text{MSE}}$
Factor C	SSC	$c-1$	MSC	$\dfrac{\text{MSC}}{\text{MSE}}$
A × B	SS(AB)	$(a-1)(b-1)$	MS(AB)	$\dfrac{\text{MS(AB)}}{\text{MSE}}$
A × C	SS(AC)	$(a-1)(c-1)$	MS(AC)	$\dfrac{\text{MS(AC)}}{\text{MSE}}$
B × C	SS(BC)	$(b-1)(c-1)$	MS(BC)	$\dfrac{\text{MS(BC)}}{\text{MSE}}$
A × B × C	SS(ABC)	$(a-1)(b-1)(c-1)$	MS(ABC)	$\dfrac{\text{MS(ABC)}}{\text{MSE}}$
Error	SSE	$abc(n-1)$	MSE	
Total	SST	$abcn-1$		

Hypothesis tests:

(1) Test for significant factor A main effect

    H_0: There is no effect due to factor A

        (Fixed effects model: $\alpha_1 = \alpha_2 = \cdots = \alpha_a = 0$)

        (Random effects model: $\sigma_\alpha^2 = 0$)

        (Mixed effects model (α): $\alpha_1 = \alpha_2 = \cdots = \alpha_a = 0$)

        (Mixed effects model (α, β): $\alpha_1 = \alpha_2 = \cdots = \alpha_a = 0$)

    H_a: There is an effect due to factor A

        (Fixed effects model: $\alpha_i \neq 0$ for some i)

        (Random effects model: $\sigma_\alpha^2 \neq 0$)

        (Mixed effects model (α): $\alpha_i \neq 0$ for some i)

        (Mixed effects model (α, β): $\alpha_i \neq 0$ for some i)

    TS: $F = S_{\text{A}}^2 / S^2$

    RR: $F \geq F_{\alpha, a-1, abc(n-1)}$

(2) Test for significant factor B main effect

H_0: There is no effect due to factor B
    (Fixed effects model: $\beta_1 = \beta_2 = \cdots = \beta_b = 0$)
    (Random effects model: $\sigma_\beta^2 = 0$)
    (Mixed effects model (α): $\sigma_\beta^2 = 0$)
    (Mixed effects model (α, β): $\beta_1 = \beta_2 = \cdots = \beta_b = 0$)

H_a: There is an effect due to factor B
    (Fixed effects model: $\beta_j \neq 0$ for some j)
    (Random effects model: $\sigma_\beta^2 \neq 0$)
    (Mixed effects model (α): $\sigma_\beta^2 \neq 0$)
    (Mixed effects model (α, β): $\beta_j \neq 0$ for some j)

TS: $F = S_{\mathrm{B}}^2 / S^2$

RR: $F \geq F_{\alpha, b-1, abc(n-1)}$

(3) Test for significant factor C main effect

H_0: There is no effect due to factor C
    (Fixed effects model: $\gamma_1 = \gamma_2 = \cdots = \gamma_c = 0$)
    (Random effects model: $\sigma_\gamma^2 = 0$)
    (Mixed effects model (α): $\sigma_\gamma^2 = 0$)
    (Mixed effects model (α, β): $\sigma_\gamma^2 = 0$)

H_a: There is an effect due to factor C
    (Fixed effects model: $\gamma_k \neq 0$ for some k)
    (Random effects model: $\sigma_\gamma^2 \neq 0$)
    (Mixed effects model (α): $\sigma_\gamma^2 \neq 0$)
    (Mixed effects model (α, β): $\sigma_\gamma^2 \neq 0$)

TS: $F = S_{\mathrm{C}}^2 / S^2$

RR: $F \geq F_{\alpha, c-1, abc(n-1)}$

(4) Test for significant AB interaction effect

H_0: There is no effect due to interaction
    (Fixed effects model: $(\alpha\beta)_{11} = \cdots = (\alpha\beta)_{ab} = 0$)
    (Random effects model: $\sigma_{\alpha\beta}^2 = 0$)
    (Mixed effects model (α): $\sigma_{\alpha\beta}^2 = 0$)
    (Mixed effects model (α, β): $(\alpha\beta)_{11} = \cdots = (\alpha\beta)_{ab} = 0$)

H_a: There is an effect due to AB interaction
    (Fixed effects model: $(\alpha\beta)_{ij} \neq 0$ for some ij)
    (Random effects model: $\sigma_{\alpha\beta}^2 \neq 0$)
    (Mixed effects model (α): $\sigma_{\alpha\beta}^2 \neq 0$)
    (Mixed effects model (α, β): $(\alpha\beta)_{ij} \neq 0$ for some ij)

TS: $F = S_{\mathrm{AB}}^2 / S^2$

RR: $F \geq F_{\alpha, (a-1)(b-1), abc(n-1)}$

(5) Test for significant AC interaction effect

H_0: There is no effect due to interaction
   (Fixed effects model: $(\alpha\gamma)_{11} = \cdots = (\alpha\gamma)_{ac} = 0$)
   (Random effects model: $\sigma^2_{\alpha\gamma} = 0$)
   (Mixed effects model (α): $\sigma^2_{\alpha\gamma} = 0$)
   (Mixed effects model (α, β): $\sigma^2_{\alpha\gamma} = 0$)

H_a: There is an effect due to AC interaction
   (Fixed effects model: $(\alpha\gamma)_{ik} \neq 0$ for some ik)
   (Random effects model: $\sigma^2_{\alpha\gamma} \neq 0$)
   (Mixed effects model (α): $\sigma^2_{\alpha\gamma} \neq 0$)
   (Mixed effects model (α, β): $\sigma^2_{\alpha\gamma} \neq 0$)

TS: $F = S^2_{\mathrm{AC}}/S^2$

RR: $F \geq F_{\alpha,(a-1)(c-1),abc(n-1)}$

(6) Test for significant BC interaction effect

H_0: There is no effect due to interaction
   (Fixed effects model: $(\beta\gamma)_{11} = \cdots = (\beta\gamma)_{bc} = 0$)
   (Random effects model: $\sigma^2_{\beta\gamma} = 0$)
   (Mixed effects model (α): $\sigma^2_{\beta\gamma} = 0$)
   (Mixed effects model (α, β): $\sigma^2_{\beta\gamma} = 0$)

H_a: There is an effect due to BC interaction
   (Fixed effects model: $(\beta\gamma)_{jk} \neq 0$ for some jk
   (Random effects model: $\sigma^2_{\beta\gamma} \neq 0$)
   (Mixed effects model (α): $\sigma^2_{\beta\gamma} \neq 0$)
   (Mixed effects model (α, β): $\sigma^2_{\beta\gamma} \neq 0$)

TS: $F = S^2_{\mathrm{BC}}/S^2$

RR: $F \geq F_{\alpha,(b-1)(c-1),abc(n-1)}$

(7) Test for significant ABC interaction effect

H_0: There is no effect due to interaction
   (Fixed effects model: $(\alpha\beta\gamma)_{111} = \cdots = (\alpha\beta\gamma)_{abc} = 0$)
   (Random effects model: $\sigma^2_{\alpha\beta\gamma} = 0$)
   (Mixed effects model (α): $\sigma^2_{\alpha\beta\gamma} = 0$)
   (Mixed effects model (α, β): $\sigma^2_{\alpha\beta\gamma} = 0$)

H_a: There is an effect due to BC interaction
   (Fixed effects model: $(\alpha\beta\gamma)_{ijk} \neq 0$ for some ijk)
   (Random effects model: $\sigma^2_{\alpha\beta\gamma} \neq 0$)
   (Mixed effects model (α): $\sigma^2_{\alpha\beta\gamma} \neq 0$)
   (Mixed effects model (α, β): $\sigma^2_{\alpha\beta\gamma} \neq 0$)

TS: $F = S^2_{\mathrm{ABC}}/S^2$

RR: $F \geq F_{\alpha,(a-1)(b-1)(c-1),abc(n-1)}$

12.4 MANOVA

Manova means *multiple anova*, used if there are multiple dependent variables to be analyzed simultaneously.

The use of repeated measurement is a subset of manova. Using multiple one-way anovas to do this will raise the probability of a Type I error. Manova controls the experiment-wide error rate. (While it may seem that several simultaneous anovas raise power, the Type I error rate increases also.)

Manova assumptions:

(a) Usual anova assumptions (normality, independence, HOV).

(b) Linearity or multicollinearity of dependent variables.

(c) Manova does not have the compound symmetry requirement that the one-factor repeated measures anova model requires.

Manova advantages:

(a) Manova is a "gateway" test. If the multivariate F test is significant, then individual univariate analyses may be considered.

(b) Manova may be used with assorted dependent variables, or with repeated measures. This is an important feature of the model if the factors cannot be collapsed because they are all essentially different.

(c) Manova may detect combined differences not found by univariate analyses if there is multicollinearity (a linear combination of the dependent variables).

Manova limitations:

(a) Manova may be very sensitive to outliers, for small sample sizes.

(b) Manova assumes a linear relationship between dependent variables.

(c) Manova cannot give the interaction effects between the main effect and a repeated factor.

12.5 FACTOR ANALYSIS

The purpose of factor analysis is to examine the covariance, or correlation, relationships among all of the variables. This technique is used to group variables that tend to move together into an unobservable, random quantity called a *factor*.

Suppose highly correlated observable variables are grouped together, i.e., grouped by correlations. Variables within a group tend to move together and have very little correlation with variables outside their group. It is possible that each group of variables may be represented by, and depend on, a single, unobserved factor. Factor analysis attempts to discover this model structure so that each factor has a large correlation with a few variables and little correlation with the remaining variables.

Let $\mathbf{X}$ be a $p \times 1$ random vector with mean $\boldsymbol{\mu}$ and variance-covariance matrix $\boldsymbol{\Sigma}$. Assume $\mathbf{X}$ is a linear function of a set of unobservable factors, $F_1, F_2, \ldots, F_m$, and p error terms, $\epsilon_1, \epsilon_2, \ldots, \epsilon_p$. The factor analysis model may be written as

$$
\begin{aligned}
X_1 - \mu_1 &= \ell_{11} F_1 + \ell_{12} F_2 + \cdots + \ell_{1m} F_m + \epsilon_1 \\
X_2 - \mu_2 &= \ell_{21} F_1 + \ell_{22} F_2 + \cdots + \ell_{2m} F_m + \epsilon_2 \\
&\vdots \qquad\qquad\qquad\qquad \vdots \\
X_p - \mu_p &= \ell_{p1} F_1 + \ell_{p2} F_2 + \cdots + \ell_{pm} F_m + \epsilon_p
\end{aligned}
\tag{12.8}
$$

In matrix notation, the factor analysis model may be written as

$$
\begin{array}{ccccc}
\mathbf{X} - \boldsymbol{\mu} = & \mathbf{L} & \mathbf{F} & + & \boldsymbol{\epsilon} \\
(p \times 1) & (p \times m) & (m \times 1) & & (p \times 1)
\end{array}
\tag{12.9}
$$

where $\mathbf{L}$ is the matrix of factor loadings and ℓ_{ij} is the loading of the ith variable on the jth factor. There are additional model assumptions involving the unobservable random vectors $\mathbf{F}$ and $\boldsymbol{\epsilon}$:

(1) $\mathbf{F}$ and $\boldsymbol{\epsilon}$ are independent.

(2) $\mathrm{E}\,[\mathbf{F}] = \mathbf{0}$, $\mathrm{Cov}\,[\mathbf{F}] = \mathbf{I}$.

(3) $\mathrm{E}\,[\boldsymbol{\epsilon}] = \mathbf{0}$, $\mathrm{Cov}\,[\boldsymbol{\epsilon}] = \boldsymbol{\Psi}$, where $\boldsymbol{\Psi}$ is a diagonal matrix.

The orthogonal factor analysis model with m common factors (equation (12.8) and these assumptions) implies the following covariance structure for the random vector $\mathbf{X}$:

(1) $\mathrm{Cov}\,[\mathbf{X}] = \mathbf{L}\mathbf{L}^{\mathrm{T}} + \boldsymbol{\Psi}$ or

$$
\mathrm{Var}\,[X_i] = \ell_{i1}^2 + \cdots + \ell_{im}^2 + \Psi_i
\tag{12.10}
$$

$$
\mathrm{Cov}\,[X_i, X_k] = \ell_{i1}\ell_{k1} + \cdots + \ell_{im}\ell_{km}
\tag{12.11}
$$

(2) $\mathrm{Cov}\,[\mathbf{X}, \mathbf{F}] = \mathbf{L}$, or $\mathrm{Cov}\,[X_i, F_j] = \ell_{ij}$

The variance of the ith variable, σ_{ii}, is the sum of two terms: the ith *communality* and the *specific variance*.

$$
\underbrace{\sigma_{ii}}_{\mathrm{Var}[X_i]} = \underbrace{\ell_{i1}^2 + \ell_{i2}^2 + \cdots + \ell_{im}^2}_{\text{communality}} + \underbrace{\Psi_i}_{\text{specific variance}}
\tag{12.12}
$$

Factor analysis is most useful when the number of unobserved factors, m, is small relative to the number of observed random variables, p. The objective of the factor analysis model is to provide a simpler explanation for the relationships in $\mathbf{X}$ rather than referring to the complete variance-covariance $\boldsymbol{\Sigma}$. A problem with this procedure is that most variance-covariance matrices cannot be written as in equation (12.11) with m much less than p.

If $m > 1$, there are additional conditions necessary in order to obtain unique estimates of $\mathbf{L}$ and $\boldsymbol{\Psi}$. The estimate of the loading matrix $\mathbf{L}$ is determined only up to an orthogonal (rotation) matrix. The rotation matrix is usually constructed so that the model may be realistically interpreted.

There are two common procedures used to estimate the parameters ℓ_{ij} and Ψ_i: the method of principal components and the method of maximum likelihood. Each of these solutions may be rotated in order to more appropriately interpret the model. See, for example, R. A. Johnson and D. W. Wichern, *Applied Multivariate Statistical Analysis*, Fourth Edition, Prentice-Hall, Inc., Upper Saddle River, NJ, 1998.

12.6   LATIN SQUARE DESIGN

12.6.1   Models and assumptions

Let $Y_{ij(k)}$ be the random observation corresponding to the ith row, the jth column, and the kth treatment. The parentheses in the subscripts are used to denote the one value k assumes for each ij combination: $i, j, k = 1, 2, \ldots, r$. It is assumed there are no interactions among these three factors.

Fixed effects experiment:

Model: $\qquad Y_{ij(k)} = \mu + \alpha_i + \beta_j + \gamma_k + \epsilon_{ij(k)}$

Assumptions: $\epsilon_{ij(k)} \overset{\text{ind}}{\sim} N(0, \sigma^2), \quad \sum_{i=1}^{r} \alpha_i = \sum_{j=1}^{r} \beta_j = \sum_{k=1}^{r} \gamma_k = 0$

Random effects experiment:

Model: $\qquad Y_{ij(k)} = \mu + A_i + B_j + C_k + \epsilon_{ij(k)}$

Assumptions:

$\epsilon_{ij(k)} \overset{\text{ind}}{\sim} N(0, \sigma^2), \quad A_i \overset{\text{ind}}{\sim} N(0, \sigma_\alpha^2), \quad B_j \overset{\text{ind}}{\sim} N(0, \sigma_\beta^2), \quad C_k \overset{\text{ind}}{\sim} N(0, \sigma_\gamma^2)$

The $\epsilon_{ij(k)}$'s are independent of the other random components.

Mixed effects experiment (γ):

Model: $\qquad Y_{ij(k)} = \mu + A_i + B_j + \gamma_k + \epsilon_{ij(k)}$

Assumptions:

$\epsilon_{ij(k)} \overset{\text{ind}}{\sim} N(0, \sigma^2), \quad A_i \overset{\text{ind}}{\sim} N(0, \sigma_\alpha^2), \quad B_j \overset{\text{ind}}{\sim} N(0, \sigma_\beta^2), \quad \sum_{k=1}^{r} \gamma_k = 0$

The $\epsilon_{ij(k)}$'s are independent of the other random components.

Mixed effects experiment (α, γ):

Model: $\qquad Y_{ij(k)} = \mu + \alpha_i + B_j + \gamma_k + \epsilon_{ij(k)}$

Assumptions:

$\epsilon_{ij(k)} \overset{\text{ind}}{\sim} N(0, \sigma^2), \quad B_j \overset{\text{ind}}{\sim} N(0, \sigma_\beta^2), \quad \sum_{i=1}^{r} \alpha_i = \sum_{k=1}^{r} \gamma_k = 0$

The $\epsilon_{ij(k)}$'s and the B_j's are independent.

12.6.2   Sum of squares

Dots in the subscript of $\overline{y}$ and T indicate the mean and sum of $y_{ij(k)}$, respectively, over the appropriate subscript(s).

$$\text{SST} = \sum_{i=1}^{r}\sum_{j=1}^{r}(y_{ij(k)} - \overline{y}_{...})^2 = \sum_{i=1}^{r}\sum_{j=1}^{r} y_{ij(k)}^2 - \frac{T_{...}^2}{r^2}$$

$$\text{SSR} = r\sum_{i=1}^{r}(\overline{y}_{i..} - \overline{y}_{...})^2 = \frac{\sum_{i=1}^{r} T_{i..}^2}{r} - \frac{T_{...}^2}{r^2}$$

$$\text{SSC} = r\sum_{j=1}^{r}(\overline{y}_{.j.} - \overline{y}_{...})^2 = \frac{\sum_{j=1}^{r} T_{.j.}^2}{r} - \frac{T_{...}^2}{r^2}$$

$$\text{SSTr} = r\sum_{k=1}^{r}(\overline{y}_{..k} - \overline{y}_{...})^2 = \frac{\sum_{k=1}^{r} T_{..k}^2}{r} - \frac{T_{...}^2}{r^2}$$

$$\text{SSE} = \sum_{i=1}^{r}\sum_{j=1}^{r}(y_{ij(k)} - \overline{y}_{i..} - \overline{y}_{.j.} - \overline{y}_{..k} + 2\overline{y}_{...})^2$$

$$= \text{SST} - \text{SSR} - \text{SSC} - \text{SSTr}$$

12.6.3   Mean squares and properties

$$\text{MSR} = \frac{\text{SSR}}{r-1} \qquad = S_{\text{R}}^2 = \text{mean square due to rows}$$

$$\text{MSC} = \frac{\text{SSC}}{r-1} \qquad = S_{\text{C}}^2 = \text{mean square due to columns}$$

$$\text{MSTr} = \frac{\text{SSTr}}{r-1} \qquad = S_{\text{Tr}}^2 = \text{mean square due to treatments}$$

$$\text{MSE} = \frac{\text{SSE}}{(r-1)(r-2)} = S^2 \quad = \text{mean square due to error}$$

Mean square			
		Expected value	
Fixed model	Random model	Mixed model (γ)	Mixed model (α, γ)
MSR			
$\sigma^2 + r\dfrac{\sum\limits_{i=1}^{r} \alpha_i^2}{r-1}$	$\sigma^2 + r\sigma_\alpha^2$	$\sigma^2 + r\sigma_\alpha^2$	$\sigma^2 + r\dfrac{\sum\limits_{i=1}^{r} \alpha_i^2}{r-1}$
MSC			
$\sigma^2 + r\dfrac{\sum\limits_{j=1}^{r} \beta_j^2}{r-1}$	$\sigma^2 + r\sigma_\beta^2$	$\sigma^2 + r\sigma_\beta^2$	$\sigma^2 + r\sigma_\beta^2$
MSTr			
$\sigma^2 + r\dfrac{\sum\limits_{k=1}^{r} \gamma_k^2}{r-1}$	$\sigma^2 + r\sigma_\gamma^2$	$\sigma^2 + r\dfrac{\sum\limits_{k=1}^{r} \gamma_k^2}{r-1}$	$\sigma^2 + r\dfrac{\sum\limits_{k=1}^{r} \gamma_k^2}{r-1}$
MSE			
σ^2	σ^2	σ^2	σ^2

(1) $F = S_{\mathrm{R}}^2/S^2$ has an F distribution with $r-1$ and $(r-1)(r-2)$ degrees of freedom.

(2) $F = S_{\mathrm{C}}^2/S^2$ has an F distribution with $r-1$ and $(r-1)(r-2)$ degrees of freedom.

(3) $F = S_{\mathrm{Tr}}^2/S^2$ has an F distribution with $r-1$ and $(r-1)(r-2)$ degrees of freedom.

12.6.4 Analysis of variance table

Source of variation	Sum of squares	Degrees of freedom	Mean square	Computed F
Rows	SSR	$r-1$	MSA	$\dfrac{\text{MSR}}{\text{MSE}}$
Columns	SSC	$r-1$	MSB	$\dfrac{\text{MSC}}{\text{MSE}}$
Treatments	SSTr	$r-1$	MSC	$\dfrac{\text{MSTr}}{\text{MSE}}$
Error	SSE	$(r-1)(r-2)$	MSE	
Total	SST	r^2-1		

Hypothesis tests:

(1) Test for significant row effect

H_0: There is no effect due to rows
    (Fixed effects model: $\alpha_1 = \alpha_2 = \cdots = \alpha_r = 0$)
    (Random effects model: $\sigma_\alpha^2 = 0$)
    (Mixed effects model (γ): $\sigma_\alpha^2 = 0$)
    (Mixed effects model (α, γ): $\alpha_1 = \alpha_2 = \cdots = \alpha_r = 0$)

H_a: There is an effect due to rows
    (Fixed effects model: $\alpha_i \neq 0$ for some i)
    (Random effects model: $\sigma_\alpha^2 \neq 0$)
    (Mixed effects model (γ): $\sigma_\alpha^2 \neq 0$)
    (Mixed effects model (α, γ): $\alpha_i \neq 0$ for some i)

TS: $F = S_R^2 / S^2$

RR: $F \geq F_{\alpha, r-1, (r-1)(r-2)}$

(2) Test for significant column effect

H_0: There is no effect due to columns
    (Fixed effects model: $\beta_1 = \beta_2 = \cdots = \beta_r = 0$)
    (Random effects model: $\sigma_\beta^2 = 0$)
    (Mixed effects model (γ): $\sigma_\beta^2 = 0$)
    (Mixed effects model (α, γ): $\sigma_\beta^2 = 0$)

H_a: There is an effect due to columns
    (Fixed effects model: $\beta_j \neq 0$ for some j)
    (Random effects model: $\sigma_\beta^2 \neq 0$)
    (Mixed effects model (γ): $\sigma_\beta^2 \neq 0$)
    (Mixed effects model (α, γ): $\sigma_\beta^2 \neq 0$)

TS: $F = S_C^2 / S^2$

RR: $F \geq F_{\alpha, r-1, (r-1)(r-2)}$

(3) Test for significant treatment effect

H_0: There is no effect due to treatments
    (Fixed effects model: $\gamma_1 = \gamma_2 = \cdots = \gamma_r = 0$)
    (Random effects model: $\sigma_\gamma^2 = 0$)
    (Mixed effects model (γ): $\gamma_1 = \gamma_2 = \cdots = \gamma_r = 0$)
    (Mixed effects model (α, γ): $\gamma_1 = \gamma_2 = \cdots = \gamma_r = 0$)

H_a: There is an effect due to treatments
    (Fixed effects model: $\gamma_k \neq 0$ for some k)
    (Random effects model: $\sigma_\gamma^2 \neq 0$)
    (Mixed effects model (γ): $\gamma_k \neq 0$ for some k)
    (Mixed effects model (α, γ): $\gamma_k \neq 0$ for some k)

TS: $F = S_{Tr}^2 / S^2$

RR: $F \geq F_{\alpha, r-1, (r-1)(r-2)}$

CHAPTER 13

Experimental Design

Contents

13.1 LATIN SQUARES

A *Latin square of order n* is an $n \times n$ array in which each cell contains a single element from an n-set, such that each element occurs exactly once in each row and exactly once in each column. A Latin square is in standard form if in the first row and column the elements occur in natural order. The number of Latin squares in standard form are:

n	1	2	3	4	5	6	7	8
number	1	1	1	4	56	9,408	16,942,080	535,281,401,856

The unique Latin squares of order 1 and 2 are $\boxed{\text{A}}$ and $\begin{array}{|cc|}\hline \text{A} & \text{B} \\ \text{B} & \text{A} \\ \hline\end{array}$. There are 4 Latin squares of order 4.

4×4

3 × 3

$$\begin{array}{|ccc|}\hline A & B & C \\ B & C & A \\ C & A & B \\ \hline\end{array}$$

a

$$\begin{array}{|cccc|}\hline A & B & C & D \\ B & A & D & C \\ C & D & B & A \\ D & C & A & B \\ \hline\end{array}$$

b

$$\begin{array}{|cccc|}\hline A & B & C & D \\ B & C & D & A \\ C & D & A & B \\ D & A & B & C \\ \hline\end{array}$$

c

$$\begin{array}{|cccc|}\hline A & B & C & D \\ B & D & A & C \\ C & A & D & B \\ D & C & B & A \\ \hline\end{array}$$

d

$$\begin{array}{|cccc|}\hline A & B & C & D \\ B & A & D & C \\ C & D & A & B \\ D & C & B & A \\ \hline\end{array}$$

323

5 × 5

$$
\begin{array}{ccccc}
A & B & C & D & E \\
B & A & E & C & D \\
C & D & A & E & B \\
D & E & B & A & C \\
E & C & D & B & A \\
\end{array}
$$

6 × 6

$$
\begin{array}{cccccc}
A & B & C & D & E & F \\
B & F & D & C & A & E \\
C & D & E & F & B & A \\
D & A & F & E & C & B \\
E & C & A & B & F & D \\
F & E & B & A & D & C \\
\end{array}
$$

7 × 7

$$
\begin{array}{ccccccc}
A & B & C & D & E & F & G \\
B & C & D & E & F & G & A \\
C & D & E & F & G & A & B \\
D & E & F & G & A & B & C \\
E & F & G & A & B & C & D \\
F & G & A & B & C & D & E \\
G & A & B & C & D & E & F \\
\end{array}
$$

8 × 8

$$
\begin{array}{cccccccc}
A & B & C & D & E & F & G & H \\
B & C & D & E & F & G & H & A \\
C & D & E & F & G & H & A & B \\
D & E & F & G & H & A & B & C \\
E & F & G & H & A & B & C & D \\
F & G & H & A & B & C & D & E \\
G & H & A & B & C & D & E & F \\
H & A & B & C & D & E & F & G \\
\end{array}
$$

9 × 9

$$
\begin{array}{ccccccccc}
A & B & C & D & E & F & G & H & I \\
B & C & D & E & F & G & H & I & A \\
C & D & E & F & G & H & I & A & B \\
D & E & F & G & H & I & A & B & C \\
E & F & G & H & I & A & B & C & D \\
F & G & H & I & A & B & C & D & E \\
G & H & I & A & B & C & D & E & F \\
H & I & A & B & C & D & E & F & G \\
I & A & B & C & D & E & F & G & H \\
\end{array}
$$

10 × 10

$$
\begin{array}{cccccccccc}
A & B & C & D & E & F & G & H & I & J \\
B & C & D & E & F & G & H & I & J & A \\
C & D & E & F & G & H & I & J & A & B \\
D & E & F & G & H & I & J & A & B & C \\
E & F & G & H & I & J & A & B & C & D \\
F & G & H & I & J & A & B & C & D & E \\
G & H & I & J & A & B & C & D & E & F \\
H & I & J & A & B & C & D & E & F & G \\
I & J & A & B & C & D & E & F & G & H \\
J & A & B & C & D & E & F & G & H & I \\
\end{array}
$$

11 × 11

$$
\begin{array}{ccccccccccc}
A & B & C & D & E & F & G & H & I & J & K \\
B & C & D & E & F & G & H & I & J & K & A \\
C & D & E & F & G & H & I & J & K & A & B \\
D & E & F & G & H & I & J & K & A & B & C \\
E & F & G & H & I & J & K & A & B & C & D \\
F & G & H & I & J & K & A & B & C & D & E \\
G & H & I & J & K & A & B & C & D & E & F \\
H & I & J & K & A & B & C & D & E & F & G \\
I & J & K & A & B & C & D & E & F & G & H \\
J & K & A & B & C & D & E & F & G & H & I \\
K & A & B & C & D & E & F & G & H & I & J \\
\end{array}
$$

13.2 GRAECO–LATIN SQUARES

Two Latin squares K and L of order n are *orthogonal* if $K(a,b) = K(c,d)$ and $L(a,b) = L(c,d)$ implies $a = c$ and $b = d$. Equivalently, all of the n^2 pairs $(K_{i,j}, L_{i,j})$ are distinct. A pair of orthogonal Latin squares are called Graeco–Latin squares. There is a pair of orthogonal Latin squares of order n for all $n > 1$ except $n = 2$ or 6.

A set of Latin squares $L_1, \ldots, L_m$ are *mutually orthogonal* if for every $1 \leq i < j \leq m$, the Latin squares L_i and L_j are orthogonal. Three mutually orthogonal Latin squares of size 4 are

$$
\begin{bmatrix}
A & B & C & D \\
B & A & D & C \\
C & D & A & B \\
D & C & B & A
\end{bmatrix}
\quad
\begin{bmatrix}
1 & 3 & 4 & 2 \\
2 & 4 & 3 & 1 \\
3 & 1 & 2 & 4 \\
4 & 2 & 1 & 3
\end{bmatrix}
\quad
\begin{bmatrix}
a & d & b & c \\
b & c & a & d \\
c & b & d & a \\
d & a & c & b
\end{bmatrix}
\tag{13.1}
$$

3 × 3

A_1	B_3	C_2
B_2	C_1	A_3
C_3	A_2	B_1

4 × 4

A_1	B_3	C_4	D_2
B_2	A_4	D_3	C_1
C_3	D_1	A_2	B_4
D_4	C_2	B_1	A_3

5 × 5

A_1	B_3	C_5	D_2	E_4
B_2	C_4	D_1	E_3	A_5
C_3	D_5	E_2	A_4	B_1
D_4	E_1	A_3	B_5	C_2
E_5	A_2	B_4	C_1	D_3

There are no 6 × 6 Graeco–Latin squares

7 × 7

A_1	B_5	C_2	D_6	E_3	F_7	G_4
B_2	C_6	D_3	E_7	F_4	G_1	A_5
C_3	D_7	E_4	F_1	G_5	A_2	B_6
D_4	E_1	F_5	G_2	A_6	B_3	C_7
E_5	F_2	G_6	A_3	B_7	C_4	D_1
F_6	G_3	A_7	B_4	C_1	D_5	E_2
G_7	A_4	B_1	C_5	D_2	E_6	F_3

8 × 8

A_1	B_5	C_2	D_3	E_7	F_4	G_8	H_6
B_2	A_8	G_1	F_7	H_3	D_6	C_5	E_4
C_3	G_4	A_7	E_1	D_2	H_5	B_6	F_8
D_4	F_3	E_6	A_5	C_8	B_1	H_7	G_2
E_5	H_1	D_8	C_4	A_6	G_3	F_2	B_7
F_6	D_7	H_4	B_8	G_5	A_2	E_3	C_1
G_7	C_6	B_3	H_2	F_1	E_8	A_4	D_5
H_8	E_2	F_5	G_6	B_4	C_7	D_1	A_3

9 × 9

A_1	B_3	C_2	D_7	E_9	F_8	G_4	H_6	I_5
B_2	C_1	A_3	E_8	F_7	D_9	H_5	I_4	G_6
C_3	A_2	B_1	F_8	D_8	E_7	I_6	G_5	H_4
D_4	E_6	F_5	G_1	H_3	I_2	A_7	B_9	C_8
E_5	F_4	D_6	H_2	I_1	G_3	B_8	C_7	A_9
F_6	D_5	E_4	I_3	G_2	H_1	C_9	A_8	B_7
G_7	H_9	I_8	A_4	B_6	A_5	D_1	E_3	F_2
H_8	I_7	G_9	B_5	C_4	A_6	E_2	F_1	D_3
I_9	G_8	H_7	C_6	A_5	B_4	F_3	D_2	E_1

10 × 10

A_1	B_8	C_9	D_{10}	E_2	F_4	G_6	H_3	I_5	J_7
G_7	H_2	A_8	B_9	C_{10}	E_3	F_5	I_4	J_6	D_1
F_6	G_1	I_3	H_8	A_9	B_{10}	E_4	J_5	D_7	C_2
E_5	F_7	G_2	J_4	I_8	H_9	A_{10}	D_6	C_1	B_3
H_{10}	E_6	F_1	G_3	D_5	J_8	I_9	C_7	B_2	A_4
J_9	I_{10}	E_7	F_2	G_4	C_6	D_8	B_1	A_3	H_5
C_8	D_9	J_{10}	E_1	F_3	G_5	B_7	A_2	H_4	I_6
I_2	J_3	D_4	C_5	B_6	A_7	H_1	G_8	F_9	E_{10}
D_3	C_4	B_5	A_6	H_7	I_1	J_2	F_{10}	E_8	G_9
B_4	A_5	H_6	I_7	J_1	D_2	C_3	E_9	G_{10}	F_8

13.3 BLOCK DESIGNS

A *balanced incomplete block design* (BIBD) is a pair $(\mathcal{V}, \mathcal{B})$ where $\mathcal{V}$ is a v-set and $\mathcal{B}$ is a collection of b subsets of $\mathcal{V}$ (each subset containing k elements) such that each element of $\mathcal{V}$ is contained in exactly r blocks and any 2-subset of $\mathcal{V}$ is contained in exactly λ blocks. The numbers v, b, r, k, λ are parameters of the BIBD.

The parameters are necessarily related by $vr = bk$ and $r(k-1) = \lambda(v-1)$. BIBDs are usually described by specifying (v, k, λ); this is a (v, k, λ)-*design*. From these values $r = \frac{\lambda(v-1)}{k-1}$ and $b = \frac{vr}{k} = \frac{v\lambda(v-1)}{k(k-1)}$.

The complement of a design for $(\mathcal{V}, \mathcal{B})$ is a design for $(\mathcal{V}, \overline{\mathcal{B}})$ where $\overline{\mathcal{B}} = (V \backslash B \mid B \in \mathcal{B})$. The complement of a design with parameters (v, b, r, k, λ) is a design with parameters $(v, b, b-r, v-k, b-2r+\lambda)$. For this reason, tables are usually given for $v \geq 2k$ (the designs for $v < 2k$ are then obtained by taking complements).

The example designs given below are from Colbourn and Dinitz. To conserve space, designs are displayed in a $k \times b$ array in which each column contains the elements forming a block.

(a) The unique (6,3,2) design is
$$\begin{array}{cccccccccc}
0 & 0 & 0 & 0 & 0 & 1 & 1 & 1 & 2 & 2 \\
1 & 1 & 2 & 3 & 4 & 2 & 3 & 4 & 3 & 3 \\
2 & 3 & 4 & 5 & 5 & 5 & 4 & 5 & 4 & 5
\end{array}$$

Here there are $v = 6$ elements (numbered $0, 1, \ldots, 5$), $k = 3$ elements per block, and each pair is in $\lambda = 2$ blocks. The other parameters are $r = 5$ and $b = 10$.

As an illustration of how to interpret this design, note that the pair (0,1) appears in columns 1 and 2, and in no other columns. Note that the pair (0,2) appears in columns 1 and 3, and in no other columns, etc.

(b) One of the 4 nonisomorphic (7,3,2) designs is
$$\begin{array}{cccccccccccccc}
0 & 0 & 0 & 0 & 0 & 0 & 1 & 1 & 1 & 1 & 2 & 2 & 2 & 2 \\
1 & 1 & 3 & 3 & 5 & 5 & 3 & 3 & 4 & 4 & 3 & 3 & 4 & 4 \\
2 & 2 & 4 & 4 & 6 & 6 & 5 & 5 & 6 & 6 & 6 & 6 & 5 & 5
\end{array}$$

(c) One of the 10 nonisomorphic (7,3,3) designs is
$$\begin{array}{ccccccccccccccccccccc}
0 & 0 & 0 & 0 & 0 & 0 & 0 & 0 & 0 & 1 & 1 & 1 & 1 & 1 & 1 & 2 & 2 & 2 & 2 & 2 & 2 \\
1 & 1 & 1 & 3 & 3 & 3 & 5 & 5 & 5 & 3 & 3 & 3 & 4 & 4 & 4 & 3 & 3 & 3 & 4 & 4 & 4 \\
2 & 2 & 2 & 4 & 4 & 4 & 6 & 6 & 6 & 5 & 5 & 5 & 6 & 6 & 6 & 6 & 6 & 6 & 5 & 5 & 5
\end{array}$$

(d) One of the 4 nonisomorphic (8,4,3) designs is
$$\begin{array}{cccccccccccccc}
0 & 0 & 0 & 0 & 0 & 0 & 0 & 1 & 1 & 1 & 1 & 2 & 2 & 2 \\
1 & 1 & 1 & 2 & 3 & 3 & 4 & 2 & 3 & 3 & 4 & 3 & 3 & 4 \\
2 & 2 & 5 & 5 & 4 & 6 & 6 & 6 & 4 & 5 & 5 & 4 & 5 & 5 \\
3 & 4 & 6 & 7 & 5 & 7 & 7 & 7 & 6 & 7 & 7 & 7 & 6 & 6
\end{array}$$

(e) The unique (9,3,1) design is
$$\begin{array}{cccccccccccc}
0 & 0 & 0 & 0 & 1 & 1 & 1 & 2 & 2 & 2 & 3 & 6 \\
1 & 3 & 4 & 5 & 3 & 4 & 5 & 3 & 4 & 5 & 4 & 7 \\
2 & 6 & 8 & 7 & 8 & 7 & 6 & 7 & 6 & 8 & 5 & 8
\end{array}$$

(f) One of the 36 nonisomorphic (9,3,2) designs is
$$\begin{array}{cccccccccccccccccccccccc}
0 & 0 & 0 & 0 & 0 & 0 & 0 & 0 & 1 & 1 & 1 & 1 & 1 & 1 & 2 & 2 & 2 & 2 & 2 & 2 & 3 & 3 & 4 & 4 \\
1 & 1 & 3 & 3 & 5 & 5 & 7 & 7 & 3 & 3 & 4 & 4 & 6 & 6 & 3 & 3 & 4 & 4 & 5 & 5 & 6 & 6 & 5 & 5 \\
2 & 2 & 4 & 4 & 6 & 6 & 8 & 8 & 5 & 5 & 7 & 7 & 8 & 8 & 8 & 8 & 6 & 6 & 7 & 7 & 7 & 7 & 8 & 8
\end{array}$$

(g) One of the 11 nonisomorphic (9,4,3) designs is

```
0 0 0 0 0 0 0 0 1 1 1 1 1 2 2 2 2 3
1 1 1 2 3 3 5 6 2 3 3 4 4 3 3 4 4 4
2 2 5 5 4 4 7 7 7 5 5 6 6 6 6 5 5 5
3 4 6 6 7 8 8 8 8 7 8 7 8 7 8 7 8 6
```

(h) One of the 3 nonisomorphic (10,4,2) designs is

```
0 0 0 0 0 0 1 1 1 1 2 2 2 3 3
1 1 2 3 5 6 2 3 4 5 3 4 5 4 4
2 4 4 7 7 8 7 6 7 6 5 8 6 5 6
3 5 6 8 9 9 8 9 9 8 9 9 7 8 7
```

(i) The unique (11,5,2) design is

```
0 0 0 0 0 1 1 1 2 2 3
1 1 2 3 4 1 3 6 3 5 4
2 4 5 6 7 4 5 7 4 6 5
3 5 8 8 9 8 9 8 6 7 7
7 6 9 a a a a 9 9 a 8
```

(j) The unique (13,4,1) design is

```
0 0 0 0 1 1 1 2 2 3 3 4 5
1 2 4 6 2 5 7 3 6 4 7 8 9
3 8 5 a 4 6 b 5 7 6 8 9 a
9 c 7 b a 8 c b 9 c a b c
```

13.4 FACTORIAL EXPERIMENTATION: 2 FACTORS

If two factors A and B are to be investigated at a levels and b levels, respectively, and if there are ab experimental conditions (treatments) corresponding to all possible combinations of the levels of the two factors, the resulting experiment is called a *complete $a \times b$ factorial experiment*. Assume the entire set of ab experimental conditions are repeated r times.

Let y_{ijk} be the observation in the kth replicate taken at the ith factor of A and the jth factor of B. The model has the form

$$y_{ijk} = \mu + \alpha_i + \beta_j + (\alpha\beta)_{ij} + \rho_k + \epsilon_{ijk} \qquad (13.2)$$

for $i = 1, 2, \ldots, a$, $j = 1, 2, \ldots, b$, and $k = 1, 2, \ldots, r$. Here

- μ is the grand mean
- α_i is the effect of the ith level of factor A
- β_j is the effect of the jth level of factor B
- $(\alpha\beta)_{ij}$ is the *interaction effect* or *joint effect* of the ith level of factor A and the jth level of factor B
- ρ_k is the effect of the kth replicate
- ϵ_{ijk} are independent normally distributed random variables with mean zero and variance σ^2

The following conditions are also imposed

$$\sum_{i=1}^{a}\alpha_i = \sum_{j=1}^{b}\beta_j = \sum_{i=1}^{a}(\alpha\beta)_{ij} = \sum_{j=1}^{b}(\alpha\beta)_{ij} = \sum_{k=1}^{r}\rho_k = 0 \qquad (13.3)$$

In the usual way

$$\overbrace{\sum_{i=1}^{a}\sum_{j=1}^{b}\sum_{k=1}^{r}(y_{ijk}-\overline{y}_{...})^2}^{\text{SST}} = r\overbrace{\sum_{i=1}^{a}\sum_{j=1}^{b}(\overline{y}_{ij.}-\overline{y}_{...})^2}^{\text{SS(Tr)}}$$

$$+ ab\sum_{k=1}^{r}(\overline{y}_{..k}-\overline{y}_{...})^2 \tag{13.4}$$

$$+ \underbrace{\sum_{i=1}^{a}\sum_{j=1}^{b}\sum_{k=1}^{r}\left(y_{ijk}-\overline{y}_{ij.}-\overline{y}_{..k}+\overline{y}_{...}\right)^2}_{\text{SSE}}$$

SST is the *total sum of squares*, SS(Tr) is the *treatment sum of squares*, SSE is the *error sum of squares*.

The distinguishing feature of a factorial experiment is that the treatment sum of squares can be further subdivided into components corresponding to the various factorial effects. Here:

$$r\overbrace{\sum_{i=1}^{a}\sum_{j=1}^{b}(y_{ij.}-\overline{y}_{...})^2}^{\text{SS(Tr)}} = rb\overbrace{\sum_{i=1}^{a}(\overline{y}_{i..}-\overline{y}_{...})^2}^{\text{SS(A)}} + ra\overbrace{\sum_{j=1}^{b}(\overline{y}_{.j.}-\overline{y}_{...})^2}^{\text{SS(B)}}$$

$$+ r\underbrace{\sum_{i=1}^{a}\sum_{j=1}^{b}(\overline{y}_{ij.}-\overline{y}_{i..}-\overline{y}_{.j.}+\overline{y}_{...})^2}_{\text{SS(AB)}} \tag{13.5}$$

SS(A) is the *factor A sum of squares*, SS(B) is the *factor B sum of squares*, SS(AB) is the *interaction sum of squares*.

13.5 2^r FACTORIAL EXPERIMENTS

A factorial experiment in which there are r factors, each at only two levels, is a 2^r factorial experiment. The two levels are often denoted as *high* and *low*, or 0 and 1. A complete 2^r factorial experiment includes observations for every combination of factor and level, for a total of 2^r observations. A 2^3 factorial experiment has 8 treatment combinations, and the model is given by

$$Y_{ijkl} = \mu + \alpha_i + \beta_j + \gamma_k + (\alpha\beta)_{ij} + (\alpha\gamma)_{ik} + (\beta\gamma)_{jk}$$
$$+ (\alpha\beta\gamma)_{ijk} + \epsilon_{ijkl} \tag{13.6}$$

where $i = 0, 1$, $j = 0, 1$, $k = 0, 1$, and $l = 1, 2, \dots, n$. The assumptions are

$$\epsilon_{ij} \overset{\text{ind}}{\sim} N(0, \sigma^2), \qquad \alpha_1 = -\alpha_0, \qquad \beta_1 = -\beta_0, \qquad \gamma_1 = -\gamma_0,$$
$$(\alpha\beta)_{10} = (\alpha\beta)_{01} = -(\alpha\beta)_{11} = -(\alpha\beta)_{00}, \quad \dots \tag{13.7}$$

A 2^n factorial experiment requires 2^n experimental conditions. These conditions are listed in a standard order using a special notation. Factor A at the low level, or level 0, is denoted by "1", at the high level, or level 1, by "a". The levels of factor B are represented by "1" and "b", etc. In a 2^3 factorial experiment, the treatment combination that consists of high levels of factors A and C, and a low level of factor B, is denoted by ac. The treatment combination of all low levels is denoted simply by 1.

The treatment combinations are given by a binary expansion of the factor levels. For $n = 2$ the standard order of combinations is $\{1, a, b, ab\}$. For $n = 3$ the standard form is:

Experimental	Level of factor		
condition	A	B	C
1	0	0	0
a	1	0	0
b	0	1	0
ab	1	1	0
c	0	0	1
ac	1	0	1
bc	0	1	1
abc	1	1	1

The symbols for the first four experimental conditions are like those for a two–factor experiment, and the second four are obtained by multiplying each of the first four symbols by c.

Define $(1), (a), (b), (ab), (c), \ldots$, to be the treatment totals corresponding to the experimental conditions $1, a, b, ab, c, \ldots$. For example, in a 2^3 factorial experiment,

$$(1) = \sum_{\ell=1}^{r} y_{000\ell} \qquad (a) = \sum_{\ell=1}^{r} y_{100\ell}$$

$$\cdots \qquad\qquad\qquad\qquad (13.8)$$

$$(bc) = \sum_{\ell=1}^{r} y_{011\ell} \qquad (abc) = \sum_{\ell=1}^{r} y_{111\ell}$$

Certain linear combinations of these totals result in the sum of squares for the main effects and interaction effects. Define the **effect total** for factor A:

$$[A] = -(1) + (a) - (b) + (ab) - (c) + (ac) - (bc) + (abc). \qquad (13.9)$$

The sum of squares due to each factor may be obtained from its effect total. $[1]$ is the total effect.

The linear combination for each experimental condition effect total may be presented in a **table of signs** (a larger table is on page 338):

| Experimental | Effect total | | | | | | | |
condition	[1]	[A]	[B]	[AB]	[C]	[AC]	[BC]	[ABC]
1	+	−	−	+	−	+	+	−
a	+	+	−	−	−	−	+	+
b	+	−	+	−	−	+	−	+
ab	+	+	+	+	−	−	−	−
c	+	−	−	+	+	−	−	+
ac	+	+	−	−	+	+	−	−
bc	+	−	+	−	+	−	+	−
abc	+	+	+	+	+	+	+	+

To compute the sum of squares: $\text{SSA} = [A]^2/(8r)$, $\text{SSB} = [B]^2/(8r)$, $\text{SSC} = [C]^2/(8r)$, $\text{SS(AB)} = [AB]^2/(8r), \ldots$.

Note: The expression for a single effect total may be found by expanding an algebraic expression. Consider the effect total $[AB]$ in a 2^3 factorial experiment. Take the expression $(a \pm 1)(b \pm 1)(c \pm 1)$ and use a "−" if the corresponding letter appears in the symbol for the main effect, and use a "+" if the letter does not appear. Expand the expression and add parentheses. For $[AB]$ the calculation is given by

$$(a - 1)(b - 1)(c + 1) = abc - ac - bc + c + ab - a - b + 1$$

$$= (1) - (a) - (b) + (ab) + (c) - (ac) - (bc) + (abc) \qquad (13.10)$$

13.6 CONFOUNDING IN 2^N FACTORIAL EXPERIMENTS

Sometimes it is impossible to run all the required experiments in a single block. When experimental conditions are distributed over several blocks, one or more of the effects may become *confounded* (i.e., inseparable) with possible block effects, that is, between-block differences.

For example, in a 2^3 factorial experiment, let 1, b, c, and bc be in one block, and let a, ab, ac, and abc be in another block. The "block effect", the difference between the two block totals, is given by

$$[(a) + (ab) + (ac) + (abc)] - [(1) + (b) + (c) + (bc)]$$

This happens to be equal to $[A]$. Hence, the main effect of A is *confounded with blocks*.

Using instead a block of a, b, c, abc and a block of 1, ab, ac, bc would result in ABC being confounded with blocks.

If the number of blocks is 2^p then a total of $2^p - 1$ effects are confounded by blocks.

13.7 TABLES FOR DESIGN OF EXPERIMENTS

The following tables present combinatorial patterns that may be used as experimental designs. The plans and plan numbers are from a more numerous set of patterns in W. G. Cochran and G. M. Cox, *Experimental Designs*, Second Edition, John Wiley & Sons, Inc, New York, 1957. Reprinted by permission of John Wiley & Sons, Inc.

Plans 6.1–6.6 are plans of factorial experiments confounded in randomized incomplete blocks. Plans 6A.1–6A.6 are plans of 2^n factorials in fractional replication. Plans 13.1–13.5 are plans of incomplete block designs.

13.7.1 Plans of factorial experiments confounded in randomized incomplete blocks

Plan 6.1: 2^3 factorial, 4 unit blocks

Rep. I, ABC confounded

abc	ab
a	ac
b	bc
c	(1)

Plan 6.2: 2^4 factorial, 8 unit blocks

Rep. I, $ABCD$ confounded

a	(1)
b	ab
c	ac
d	bc
abc	ad
abd	bd
acd	cd
bcd	abcd

Plan 6.3: 2^6 factorial, 16 unit blocks

Rep. I, $ABCD$, $ABEF$, $CDEF$ confounded

a	c	ab	ac
b	d	cd	ad
acd	abc	(1)	bc
bcd	abd	abcd	bd
ce	ae	ace	abe
de	be	ade	cde
abce	acde	bce	e
abde	bcde	bde	abcde
cf	af	acf	abf
df	bf	adf	cdf
abcf	acdf	bcf	f
abdf	bcdf	bdf	abcdf
aef	cef	abef	acef
bef	def	cdef	adef
acdef	abcef	ef	bcef
bcdef	abdef	abcdef	bdef

Plan 6.4: Balanced group of sets for 2^4 factorial, 4 unit blocks

Two–factor interactions are confounded in 1 replication and three–factor interactions are confounded in 3 replications. The columns are the blocks.

Rep. I, AB, ACD,
    BCD confounded

(1)	ab	a	b
abc	c	bc	ac
abd	d	bd	ad
cd	abcd	acd	bcd

Rep. II, AC, ABD, BCD

(1)	ac	a	c
abc	b	bc	ab
acd	d	cd	ad
bd	abcd	abd	bcd

Rep. III, AD, ABC, BCD

(1)	ad	a	d
abd	b	bd	ab
acd	c	cd	ac
bc	abcd	abc	bcd

Rep. IV, BC, ABD, ACD

(1)	bc	b	c
abc	a	ac	ab
bcd	d	cd	bd
ad	abcd	abd	acd

Rep. V, BD, ABC, ACD

(1)	bd	b	d
abd	a	ad	ab
bcd	c	cd	bc
ac	abcd	abc	acd

Rep. VI, CD, ABC, ABD

(1)	cd	c	d
acd	a	ad	ac
bcd	b	bd	bc
ab	abcd	abc	abd

Plan 6.5: Balanced group of sets for 2^5 factorial, 8 unit blocks

Three– and four–factor interactions are confounded in 1 replication.

Rep. I, ABC, ADE, $BCDE$ confounded

(1)	ab	a	b
bc	ac	abc	c
abd	d	bd	ad
acd	bcd	be	abcd
abe	e	ce	ae
ace	bce	ade	abce
de	abde	abcde	bde
bcde	acde	cd	cde

Rep. II, ABD, BCE, $ACDE$

(1)	ab	a	b
ad	bd	d	abd
abc	c	bc	ac
bcd	acd	abcd	cd
abe	e	be	ae
bde	ade	abde	de
ce	abce	ace	bce
acde	bcde	cde	abcde

Rep. III, ACE, BCD, $ABDE$

(1)	ac	a	c
ae	ce	e	ace
abc	b	bc	ab
bce	abe	abce	be
acd	d	cd	ad
cde	ade	acde	de
bd	abcd	abd	bcd
abde	bcde	bde	abcde

Rep. IV, ACD, BDE, $ABCE$

(1)	ad	a	d
ac	cd	c	acd
abd	b	bd	ab
bcd	abc	abcd	bc
ade	e	de	ae
cde	ace	acde	ce
be	abde	abe	bde
abce	bcde	bce	abcde

Rep. V, ABE, CDE, $ABCD$

(1)	ae	a	e
ab	be	b	abe
ace	c	ce	ac
bce	abc	abce	bc
ade	d	de	ad
bde	abd	abde	bd
cd	acde	acd	cde
abcd	bcde	bcd	abcde

Plan 6.6: Balanced group of sets for 2^6 factorial, 8 unit blocks

All three– and four–factor interactions are confounded in 2 replications.

Rep. I, ABC, CDE, ADF, BEF, $ABDE$, $BCDF$, $ACEF$ confounded

abc	a	b	(1)	bc	ac	c	ab
bd	cd	abcd	acd	abd	d	ad	bcd
ae	abce	ce	bce	e	abe	be	ace
cde	bde	ade	abde	acde	bcde	abcde	de
cf	bf	af	abf	acf	bcf	abcf	f
adf	abcdf	cdf	bcdf	df	abdf	bdf	acdf
bef	cef	abcef	acef	abef	ef	aef	bcef
abcdef	adef	bdef	def	bcdef	acdef	cdef	abdef

Rep. II, ABD, DEF, BCF, ACE, $ABEF$, $ACDF$, $BCDE$

abd	b	a	(1)	ad	bd	d	ab
cd	ac	bc	abc	bcd	acd	abcd	c
be	abde	de	ade	e	abe	ae	bde
ace	cde	abcde	bcde	abce	ce	bce	acde
af	df	abdf	bdf	abf	f	bf	adf
bcf	abcdf	cdf	acdf	cf	abcf	acf	bcdf
def	aef	bef	abef	bdef	adef	abdef	ef
abcdef	bcef	acef	cef	acdef	bcdef	cdef	abcef

Rep. III, ABE, BDF, ACD, CEF, $ADEF$, $BCDE$, $ABCF$

bc	a	ac	(1)	abc	ab	b	c
acd	bd	bcd	abd	cd	d	ad	abcd
abe	cd	d	ace	be	bce	abce	ae
de	abcde	abde	bcde	ade	acde	cde	bde
af	bcf	bf	abcf	f	cf	acf	abf
bdf	acdf	adf	cdf	abdf	abcdf	bcdf	df
cef	abef	abcef	bef	acef	aef	ef	bcef
abcdef	def	cdef	adef	bcdef	bdef	abdef	acdef

Rep. IV, ABF, CDF, ADE, BCE, $ABCD$, $BDEF$, $ACEF$

ac	a	b	(1)	c	abc	bc	ab
bd	bcd	acd	abcd	abd	d	ad	cd
bce	be	ae	abe	abce	ce	ace	e
ade	acde	bcde	cde	de	abde	bde	abcde
abf	abcf	cf	bcf	bf	af	f	acf
cdf	df	abdf	adf	acdf	bcdf	abcdf	bdf
ef	cef	abcef	acef	aef	bef	abef	bcef
abcdef	abdef	def	bdef	bcdef	acdef	cdef	adef

Rep. V, ACF, BCD, ADE, BEF, $ABDF$, $CDEF$, $ABCE$

ab	a	bc	(1)	b	ac	c	abc
bcd	cd	abd	acd	abcd	d	ad	bd
ce	bce	ae	abce	ace	be	abe	e
ade	abde	cde	bde	de	abcde	bcde	acde
acf	abcf	f	bcf	cf	abf	bf	af
df	bdf	acdf	abdf	adf	bcdf	abcdf	cdf
bef	ef	abcef	aef	abef	cef	acef	bcef
abcdef	acdef	bdef	cdef	bcdef	adef	def	abdef

Note:

(1) Replication VI, $ABC, BDE, ADF, CEF, ACDE, BCDF, ABEF$.
    Interchange B and C in replication I.

(2) Replication VII, $ABF, DEF, BCD, ACE, ABDE, ACDF, BCEF$.
    Interchange F and D in replication II.

(3) Replication VIII, $ABE, BDF, CDE, ACF, ADEF, ABCD, BCEF$.
    Interchange A and E in replication III.

(4) Replication IX, $ABD, CDF, AEF, BCE, ABCF, BDEF, ACDE$.
    Interchange F and D in replication IV.

(5) Replication X, $AEF, BDE, ACD, BCF, ABDF, CDEF, ABCE$.
    Interchange E and C in replication V.

13.7.2   Plans of 2^n factorials in fractional replication

Plan 6A.1: 2^4 factorial in 8 units ($\frac{1}{2}$ replicate)

Defining contrast: $ABCD$

Estimable 2-factor interactions: $AB = CD$, $AC = BD$, $AD = BC$

(1)
ab
ac
ad
bc
bd
cd
abcd

Effect	df
Main	4
2-factor	3
Total	7

Plan 6A.2: 2^5 factorial in 8 units ($\frac{1}{4}$ replicate)

Defining contrast: $ABE, CDE, ABCD$

Main effects have 2-factors as aliases. The only estimable 2-factors are $AC = BD$ and $AD = BC$.

(1)
ab
cd
ace
bce
ade
bde
abcd

Effect	df
Main	5
2-factor	2
Total	7

Plan 6A.3: 2^5 factorial in 16 units ($\frac{1}{2}$ replicate)

Defining contrast: $ABCDE$

1. *Blocks of 4 units*
   Estimatable 2-factors: All except CD, CE, DE (confounded with blocks)

Blocks	(1)	(2)	(3)	(4)
	(1)	ac	ae	ad
	ab	bc	be	bd
	acde	de	cd	ce
	bcde	abde	abcd	abce

CD, CE, DE confounded

Effect	df
Block	3
Main	5
2-factor	7
Total	15

2. *Blocks of 8 units*

(a) Estimatable 2-factors: All except DE

(b) Combine blocks 1 and 2; and blocks 3 and 4. DE confounded.

Effect	df
Block	1
Main	5
2-factor	9
Total	15

3. *Blocks of 16 units*

(a) Estimatable 2-factors: All

(b) Combine blocks 1–4

Effect	df
Main	5
2-factor	10
Total	15

Plan 6A.4: 2^6 factorial in 8 units ($1/8$ replicate)

Defining contrasts: $ACE, ADF, BCF, BDE, ABCD, ABEF, CDEF$

Main effects have 2-factors as aliases. The only estimable 2-factor is the set $AB = CD = EF$

> (1)
> acf
> ade
> bce
> bdf
> $abcd$
> $abef$
> $cdef$

Effect	df
Main	6
2-factor ($AB = CD = EF$)	1
Total	7

Plan 6A.5: 2^6 factorial in 16 units ($1/4$ replicate)

Defining contrasts: $ABCE, ABDF, CDEF$

1. *Blocks of 4 units*

Estimatable 2-factors: The alias sets $AC = BD$, $AD = BF$, $AE = BC$, $AF = BD$, $CD = EF$, $CF = DE$

Blocks	(1)	(2)	(3)	(4)
	(1)	acd	ab	acf
	$abce$	aef	ce	ade
	$abdf$	bcf	df	bcd
	$cdef$	bde	$abcdef$	bef

AB, ACF, BCF confounded

Effect	df
Block	3
Main	6
2-factor	6
Total	15

2. *Blocks of 8 units*

(a) Estimatable 2-factors: Same as in blocks of 4 units, plus the set $AB = CE = DF$.

(b) Combine blocks 1 and 2; and blocks 3 and 4. ACF confounded.

Effect	df
Block	1
Main	6
2-factor	7
3-factor	1
Total	15

3. *Blocks of 16 units*

(a) Estimatable 2-factors: Same as in blocks of 8 units

(b) Combine blocks 1–4

Effect	df
Main	6
2-factor	7
3-factor	2
Total	15

Plan 6A.6: 2^6 factorial in 32 units ($1/2$ replicate)

Defining contrast: $ABCDEF$

1. *Blocks of 4 units*

   Estimatable 2-factors: All except AE, BF, and CD (confounded with blocks)

Blocks	(1)	(2)	(3)	(4)	(5)	(6)	(7)	(8)
	(1)	ab	ac	bc	ae	af	ad	bd
	$abef$	ef	de	df	bf	be	ce	cf
	$acde$	$acdf$	$abdf$	$acef$	cd	$abcd$	$abcf$	$abce$
	$bcdf$	$bcde$	$bcef$	$abde$	$abcdef$	$cdef$	$bdef$	$adef$

AE, BF, CD, ABD, ACF, ADF confounded

Effect	df
Block	7
Main	6
2-factor	12
Higher order	6
Total	31

2. *Blocks of 8 units*

   (a) Estimatable 2-factors: All except CD

   (b) Combine blocks 1 and 2; blocks 3 and 4; blocks 5 and 6; and blocks 7 and 8; CD, ABC, ABD confounded.

Effect	df
Block	3
Main	6
2-factor	14
Higher order	8
Total	31

3. *Blocks of 16 units*

   (a) Estimatable 2-factors: All

   (b) Estimatable 3-factors: $ABC = DEF$ is lost by confounding. The others are in alias pairs, e.g., $ABD = CEF$.

   (c) Combine blocks 1–4; and blocks 5–8. ABC confounded.

Effect	df
Block	1
Main	6
2-factor	15
3-factor	9
Total	31

4. *Blocks of 32 units*

   (a) Estimatable 2-factors: All

   (b) Estimatable 3-factors: These are arranged in 10 alias pairs.

   (c) Combine blocks 1–8.

Effect	df
Main	6
2-factor	15
3-factor	10
Total	31

13.7.3   Plans of incomplete block designs

Plan 13.1: $t = 7$, $k = 3$, $r = 3$, $b = 7$, $\lambda = 1$ $E = .78$, Type II

Block	Reps. I	II	III
(1)	7	1	3
(2)	1	2	4
(3)	2	3	5
(4)	3	4	6

Block	Reps. I	II	III
(5)	4	5	7
(6)	5	6	1
(7)	6	7	2

Plan 13.2: $t = 7, k = 4, r = 4, b = 7, \lambda = 2\ E = .88$, Type II

Block	Reps. I	II	III	IV
(1)	3	5	6	7
(2)	4	6	7	1
(3)	5	7	1	2
(4)	6	1	2	3

Block	Reps. I	II	III	IV
(5)	7	2	3	4
(6)	1	3	4	5
(7)	2	4	5	6

Plan 13.3: $t = 11, k = 5, r = 5, b = 11, \lambda = 2\ E = .88$, Type I

Block	Reps. I	II	III	IV	V
(1)	1	2	3	4	5
(2)	7	1	6	10	3
(3)	9	8	1	6	2
(4)	11	9	7	1	4
(5)	10	11	5	8	1
(6)	8	7	2	3	11

Block	Reps. I	II	III	IV	V
(7)	2	6	4	11	10
(8)	6	3	11	5	9
(9)	3	4	10	9	8
(10)	5	10	9	2	7
(11)	4	5	8	7	6

Plan 13.4: $t = 11, k = 6, r = 6, b = 11, \lambda = 3\ E = .92$, Type I

Block	Reps. I	II	III	IV	V	VI
(1)	6	7	8	9	10	11
(2)	5	8	4	11	2	9
(3)	4	5	7	3	11	10
(4)	3	10	2	6	5	8
(5)	2	3	9	7	4	6
(6)	1	6	10	4	9	5

Block	Reps. I	II	III	IV	V	VI
(7)	9	1	3	5	8	7
(8)	8	2	1	10	7	4
(9)	7	11	5	1	6	2
(10)	11	4	6	8	1	3
(11)	10	9	11	2	3	1

Plan 13.5: $t = 13, k = 4, r = 4, b = 13, \lambda = 1\ E = .81$, Type I

Block	Reps. I	II	III	IV
(1)	13	1	3	9
(2)	1	2	4	10
(3)	2	3	5	11
(4)	3	4	6	12
(5)	4	5	7	13
(6)	5	6	8	1
(7)	6	7	9	2

Block	Reps. I	II	III	IV
(8)	7	8	10	3
(9)	8	9	11	4
(10)	9	10	12	5
(11)	10	11	13	6
(12)	11	12	1	7
(13)	12	13	2	8

13.7.4   Main effect and interactions in factorial designs

Main effect and interactions in 2^2, 2^3, 2^4, 2^5, and 2^6 factorial designs

	(T)	A	B	AB	C	AC	BC	ABC	D	AD	BD	ABD	CD	ACD	BCD	ABCD	E	AE	BE	ABE	CE	ACE	BCE	ABCE	DE	ADE	BDE	ABDE	CDE	ACDE	BCDE	ABCDE
(1)	+	−	−	+	−	+	+	−	−	+	+	−	+	−	−	+	−	+	+	−	+	−	−	+	+	−	−	+	−	+	+	−
a	+	+	−	−	−	−	+	+	−	−	+	+	+	+	−	−	−	−	+	+	+	+	−	−	+	+	−	−	−	−	+	+
b	+	−	+	−	−	+	−	+	−	+	−	+	+	−	+	−	−	+	−	+	+	−	+	−	+	−	+	−	−	+	−	+
ab	+	+	+	+	−	−	−	−	−	−	−	−	+	+	+	+	−	−	−	−	+	+	+	+	+	+	+	+	−	−	−	−
c	+	−	−	+	+	−	−	+	−	+	+	−	−	+	+	−	−	+	+	−	−	+	+	−	+	−	−	+	+	−	−	+
ac	+	+	−	−	+	+	−	−	−	−	+	+	−	−	+	+	−	−	+	+	−	−	+	+	+	+	−	−	+	+	−	−
bc	+	−	+	−	+	−	+	−	−	+	−	+	−	+	−	+	−	+	−	+	−	+	−	+	+	−	+	−	+	−	+	−
abc	+	+	+	+	+	+	+	+	−	−	−	−	−	−	−	−	−	−	−	−	−	−	−	−	+	+	+	+	+	+	+	+
d	+	−	−	+	−	+	+	−	+	−	−	+	−	+	+	−	−	+	+	−	+	−	−	+	−	+	+	−	+	−	−	+
ad	+	+	−	−	−	−	+	+	+	+	−	−	−	−	+	+	−	−	+	+	+	+	−	−	−	−	+	+	+	+	−	−
bd	+	−	+	−	−	+	−	+	+	−	+	−	−	+	−	+	−	+	−	+	+	−	+	−	−	+	−	+	+	−	+	−
abd	+	+	+	+	−	−	−	−	+	+	+	+	−	−	−	−	−	−	−	−	+	+	+	+	−	−	−	−	+	+	+	+
cd	+	−	−	+	+	−	−	+	+	−	−	+	+	−	−	+	−	+	+	−	−	+	+	−	−	+	+	−	−	+	+	−
acd	+	+	−	−	+	+	−	−	+	+	−	−	+	+	−	−	−	−	+	+	−	−	+	+	−	−	+	+	−	−	+	+
bcd	+	−	+	−	+	−	+	−	+	−	+	−	+	−	+	−	−	+	−	+	−	+	−	+	−	+	−	+	−	+	−	+
abcd	+	+	+	+	+	+	+	+	+	+	+	+	+	+	+	+	−	−	−	−	−	−	−	−	−	−	−	−	−	−	−	−
e	+	−	−	+	−	+	+	−	−	+	+	−	+	−	−	+	+	−	−	+	−	+	+	−	−	+	+	−	+	−	−	+
ae	+	+	−	−	−	−	+	+	−	−	+	+	+	+	−	−	+	+	−	−	−	−	+	+	−	−	+	+	+	+	−	−
be	+	−	+	−	−	+	−	+	−	+	−	+	+	−	+	−	+	−	+	−	−	+	−	+	−	+	−	+	+	−	+	−
abe	+	+	+	+	−	−	−	−	−	−	−	−	+	+	+	+	+	+	+	+	−	−	−	−	−	−	−	−	+	+	+	+
ce	+	−	−	+	+	−	−	+	−	+	+	−	−	+	+	−	+	−	−	+	+	−	−	+	−	+	+	−	−	+	+	−
ace	+	+	−	−	+	+	−	−	−	−	+	+	−	−	+	+	+	+	−	−	+	+	−	−	−	−	+	+	−	−	+	+
bce	+	−	+	−	+	−	+	−	−	+	−	+	−	+	−	+	+	−	+	−	+	−	+	−	−	+	−	+	−	+	−	+
abce	+	+	+	+	+	+	+	+	−	−	−	−	−	−	−	−	+	+	+	+	+	+	+	+	−	−	−	−	−	−	−	−
de	+	−	−	+	−	+	+	−	+	−	−	+	−	+	+	−	+	−	−	+	−	+	+	−	+	−	−	+	−	+	+	−
ade	+	+	−	−	−	−	+	+	+	+	−	−	−	−	+	+	+	+	−	−	−	−	+	+	+	+	−	−	−	−	+	+
bde	+	−	+	−	−	+	−	+	+	−	+	−	−	+	−	+	+	−	+	−	−	+	−	+	+	−	+	−	−	+	−	+
abde	+	+	+	+	−	−	−	−	+	+	+	+	−	−	−	−	+	+	+	+	−	−	−	−	+	+	+	+	−	−	−	−
cde	+	−	−	+	+	−	−	+	+	−	−	+	+	−	−	+	+	−	−	+	+	−	−	+	+	−	−	+	+	−	−	+
acde	+	+	−	−	+	+	−	−	+	+	−	−	+	+	−	−	+	+	−	−	+	+	−	−	+	+	−	−	+	+	−	−
bcde	+	−	+	−	+	−	+	−	+	−	+	−	+	−	+	−	+	−	+	−	+	−	+	−	+	−	+	−	+	−	+	−
abcde	+	+	+	+	+	+	+	+	+	+	+	+	+	+	+	+	+	+	+	+	+	+	+	+	+	+	+	+	+	+	+	+

Main effect and interactions in 2^2, 2^3, 2^4, 2^5, and 2^6 factorial designs

	F	AF	BF	ABF	CF	ACF	BCF	ABCF	DF	ADF	BDF	ABDF	CDF	ACDF	BCDF	ABCDF	EF	AEF	BEF	ABEF	CEF	ACEF	BCEF	ABCEF	DEF	ADEF	BDEF	ABDEF	CDEF	ACDEF	BCDEF	ABCDEF
(1)	−	+	+	−	+	−	−	+	+	−	−	+	−	+	+	−	+	−	−	+	−	+	+	−	−	+	+	−	+	−	−	+
a	−	−	+	+	+	+	−	−	+	+	−	−	−	−	+	+	+	+	−	−	−	−	+	+	−	−	+	+	+	+	−	−
b	−	+	−	+	+	−	+	−	+	−	+	−	−	+	−	+	+	−	+	−	−	+	−	+	−	+	−	+	+	−	+	−
ab	−	−	−	−	+	+	+	+	+	+	+	+	−	−	−	−	+	+	+	+	−	−	−	−	−	−	−	−	+	+	+	+
c	−	+	+	−	−	+	+	−	+	−	−	+	+	−	−	+	+	−	−	+	+	−	−	+	−	+	+	−	−	+	+	−
ac	−	−	+	+	−	−	+	+	+	+	−	−	+	+	−	−	+	+	−	−	+	+	−	−	−	−	+	+	−	−	+	+
bc	−	+	−	+	−	+	−	+	+	−	+	−	+	−	+	−	+	−	+	−	+	−	+	−	−	+	−	+	−	+	−	+
abc	−	−	−	−	−	−	−	−	+	+	+	+	+	+	+	+	+	+	+	+	+	+	+	+	−	−	−	−	−	−	−	−
d	−	+	+	−	+	−	−	+	−	+	+	−	+	−	−	+	+	−	−	+	−	+	+	−	+	−	−	+	−	+	+	−
ad	−	−	+	+	+	+	−	−	−	−	+	+	+	+	−	−	+	+	−	−	−	−	+	+	+	+	−	−	−	−	+	+
bd	−	+	−	+	+	−	+	−	−	+	−	+	+	−	+	−	+	−	+	−	−	+	−	+	+	−	+	−	−	+	−	+
abd	−	−	−	−	+	+	+	+	−	−	−	−	+	+	+	+	+	+	+	+	−	−	−	−	+	+	+	+	−	−	−	−
cd	−	+	+	−	−	+	+	−	−	+	+	−	−	+	+	−	+	−	−	+	+	−	−	+	+	−	−	+	+	−	−	+
acd	−	−	+	+	−	−	+	+	−	−	+	+	−	−	+	+	+	+	−	−	+	+	−	−	+	+	−	−	+	+	−	−
bcd	−	+	−	+	−	+	−	+	−	+	−	+	−	+	−	+	+	−	+	−	+	−	+	−	+	−	+	−	+	−	+	−
abcd	−	−	−	−	−	−	−	−	−	−	−	−	−	−	−	−	+	+	+	+	+	+	+	+	+	+	+	+	+	+	+	+
e	−	+	+	−	+	−	−	+	+	−	−	+	−	+	+	−	−	+	+	−	+	−	−	+	+	−	−	+	−	+	+	−
ae	−	−	+	+	+	+	−	−	+	+	−	−	−	−	+	+	−	−	+	+	+	+	−	−	+	+	−	−	−	−	+	+
be	−	+	−	+	+	−	+	−	+	−	+	−	−	+	−	+	−	+	−	+	+	−	+	−	+	−	+	−	−	+	−	+
abe	−	−	−	−	+	+	+	+	+	+	+	+	−	−	−	−	−	−	−	−	+	+	+	+	+	+	+	+	−	−	−	−
ce	−	+	+	−	−	+	+	−	+	−	−	+	+	−	−	+	−	+	+	−	−	+	+	−	+	−	−	+	+	−	−	+
ace	−	−	+	+	−	−	+	+	+	+	−	−	+	+	−	−	−	−	+	+	−	−	+	+	+	+	−	−	+	+	−	−
bce	−	+	−	+	−	+	−	+	+	−	+	−	+	−	+	−	−	+	−	+	−	+	−	+	+	−	+	−	+	−	+	−
abce	−	−	−	−	−	−	−	−	+	+	+	+	+	+	+	+	−	−	−	−	−	−	−	−	+	+	+	+	+	+	+	+
de	−	+	+	−	+	−	−	+	−	+	+	−	+	−	−	+	−	+	+	−	+	−	−	+	−	+	+	−	+	−	−	+
ade	−	−	+	+	+	+	−	−	−	−	+	+	+	+	−	−	−	−	+	+	+	+	−	−	−	−	+	+	+	+	−	−
bde	−	+	−	+	+	−	+	−	−	+	−	+	+	−	+	−	−	+	−	+	+	−	+	−	−	+	−	+	+	−	+	−
abde	−	−	−	−	+	+	+	+	−	−	−	−	+	+	+	+	−	−	−	−	+	+	+	+	−	−	−	−	+	+	+	+
cde	−	+	+	−	−	+	+	−	−	+	+	−	−	+	+	−	−	+	+	−	−	+	+	−	−	+	+	−	−	+	+	−
acde	−	−	+	+	−	−	+	+	−	−	+	+	−	−	+	+	−	−	+	+	−	−	+	+	−	−	+	+	−	−	+	+
bcde	−	+	−	+	−	+	−	+	−	+	−	+	−	+	−	+	−	+	−	+	−	+	−	+	−	+	−	+	−	+	−	+
abcde	−	−	−	−	−	−	−	−	−	−	−	−	−	−	−	−	−	−	−	−	−	−	−	−	−	−	−	−	−	−	−	−

Main effect and interactions in 2^2, 2^3, 2^4, 2^5, and 2^6 factorial designs

	(T)	A	B	AB	C	AC	BC	ABC	D	AD	BD	ABD	CD	ACD	BCD	ABCD	E	AE	BE	ABE	CE	ACE	BCE	ABCE	DE	ADE	BDE	ABDE	CDE	ACDE	BCDE	ABCDE
f	+	−	−	+	−	+	+	−	−	+	+	−	+	−	−	+	−	+	+	−	+	−	−	+	+	−	−	+	−	+	+	−
af	+	+	−	−	−	−	+	+	−	−	+	+	+	+	−	−	−	−	+	+	+	+	−	−	+	+	−	−	−	−	+	+
bf	+	−	+	−	−	+	−	+	−	+	−	+	+	−	+	−	−	+	−	+	+	−	+	−	+	−	+	−	−	+	−	+
abf	+	+	+	+	−	−	−	−	−	−	−	−	+	+	+	+	−	−	−	−	+	+	+	+	+	+	+	+	−	−	−	−
cf	+	−	−	+	+	−	−	+	−	+	+	−	−	+	+	−	−	+	+	−	−	+	+	−	+	−	−	+	+	−	−	+
acf	+	+	−	−	+	+	−	−	−	−	+	+	−	−	+	+	−	−	+	+	−	−	+	+	+	+	−	−	+	+	−	−
bcf	+	−	+	−	+	−	+	−	−	+	−	+	−	+	−	+	−	+	−	+	−	+	−	+	+	−	+	−	+	−	+	−
abcf	+	+	+	+	+	+	+	+	−	−	−	−	−	−	−	−	−	−	−	−	−	−	−	−	+	+	+	+	+	+	+	+
df	+	−	−	+	−	+	+	−	+	−	−	+	+	−	−	+	−	+	+	−	+	−	−	+	−	+	+	−	+	−	−	+
adf	+	+	−	−	−	−	+	+	+	+	−	−	−	−	+	+	−	−	+	+	+	+	−	−	−	−	+	+	+	+	−	−
bdf	+	−	+	−	−	+	−	+	+	−	+	−	−	+	−	+	−	+	−	+	+	−	+	−	−	+	−	+	+	−	+	−
abdf	+	+	+	+	−	−	−	−	+	+	+	+	−	−	−	−	−	−	−	−	+	+	+	+	−	−	−	−	+	+	+	+
cdf	+	−	−	+	+	−	−	+	+	−	−	+	+	−	−	+	−	+	+	−	−	+	+	−	−	+	+	−	−	+	+	−
acdf	+	+	−	−	+	+	−	−	+	+	−	−	+	+	−	−	−	−	+	+	−	−	+	+	−	−	+	+	−	−	+	+
bcdf	+	−	+	−	+	−	+	−	+	−	+	−	+	−	+	−	−	+	−	+	−	+	−	+	−	+	−	+	−	+	−	+
abcdf	+	+	+	+	+	+	+	+	+	+	+	+	+	+	+	+	−	−	−	−	−	−	−	−	−	−	−	−	−	−	−	−
ef	+	−	−	+	−	+	+	−	−	+	+	−	+	−	−	+	+	−	−	+	−	+	+	−	−	+	+	−	+	−	−	+
aef	+	+	−	−	−	−	+	+	−	−	+	+	+	+	−	−	+	+	−	−	−	−	+	+	+	+	−	−	−	−	+	+
bef	+	−	+	−	−	+	−	+	−	+	−	+	+	−	+	−	+	−	+	−	−	+	−	+	−	+	−	+	+	−	+	−
abef	+	+	+	+	−	−	−	−	−	−	−	−	+	+	+	+	+	+	+	+	−	−	−	−	−	−	−	−	+	+	+	+
cef	+	−	−	+	+	−	−	+	−	+	+	−	−	+	+	−	+	−	−	+	+	−	−	+	−	+	+	−	−	+	+	−
acef	+	+	−	−	+	+	−	−	−	−	+	+	−	−	+	+	+	+	−	−	+	+	−	−	−	−	+	+	−	−	+	+
bcef	+	−	+	−	+	−	+	−	−	+	−	+	−	+	−	+	+	−	+	−	+	−	+	−	−	+	−	+	−	+	−	+
abcef	+	+	+	+	+	+	+	+	−	−	−	−	−	−	−	−	+	+	+	+	+	+	+	+	−	−	−	−	−	−	−	−
def	+	−	−	+	−	+	+	−	+	−	−	+	+	−	−	+	+	−	−	+	−	+	+	−	+	−	−	+	+	−	−	+
adef	+	+	−	−	−	−	+	+	+	+	−	−	−	−	+	+	+	+	−	−	−	−	+	+	+	+	−	−	−	−	+	+
bdef	+	−	+	−	−	+	−	+	+	−	+	−	−	+	−	+	+	−	+	−	−	+	−	+	+	−	+	−	−	+	−	+
abdef	+	+	+	+	−	−	−	−	+	+	+	+	−	−	−	−	+	+	+	+	−	−	−	−	+	+	+	+	−	−	−	−
cdef	+	−	−	+	+	−	−	+	+	−	−	+	+	−	−	+	+	−	−	+	+	−	−	+	+	−	−	+	+	−	−	+
acdef	+	+	−	−	+	+	−	−	+	+	−	−	+	+	−	−	+	+	−	−	+	+	−	−	+	+	−	−	+	+	−	−
bcdef	+	−	+	−	+	−	+	−	+	−	+	−	+	−	+	−	+	−	+	−	+	−	+	−	+	−	+	−	+	−	+	−
abcdef	+	+	+	+	+	+	+	+	+	+	+	+	+	+	+	+	+	+	+	+	+	+	+	+	+	+	+	+	+	+	+	+

Main effect and interactions in 2^2, 2^3, 2^4, 2^5, and 2^6 factorial designs

	F	AF	BF	ABF	CF	ACF	BCF	ABCF	DF	ADF	BDF	ABDF	CDF	ACDF	BCDF	ABCDF	EF	AEF	BEF	ABEF	CEF	ACEF	BCEF	ABCEF	DEF	ADEF	BDEF	ABDEF	CDEF	ACDEF	BCDEF	ABCDEF
f	+	−	−	+	−	+	+	−	−	+	+	−	+	−	−	+	−	+	+	−	+	−	−	+	+	−	−	+	−	+	+	−
af	+	+	−	−	−	−	+	+	−	−	+	+	+	+	−	−	−	−	+	+	+	+	−	−	+	+	−	−	−	−	+	+
bf	+	−	+	−	−	+	−	+	−	+	−	+	+	−	+	−	−	+	−	+	+	−	+	−	+	−	+	−	−	+	−	+
abf	+	+	+	+	−	−	−	−	−	−	−	−	+	+	+	+	−	−	−	−	+	+	+	+	+	+	+	+	−	−	−	−
cf	+	−	−	+	+	−	−	+	−	+	+	−	−	+	+	−	−	+	+	−	−	+	+	−	+	−	−	+	+	−	−	+
acf	+	+	−	−	+	+	−	−	−	−	+	+	−	−	+	+	−	−	+	+	−	−	+	+	+	+	−	−	+	+	−	−
bcf	+	−	+	−	+	−	+	−	−	+	−	+	−	+	−	+	−	+	−	+	−	+	−	+	+	−	+	−	+	−	+	−
abcf	+	+	+	+	+	+	+	+	−	−	−	−	−	−	−	−	−	−	−	−	−	−	−	−	+	+	+	+	+	+	+	+
df	+	−	−	+	−	+	+	−	+	−	−	+	−	+	+	−	−	+	+	−	+	−	−	+	−	+	+	−	+	−	−	+
adf	+	+	−	−	−	−	+	+	+	+	−	−	−	−	+	+	−	−	+	+	+	+	−	−	−	−	+	+	+	+	−	−
bdf	+	−	+	−	−	+	−	+	+	−	+	−	−	+	−	+	−	+	−	+	+	−	+	−	−	+	−	+	+	−	+	−
abdf	+	+	+	+	−	−	−	−	+	+	+	+	−	−	−	−	−	−	−	−	+	+	+	+	−	−	−	−	+	+	+	+
cdf	+	−	−	+	+	−	−	+	+	−	−	+	+	−	−	+	−	+	+	−	−	+	+	−	−	+	+	−	−	+	+	−
acdf	+	+	−	−	+	+	−	−	+	+	−	−	+	+	−	−	−	−	+	+	−	−	+	+	−	−	+	+	−	−	+	+
bcdf	+	−	+	−	+	−	+	−	+	−	+	−	+	−	+	−	−	+	−	+	−	+	−	+	−	+	−	+	−	+	−	+
abcdf	+	+	+	+	+	+	+	+	+	+	+	+	+	+	+	+	−	−	−	−	−	−	−	−	−	−	−	−	−	−	−	−
ef	+	−	−	+	−	+	+	−	−	+	+	−	+	−	−	+	+	−	−	+	−	+	+	−	−	+	+	−	+	−	−	+
aef	+	+	−	−	−	−	+	+	−	−	+	+	+	+	−	−	+	+	−	−	−	−	+	+	−	−	+	+	+	+	−	−
bef	+	−	+	−	−	+	−	+	−	+	−	+	+	−	+	−	+	−	+	−	−	+	−	+	−	+	−	+	+	−	+	−
abef	+	+	+	+	−	−	−	−	−	−	−	−	+	+	+	+	+	+	+	+	−	−	−	−	−	−	−	−	+	+	+	+
cef	+	−	−	+	+	−	−	+	−	+	+	−	−	+	+	−	+	−	−	+	+	−	−	+	−	+	+	−	−	+	+	−
acef	+	+	−	−	+	+	−	−	−	−	+	+	−	−	+	+	+	+	−	−	+	+	−	−	−	−	+	+	−	−	+	+
bcef	+	−	+	−	+	−	+	−	−	+	−	+	−	+	−	+	+	−	+	−	+	−	+	−	−	+	−	+	−	+	−	+
abcef	+	+	+	+	+	+	+	+	−	−	−	−	−	−	−	−	+	+	+	+	+	+	+	+	−	−	−	−	−	−	−	−
def	+	−	−	+	−	+	+	−	+	−	−	+	−	+	+	−	+	−	−	+	−	+	+	−	+	−	−	+	−	+	+	−
adef	+	+	−	−	−	−	+	+	+	+	−	−	−	−	+	+	+	+	−	−	−	−	+	+	+	+	−	−	−	−	+	+
bdef	+	−	+	−	−	+	−	+	+	−	+	−	−	+	−	+	+	−	+	−	−	+	−	+	+	−	+	−	−	+	−	+
abdef	+	+	+	+	−	−	−	−	+	+	+	+	−	−	−	−	+	+	+	+	−	−	−	−	+	+	+	+	−	−	−	−
cdef	+	−	−	+	+	−	−	+	+	−	−	+	+	−	−	+	+	−	−	+	+	−	−	+	+	−	−	+	+	−	−	+
acdef	+	+	−	−	+	+	−	−	+	+	−	−	+	+	−	−	+	+	−	−	+	+	−	−	+	+	−	−	+	+	−	−
bcdef	+	−	+	−	+	−	+	−	+	−	+	−	+	−	+	−	+	−	+	−	+	−	+	−	+	−	+	−	+	−	+	−
abcdef	+	+	+	+	+	+	+	+	+	+	+	+	+	+	+	+	+	+	+	+	+	+	+	+	+	+	+	+	+	+	+	+

13.8 REFERENCES

1. W. G. Cochran and G. M. Cox, *Experimental Designs*, Second Edition, John Wiley & Sons, Inc, New York, 1957.

2. C. J. Colbourn and J. H. Dinitz, *CRC Handbook of Combinatorial Designs*, CRC Press, Boca Raton, FL, 1996, pages 578–581.

CHAPTER 14

Nonparametric Statistics

Contents

Nonparametric, or **distribution–free**, statistical procedures generally assume very little about the underlying population(s). The test statistic used in each procedure is usually easy to compute and may involve qualitative measurements or measurements made on an ordinal scale.

If both a parametric and nonparametric test are applicable, the nonparametric test is less efficient because it does not utilize all of the information in the sample. A larger sample size is required in order for the nonparametric test to have the same probability of a type II error.

14.1   FRIEDMAN TEST FOR RANDOMIZED BLOCK DESIGN

Assumptions: Let there be k independent random samples (treatments) from continuous distributions and b blocks.

Hypothesis test:

H_0: the k samples are from identical populations.

H_a: at least two of the populations differ in location.

Rank each observation from 1 (smallest) to k (largest) within each block. Equal observations are assigned the mean rank for their positions. Let R_i be the rank sum of the ith sample (treatment).

TS: $F_r = \left(\dfrac{12}{bk(k+1)} \displaystyle\sum_{i=1}^{k} R_i^2 \right) - 3b(k+1)$

RR: $F_r \geq \chi^2_{\alpha, k-1}$

14.2   KENDALL'S RANK CORRELATION COEFFICIENT

Given two sets containing ranked elements of the same size, consider each of the $\binom{n}{2} = \frac{n(n-1)}{2}$ pairs of elements from within each set. Associate with each pair (a) a score of $+1$ if the relative ranking of both samples is the same, or (b) a score of -1 if the relative rankings are different. Kendall's score, S_t, is defined as the total of these $\binom{n}{2}$ individual scores. S_t will have a maximum value of $\frac{n(n-1)}{2}$ if the two rankings are identical and a minimum value of $-\frac{n(n-1)}{2}$ if the sets are ranked in exactly the opposite order. Kendall's Tau is defined as

$$\tau = S_t \Big/ \left(\frac{n(n-1)}{2} \right) \tag{14.1}$$

and has the range $-1 \leq \tau \leq 1$.

The table on page 345 may be used to determine the exact probability associated with an occurrence (one-tailed) of a specific value of S_t. In this case the null hypothesis is the existence of an association between the two sets as extreme as an observed S_t. The tabled value is the probability that S_t is equaled or exceeded.

Example 14.68:   Consider the sets of ranked elements: $\mathbf{a} = \{4, 12, 6, 10\}$ and $\mathbf{b} = \{8, 7, 16, 2\}$. Kendall's score is $S_t = 0$ for these sets since

For the $(1, 2)$ term (with $a_1 < a_2$ and $b_1 > b_2$) score is -1
For the $(1, 3)$ term (with $a_1 < a_3$ and $b_1 < b_3$) score is $+1$
For the $(1, 4)$ term (with $a_1 < a_4$ and $b_1 > b_4$) score is -1
For the $(2, 3)$ term (with $a_2 > a_3$ and $b_2 > b_3$) score is $+1$
For the $(2, 4)$ term (with $a_2 > a_4$ and $b_2 > b_4$) score is $+1$
For the $(3, 4)$ term (with $a_3 < a_4$ and $b_3 > b_4$) score is -1

Total is $\overline{\quad 0 \quad}$

Using the table on page 345 with $n = 4$ we find $\text{Prob}\,[S_t \geq 0] = .625$. That is, an S_t value of 0 or larger would be expected 62.5% of the time.

14.2.1 Tables for Kendall rank correlation coefficient

The following table may be used to determine the exact probability associated with an occurrence (one-tailed) of a specific value of S_t. In this case the null hypothesis is the existence of an association between the two sets as extreme as an observed S_t. The tabled value is the probability that S_t is equaled or exceeded.

Distribution of Kendall's rank correlation coefficient in random rankings

S_t	$n = 3$	4	5	6	7	8	9	10
0	0.5000	0.6250	0.5917	0.5000	0.5000	0.5476	0.5403	0.5000
2	0.1667	0.3750	0.4083	0.3597	0.3863	0.4524	0.4597	0.4309
4		0.1667	0.2417	0.2347	0.2810	0.3598	0.3807	0.3637
6		0.0417	0.1167	0.1361	0.1907	0.2742	0.3061	0.3003
8			0.0417	0.0681	0.1194	0.1994	0.2384	0.2422
10			0.0083	0.0278	0.0681	0.1375	0.1792	0.1904
12				0.0083	0.0345	0.0894	0.1298	0.1456
14				0.0014	0.0151	0.0543	0.0901	0.1082
16					0.0054	0.0305	0.0597	0.0779
18					0.0014	0.0156	0.0376	0.0542
20					0.0002	0.0071	0.0223	0.0363
22						0.0028	0.0124	0.0233
24						0.0009	0.0063	0.0143
26						0.0002	0.0029	0.0083
28							0.0012	0.0046
30							0.0004	0.0023
32							0.0001	0.0011
34								0.0005
36								0.0002
38								0.0001

Note that each distribution is symmetric about $S_t = 0$: e.g., for $n = 4$, $\text{Prob}\,[S_t = 2] = \text{Prob}\,[S_t = -2] = 0.375$. Note also that S_t can only assume values with the same parity as n (for example, if n is even then $\text{Prob}\,[S_t = \text{odd}] = 0$); e.g., for $n = 4$, $\text{Prob}\,[S_t = \pm 1] = \text{Prob}\,[S_t = \pm 3] = \text{Prob}\,[S_t = \pm 5] = 0$.

14.3   KOLMOGOROV–SMIRNOFF TESTS

A *one-sample Kolmogorov–Smirnoff* test is used to compare an observed cumulative distribution function (computed from a sample) to a specific continuous distribution function. This is a special test of goodness of fit. A *two-sample Kolmogorov–Smirnoff* test is used to compare two observed cumulative distribution functions; the null hypothesis is that the two independent samples come from identical continuous distributions.

14.3.1   One-sample Kolmogorov–Smirnoff test

Suppose a sample of size n is drawn from a population with known cumulative distribution function $F(x)$. The empirical distribution function, $F_n(x)$, is defined by the sample and is a step function given by

$$F_n(x) = \frac{k}{n} \quad \text{when} \quad x_{(i)} \le x < x_{(i+1)} \tag{14.2}$$

where k is the number of observations less than or equal to x and the $\{x_{(i)}\}$ are the order statistics. If the sample is drawn from the hypothesized distribution, then the empirical distribution function, $F_n(x)$, should be close to $F(x)$. Define the maximum absolute difference between the two distributions to be

$$D = \max \left| F_n(x) - F(x) \right| \tag{14.3}$$

For a two-tailed test the table on page 348 gives critical values for the sampling distribution of D under the null hypothesis. One should reject the hypothetical distribution $F(x)$ if the value D exceeds the tabulated value.

A corresponding one-tailed test is provided by the statistic

$$D^+ = \max \left(F_n(x) - F(x) \right) \tag{14.4}$$

Example 14.69: The values $\{.5, .75, .9, .1\}$ are observed from data that are presumed to be uniformly distributed on the interval $(0, 1)$. Since the presumed distribution is uniform, we have $F(x) = x$. Figure 14.1 shows $F_n(x)$ and $F(x)$. To determine D, only the values of $|F_n(x) - F(x)|$ for x at the endpoints ($x = 0$ and $x = 1$) and on each side of the sample values (since $F_n(x)$ has discontinuities at the sample values) need to be considered. Constructing Table 14.1 results in $D = .25$. If $\alpha = .05$ and $n = 4$ the table on page 348 yields a critical value of $c = .624$. Since $D < c$, the null hypothesis is not rejected.

14.3.2   Two-sample Kolmogorov–Smirnoff test

Suppose two independent samples of sizes n_1 and n_2 are drawn from a population with cumulative distribution function $F(x)$. For each sample j an

x	$F_n(x)$	$F(x) = x$	$\|F_n(x) - F(x)\|$
$x = 0$	0	0	$\|0 - 0\| = 0$
$x = .1-$	0	.10	$\|0 - .10\| = .10$
$x = .1+$	.25	.10	$\|.25 - .10\| = .15$
$x = .5-$	.25	.50	$\|.25 - .50\| = .25$
$x = .5+$	.50	.50	$\|.50 - .50\| = 0$
$x = .75-$	.50	.75	$\|.50 - .75\| = .25$
$x = .75+$	.75	.75	$\|.75 - .75\| = 0$
$x = .90-$	.75	.90	$\|.75 - .90\| = .15$
$x = .90+$	.90	.90	$\|.90 - .90\| = 0$
$x = 1$	1	1	$\|1 - 1\| = 0$

Table 14.1: Table for Kolmogorov–Smirnoff computation.

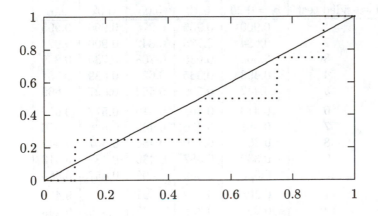

Figure 14.1: Comparison of the sample distribution function (dotted curve) with the distribution function (solid line) for a uniform random variable on the interval $(0, 1)$.

empirical distribution function $F_{n_j}(x)$ is given by the step function,

$$F_{n_j}(x) = \frac{k}{n} \quad \text{when} \quad x_{(i)}^{\text{sample } j} \leq x < x_{(i+1)}^{\text{sample } j} \tag{14.5}$$

where k is the number of observations less than or equal to x and the $\{x_{(i)}^{\text{sample } j}\}$ are the order statistics for the jth sample (for $j = 1$ or $j = 2$).

If the two samples have been drawn from the same population, or from populations with the same distribution (the null hypothesis), then $F_{n_1}(x)$ should be close to $F_{n_2}(x)$. Define the maximum absolute difference between the two empirical distributions to be

$$D = \max \left| F_{n_1}(x) - F_{n_2}(x) \right| \tag{14.6}$$

For a two-tailed test the table on page 350 gives critical values for the sampling distribution of D under the null hypothesis. The null hypothesis is rejected if the value of D exceeds the tabulated value.

A corresponding one-tailed test is provided by the statistic

$$D^+ = \max \left(F_{n_1}(x) - F_{n_2}(x) \right) \qquad (14.7)$$

14.3.3   Tables for Kolmogorov–Smirnoff tests

14.3.3.1   Critical values, one-sample Kolmogorov–Smirnoff test

Critical values, one-sample Kolmogorov–Smirnov test

One-sided test	$\alpha = 0.10$	0.05	0.025	0.01	0.005
Two-sided test	$\alpha = 0.20$	0.10	0.05	0.02	0.01
$n = 1$	0.900	0.950	0.975	0.990	0.995
2	0.684	0.776	0.842	0.900	0.929
3	0.565	0.636	0.708	0.785	0.829
4	0.493	0.565	0.624	0.689	0.734
5	0.447	0.509	0.563	0.627	0.669
6	0.410	0.468	0.519	0.577	0.617
7	0.381	0.436	0.483	0.538	0.576
8	0.358	0.410	0.454	0.507	0.542
9	0.339	0.387	0.430	0.480	0.513
10	0.323	0.369	0.409	0.457	0.489
11	0.308	0.352	0.391	0.437	0.468
12	0.296	0.338	0.375	0.419	0.449
13	0.285	0.325	0.361	0.404	0.432
14	0.275	0.314	0.349	0.390	0.418
15	0.266	0.304	0.338	0.377	0.404
16	0.258	0.295	0.327	0.366	0.392
17	0.250	0.286	0.318	0.355	0.381
18	0.244	0.279	0.309	0.346	0.371
19	0.237	0.271	0.301	0.337	0.361
20	0.232	0.265	0.294	0.329	0.352
21	0.226	0.259	0.287	0.321	0.344
22	0.221	0.253	0.281	0.314	0.337
23	0.216	0.247	0.275	0.307	0.330
24	0.212	0.242	0.269	0.301	0.323
25	0.208	0.238	0.264	0.295	0.317

Critical values, one-sample Kolmogorov–Smirnov test

One-sided test	$\alpha = 0.10$	0.05	0.025	0.01	0.005
Two-sided test	$\alpha = 0.20$	0.10	0.05	0.02	0.01
26	0.204	0.233	0.259	0.290	0.311
27	0.200	0.229	0.254	0.284	0.305
28	0.197	0.225	0.250	0.279	0.300
29	0.193	0.221	0.246	0.275	0.295
30	0.190	0.218	0.242	0.270	0.290
31	0.187	0.214	0.238	0.266	0.285
32	0.184	0.211	0.234	0.262	0.281
33	0.182	0.208	0.231	0.258	0.277
34	0.179	0.205	0.227	0.254	0.273
35	0.177	0.202	0.224	0.251	0.269
36	0.174	0.199	0.221	0.247	0.265
37	0.172	0.196	0.218	0.244	0.262
38	0.170	0.194	0.215	0.241	0.258
39	0.168	0.191	0.213	0.238	0.255
40	0.165	0.189	0.210	0.235	0.252
Approximation for $n > 40$:	$\dfrac{1.07}{\sqrt{n}}$	$\dfrac{1.22}{\sqrt{n}}$	$\dfrac{1.36}{\sqrt{n}}$	$\dfrac{1.52}{\sqrt{n}}$	$\dfrac{1.63}{\sqrt{n}}$

14.3.3.2   Critical values, two-sample Kolmogorov–Smirnoff test

Given the null hypothesis that the two distributions are the same (H_0: $F_1(x) = F_2(x)$), compute $D = \max |F_{n_1}(x) - F_{n_2}(x)|$.

(a) Reject H_0 if D exceeds the value in the table on page 350.

(b) Where $*$ appears in the table on page 350, do not reject H_0 at the given significance level.

(c) For large values of n_1 and n_2, and various values of α, the approximate critical value of D is given in the table below.

Level of significance	Approximate critical value
$\alpha = 0.10$	$1.22\sqrt{\dfrac{n_1 + n_2}{n_1 n_2}}$
$\alpha = 0.05$	$1.36\sqrt{\dfrac{n_1 + n_2}{n_1 n_2}}$
$\alpha = 0.025$	$1.48\sqrt{\dfrac{n_1 + n_2}{n_1 n_2}}$
$\alpha = 0.01$	$1.63\sqrt{\dfrac{n_1 + n_2}{n_1 n_2}}$
$\alpha = 0.005$	$1.73\sqrt{\dfrac{n_1 + n_2}{n_1 n_2}}$
$\alpha = 0.001$	$1.95\sqrt{\dfrac{n_1 + n_2}{n_1 n_2}}$

The entries in the following table are expressed as rational numbers since all critical values of D are an integer divided by $n_1 n_2$. For example, if $n_1 = 6$ and $n_2 = 5$, then

$$.108225 = \text{Prob}\left[D \geq \frac{20}{30}\right] = \text{Prob}\left[D \geq \frac{21}{30}\right] = \text{Prob}\left[D \geq \frac{22}{30}\right] = \text{Prob}\left[D \geq \frac{23}{30}\right]$$

$$.047619 = \text{Prob}\left[D \geq \frac{24}{30}\right] \quad \text{(least value of } D \text{ for which } \alpha < 0.05)$$

$$.025974 = \text{Prob}\left[D \geq \frac{25}{30}\right] = \text{Prob}\left[D \geq \frac{26}{30}\right] \tag{14.8}$$

$$= \text{Prob}\left[D \geq \frac{27}{30}\right] = \text{Prob}\left[D \geq \frac{28}{30}\right] = \text{Prob}\left[D \geq \frac{29}{30}\right]$$

$$.004329 = \text{Prob}\left[D \geq \frac{30}{30}\right] \quad \text{(least value of } D \text{ for which } \alpha < 0.01)$$

See P. J. Kim and R. I. Jennrich, Tables of the exact sampling distribution of the two-sample Kolmogorov–Smirnov criterion, D_{mn}, $m \leq n$, pages 79–170, in H. L. Harter and D. B. Brown (ed.), *Selected Tables in Mathematical Statistics*, Volume 1, American Mathematical Society, Providence, RI, 1973.

Critical values for the Kolmogorov–Smirnov test of $F_1(x) = F_2(x)$
(upper value for $\alpha \leq .05$, lower value for $\alpha \leq .01$)

Sample size n_2	Sample size n_1									
	3	4	5	6	7	8	9	10	11	12
1	*	*	*	*	*	*	*	*	*	*
	*	*	*	*	*	*	*	*	*	*
2	*	*	*	*	*	16/16	18/18	20/20	22/22	24/24
	*	*	*	*	*	*	*	*	*	*
3	*	*	15/15	18/18	21/21	21/24	24/27	27/30	30/33	30/36
	*	*	*	*	*	24/24	27/27	30/30	33/33	36/35
4		16/16	20/20	20/24	24/28	28/32	28/36	30/40	33/44	36/48
		*	*	24/24	28/28	32/32	32/36	36/40	40/44	44/48
5			*	24/30	30/35	30/40	35/45	40/50	39/55	43/60
			*	30/30	35/35	35/40	40/45	45/50	45/55	50/60
6				30/36	30/42	34/48	39/54	40/60	43/66	48/72
				36/36	36/42	40/48	45/54	48/60	54/66	60/72
7					42/49	40/56	42/63	46/70	48/77	53/84
					42/49	48/56	49/63	53/70	59/77	60/84
8						48/64	46/72	48/80	53/88	60/96
						56/64	55/72	60/80	64/88	68/96
9							54/81	53/90	59/99	63/108
							63/81	70/90	70/99	75/108
10								70/100	60/110	66/120
								80/100	77/110	80/120
11									77/121	72/132
									88/121	86/132
12										96/144
										84/144

14.4 KRUSKAL–WALLIS TEST

Assumptions: Suppose there are $k > 2$ independent random samples from continuous distributions, let n_i (for $i = 1, 2, \ldots, k$) be the number of observations in each sample, and let $n = n_1 + n_2 + \cdots + n_k$.

Hypothesis test:

H_0: the k samples are from identical populations.

H_a: at least two of the populations differ.

Rank all n observations from 1 (smallest) to n (largest). Equal observations are assigned the mean rank for their positions. Let R_{ij} be the rank assigned to the jth observation in the ith sample, and let R_i be the total of the ranks in the ith sample.

TS: $H = \left[\dfrac{12}{n(n+1)} \displaystyle\sum_{i=1}^{k} \dfrac{R_i^2}{n_i} \right] - 3(n+1)$

RR: $H \geq h$

where h is the critical value for the Kruskal–Wallis statistic (see table on page 352) such that $\text{Prob}[H \geq h] \approx \alpha$.

Note:

(1) The Kruskal–Wallis procedure is equivalent to an analysis of variance of the ranks. Define the variance ratio as

$$\text{VR} = \frac{\displaystyle\sum_{i=1}^{k} n_i \frac{\left(\overline{R}_i - \overline{\overline{R}}\right)^2}{k-1}}{\displaystyle\sum_{i=1}^{k} \sum_{j=1}^{n_i} \frac{\left(R_{ij} - \overline{R}_i\right)^2}{n-k}} \tag{14.9}$$

where $\overline{R}_i = R_i/n_i$ is the mean of the ranks assigned to the ith sample and $\overline{\overline{R}} = (n+1)/2$ is the overall mean. The Kruskal–Wallis test statistic, H, and VR are related by the equations

$$\text{VR} = \frac{H(n-k)}{(k-1)(n-1-H)}, \qquad H = \frac{(n-1)(k-1)\text{VR}}{(n-k) + (k-1)\text{VR}}. \tag{14.10}$$

(2) As $n \to \infty$ and each $n_i/n \to \lambda_i > 0$, H has approximately a chi–square distribution with $k - 1$ degrees of freedom. Practically, if H_0 is true, and either

(a) $k = 3$, $\quad n_i \geq 6$, $\quad i = 1, 2, 3$ or

(b) $k > 3$, $\quad n_i \geq 5$, $\quad i = 1, 2, \ldots, k$

then H has a chi–square distribution with $k - 1$ degrees of freedom.

(3) The variance ratio, VR, has approximately an F distribution with $k - 1$ and $n - k$ degrees of freedom.

Example 14.70: Suppose that $k = 3$ treatments (A, B, and C) result in the following observations $\{1.2, 1.8, 1.7\}$, $\{0.9, 0.7\}$, and $\{1.0, 0.8\}$. (Therefore, $n_1 = 3$, $n_2 = 2$, $n_3 = 2$, $n = 7$.) Ranking these values:

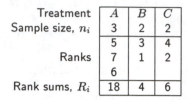

Treatment	A	B	C
Sample size, n_i	3	2	2
	5	3	4
Ranks	7	1	2
	6		
Rank sums, R_i	18	4	6

Hence, $H = \frac{12}{7(8)} \left(\frac{18^2}{3} + \frac{4^2}{2} + \frac{6^2}{2} \right) - 3(8) = \frac{33}{7} \approx 4.714$. From the table on page 352 with $\{n_i\} = \{3, 2, 2\}$, we observe that Prob $[H \geq 4.714] = .0476$. At the $\alpha = .05$ level of significance, there is evidence to suggest at least two of the populations differ.

See R. L. Iman, D. Quade, and D. A. Alexander, Exact probability levels for the Kruskal–Wallis test, *Selected Tables in Mathematical Statistics*, Volume 3, American Mathematical Society, Providence, RI, 1975.

14.4.1 Tables for Kruskal–Wallis test

$\{n_i\} = \{2,1,1\}$		$\{n_i\} = \{2,2,1\}$		$\{n_i\} = \{2,2,2\}$		$\{n_i\} = \{3,2,1\}$	
h	$P(H \geq h)$	h	$P(H \geq h)$	h	$P(H \geq h)$	h	$P(H \geq h)$
2.700	0.5000	3.600	0.2000	4.571	0.0667	4.286	0.1000
				3.714	0.2000	3.857	0.1333

$\{n_i\} = \{3,2,2\}$		$\{n_i\} = \{3,3,1\}$		$\{n_i\} = \{3,3,2\}$		$\{n_i\} = \{3,3,3\}$	
h	$P(H \geq h)$	h	$P(H \geq h)$	h	$P(H \geq h)$	h	$P(H \geq h)$
5.357	0.0286	5.143	0.0429	6.250	0.0107	7.200	0.0036
4.714	0.0476	4.571	0.1000	5.556	0.0250	6.489	0.0107
4.500	0.0667	4.000	0.1286	5.361	0.0321	5.956	0.0250
4.464	0.1048	3.286	0.1571	5.139	0.0607	5.689	0.0286
3.929	0.1810	3.143	0.2429	5.000	0.0750	5.600	0.0500
3.750	0.2190	2.571	0.3286	4.694	0.0929	5.067	0.0857
3.607	0.2381	2.286	0.4857	4.556	0.1000	4.622	0.1000

$\{n_i\} = \{4,2,1\}$		$\{n_i\} = \{4,2,2\}$		$\{n_i\} = \{4,3,1\}$		$\{n_i\} = \{4,3,2\}$	
h	$P(H \geq h)$	h	$P(H \geq h)$	h	$P(H \geq h)$	h	$P(H \geq h)$
4.821	0.0571	6.000	0.0143	5.833	0.0214	7.000	0.0048
4.500	0.0762	5.500	0.0238	5.389	0.0357	6.444	0.0079
4.018	0.1143	5.333	0.0333	5.208	0.0500	6.300	0.0111
3.750	0.1333	5.125	0.0524	5.000	0.0571	6.111	0.0206
3.696	0.1714	4.500	0.0905	4.764	0.0714	5.800	0.0302
3.161	0.1905	4.458	0.1000	4.208	0.0786	5.500	0.0397
2.893	0.2667	4.167	0.1048	4.097	0.0857	5.400	0.0508
2.786	0.2857	4.125	0.1524	4.056	0.0929	4.444	0.1016

$\{n_i\} = \{4,3,3\}$		$\{n_i\} = \{4,4,1\}$		$\{n_i\} = \{4,4,2\}$		$\{n_i\} = \{4,4,3\}$	
h	$P(H \geq h)$	h	$P(H \geq h)$	h	$P(H \geq h)$	h	$P(H \geq h)$
8.018	0.0014	6.667	0.0095	7.855	0.0019	8.909	0.0005
7.000	0.0062	6.167	0.0222	6.873	0.0108	7.144	0.0097
6.745	0.0100	6.000	0.0286	6.545	0.0203	7.136	0.0107
6.564	0.0171	5.667	0.0349	5.945	0.0279	6.659	0.0201
6.018	0.0267	5.100	0.0413	5.645	0.0394	6.182	0.0296
5.982	0.0343	4.967	0.0476	5.236	0.0521	6.167	0.0306
5.727	0.0505	4.867	0.0540	4.991	0.0648	6.000	0.0400
5.436	0.0619	4.267	0.0698	4.691	0.0800	5.576	0.0507
5.064	0.0705	4.167	0.0825	4.555	0.0978	4.712	0.0902
4.845	0.0810	4.067	0.1016	4.445	0.1029	4.477	0.1022
4.700	0.1010	3.900	0.1079				

14.5   THE RUNS TEST

A **run** is a maximal subsequence of elements with a common property.

Hypothesis test:

H_0: the sequence is random.

H_a: the sequence is not random.

TS: $V = $ the total number of runs

RR: $V \geq v_1$ or $V \leq v_2$

> where v_1 and v_2 are critical values for the runs test (see page 354) such that $\text{Prob}[V \geq v_1] \approx \alpha/2$ and $\text{Prob}[V \leq v_2] \approx \alpha/2$.

The normal approximation: Let m be the number of elements with the property that occurs least and n be the number of elements with the other property. As m and n increase, V has approximately a normal distribution with

$$\mu_V = \frac{2mn}{m+n} + 1 \quad \text{and} \quad \sigma_V^2 = \frac{2mn(2mn - m - n)}{(m+n)^2(m+n+1)}. \tag{14.11}$$

The random variable

$$Z = \frac{V - \mu_V}{\sigma_V} \tag{14.12}$$

has approximately a standard normal distribution.

Example 14.71:   Suppose the following sequence of heads (H) and tails (T) was obtained from flipping a coin: $\{H, H, T, T, H, T, H, T, T, T, T, H\}$. Is there any evidence to suggest the coin is biased?

Solution:

(S1) Place vertical bars at the end of each run. The data set may be written to easily count the number of runs.

$$HH \mid TT \mid H \mid T \mid H \mid TTTT \mid H \mid$$

(S2) Using this notation, there are 5 H's, 7 T's, and 7 runs.

(S3) The table on page 356 (using $m = 5$ and $n = 7$) indicates that 65% of the time one would expect there to be 7 runs or fewer.

(S4) The table on page 356 (using $m = 5$ and $n = 6$) indicates that 42% of the time one would expect there be 6 runs or fewer. Alternatively, 58% (since $1 - 0.42 = 0.58$) of the time there would be 7 runs or more.

(S5) In neither case is there any evidence to suggest the coin is biased.

14.5.1   Tables for the runs test

Runs can be used to test data for randomness or to test the hypothesis that two samples come from the same distribution. A *run* is defined as a succession of identical elements which are followed and preceded by different elements or by no elements at all. Let m be the number of elements of one kind and n be the number of elements of the other kind. Let v equal the total number of runs among the $n + m$ elements. The probability that exactly v runs occur is given by

$$
\text{Prob}\,[v \text{ runs}\,] =
\begin{cases}
\dfrac{2 \binom{n-1}{(k-2)/2}\binom{m-1}{(k-2)/2}}{\binom{n+m}{n}} & \text{if } k \text{ is even} \\[4mm]
\dfrac{\binom{n-1}{(k-3)/2}\binom{m-1}{(k-1)/2} + \binom{n-1}{(k-1)/2}\binom{m-1}{(k-3)/2}}{\binom{n+m}{n}} & \text{if } k \text{ is odd}
\end{cases}
\tag{14.13}
$$

The following tables give the sampling distribution for v for values of m and n less than or equal to 20. That is, the values listed in this table give the probability that v or fewer runs will occur.

The table on page 364 gives percentage points of the distribution for larger sample sizes when $m = n$. The columns headed with 0.5%, 1%, 2.5%, 5% indicate the values of v such that v or fewer runs occur with probability less than that indicated; the columns headed with 97.5%, 99%, 99.5% indicate values of v for which the probability of v or more runs is less than 2.5%, 1%, 0.5%. For large values of m and n, particularly for $m = n$ greater than 10, a normal approximation may be used, with the parameters given in equation (14.11).

Distribution of total number of runs v in samples of size (m, n)

m, n	$v = 2$	3	4	5	6	7	8	9	10
2, 2	0.3333	0.6667	1.0000						
2, 3	0.2000	0.5000	0.9000	1.0000					
2, 4	0.1333	0.4000	0.8000	1.0000					
2, 5	0.0952	0.3333	0.7143	1.0000					
2, 6	0.0714	0.2857	0.6429	1.0000					
2, 7	0.0556	0.2500	0.5833	1.0000					
2, 8	0.0444	0.2222	0.5333	1.0000					
2, 9	0.0364	0.2000	0.4909	1.0000					
2, 10	0.0303	0.1818	0.4545	1.0000					
2, 11	0.0256	0.1667	0.4231	1.0000					
2, 12	0.0220	0.1538	0.3956	1.0000					
2, 13	0.0190	0.1429	0.3714	1.0000					
2, 14	0.0167	0.1333	0.3500	1.0000					
2, 15	0.0147	0.1250	0.3309	1.0000					
2, 16	0.0131	0.1176	0.3137	1.0000					
2, 17	0.0117	0.1111	0.2982	1.0000					
2, 18	0.0105	0.1053	0.2842	1.0000					
2, 19	0.0095	0.1000	0.2714	1.0000					
2, 20	0.0087	0.0952	0.2597	1.0000					
3, 3	0.1000	0.3000	0.7000	0.9000	1.0000				
3, 4	0.0571	0.2000	0.5429	0.8000	0.9714	1.0000			
3, 5	0.0357	0.1429	0.4286	0.7143	0.9286	1.0000			
3, 6	0.0238	0.1071	0.3452	0.6429	0.8810	1.0000			
3, 7	0.0167	0.0833	0.2833	0.5833	0.8333	1.0000			
3, 8	0.0121	0.0667	0.2364	0.5333	0.7879	1.0000			
3, 9	0.0091	0.0545	0.2000	0.4909	0.7455	1.0000			
3, 10	0.0070	0.0455	0.1713	0.4545	0.7063	1.0000			
3, 11	0.0055	0.0385	0.1484	0.4231	0.6703	1.0000			
3, 12	0.0044	0.0330	0.1297	0.3956	0.6374	1.0000			
3, 13	0.0036	0.0286	0.1143	0.3714	0.6071	1.0000			
3, 14	0.0029	0.0250	0.1015	0.3500	0.5794	1.0000			
3, 15	0.0025	0.0221	0.0907	0.3309	0.5539	1.0000			
3, 16	0.0021	0.0196	0.0815	0.3137	0.5304	1.0000			
3, 17	0.0018	0.0175	0.0737	0.2982	0.5088	1.0000			
3, 18	0.0015	0.0158	0.0669	0.2842	0.4887	1.0000			
3, 19	0.0013	0.0143	0.0610	0.2714	0.4701	1.0000			
3, 20	0.0011	0.0130	0.0559	0.2597	0.4529	1.0000			

Distribution of total number of runs v in samples of size (m, n)

m, n	$v = 2$	3	4	5	6	7	8	9	10
4, 4	0.0286	0.1143	0.3714	0.6286	0.8857	0.9714	1.0000		
4, 5	0.0159	0.0714	0.2619	0.5000	0.7857	0.9286	0.9921	1.0000	
4, 6	0.0095	0.0476	0.1905	0.4048	0.6905	0.8810	0.9762	1.0000	
4, 7	0.0061	0.0333	0.1424	0.3333	0.6061	0.8333	0.9545	1.0000	
4, 8	0.0040	0.0242	0.1091	0.2788	0.5333	0.7879	0.9293	1.0000	
4, 9	0.0028	0.0182	0.0853	0.2364	0.4713	0.7455	0.9021	1.0000	
4, 10	0.0020	0.0140	0.0679	0.2028	0.4186	0.7063	0.8741	1.0000	
4, 11	0.0015	0.0110	0.0549	0.1758	0.3736	0.6703	0.8462	1.0000	
4, 12	0.0011	0.0088	0.0451	0.1538	0.3352	0.6374	0.8187	1.0000	
4, 13	0.0008	0.0071	0.0374	0.1357	0.3021	0.6071	0.7920	1.0000	
4, 14	0.0007	0.0059	0.0314	0.1206	0.2735	0.5794	0.7663	1.0000	
4, 15	0.0005	0.0049	0.0266	0.1078	0.2487	0.5539	0.7417	1.0000	
4, 16	0.0004	0.0041	0.0227	0.0970	0.2270	0.5304	0.7183	1.0000	
4, 17	0.0003	0.0035	0.0195	0.0877	0.2080	0.5088	0.6959	1.0000	
4, 18	0.0003	0.0030	0.0170	0.0797	0.1913	0.4887	0.6746	1.0000	
4, 19	0.0002	0.0026	0.0148	0.0727	0.1764	0.4701	0.6544	1.0000	
4, 20	0.0002	0.0023	0.0130	0.0666	0.1632	0.4529	0.6352	1.0000	
5, 5	0.0079	0.0397	0.1667	0.3571	0.6429	0.8333	0.9603	0.9921	1.0000
5, 6	0.0043	0.0238	0.1104	0.2619	0.5216	0.7381	0.9113	0.9762	0.9978
5, 7	0.0025	0.0152	0.0758	0.1970	0.4242	0.6515	0.8535	0.9545	0.9924
5, 8	0.0016	0.0101	0.0536	0.1515	0.3473	0.5758	0.7933	0.9293	0.9837
5, 9	0.0010	0.0070	0.0390	0.1189	0.2867	0.5105	0.7343	0.9021	0.9720
5, 10	0.0007	0.0050	0.0290	0.0949	0.2388	0.4545	0.6783	0.8741	0.9580
5, 11	0.0005	0.0037	0.0220	0.0769	0.2005	0.4066	0.6264	0.8462	0.9423
5, 12	0.0003	0.0027	0.0170	0.0632	0.1698	0.3654	0.5787	0.8187	0.9253
5, 13	0.0002	0.0021	0.0133	0.0525	0.1450	0.3298	0.5352	0.7920	0.9076
5, 14	0.0002	0.0016	0.0106	0.0441	0.1246	0.2990	0.4958	0.7663	0.8893
5, 15	0.0001	0.0013	0.0085	0.0374	0.1078	0.2722	0.4600	0.7417	0.8709
5, 16	$.0^4983$	0.0010	0.0069	0.0320	0.0939	0.2487	0.4276	0.7183	0.8524
5, 17	$.0^4759$	0.0008	0.0057	0.0276	0.0823	0.2281	0.3982	0.6959	0.8341
5, 18	$.0^4594$	0.0007	0.0047	0.0239	0.0724	0.2098	0.3715	0.6746	0.8161
5, 19	$.0^4471$	0.0006	0.0040	0.0209	0.0641	0.1937	0.3473	0.6544	0.7984
5, 20	$.0^4376$	0.0005	0.0033	0.0184	0.0570	0.1793	0.3252	0.6352	0.7811
6, 6	0.0022	0.0130	0.0671	0.1753	0.3918	0.6082	0.8247	0.9329	0.9870
6, 7	0.0012	0.0076	0.0425	0.1212	0.2960	0.5000	0.7331	0.8788	0.9662
6, 8	0.0007	0.0047	0.0280	0.0862	0.2261	0.4126	0.6457	0.8205	0.9371
6, 9	0.0004	0.0030	0.0190	0.0629	0.1748	0.3427	0.5664	0.7622	0.9021
6, 10	0.0002	0.0020	0.0132	0.0470	0.1369	0.2867	0.4965	0.7063	0.8636
6, 11	0.0002	0.0014	0.0095	0.0357	0.1084	0.2418	0.4357	0.6538	0.8235
6, 12	0.0001	0.0010	0.0069	0.0276	0.0869	0.2054	0.3832	0.6054	0.7831
6, 13	$.0^4737$	0.0007	0.0051	0.0217	0.0704	0.1758	0.3379	0.5609	0.7434
6, 14	$.0^4516$	0.0005	0.0039	0.0173	0.0575	0.1514	0.2990	0.5204	0.7049
6, 15	$.0^4369$	0.0004	0.0030	0.0139	0.0475	0.1313	0.2655	0.4835	0.6680
6, 16	$.0^4268$	0.0003	0.0023	0.0114	0.0395	0.1146	0.2365	0.4499	0.6329
6, 17	$.0^4198$	0.0002	0.0018	0.0093	0.0331	0.1005	0.2114	0.4195	0.5998
6, 18	$.0^4149$	0.0002	0.0014	0.0078	0.0280	0.0886	0.1896	0.3917	0.5685
6, 19	$.0^4113$	0.0001	0.0012	0.0065	0.0238	0.0785	0.1706	0.3665	0.5392
6, 20	$.0^5869$	0.0001	0.0009	0.0055	0.0203	0.0698	0.1540	0.3434	0.5118

Distribution of total number of runs v in samples of size (m, n)

m, n	$v = 11$	12	13	14	15	16	17	18	19	20	21
4, 4											
4, 5											
4, 6											
4, 7											
4, 8											
4, 9											
4, 10											
4, 11											
4, 12											
4, 13											
4, 14											
4, 15											
4, 16											
4, 17											
4, 18											
4, 19											
4, 20											
5, 5											
5, 6	1.0000										
5, 7	1.0000										
5, 8	1.0000										
5, 9	1.0000										
5, 10	1.0000										
5, 11	1.0000										
5, 12	1.0000										
5, 13	1.0000										
5, 14	1.0000										
5, 15	1.0000										
5, 16	1.0000										
5, 17	1.0000										
5, 18	1.0000										
5, 19	1.0000										
5, 20	1.0000										
6, 6	0.9978	1.0000									
6, 7	0.9924	0.9994	1.0000								
6, 8	0.9837	0.9977	1.0000								
6, 9	0.9720	0.9944	1.0000								
6, 10	0.9580	0.9895	1.0000								
6, 11	0.9423	0.9830	1.0000								
6, 12	0.9253	0.9751	1.0000								
6, 13	0.9076	0.9659	1.0000								
6, 14	0.8893	0.9557	1.0000								
6, 15	0.8709	0.9447	1.0000								
6, 16	0.8524	0.9329	1.0000								
6, 17	0.8341	0.9207	1.0000								
6, 18	0.8161	0.9081	1.0000								
6, 19	0.7984	0.8952	1.0000								
6, 20	0.7811	0.8822	1.0000								

Distribution of total number of runs v in samples of size (m, n)

m, n	$v = 2$	3	4	5	6	7	8	9	10
7, 7	0.0006	0.0041	0.0251	0.0775	0.2086	0.3834	0.6166	0.7914	0.9225
7, 8	0.0003	0.0023	0.0154	0.0513	0.1492	0.2960	0.5136	0.7040	0.8671
7, 9	0.0002	0.0014	0.0098	0.0350	0.1084	0.2308	0.4266	0.6224	0.8059
7, 10	0.0001	0.0009	0.0064	0.0245	0.0800	0.1818	0.3546	0.5490	0.7433
7, 11	$.0^4628$	0.0006	0.0043	0.0175	0.0600	0.1448	0.2956	0.4842	0.6821
7, 12	$.0^4397$	0.0004	0.0030	0.0128	0.0456	0.1165	0.2475	0.4276	0.6241
7, 13	$.0^4258$	0.0003	0.0021	0.0095	0.0351	0.0947	0.2082	0.3785	0.5700
7, 14	$.0^4172$	0.0002	0.0015	0.0072	0.0273	0.0777	0.1760	0.3359	0.5204
7, 15	$.0^4117$	0.0001	0.0011	0.0055	0.0216	0.0642	0.1496	0.2990	0.4751
7, 16	$.0^5816$	$.0^4938$	0.0008	0.0043	0.0172	0.0536	0.1278	0.2670	0.4340
7, 17	$.0^5578$	$.0^4693$	0.0006	0.0034	0.0138	0.0450	0.1097	0.2392	0.3969
7, 18	$.0^5416$	$.0^4520$	0.0005	0.0027	0.0112	0.0381	0.0947	0.2149	0.3634
7, 19	$.0^5304$	$.0^4395$	0.0004	0.0022	0.0092	0.0324	0.0820	0.1937	0.3332
7, 20	$.0^5225$	$.0^4304$	0.0003	0.0018	0.0075	0.0278	0.0714	0.1751	0.3060
8, 8	0.0002	0.0012	0.0089	0.0317	0.1002	0.2145	0.4048	0.5952	0.7855
8, 9	$.0^4823$	0.0007	0.0053	0.0203	0.0687	0.1573	0.3186	0.5000	0.7016
8, 10	$.0^4457$	0.0004	0.0033	0.0134	0.0479	0.1170	0.2514	0.4194	0.6209
8, 11	$.0^4265$	0.0003	0.0021	0.0090	0.0341	0.0882	0.1994	0.3522	0.5467
8, 12	$.0^4159$	0.0002	0.0014	0.0063	0.0246	0.0674	0.1591	0.2966	0.4800
8, 13	$.0^5983$	0.0001	0.0009	0.0044	0.0181	0.0521	0.1278	0.2508	0.4211
8, 14	$.0^5625$	$.0^4688$	0.0006	0.0032	0.0134	0.0408	0.1034	0.2129	0.3695
8, 15	$.0^5408$	$.0^4469$	0.0004	0.0023	0.0101	0.0322	0.0842	0.1816	0.3245
8, 16	$.0^5272$	$.0^4326$	0.0003	0.0017	0.0077	0.0257	0.0690	0.1556	0.2856
8, 17	$.0^5185$	$.0^4231$	0.0002	0.0013	0.0060	0.0207	0.0570	0.1340	0.2518
8, 18	$.0^5128$	$.0^4166$	0.0002	0.0010	0.0047	0.0169	0.0473	0.1159	0.2225
8, 19	$.0^6901$	$.0^4122$	0.0001	0.0008	0.0037	0.0138	0.0395	0.1006	0.1971
8, 20	$.0^6643$	$.0^5901$	$.0^4946$	0.0006	0.0029	0.0114	0.0332	0.0878	0.1751
9, 9	$.0^4411$	0.0004	0.0030	0.0122	0.0445	0.1090	0.2380	0.3992	0.6008
9, 10	$.0^4217$	0.0002	0.0018	0.0076	0.0294	0.0767	0.1786	0.3186	0.5095
9, 11	$.0^4119$	0.0001	0.0011	0.0049	0.0199	0.0549	0.1349	0.2549	0.4300
9, 12	$.0^5680$	$.0^4714$	0.0007	0.0032	0.0137	0.0399	0.1028	0.2049	0.3621
9, 13	$.0^5402$	$.0^4442$	0.0004	0.0022	0.0096	0.0294	0.0789	0.1656	0.3050
9, 14	$.0^5245$	$.0^4281$	0.0003	0.0015	0.0068	0.0220	0.0612	0.1347	0.2572
9, 15	$.0^5153$	$.0^4184$	0.0002	0.0010	0.0049	0.0166	0.0478	0.1102	0.2174
9, 16	$.0^6979$	$.0^4122$	0.0001	0.0007	0.0036	0.0127	0.0377	0.0907	0.1842
9, 17	$.0^6640$	$.0^5832$	$.0^4903$	0.0005	0.0027	0.0099	0.0299	0.0751	0.1566
9, 18	$.0^6427$	$.0^5576$	$.0^4638$	0.0004	0.0020	0.0077	0.0240	0.0626	0.1336
9, 19	$.0^6290$	$.0^5405$	$.0^4458$	0.0003	0.0015	0.0061	0.0193	0.0524	0.1144
9, 20	$.0^6200$	$.0^5290$	$.0^4333$	0.0002	0.0012	0.0048	0.0157	0.0441	0.0983
10, 10	$.0^4108$	0.0001	0.0010	0.0045	0.0185	0.0513	0.1276	0.2422	0.4141
10, 11	$.0^5567$	$.0^4595$	0.0006	0.0027	0.0119	0.0349	0.0920	0.1849	0.3350
10, 12	$.0^5309$	$.0^4340$	0.0003	0.0017	0.0078	0.0242	0.0670	0.1421	0.2707
10, 13	$.0^5175$	$.0^4201$	0.0002	0.0011	0.0053	0.0170	0.0493	0.1099	0.2189
10, 14	$.0^5102$	$.0^4122$	0.0001	0.0007	0.0036	0.0122	0.0367	0.0857	0.1775
10, 15	$.0^6612$	$.0^5765$	$.0^4847$	0.0005	0.0025	0.0088	0.0275	0.0673	0.1445
10, 16	$.0^6377$	$.0^5489$	$.0^4557$	0.0003	0.0018	0.0065	0.0209	0.0533	0.1180
10, 17	$.0^6237$	$.0^5320$	$.0^4373$	0.0002	0.0013	0.0048	0.0160	0.0425	0.0968
10, 18	$.0^6152$	$.0^5213$	$.0^4255$	0.0002	0.0009	0.0036	0.0124	0.0341	0.0798
10, 19	$.0^7999$	$.0^5145$	$.0^4176$	0.0001	0.0007	0.0028	0.0096	0.0276	0.0661
10, 20	$.0^7666$	$.0^6999$	$.0^4124$	$.0^4864$	0.0005	0.0021	0.0076	0.0225	0.0550

Distribution of total number of runs v in samples of size (m, n)

m, n	$v = 11$	12	13	14	15	16	17	18	19	20	21
7, 7	0.9749	0.9959	0.9994	1.0000							
7, 8	0.9487	0.9879	0.9977	0.9998	1.0000						
7, 9	0.9161	0.9748	0.9944	0.9993	1.0000						
7, 10	0.8794	0.9571	0.9895	0.9981	1.0000						
7, 11	0.8405	0.9355	0.9830	0.9962	1.0000						
7, 12	0.8009	0.9109	0.9751	0.9935	1.0000						
7, 13	0.7616	0.8842	0.9659	0.9898	1.0000						
7, 14	0.7233	0.8561	0.9557	0.9852	1.0000						
7, 15	0.6864	0.8273	0.9447	0.9799	1.0000						
7, 16	0.6512	0.7982	0.9329	0.9738	1.0000						
7, 17	0.6178	0.7692	0.9207	0.9669	1.0000						
7, 18	0.5862	0.7407	0.9081	0.9595	1.0000						
7, 19	0.5565	0.7128	0.8952	0.9516	1.0000						
7, 20	0.5286	0.6857	0.8822	0.9433	1.0000						
8, 8	0.8998	0.9683	0.9911	0.9988	0.9998	1.0000					
8, 9	0.8427	0.9394	0.9797	0.9958	0.9993	1.0000	1.0000				
8, 10	0.7822	0.9031	0.9636	0.9905	0.9981	0.9998	1.0000				
8, 11	0.7217	0.8618	0.9434	0.9823	0.9962	0.9994	1.0000				
8, 12	0.6634	0.8174	0.9201	0.9714	0.9935	0.9987	1.0000				
8, 13	0.6084	0.7718	0.8944	0.9580	0.9898	0.9976	1.0000				
8, 14	0.5573	0.7263	0.8672	0.9423	0.9852	0.9960	1.0000				
8, 15	0.5103	0.6818	0.8390	0.9248	0.9799	0.9939	1.0000				
8, 16	0.4674	0.6389	0.8104	0.9057	0.9738	0.9913	1.0000				
8, 17	0.4285	0.5981	0.7818	0.8855	0.9669	0.9881	1.0000				
8, 18	0.3931	0.5595	0.7536	0.8645	0.9595	0.9844	1.0000				
8, 19	0.3611	0.5232	0.7258	0.8429	0.9516	0.9803	1.0000				
8, 20	0.3322	0.4893	0.6988	0.8210	0.9433	0.9757	1.0000				
9, 9	0.7620	0.8910	0.9555	0.9878	0.9970	0.9996	1.0000	1.0000			
9, 10	0.6814	0.8342	0.9233	0.9742	0.9924	0.9986	0.9998	1.0000	1.0000		
9, 11	0.6050	0.7731	0.8851	0.9551	0.9851	0.9965	0.9994	0.9999	1.0000		
9, 12	0.5350	0.7111	0.8431	0.9311	0.9751	0.9931	0.9987	0.9998	1.0000		
9, 13	0.4721	0.6505	0.7991	0.9031	0.9625	0.9880	0.9976	0.9996	1.0000		
9, 14	0.4164	0.5928	0.7545	0.8721	0.9477	0.9813	0.9960	0.9991	1.0000		
9, 15	0.3674	0.5389	0.7104	0.8390	0.9309	0.9729	0.9939	0.9985	1.0000		
9, 16	0.3245	0.4892	0.6675	0.8047	0.9125	0.9629	0.9913	0.9976	1.0000		
9, 17	0.2871	0.4437	0.6264	0.7699	0.8929	0.9515	0.9881	0.9963	1.0000		
9, 18	0.2545	0.4024	0.5872	0.7351	0.8724	0.9388	0.9844	0.9948	1.0000		
9, 19	0.2261	0.3650	0.5503	0.7008	0.8513	0.9250	0.9803	0.9930	1.0000		
9, 20	0.2013	0.3313	0.5155	0.6672	0.8298	0.9103	0.9757	0.9908	1.0000		
10, 10	0.5859	0.7578	0.8724	0.9487	0.9815	0.9955	0.9990	0.9999	1.0000	1.0000	
10, 11	0.5000	0.6800	0.8151	0.9151	0.9651	0.9896	0.9973	0.9996	0.9999	1.0000	
10, 12	0.4250	0.6050	0.7551	0.8751	0.9437	0.9804	0.9942	0.9988	0.9998	1.0000	1.0000
10, 13	0.3607	0.5351	0.6950	0.8307	0.9180	0.9678	0.9896	0.9974	0.9996	0.9999	1.0000
10, 14	0.3062	0.4715	0.6369	0.7839	0.8889	0.9519	0.9834	0.9952	0.9991	0.9999	1.0000
10, 15	0.2602	0.4146	0.5818	0.7361	0.8574	0.9330	0.9755	0.9920	0.9985	0.9997	1.0000
10, 16	0.2216	0.3641	0.5303	0.6886	0.8243	0.9115	0.9661	0.9879	0.9976	0.9994	1.0000
10, 17	0.1893	0.3197	0.4828	0.6423	0.7904	0.8880	0.9551	0.9826	0.9963	0.9991	1.0000
10, 18	0.1621	0.2809	0.4393	0.5978	0.7562	0.8629	0.9429	0.9763	0.9948	0.9985	1.0000
10, 19	0.1392	0.2470	0.3997	0.5554	0.7223	0.8367	0.9296	0.9689	0.9930	0.9978	1.0000
10, 20	0.1200	0.2175	0.3638	0.5155	0.6889	0.8097	0.9153	0.9606	0.9908	0.9969	1.0000

Distribution of total number of runs v in samples of size (m, n)

m, n	$v = 2$	3	4	5	6	7	8	9	10
11, 11	$.0^5284$	$.0^4312$	0.0003	0.0016	0.0073	0.0226	0.0635	0.1349	0.2599
11, 12	$.0^5148$	$.0^4170$	0.0002	0.0010	0.0046	0.0150	0.0443	0.0992	0.2017
11, 13	$.0^6801$	$.0^5961$	0.0001	0.0006	0.0030	0.0101	0.0313	0.0736	0.1569
11, 14	$.0^6449$	$.0^5561$	$.0^4639$	0.0004	0.0019	0.0069	0.0223	0.0551	0.1224
11, 15	$.0^6259$	$.0^5337$	$.0^4396$	0.0002	0.0013	0.0048	0.0161	0.0416	0.0960
11, 16	$.0^6153$	$.0^5207$	$.0^4251$	0.0002	0.0009	0.0034	0.0118	0.0317	0.0757
11, 17	$.0^7931$	$.0^5130$	$.0^4162$	0.0001	0.0006	0.0025	0.0087	0.0244	0.0600
11, 18	$.0^7578$	$.0^6838$	$.0^4107$	$.0^4721$	0.0004	0.0018	0.0065	0.0189	0.0478
11, 19	$.0^7366$	$.0^6549$	$.0^5714$	$.0^4500$	0.0003	0.0013	0.0049	0.0148	0.0383
11, 20	$.0^7236$	$.0^6366$	$.0^5485$	$.0^4351$	0.0002	0.0010	0.0037	0.0116	0.0308
12, 12	$.0^6740$	$.0^5888$	$.0^4984$	0.0005	0.0028	0.0095	0.0296	0.0699	0.1504
12, 13	$.0^6385$	$.0^5481$	$.0^4556$	0.0003	0.0017	0.0061	0.0201	0.0498	0.1126
12, 14	$.0^6207$	$.0^5269$	$.0^4323$	0.0002	0.0011	0.0040	0.0138	0.0358	0.0847
12, 15	$.0^6115$	$.0^5155$	$.0^4193$	0.0001	0.0007	0.0027	0.0096	0.0260	0.0640
12, 16	$.0^7657$	$.0^6920$	$.0^4118$	$.0^4769$	0.0005	0.0018	0.0068	0.0191	0.0487
12, 17	$.0^7385$	$.0^6559$	$.0^5734$	$.0^4497$	0.0003	0.0013	0.0048	0.0142	0.0373
12, 18	$.0^7231$	$.0^6347$	$.0^5467$	$.0^4328$	0.0002	0.0009	0.0035	0.0106	0.0288
12, 19	$.0^7142$	$.0^6220$	$.0^5303$	$.0^4220$	0.0001	0.0006	0.0025	0.0080	0.0223
12, 20	$.0^8886$	$.0^6142$	$.0^5199$	$.0^4150$	$.0^4983$	0.0005	0.0019	0.0061	0.0175
13, 13	$.0^6192$	$.0^5250$	$.0^4302$	0.0002	0.0010	0.0038	0.0131	0.0341	0.0812
13, 14	$.0^7997$	$.0^5135$	$.0^4169$	0.0001	0.0006	0.0024	0.0087	0.0236	0.0589
13, 15	$.0^7534$	$.0^6748$	$.0^5972$	$.0^4636$	0.0004	0.0016	0.0058	0.0165	0.0430
13, 16	$.0^7295$	$.0^6427$	$.0^5573$	$.0^4389$	0.0002	0.0010	0.0040	0.0117	0.0316
13, 17	$.0^7167$	$.0^6251$	$.0^5346$	$.0^4243$	0.0002	0.0007	0.0027	0.0084	0.0234
13, 18	$.0^8970$	$.0^6150$	$.0^5213$	$.0^4155$	0.0001	0.0005	0.0019	0.0061	0.0175
13, 19	$.0^8576$	$.0^7921$	$.0^5134$	$.0^4100$	$.0^4682$	0.0003	0.0014	0.0045	0.0132
13, 20	$.0^8349$	$.0^7576$	$.0^6853$	$.0^5662$	$.0^4460$	0.0002	0.0010	0.0033	0.0100
14, 14	$.0^7499$	$.0^6698$	$.0^5912$	$.0^4597$	0.0004	0.0015	0.0056	0.0157	0.0412
14, 15	$.0^7258$	$.0^6374$	$.0^5507$	$.0^4344$	0.0002	0.0009	0.0036	0.0107	0.0291
14, 16	$.0^7138$	$.0^6206$	$.0^5289$	$.0^4203$	0.0001	0.0006	0.0024	0.0073	0.0207
14, 17	$.0^8754$	$.0^6117$	$.0^5169$	$.0^4123$	$.0^4829$	0.0004	0.0016	0.0051	0.0149
14, 18	$.0^8424$	$.0^7679$	$.0^5101$	$.0^5757$	$.0^4526$	0.0002	0.0011	0.0035	0.0108
14, 19	$.0^8244$	$.0^7403$	$.0^6612$	$.0^5476$	$.0^4339$	0.0002	0.0007	0.0025	0.0079
14, 20	$.0^8144$	$.0^7244$	$.0^6379$	$.0^5304$	$.0^4222$	0.0001	0.0005	0.0018	0.0058
15, 15	$.0^7129$	$.0^6193$	$.0^5272$	$.0^4191$	0.0001	0.0006	0.0023	0.0070	0.0199
15, 16	$.0^8665$	$.0^6103$	$.0^5150$	$.0^4109$	$.0^4745$	0.0003	0.0014	0.0046	0.0137
15, 17	$.0^8354$	$.0^7566$	$.0^6848$	$.0^5639$	$.0^4450$	0.0002	0.0009	0.0031	0.0095
15, 18	$.0^8193$	$.0^7318$	$.0^6491$	$.0^5382$	$.0^4277$	0.0001	0.0006	0.0021	0.0067
15, 19	$.0^8108$	$.0^7183$	$.0^6290$	$.0^5233$	$.0^4173$	$.0^4873$	0.0004	0.0014	0.0047
15, 20	$.0^9616$	$.0^7108$	$.0^6175$	$.0^5144$	$.0^4110$	$.0^4573$	0.0003	0.0010	0.0034
16, 16	$.0^8333$	$.0^7532$	$.0^6802$	$.0^5604$	$.0^4427$	0.0002	0.0009	0.0030	0.0092
16, 17	$.0^8171$	$.0^7283$	$.0^6440$	$.0^5342$	$.0^4250$	0.0001	0.0006	0.0019	0.0062
16, 18	$.0^9907$	$.0^7154$	$.0^6247$	$.0^5198$	$.0^4149$	$.0^4754$	0.0004	0.0013	0.0042
16, 19	$.0^9493$	$.0^8862$	$.0^6142$	$.0^5117$	$.0^5909$	$.0^4473$	0.0002	0.0008	0.0029
16, 20	$.0^9274$	$.0^8493$	$.0^7829$	$.0^6707$	$.0^5562$	$.0^4302$	0.0002	0.0006	0.0020
17, 17	$.0^9857$	$.0^7146$	$.0^6234$	$.0^5188$	$.0^4142$	$.0^4718$	0.0003	0.0012	0.0041
17, 18	$.0^9441$	$.0^8771$	$.0^6128$	$.0^5106$	$.0^5825$	$.0^4430$	0.0002	0.0008	0.0027
17, 19	$.0^9233$	$.0^8419$	$.0^7712$	$.0^6607$	$.0^5488$	$.0^4262$	0.0001	0.0005	0.0018
17, 20	$.0^9126$	$.0^8233$	$.0^7406$	$.0^6356$	$.0^5294$	$.0^4163$	$.0^4845$	0.0003	0.0012
18, 18	$.0^9220$	$.0^8397$	$.0^7677$	$.0^6577$	$.0^5465$	$.0^4250$	0.0001	0.0005	0.0017
18, 19	$.0^9113$	$.0^8209$	$.0^7367$	$.0^6322$	$.0^5268$	$.0^4148$	$.0^4776$	0.0003	0.0011
18, 20	$.0^{10}596$	$.0^8113$	$.0^7204$	$.0^6184$	$.0^5157$	$.0^5896$	$.0^4482$	0.0002	0.0007
19, 19	$.0^{10}566$	$.0^8108$	$.0^7194$	$.0^6175$	$.0^5150$	$.0^5856$	$.0^4462$	0.0002	0.0007
19, 20	$.0^{10}290$	$.0^9566$	$.0^7105$	$.0^7973$	$.0^6857$	$.0^5503$	$.0^4280$	0.0001	0.0005
20, 20	$.0^{10}145$	$.0^9290$	$.0^8553$	$.0^7527$	$.0^6477$	$.0^5288$	$.0^4165$	$.0^4710$	0.0003

Distribution of total number of runs v in samples of size (m, n)

m, n	$v = 11$	12	13	14	15	16	17	18	19	20	21
11, 11	0.4100	0.5900	0.7401	0.8651	0.9365	0.9774	0.9927	0.9984	0.9997	1.0000	1.0000
11, 12	0.3350	0.5072	0.6650	0.8086	0.9008	0.9594	0.9850	0.9960	0.9990	0.9999	1.0000
11, 13	0.2735	0.4334	0.5933	0.7488	0.8598	0.9360	0.9740	0.9919	0.9978	0.9996	0.9999
11, 14	0.2235	0.3690	0.5267	0.6883	0.8154	0.9078	0.9598	0.9857	0.9958	0.9991	0.9999
11, 15	0.1831	0.3137	0.4660	0.6293	0.7692	0.8758	0.9424	0.9774	0.9930	0.9981	0.9997
11, 16	0.1504	0.2665	0.4116	0.5728	0.7225	0.8410	0.9224	0.9669	0.9891	0.9967	0.9994
11, 17	0.1240	0.2265	0.3632	0.5199	0.6765	0.8043	0.9002	0.9542	0.9841	0.9948	0.9991
11, 18	0.1027	0.1928	0.3205	0.4708	0.6317	0.7666	0.8763	0.9395	0.9781	0.9922	0.9985
11, 19	0.0853	0.1644	0.2830	0.4257	0.5888	0.7286	0.8510	0.9231	0.9711	0.9889	0.9978
11, 20	0.0712	0.1404	0.2500	0.3846	0.5480	0.6908	0.8247	0.9051	0.9631	0.9849	0.9969
12, 12	0.2632	0.4211	0.5789	0.7368	0.8496	0.9301	0.9704	0.9905	0.9972	0.9995	0.9999
12, 13	0.2068	0.3475	0.5000	0.6642	0.7932	0.8937	0.9502	0.9816	0.9939	0.9985	0.9997
12, 14	0.1628	0.2860	0.4296	0.5938	0.7345	0.8518	0.9251	0.9691	0.9886	0.9968	0.9992
12, 15	0.1286	0.2351	0.3681	0.5277	0.6759	0.8062	0.8958	0.9528	0.9813	0.9940	0.9984
12, 16	0.1020	0.1933	0.3149	0.4669	0.6189	0.7585	0.8632	0.9330	0.9718	0.9899	0.9971
12, 17	0.0813	0.1591	0.2693	0.4118	0.5646	0.7101	0.8283	0.9101	0.9602	0.9844	0.9953
12, 18	0.0651	0.1312	0.2304	0.3626	0.5137	0.6621	0.7919	0.8847	0.9465	0.9774	0.9929
12, 19	0.0524	0.1085	0.1973	0.3189	0.4665	0.6153	0.7548	0.8572	0.9311	0.9690	0.9898
12, 20	0.0424	0.0900	0.1693	0.2803	0.4231	0.5703	0.7176	0.8281	0.9140	0.9590	0.9860
13, 13	0.1566	0.2772	0.4179	0.5821	0.7228	0.8434	0.9188	0.9659	0.9869	0.9962	0.9990
13, 14	0.1189	0.2205	0.3475	0.5056	0.6525	0.7880	0.8811	0.9446	0.9764	0.9921	0.9976
13, 15	0.0906	0.1753	0.2883	0.4365	0.5847	0.7299	0.8388	0.9182	0.9623	0.9858	0.9952
13, 16	0.0695	0.1396	0.2389	0.3751	0.5212	0.6714	0.7934	0.8873	0.9446	0.9771	0.9917
13, 17	0.0535	0.1113	0.1980	0.3215	0.4628	0.6141	0.7465	0.8529	0.9238	0.9658	0.9868
13, 18	0.0415	0.0890	0.1643	0.2752	0.4098	0.5592	0.6992	0.8159	0.9001	0.9520	0.9805
13, 19	0.0324	0.0714	0.1365	0.2353	0.3623	0.5074	0.6525	0.7772	0.8742	0.9358	0.9728
13, 20	0.0254	0.0575	0.1138	0.2012	0.3200	0.4592	0.6072	0.7377	0.8465	0.9174	0.9635
14, 14	0.0871	0.1697	0.2798	0.4266	0.5734	0.7202	0.8303	0.9129	0.9588	0.9843	0.9944
14, 15	0.0642	0.1306	0.2247	0.3576	0.5000	0.6519	0.7753	0.8749	0.9358	0.9727	0.9893
14, 16	0.0476	0.1007	0.1804	0.2986	0.4336	0.5854	0.7183	0.8322	0.9081	0.9574	0.9820
14, 17	0.0355	0.0779	0.1450	0.2487	0.3745	0.5226	0.6614	0.7863	0.8765	0.9382	0.9721
14, 18	0.0266	0.0604	0.1167	0.2068	0.3227	0.4643	0.6058	0.7386	0.8418	0.9155	0.9598
14, 19	0.0202	0.0471	0.0942	0.1720	0.2776	0.4110	0.5527	0.6903	0.8049	0.8898	0.9450
14, 20	0.0153	0.0368	0.0763	0.1432	0.2387	0.3630	0.5027	0.6425	0.7667	0.8616	0.9281
15, 15	0.0457	0.0974	0.1749	0.2912	0.4241	0.5759	0.7088	0.8251	0.9026	0.9543	0.9801
15, 16	0.0328	0.0728	0.1362	0.2362	0.3576	0.5046	0.6424	0.7710	0.8638	0.9305	0.9672
15, 17	0.0237	0.0546	0.1061	0.1912	0.3005	0.4393	0.5781	0.7147	0.8210	0.9020	0.9505
15, 18	0.0173	0.0412	0.0830	0.1546	0.2519	0.3806	0.5174	0.6581	0.7754	0.8693	0.9303
15, 19	0.0127	0.0312	0.0650	0.1251	0.2109	0.3286	0.4610	0.6026	0.7285	0.8334	0.9068
15, 20	0.0094	0.0237	0.0512	0.1014	0.1766	0.2831	0.4095	0.5493	0.6813	0.7952	0.8806
16, 16	0.0228	0.0528	0.1028	0.1862	0.2933	0.4311	0.5689	0.7067	0.8138	0.8972	0.9472
16, 17	0.0160	0.0385	0.0778	0.1465	0.2397	0.3659	0.5000	0.6420	0.7603	0.8584	0.9222
16, 18	0.0113	0.0282	0.0591	0.1153	0.1956	0.3091	0.4369	0.5789	0.7051	0.8155	0.8928
16, 19	0.0080	0.0207	0.0450	0.0908	0.1594	0.2603	0.3801	0.5188	0.6498	0.7697	0.8596
16, 20	0.0058	0.0153	0.0345	0.0716	0.1300	0.2188	0.3297	0.4628	0.5959	0.7224	0.8237
17, 17	0.0109	0.0272	0.0572	0.1122	0.1907	0.3028	0.4290	0.5710	0.6972	0.8093	0.8878
17, 18	0.0075	0.0194	0.0422	0.0859	0.1514	0.2495	0.3659	0.5038	0.6341	0.7567	0.8486
17, 19	0.0052	0.0139	0.0313	0.0659	0.1202	0.2049	0.3108	0.4418	0.5728	0.7022	0.8057
17, 20	0.0036	0.0100	0.0233	0.0506	0.0955	0.1680	0.2631	0.3854	0.5146	0.6474	0.7604
18, 18	0.0050	0.0134	0.0303	0.0640	0.1171	0.2004	0.3046	0.4349	0.5651	0.6954	0.7996
18, 19	0.0034	0.0094	0.0219	0.0479	0.0906	0.1606	0.2525	0.3729	0.5000	0.6338	0.7475
18, 20	0.0023	0.0066	0.0159	0.0359	0.0701	0.1285	0.2088	0.3182	0.4398	0.5736	0.6940
19, 19	0.0022	0.0064	0.0154	0.0349	0.0683	0.1256	0.2044	0.3127	0.4331	0.5669	0.6873
19, 20	0.0015	0.0044	0.0109	0.0255	0.0516	0.0981	0.1650	0.2610	0.3729	0.5033	0.6271
20, 20	0.0009	0.0029	0.0075	0.0182	0.0380	0.0748	0.1301	0.2130	0.3143	0.4381	0.5619

Distribution of total number of runs v in samples of size (m, n)

m, n	$v = 22$	23	24	25	26	27	28	29
11, 11								
11, 12	1.0000							
11, 13	1.0000							
11, 14	1.0000	1.0000						
11, 15	1.0000	1.0000						
11, 16	0.9999	1.0000						
11, 17	0.9998	1.0000						
11, 18	0.9996	1.0000						
11, 19	0.9994	1.0000						
11, 20	0.9991	1.0000						
12, 12	1.0000							
12, 13	1.0000	1.0000						
12, 14	0.9999	1.0000	1.0000					
12, 15	0.9997	1.0000	1.0000					
12, 16	0.9993	0.9999	1.0000	1.0000				
12, 17	0.9987	0.9998	1.0000	1.0000				
12, 18	0.9978	0.9996	0.9999	1.0000				
12, 19	0.9966	0.9994	0.9999	1.0000				
12, 20	0.9950	0.9991	0.9998	1.0000				
13, 13	0.9998	1.0000	1.0000					
13, 14	0.9995	0.9999	1.0000	1.0000				
13, 15	0.9988	0.9997	1.0000	1.0000				
13, 16	0.9975	0.9994	0.9999	1.0000	1.0000			
13, 17	0.9957	0.9989	0.9997	1.0000	1.0000			
13, 18	0.9930	0.9981	0.9995	0.9999	1.0000	1.0000		
13, 19	0.9894	0.9969	0.9991	0.9999	1.0000	1.0000		
13, 20	0.9848	0.9954	0.9986	0.9998	1.0000	1.0000		
14, 14	0.9985	0.9996	0.9999	1.0000				
14, 15	0.9967	0.9991	0.9998	1.0000	1.0000			
14, 16	0.9938	0.9981	0.9995	0.9999	1.0000	1.0000		
14, 17	0.9894	0.9965	0.9990	0.9998	1.0000	1.0000		
14, 18	0.9834	0.9941	0.9982	0.9996	0.9999	1.0000	1.0000	
14, 19	0.9756	0.9909	0.9970	0.9992	0.9998	1.0000	1.0000	
14, 20	0.9660	0.9867	0.9952	0.9987	0.9997	1.0000	1.0000	
15, 15	0.9930	0.9977	0.9994	0.9999	1.0000	1.0000		
15, 16	0.9872	0.9954	0.9987	0.9997	0.9999	1.0000	1.0000	
15, 17	0.9789	0.9918	0.9974	0.9992	0.9998	1.0000	1.0000	
15, 18	0.9678	0.9866	0.9953	0.9985	0.9996	0.9999	1.0000	1.0000
15, 19	0.9540	0.9798	0.9923	0.9975	0.9993	0.9998	1.0000	1.0000
15, 20	0.9375	0.9712	0.9881	0.9959	0.9987	0.9997	0.9999	1.0000
16, 16	0.9772	0.9908	0.9970	0.9991	0.9998	1.0000	1.0000	
16, 17	0.9634	0.9840	0.9942	0.9981	0.9995	0.9999	1.0000	1.0000
16, 18	0.9457	0.9747	0.9900	0.9964	0.9989	0.9997	0.9999	1.0000
16, 19	0.9244	0.9626	0.9840	0.9938	0.9980	0.9994	0.9999	1.0000
16, 20	0.8996	0.9479	0.9761	0.9903	0.9965	0.9989	0.9997	0.9999
17, 17	0.9428	0.9728	0.9891	0.9959	0.9988	0.9997	0.9999	1.0000
17, 18	0.9172	0.9578	0.9816	0.9925	0.9975	0.9992	0.9998	1.0000
17, 19	0.8872	0.9391	0.9714	0.9876	0.9954	0.9985	0.9996	0.9999
17, 20	0.8534	0.9168	0.9584	0.9808	0.9924	0.9972	0.9992	0.9998
18, 18	0.8829	0.9360	0.9697	0.9866	0.9950	0.9983	0.9995	0.9999
18, 19	0.8438	0.9094	0.9540	0.9781	0.9911	0.9966	0.9990	0.9997
18, 20	0.8010	0.8788	0.9345	0.9670	0.9856	0.9941	0.9980	0.9994
19, 19	0.7956	0.8744	0.9317	0.9651	0.9846	0.9936	0.9978	0.9993
19, 20	0.7444	0.8350	0.9048	0.9484	0.9756	0.9891	0.9959	0.9985
20, 20	0.6857	0.7870	0.8699	0.9252	0.9620	0.9818	0.9925	0.9971

Distribution of total number of runs v in samples of size (m, n)

m, n	$v = 30$	31	32	33	34	35	36	37
11, 11								
11, 12								
11, 13								
11, 14								
11, 15								
11, 16								
11, 17								
11, 18								
11, 19								
11, 20								
12, 12								
12, 13								
12, 14								
12, 15								
12, 16								
12, 17								
12, 18								
12, 19								
12, 20								
13, 13								
13, 14								
13, 15								
13, 16								
13, 17								
13, 18								
13, 19								
13, 20								
14, 14								
14, 15								
14, 16								
14, 17								
14, 18								
14, 19								
14, 20								
15, 15								
15, 16								
15, 17								
15, 18								
15, 19								
15, 20								
16, 16								
16, 17								
16, 18	1.0000							
16, 19	1.0000							
16, 20	1.0000	1.0000						
17, 17	1.0000							
17, 18	1.0000							
17, 19	1.0000	1.0000						
17, 20	0.9999	1.0000						
18, 18	1.0000	1.0000						
18, 19	0.9999	1.0000	1.0000					
18, 20	0.9998	1.0000	1.0000					
19, 19	0.9998	1.0000	1.0000					
19, 20	0.9996	0.9999	1.0000	1.0000				
20, 20	0.9991	0.9997	0.9999	1.0000	1.0000			

The values listed in the previous tables indicate the probability that v or fewer runs will occur. For example, for two samples of size 4, the probability of three or fewer runs is 0.114. For sample size $m = n$, and m larger than 10, the following table can be used. The columns headed 0.5, 1, 2.5, and 5 give values of v such that v or fewer runs occur with probability less than the indicated percentage. For example, for $m = n = 12$, the probability of 8 or fewer runs is approximately 5%. The columns headed 95, 97.5, 99, and 99.5 give values of v for which the probability of v or more runs is less than 5, 2.5, 1, or 0.5 percent.

Distribution of the total number of runs v in samples of size $m = n$.

$m = n$	0.5	1.0	2.5	5.0	95.0	97.5	99.0	99.5	mean	var (σ^2)	s.d. (σ)
11	5	6	7	7	16	16	17	18	12	5.24	2.29
12	6	7	7	8	17	18	18	19	13	5.74	2.40
13	7	7	8	9	18	19	20	20	14	6.24	2.50
14	7	8	9	10	19	20	21	22	15	6.74	2.60
15	8	9	10	11	20	21	22	23	16	7.24	2.69
16	9	10	11	11	22	22	23	24	17	7.74	2.78
17	10	10	11	12	23	24	25	25	18	8.24	2.87
18	11	11	12	13	24	25	26	26	19	8.74	2.96
19	11	12	13	14	25	26	27	28	20	9.24	3.04
20	12	13	14	15	26	27	28	29	21	9.74	3.12
25	16	17	18	19	32	33	34	35	26	12.24	3.50
30	20	21	23	24	37	38	40	41	31	14.75	3.84
35	24	25	27	28	43	44	46	47	36	17.25	4.15
40	29	30	31	33	48	50	51	52	41	19.75	4.44
45	33	34	36	37	54	55	57	58	46	22.25	4.72
50	37	38	40	42	59	61	63	64	51	24.75	4.97
55	42	43	45	46	65	66	68	69	56	27.25	5.22
60	46	47	49	51	70	72	74	75	61	29.75	5.45
65	50	52	54	56	75	77	79	81	66	32.25	5.68
70	55	56	58	60	81	83	85	86	71	34.75	5.89
75	59	61	63	65	86	88	90	92	76	37.25	6.10
80	64	65	68	70	91	93	96	97	81	39.75	6.30
85	68	70	72	74	97	99	101	103	86	42.25	6.50
90	73	74	77	79	102	104	107	108	91	44.75	6.69
95	77	79	82	84	107	109	112	114	96	47.25	6.87
100	82	84	86	88	113	115	117	119	101	49.75	7.05

14.6 THE SIGN TEST

Assumptions: Let $X_1, X_2, \ldots, X_n$ be a random sample from a continuous distribution.

Hypothesis test:

H_0: $\tilde{\mu} = \tilde{\mu}_0$

H_a: $\tilde{\mu} > \tilde{\mu}_0$, $\quad \tilde{\mu} < \tilde{\mu}_0$, $\quad \tilde{\mu} \neq \tilde{\mu}_0$

TS: Y = the number of X_i's greater than $\tilde{\mu}_0$.

Under the null hypothesis, Y has a binomial distribution with parameters n and $p = .5$.

RR: $Y \geq c_1$, $\quad Y \leq c_2$, $\quad Y \geq c$ or $Y \leq n - c$

The critical values c_1, c_2, and c are obtained from the binomial distribution with parameters n and $p = .5$ to yield the desired significance level α. (See the table on page 366.)

Sample values equal to $\tilde{\mu}_0$ are excluded from the analysis and the sample size is reduced accordingly.

The normal approximation: When $n \geq 10$ and $np \geq 5$ the binomial distribution can be approximated by a normal distribution with

$$\mu_Y = np \quad \text{and} \quad \sigma_Y^2 = np(1 - p) \tag{14.14}$$

The random variable

$$Z = \frac{Y - \mu_Y}{\sigma_Y} = \frac{Y - np}{\sqrt{np(1 - p)}} \tag{14.15}$$

has approximately a standard normal distribution when H_0 is true and $n \geq 10$ and $np \geq 5$.

14.6.1 Table of critical values for the sign test

Let $X_1, X_2, \ldots, X_n$ be a random sample from a continuous distribution with hypothesized median $\tilde{\mu}_0$. The test statistic is Y, the number of X_i's greater than $\tilde{\mu}_0$. If the null hypothesis is true, the probability X_i is greater than the median is $1/2$. The probability distribution for Y is given by the binomial probability function

$$\text{Prob}[Y = y] = f(y) = \binom{n}{y} \left(\frac{1}{2}\right)^n. \tag{14.16}$$

The following table contains critical values k such that

$$\text{Prob}[Y \leq k] = \sum_{y=0}^{k} \binom{n}{y} \left(\frac{1}{2}\right)^n < \frac{\alpha}{2}. \tag{14.17}$$

For a one-tailed test with significance level α, enter the table in the column headed by 2α.

For larger values of n, approximate critical values may be found using equation (14.15).

$$k = \left\lfloor np + \sqrt{np(1-p)}z_{\alpha/2} \right\rfloor \qquad (14.18)$$

where z_α is the critical value for the normal distribution.

Critical values for the sign test

n	.01	.025	.05	.10	n	.01	.025	.05	.10
1	0	0	0	0	26	6	6	7	8
2	0	0	0	0	27	6	7	7	8
3	0	0	0	0	28	6	7	8	9
4	0	0	0	0	29	7	8	8	9
5	0	0	0	0	30	7	8	9	10
6	0	0	0	0	31	7	8	9	10
7	0	0	0	0	32	8	9	9	10
8	0	0	0	1	33	8	9	10	11
9	0	0	1	1	34	9	10	10	11
10	0	1	1	1	35	9	10	11	12
11	0	1	1	2	36	9	10	11	12
12	1	1	2	2	37	10	11	12	13
13	1	2	2	3	38	10	11	12	13
14	1	2	2	3	39	11	12	12	13
15	2	2	3	3	40	11	12	13	14
16	2	3	3	4	41	11	12	13	14
17	2	3	4	4	42	12	13	14	15
18	3	3	4	5	43	12	13	14	15
19	3	4	4	5	44	13	14	15	16
20	3	4	5	5	45	13	14	15	16
21	4	4	5	6	46	13	14	15	16
22	4	5	5	6	47	14	15	16	17
23	4	5	6	7	48	14	15	16	17
24	5	6	6	7	49	15	16	17	18
25	5	6	7	7	50	15	16	17	18

14.7 SPEARMAN'S RANK CORRELATION COEFFICIENT

Suppose there are n pairs of observations from continuous distributions. Rank the observations in the two samples separately from smallest to largest. Equal observations are assigned the mean rank for their positions. Let u_i be the rank of the ith observation in the first sample and let v_i be the rank of the ith observation in the second sample. Spearman's rank correlation coefficient, r_S, is a measure of the correlation between ranks, calculated by using the ranks in place of the actual observations in the formula for the correlation coefficient r.

$$r_S = \frac{SS_{uv}}{\sqrt{SS_{uu} \, SS_{vv}}} = \frac{n \sum_{i=1}^{n} u_i v_i - \left(\sum_{i=1}^{n} u_i\right)\left(\sum_{i=1}^{n} v_i\right)}{\sqrt{\left[n \sum_{i=1}^{n} u_i^2 - \left(\sum_{i=1}^{n} u_i\right)^2\right]\left[n \sum_{i=1}^{n} v_i^2 - \left(\sum_{i=1}^{n} v_i\right)^2\right]}}$$

$$= 1 - \frac{6 \sum_{i=1}^{n} d_i^2}{n(n^2 - 1)} \quad \text{where } d_i = u_i - v_i.$$

(14.19)

The shortcut formula for r_S that only uses the differences $\{d_i\}$ is not exact when there are tied measurements. The approximation is good when the number of ties is small in comparison to n.

Hypothesis test:

H_0: $\rho_S = 0$ (no population correlation between ranks)

H_a: $\rho_S > 0$, $\rho_S < 0$, $\rho_S \neq 0$

TS: r_S

RR: $r_S \geq r_{S,\alpha}$, $r_S \leq -r_{S,\alpha}$, $|r_S| \geq r_{S,\alpha/2}$

where $r_{S,\alpha}$ is a critical value for Spearman's rank correlation coefficient test (see page 367).

The normal approximation: When H_0 is true r_S has approximately a normal distribution with

$$\mu_{r_S} = 0 \quad \text{and} \quad \sigma_{r_S}^2 = \frac{1}{n - 1}.$$

(14.20)

The random variable

$$Z = \frac{r_S - \mu_{r_S}}{\sigma_{r_S}} = \frac{r_S - 0}{1/\sqrt{n-1}} = r_S \sqrt{n - 1}$$

(14.21)

has approximately a standard normal distribution as n increases.

14.7.1 Tables for Spearman's rank correlation coefficient

Spearman's coefficient of rank correlation, r_S, measures the correspondence between two rankings; see equation (14.19). The table below gives critical values for r_S assuming the samples are independent; their derivation comes from the subsequent table.

Critical values of Spearman's rank correlation coefficient

n	$\alpha = 0.10$	$\alpha = 0.05$	$\alpha = 0.01$	$\alpha = 0.001$
4	0.8000	0.8000	–	–
5	0.7000	0.8000	0.9000	–
6	0.6000	0.7714	0.8857	–
7	0.5357	0.6786	0.8571	0.9643
8	0.5000	0.6190	0.8095	0.9286
9	0.4667	0.5833	0.7667	0.9000
10	0.4424	0.5515	0.7333	0.8667
11	0.4182	0.5273	0.7000	0.8364
12	0.3986	0.4965	0.6713	0.8182
13	0.3791	0.4780	0.6429	0.7912
14	0.3626	0.4593	0.6220	0.7670
15	0.3500	0.4429	0.6000	0.7464
20	0.2977	0.3789	0.5203	0.6586
25	0.2646	0.3362	0.4654	0.5962
30	0.2400	0.3059	0.4251	0.5479

Let $\sum m$ represents the mean value of the sum of squares. Then the following tables give the probability that $\sum d^2 \geq S$ for $S \geq \sum m$, or that $\sum d^2 \leq S$ for $S \leq \sum m$. The tables for $n = 9$ and $n = 10$ can be completed by symmetry.

The values in the next table create the critical values in the last table. For example, taking $n = 9$ we note that (a) $S = 26$ (corresponding to a Spearman rank correlation coefficient of $1 - \frac{26}{120} \approx 0.7833$) has a probability of 0.0086; and (b) $S = 28$ (corresponding a Spearman rank correlation coefficient of $1 - \frac{28}{120} \approx 0.7667$) has a probability of 0.0107. Hence, the critical value for $n = 9$ and $\alpha = 0.01$, the least value of the coefficient whose probability is less that 0.01, is 0.7667.

Exact values for Spearman's rank correlation coefficient

S	$n = 2$ $\sum m = 1$	3 4	4 10	5 20	6 35	7 56	8 84	9 120	10 165
0	0.5000	0.1667	0.0417	0.0083	0.0014	0.0002	0.0000	0.0000	0.0000
2	0.5000	0.5000	0.1667	0.0417	0.0083	0.0014	0.0002	0.0000	0.0000
4		0.5000	0.2083	0.0667	0.0167	0.0034	0.0006	0.0001	0.0000
6		0.5000	0.3750	0.1167	0.0292	0.0062	0.0011	0.0002	0.0000
8		0.1667	0.4583	0.1750	0.0514	0.0119	0.0023	0.0004	0.0001
10			0.5417	0.2250	0.0681	0.0171	0.0036	0.0007	0.0001
12			0.4583	0.2583	0.0875	0.0240	0.0054	0.0010	0.0002
14			0.3750	0.3417	0.1208	0.0331	0.0077	0.0015	0.0003
16			0.2083	0.3917	0.1486	0.0440	0.0109	0.0023	0.0004
18			0.1667	0.4750	0.1778	0.0548	0.0140	0.0030	0.0006
20			0.0417	0.5250	0.2097	0.0694	0.0184	0.0041	0.0008
22				0.4750	0.2486	0.0833	0.0229	0.0054	0.0011
24				0.3917	0.2819	0.1000	0.0288	0.0069	0.0014
26				0.3417	0.3292	0.1179	0.0347	0.0086	0.0019
28				0.2583	0.3569	0.1333	0.0415	0.0107	0.0024
30				0.2250	0.4014	0.1512	0.0481	0.0127	0.0029
32				0.1750	0.4597	0.1768	0.0575	0.0156	0.0036
34				0.1167	0.5000	0.1978	0.0661	0.0184	0.0044
36				0.0667	0.5000	0.2222	0.0756	0.0216	0.0053
38				0.0417	0.5000	0.2488	0.0855	0.0252	0.0063
40				0.0083	0.4597	0.2780	0.0983	0.0294	0.0075
42					0.4014	0.2974	0.1081	0.0333	0.0087
44					0.3569	0.3308	0.1215	0.0380	0.0101
46					0.3292	0.3565	0.1337	0.0429	0.0117
48					0.2819	0.3913	0.1496	0.0484	0.0134
50					0.2486	0.4198	0.1634	0.0540	0.0153
52					0.2097	0.4532	0.1799	0.0603	0.0173
54					0.1778	0.4817	0.1947	0.0664	0.0195
56					0.1486	0.5183	0.2139	0.0738	0.0219
58					0.1208	0.4817	0.2309	0.0809	0.0245
60					0.0875	0.4532	0.2504	0.0888	0.0272
62					0.0681	0.4198	0.2682	0.0969	0.0302
64					0.0514	0.3913	0.2911	0.1063	0.0334
66					0.0292	0.3565	0.3095	0.1149	0.0367
68					0.0167	0.3308	0.3323	0.1250	0.0403
70					0.0083	0.2974	0.3517	0.1348	0.0441
72					0.0014	0.2780	0.3760	0.1456	0.0481
74						0.2488	0.3965	0.1563	0.0524
76						0.2222	0.4201	0.1681	0.0569
78						0.1978	0.4410	0.1793	0.0616
80						0.1768	0.4674	0.1927	0.0667

Exact values for Spearman's rank correlation coefficient

S	$n = 2$	3	4	5	6	7	8	9	10
	$\sum m = 1$	4	10	20	35	56	84	120	165
80						0.1768	0.4674	0.1927	0.0667
82						0.1512	0.4884	0.2050	0.0720
84						0.1333	0.5116	0.2183	0.0774
86						0.1179	0.4884	0.2315	0.0831
88						0.1000	0.4674	0.2467	0.0893
90						0.0833	0.4410	0.2603	0.0956
92						0.0694	0.4201	0.2759	0.1022
94						0.0548	0.3965	0.2905	0.1091
96						0.0440	0.3760	0.3067	0.1163
98						0.0331	0.3517	0.3218	0.1237
100						0.0240	0.3323	0.3389	0.1316
102						0.0171	0.3095	0.3540	0.1394
104						0.0119	0.2911	0.3718	0.1478
106						0.0062	0.2682	0.3878	0.1564
108						0.0034	0.2504	0.4050	0.1652
110						0.0014	0.2309	0.4216	0.1744
112						0.0002	0.2139	0.4400	0.1839
114							0.1947	0.4558	0.1935
116							0.1799	0.4742	0.2035
118							0.1634	0.4908	0.2135
120							0.1496	0.5092	0.2241
122							0.1337	0.4908	0.2349
124							0.1215	0.4742	0.2459
126							0.1081	0.4558	0.2567
128							0.0983	0.4400	0.2683
130							0.0855	0.4216	0.2801
132							0.0756	0.4050	0.2918
134							0.0661	0.3878	0.3037
136							0.0575	0.3718	0.3161
138							0.0481	0.3540	0.3284
140							0.0415	0.3389	0.3410
142							0.0347	0.3218	0.3536
144							0.0288	0.3067	0.3665
146							0.0229	0.2905	0.3795
148							0.0184	0.2759	0.3925
150							0.0140	0.2603	0.4056
152							0.0109	0.2467	0.4191
154							0.0077	0.2315	0.4326
156							0.0054	0.2183	0.4458
158							0.0036	0.2050	0.4592
160							0.0023	0.1927	0.4730

14.8 WILCOXON MATCHED-PAIRS SIGNED-RANKS TEST

Assume we have a matched set of n observations $\{x_i, y_i\}$. Let d_i denote the differences $d_i = x_i - y_i$.

Hypothesis test:

H_0: there is no difference in the distribution of the x_i's and the y_i's

H_a: there is a difference

Rank all of the d_i's without regard to sign: the least value of $|d_i|$ gets rank 1, the next largest value gets rank 2, etc. After determining the ranking, affix the signs of the differences to each rank.

TS: $T =$ the smaller sum of the like-signed ranks.

RR: $T \geq c$

where c is found from the table on page 372.

Example 14.72: Suppose $n = 10$ values are as shown in the first two columns of the following table:

x_i	y_i	$d_i = x_i - y_i$	rank of $|d_i|$	signed rank of $|d_i|$
9	8	1	2	2
2	2	0	–	–
1	3	−2	4.5	−4.5
4	2	2	4.5	4.5
6	3	3	7	7
4	0	4	9	9
7	4	3	7	7
8	5	3	7	7
5	4	1	2	2
1	0	1	2	2

$$\sum R^+ = 40.5$$
$$\sum R^- = -4.5$$

The subsequent columns show the differences, the ranks (note how ties are handled), and the signed ranks. The smaller of the two sums is $T = 4.5$. From the following table (with $n = 10$) we conclude that there is evidence of a difference in distributions at the .005 significance level.

See D. J. Sheskin, *Handbook of Parametric and Nonparametric Statistical Procedures*, CRC Press LLC, Boca Raton, FL, 1997, pages 291–301, 681.

Critical values for the Wilcoxon signed-ranks test
and the matched-pairs signed-ranks test

One sided	Two sided	$n = 5$	6	7	8	9	10	11	12	13	14
$\alpha = .05$	$\alpha = .10$	0	2	3	5	8	10	13	17	21	25
$\alpha = .025$	$\alpha = .05$		0	2	3	5	8	10	13	17	21
$\alpha = .01$	$\alpha = .02$			0	1	3	5	7	9	12	15
$\alpha = .005$	$\alpha = .01$				0	1	3	5	7	9	12

One sided	Two sided	$n = 15$	16	17	18	19	20	21	22	23	24
$\alpha = .05$	$\alpha = .10$	30	35	41	47	53	60	67	75	83	91
$\alpha = .025$	$\alpha = .05$	25	29	34	40	46	52	58	65	73	81
$\alpha = .01$	$\alpha = .02$	19	23	27	32	37	43	49	55	62	69
$\alpha = .005$	$\alpha = .01$	15	19	23	27	32	37	42	48	54	61

One sided	Two sided	$n = 25$	26	27	28	29	30	31	32	33	34
$\alpha = .05$	$\alpha = .10$	100	110	119	130	140	151	163	175	187	200
$\alpha = .025$	$\alpha = .05$	89	98	107	116	126	137	147	159	170	182
$\alpha = .01$	$\alpha = .02$	76	84	92	101	110	120	130	140	151	162
$\alpha = .005$	$\alpha = .01$	68	75	83	91	100	109	118	128	138	148

14.9 WILCOXON RANK–SUM (MANN–WHITNEY) TEST

Assumptions: Let $X_1, X_2, \ldots, X_m$ and $Y_1, Y_2, \ldots, Y_n$ (with $m \leq n$) be independent random samples from continuous distributions.

Hypothesis test:

H_0: $\tilde{\mu}_1 - \tilde{\mu}_2 = \Delta_0$

H_a: $\tilde{\mu}_1 - \tilde{\mu}_2 > \Delta_0$, $\tilde{\mu}_1 - \tilde{\mu}_2 < \Delta_0$, $\tilde{\mu}_1 - \tilde{\mu}_2 \neq \Delta_0$

Subtract Δ_0 from each X_i. Combine the $(X_i - \Delta_0)$'s and the Y_j's into one sample and rank all of the observations. Equal differences are assigned the mean rank for their positions.

TS: $W = \sum_{i=1}^{m} R_i$

where R_i is the rank of $(X_i - \Delta_0)$ in the combined sample.

RR: $W \geq c_1$, $W \leq c_2$, $W \geq c$ or $W \leq m(m + n + 1) - c$

where c_1, c_2, and c are critical values for the Wilcoxon rank–sum statistic such that $\text{Prob}\,[W \geq c_1] \approx \alpha$, $\text{Prob}\,[W \leq c_2] \approx \alpha$, and $\text{Prob}\,[W \geq c] \approx \alpha/2$. (In practice, we convert from W to U via equation (14.24) and look up U values.)

The normal approximation: When both m and n are greater than 8, W has approximately a normal distribution with

$$\mu_W = \frac{m(m + n + 1)}{2} \quad \text{and} \quad \sigma_W^2 = \frac{mn(m + n + 1)}{12}. \tag{14.22}$$

The random variable

$$Z = \frac{W - \mu_W}{\sigma_W} \qquad (14.23)$$

has approximately a standard normal distribution.

The Mann–Whitney U statistic: The rank–sum test may also be based on the test statistic

$$U = \frac{m(m + 2n + 1)}{2} - W. \qquad (14.24)$$

When both m and n are greater than 8, U has approximately a normal distribution with

$$\mu_U = \frac{mn}{2} \quad \text{and} \quad \sigma_U^2 = \frac{mn(m + n + 1)}{12}. \qquad (14.25)$$

The random variable

$$Z = \frac{U - \mu_U}{\sigma_U} \qquad (14.26)$$

has approximately a standard normal distribution.

Note that there are two tests commonly called the Mann–Whitney U test: one developed by Mann and Whitney and one developed by Wilcoxon. Although they employ different equations and different tables, the two versions yield comparable results.

Example 14.73: The Pennsylvania State Police theorize that cars travel faster during the evening rush hour versus the morning rush hour. Randomly selected cars were selected during each rush hour and there speeds were computed using radar. The data is given in the table below.

Morning:	68	65	80	61	64	64
	63	73	75	71		
Evening:	70	70	71	72	72	71
	75	74	81	72	74	71

Use the Mann–Whitney U test to determine if there is any evidence to suggest the median speeds are different. Use $\alpha = .05$.

Solution:

(S1) Computations:

$m = 10$, $n = 12$, $W = 88$, $U = 87$

(S2) Using the tables, the critical value for a two sided test with $\alpha = .05$ is 29.

(S3) The value of the test statistic is not in the rejection region. There is no evidence to suggest the median speeds are different.

14.9.1   Tables for Wilcoxon (Mann–Whitney) U statistic

Given two sample of sizes m and n (with $m \leq n$) the Mann–Whitney U-statistic (see equation (14.24)) is used to test the hypothesis that the two samples are from populations with the same median. Rank all of the observations in ascending order of magnitude. Let W be the sum of the ranks assigned to the sample of size m. Then U is defined as

$$U = \frac{m(m + 2n + 1)}{2} - W \qquad (14.27)$$

The following tables present cumulative probability and are used to determine exact probabilities associated with this test statistic. If the null hypothesis is true, the body of the tables contains probabilities such that $\text{Prob}\,[U \le u]$.

Only *small* values of u are shown in the tables since the probability distribution for U is symmetric. For example, for $n = 3$ and $m = 2$ the probability distribution of U values is:

$$\text{Prob}\,[U = 0] = \text{Prob}\,[U = 1] = \text{Prob}\,[U = 5] = \text{Prob}\,[U = 6] = 0.1$$
$$\text{Prob}\,[U = 2] = \text{Prob}\,[U = 3] = \text{Prob}\,[U = 4] = 0.2$$

so that the distribution function is:

$$\text{Prob}\,[U \le 0] = 0.1, \qquad \text{Prob}\,[U \le 1] = 0.2, \qquad \text{Prob}\,[U \le 2] = 0.4,$$
$$\text{Prob}\,[U \le 3] = 0.6, \qquad \text{Prob}\,[U \le 4] = 0.8,$$
$$\text{Prob}\,[U \le 5] = 0.9, \qquad \text{Prob}\,[U \le 6] = 1$$

Example 14.74:   Consider the two samples $\{13, 9\}$ $(m = 2)$ and $\{12, 16, 14\}$ $(n = 3)$. Arrange the combined samples in rank order and box the values from the first sample:

$$\boxed{9} \quad 12 \quad \boxed{13} \quad 14 \quad 16$$
$$\text{rank} \quad 1 \quad\; 2 \quad\; 3 \quad\;\; 4 \quad\; 5 \qquad (14.28)$$

Compute the U statistic:

(a) $W = 1 + 3 = 4$

(b) $U = \dfrac{2(2 + 2 \cdot 3 + 1)}{2} - 4 = 5$

Using the tables below (and the comment above): $\text{Prob}\,[U \le 5] = .9$. There is little evidence to suggest the medians are different.

	$n = 3$		
u	$m = 1$	2	3
0	0.250	0.100	0.050
1	0.500	0.200	0.100
2	0.750	0.400	0.200
3		0.600	0.350
4			0.500
5			0.650

	$n = 4$			
u	$m = 1$	2	3	4
0	0.200	0.067	0.029	0.014
1	0.400	0.133	0.057	0.029
2	0.600	0.267	0.114	0.057
3		0.400	0.200	0.100
4		0.600	0.314	0.171
5			0.429	0.243
6			0.571	0.343
7				0.443
8				0.557

		$n = 5$			
u	$m = 1$	2	3	4	5
0	0.167	0.048	0.018	0.008	0.004
1	0.333	0.095	0.036	0.016	0.008
2	0.500	0.190	0.071	0.032	0.016
3	0.667	0.286	0.125	0.056	0.028
4		0.429	0.196	0.095	0.048
5		0.571	0.286	0.143	0.075
6			0.393	0.206	0.111
7			0.500	0.278	0.155
8			0.607	0.365	0.210
9				0.452	0.274
10				0.548	0.345
11					0.421
12					0.500
13					0.579

			$n = 6$			
u	$m = 1$	2	3	4	5	6
0	0.143	0.036	0.012	0.005	0.002	0.001
1	0.286	0.071	0.024	0.010	0.004	0.002
2	0.429	0.143	0.048	0.019	0.009	0.004
3	0.571	0.214	0.083	0.033	0.015	0.008
4		0.321	0.131	0.057	0.026	0.013
5		0.429	0.190	0.086	0.041	0.021
6		0.571	0.274	0.129	0.063	0.032
7			0.357	0.176	0.089	0.047
8			0.452	0.238	0.123	0.066
9			0.548	0.305	0.165	0.090
10				0.381	0.214	0.120
11				0.457	0.268	0.155
12				0.543	0.331	0.197
13					0.396	0.242
14					0.465	0.294
15					0.535	0.350
16						0.409
17						0.469
18						0.531

u	$m = 1$	2	3	4	5	6	7
				$n = 7$			
0	0.125	0.028	0.008	0.003	0.001	0.001	0.000
1	0.250	0.056	0.017	0.006	0.003	0.001	0.001
2	0.375	0.111	0.033	0.012	0.005	0.002	0.001
3	0.500	0.167	0.058	0.021	0.009	0.004	0.002
4	0.625	0.250	0.092	0.036	0.015	0.007	0.003
5		0.333	0.133	0.055	0.024	0.011	0.006
6		0.444	0.192	0.082	0.037	0.017	0.009
7		0.556	0.258	0.115	0.053	0.026	0.013
8			0.333	0.158	0.074	0.037	0.019
9			0.417	0.206	0.101	0.051	0.027
10			0.500	0.264	0.134	0.069	0.036
11			0.583	0.324	0.172	0.090	0.049
12				0.394	0.216	0.117	0.064
13				0.464	0.265	0.147	0.082
14				0.536	0.319	0.183	0.104
15					0.378	0.223	0.130
16					0.438	0.267	0.159
17					0.500	0.314	0.191
18					0.562	0.365	0.228
19						0.418	0.267
20						0.473	0.310
21						0.527	0.355
22							0.402
23							0.451
24							0.500
25							0.549

| $n = 8$ | | | | | | | | |
u	$m = 1$	2	3	4	5	6	7	8
0	0.111	0.022	0.006	0.002	0.001	0.000	0.000	0.000
1	0.222	0.044	0.012	0.004	0.002	0.001	0.000	0.000
2	0.333	0.089	0.024	0.008	0.003	0.001	0.001	0.000
3	0.444	0.133	0.042	0.014	0.005	0.002	0.001	0.001
4	0.556	0.200	0.067	0.024	0.009	0.004	0.002	0.001
5		0.267	0.097	0.036	0.015	0.006	0.003	0.001
6		0.356	0.139	0.055	0.023	0.010	0.005	0.002
7		0.444	0.188	0.077	0.033	0.015	0.007	0.003
8		0.556	0.248	0.107	0.047	0.021	0.010	0.005
9			0.315	0.141	0.064	0.030	0.014	0.007
10			0.388	0.184	0.085	0.041	0.020	0.010
11			0.461	0.230	0.111	0.054	0.027	0.014
12			0.539	0.285	0.142	0.071	0.036	0.019
13				0.341	0.177	0.091	0.047	0.025
14				0.404	0.218	0.114	0.060	0.032
15				0.467	0.262	0.141	0.076	0.041
16				0.533	0.311	0.172	0.095	0.052
17					0.362	0.207	0.116	0.065
18					0.416	0.245	0.140	0.080
19					0.472	0.286	0.168	0.097
20					0.528	0.331	0.198	0.117
21						0.377	0.232	0.139
22						0.426	0.268	0.164
23						0.475	0.306	0.191
24						0.525	0.347	0.221
25							0.389	0.253
26							0.433	0.287
27							0.478	0.323
28							0.522	0.360
29								0.399
30								0.439
31								0.480
32								0.520

14.9.2   Critical values for Wilcoxon (Mann–Whitney) statistic

The following tables give critical values for U for significance levels of 0.00005, 0.0001, 0.005, 0.01, 0.025, 0.05, and 0.10 for a one-tailed test. For a two-tailed test, the significance levels are doubled. If an observed U is equal to or less than the tabular value, the null hypothesis may be rejected at the level of significance indicated at the head of the table.

Critical values of U in the Mann–Whitney test
Critical values for the $\alpha = 0.10$ level of significance

m	n=1	2	3	4	5	6	7	8	9	10	11	12	13	14	15	16	17	18	19	20
1																				
2			0	0	1	1	1	2	2	3	3	4	4	5	5	5	6	6	7	7
3		0	1	1	2	3	4	5	5	6	7	8	9	10	10	11	12	13	14	15
4		0	1	3	4	5	6	7	9	10	11	12	13	15	16	17	18	20	21	22
5		1	2	4	5	7	8	10	12	13	15	17	18	20	22	23	25	27	28	30
6		1	3	5	7	9	11	13	15	17	19	21	23	25	27	29	31	34	36	38
7		1	4	6	8	11	13	16	18	21	23	26	28	31	33	36	38	41	43	46
8		2	5	7	10	13	16	19	22	24	27	30	33	36	39	42	45	48	51	54
9		2	5	9	12	15	18	22	25	28	31	35	38	41	45	48	52	55	58	62
10		3	6	10	13	17	21	24	28	32	36	39	43	47	51	54	58	62	66	70
11		3	7	11	15	19	23	27	31	36	40	44	48	52	57	61	65	69	73	78
12		4	8	12	17	21	26	30	35	39	44	49	53	58	63	67	72	77	81	86
13		4	9	13	18	23	28	33	38	43	48	53	58	63	68	74	79	84	89	94
14		5	10	15	20	25	31	36	41	47	52	58	63	69	74	80	85	91	97	102
15		5	10	16	22	27	33	39	45	51	57	63	68	74	80	86	92	98	104	110
16		5	11	17	23	29	36	42	48	54	61	67	74	80	86	93	99	106	112	119
17		6	12	18	25	31	38	45	52	58	65	72	79	85	92	99	106	113	120	127
8		6	13	20	27	34	41	48	55	62	69	77	84	91	98	106	113	120	128	135
19		7	14	21	28	36	43	51	58	66	73	81	89	97	104	112	120	128	135	143
20		7	15	22	30	38	46	54	62	70	78	86	94	102	110	119	127	135	143	151

Critical values of U in the Mann–Whitney test
Critical values for the $\alpha = 0.05$ level of significance

m	n=1	2	3	4	5	6	7	8	9	10	11	12	13	14	15	16	17	18	19	20
1																				
2					0	0	0	1	1	1	1	2	2	3	3	3	3	4	4	4
3			0	0	1	2	2	3	4	4	5	5	6	7	7	8	9	9	10	11
4			0	1	2	3	4	5	6	7	8	9	10	11	12	14	15	16	17	18
5		0	1	2	4	5	6	8	9	11	12	13	15	16	18	19	20	22	23	25
6		0	2	3	5	7	8	10	12	14	16	17	19	21	23	25	26	28	30	32
7		0	2	4	6	8	11	13	15	17	19	21	24	26	28	30	33	35	37	39
8		1	3	5	8	10	13	15	18	20	23	26	28	31	33	36	39	41	44	47
9		1	4	6	9	12	15	18	21	24	27	30	33	36	39	42	45	48	51	54
10		1	4	7	11	14	17	20	24	27	31	34	37	41	44	48	51	55	58	62
11		1	5	8	12	16	19	23	27	31	34	38	42	46	50	54	57	61	65	69
12		2	5	9	13	17	21	26	30	34	38	42	47	51	55	60	64	68	72	77
13		2	6	10	15	19	24	28	33	37	42	47	51	56	61	65	70	75	80	84
14		3	7	11	16	21	26	31	36	41	46	51	56	61	66	71	77	82	87	92
15		3	7	12	18	23	28	33	39	44	50	55	61	66	72	77	83	88	94	100
16		3	8	14	19	25	30	36	42	48	54	60	65	71	77	83	89	95	101	107
17		3	9	15	20	26	33	39	45	51	57	64	70	77	83	89	96	102	109	115
18		4	9	16	22	28	35	41	48	55	61	68	75	82	88	95	102	109	116	123
19		4	10	17	23	30	37	44	51	58	65	72	80	87	94	101	109	116	123	130
20		4	11	18	25	32	39	47	54	62	69	77	84	92	100	107	115	123	130	138

Critical values of U in the Mann–Whitney test
Critical values for the $\alpha = 0.025$ level of significance

m	$n=1$	2	3	4	5	6	7	8	9	10	11	12	13	14	15	16	17	18	19	20
1																				
2								0	0	0	0	1	1	1	1	1	2	2	2	2
3					0	1	1	2	2	3	3	4	4	5	5	6	6	7	7	8
4				0	1	2	3	4	4	5	6	7	8	9	10	11	11	12	13	14
5			0	1	2	3	5	6	7	8	9	11	12	13	14	15	17	18	19	20
6			1	2	3	5	6	8	10	11	13	14	16	17	19	21	22	24	25	27
7			1	3	5	6	8	10	12	14	16	18	20	22	24	26	28	30	32	34
8		0	2	4	6	8	10	13	15	17	19	22	24	26	29	31	34	36	38	41
9		0	2	4	7	10	12	15	17	20	23	26	28	31	34	37	39	42	45	48
10		0	3	5	8	11	14	17	20	23	26	29	33	36	39	42	45	48	52	55
11		0	3	6	9	13	16	19	23	26	30	33	37	40	44	47	51	55	58	62
12		1	4	7	11	14	18	22	26	29	33	37	41	45	49	53	57	61	65	69
13		1	4	8	12	16	20	24	28	33	37	41	45	50	54	59	63	67	72	76
14		1	5	9	13	17	22	26	31	36	40	45	50	55	59	64	69	74	78	83
15		1	5	10	14	19	24	29	34	39	44	49	54	59	64	70	75	80	85	90
16		1	6	11	15	21	26	31	37	42	47	53	59	64	70	75	81	86	92	98
17		2	6	11	17	22	28	34	39	45	51	57	63	69	75	81	87	93	99	105
18		2	7	12	18	24	30	36	42	48	55	61	67	74	80	86	93	99	106	112
19		2	7	13	19	25	32	38	45	52	58	65	72	78	85	92	99	106	113	119
20		2	8	14	20	27	34	41	48	55	62	69	76	83	90	98	105	112	119	127

Critical values of U in the Mann–Whitney test
Critical values for the $\alpha = 0.01$ level of significance

m	$n=1$	2	3	4	5	6	7	8	9	10	11	12	13	14	15	16	17	18	19	20
1																				
2													0	0	0	0	0	0	1	1
3							0	0	1	1	1	2	2	2	3	3	4	4	4	5
4					0	1	1	2	3	3	4	5	5	6	7	7	8	9	9	10
5				0	1	2	3	4	5	6	7	8	9	10	11	12	13	14	15	16
6				1	2	3	4	6	7	8	9	11	12	13	15	16	18	19	20	22
7			0	1	3	4	6	7	9	11	12	14	16	17	19	21	23	24	26	28
8			0	2	4	6	7	9	11	13	15	17	20	22	24	26	28	30	32	34
9			1	3	5	7	9	11	14	16	18	21	23	26	28	31	33	36	38	40
10			1	3	6	8	11	13	16	19	22	24	27	30	33	36	38	41	44	47
11			1	4	7	9	12	15	18	22	25	28	31	34	37	41	44	47	50	53
12			2	5	8	11	14	17	21	24	28	31	35	38	42	46	49	53	56	60
13		0	2	5	9	12	16	20	23	27	31	35	39	43	47	51	55	59	63	67
14		0	2	6	10	13	17	22	26	30	34	38	43	47	51	56	60	65	69	73
15		0	3	7	11	15	19	24	28	33	37	42	47	51	56	61	66	70	75	80
16		0	3	7	12	16	21	26	31	36	41	46	51	56	61	66	71	76	82	87
17		0	4	8	13	18	23	28	33	38	44	49	55	60	66	71	77	82	88	93
18		0	4	9	14	19	24	30	36	41	47	53	59	65	70	76	82	88	94	100
19		1	4	9	15	20	26	32	38	44	50	56	63	69	75	82	88	94	101	107
20		1	5	10	16	22	28	34	40	47	53	60	67	73	80	87	93	100	107	114

Critical values of U in the Mann–Whitney test

Critical values for the $\alpha = 0.005$ level of significance

m	1	2	3	4	5	6	7	8	9	10	11	12	13	14	15	16	17	18	19	20
1																				
2																			0	0
3									0	0	0	1	1	1	2	2	2	2	3	3
4						0	0	1	1	2	2	3	3	4	5	5	6	6	7	8
5					0	1	1	2	3	4	5	6	7	7	8	9	10	11	12	13
6				0	1	2	3	4	5	6	7	9	10	11	12	13	15	16	17	18
7				0	1	3	4	6	7	9	10	12	13	15	16	18	19	21	22	24
8				1	2	4	6	7	9	11	13	15	17	18	20	22	24	26	28	30
9			0	1	3	5	7	9	11	13	16	18	20	22	24	27	29	31	33	36
10			0	2	4	6	9	11	13	16	18	21	24	26	29	31	34	37	39	42
11			0	2	5	7	10	13	16	18	21	24	27	30	33	36	39	42	45	48
12			1	3	6	9	12	15	18	21	24	27	31	34	37	41	44	47	51	54
13			1	3	7	10	13	17	20	24	27	31	34	38	42	45	49	53	57	60
14			1	4	7	11	15	18	22	26	30	34	38	42	46	50	54	58	63	67
15			2	5	8	12	16	20	24	29	33	37	42	46	51	55	60	64	69	73
16			2	5	9	13	18	22	27	31	36	41	45	50	55	60	65	70	74	79
17			2	6	10	15	19	24	29	34	39	44	49	54	60	65	70	75	81	86
18			2	6	11	16	21	26	31	37	42	47	53	58	64	70	75	81	87	92
19		0	3	7	12	17	22	28	33	39	45	51	57	63	69	74	81	87	93	99
20		0	3	8	13	18	24	30	36	42	48	54	60	67	73	79	86	92	99	105

Critical values of U in the Mann–Whitney test

Critical values for the $\alpha = 0.001$ level of significance

m	1	2	3	4	5	6	7	8	9	10	11	12	13	14	15	16	17	18	19	20
1																				
2																				
3																	0	0	0	0
4										0	0	0	1	1	1	2	2	3	3	3
5								0	1	1	2	2	3	3	4	5	5	6	7	7
6							0	1	2	3	4	4	5	6	7	8	9	10	11	12
7						0	1	2	3	5	6	7	8	9	10	11	13	14	15	16
8					0	1	2	4	5	6	8	9	11	12	14	15	17	18	20	21
9					1	2	3	5	7	8	10	12	14	15	17	19	21	23	25	26
10				0	1	3	5	6	8	10	12	14	17	19	21	23	25	27	29	32
11				0	2	4	6	8	10	12	15	17	20	22	24	27	29	32	34	37
12				0	2	4	7	9	12	14	17	20	23	25	28	31	34	37	40	42
13				1	3	5	8	11	14	17	20	23	26	29	32	35	38	42	45	48
14				1	3	6	9	12	15	19	22	25	29	32	36	39	43	46	50	54
15				1	4	7	10	14	17	21	24	28	32	36	40	43	47	51	55	59
16				2	5	8	11	15	19	23	27	31	35	39	43	48	52	56	60	65
17			0	2	5	9	13	17	21	25	29	34	38	43	47	52	57	61	66	70
18			0	3	6	10	14	18	23	27	32	37	42	46	51	56	61	66	71	76
19			0	3	7	11	15	20	25	29	34	40	45	50	55	60	66	71	77	82
20			0	3	7	12	16	21	26	32	37	42	48	54	59	65	70	76	82	88

Critical values of U in the Mann–Whitney test
Critical values for the $\alpha = 0.0005$ level of significance

m	$n=1$	2	3	4	5	6	7	8	9	10	11	12	13	14	15	16	17	18	19	20
1																				
2																				
3																				
4													0	0	0	1	1	1	2	2
5									0	0	1	1	2	2	3	3	4	4	5	5
6								0	1	2	2	3	4	5	5	6	7	8	8	9
7							0	1	2	3	4	5	6	7	8	9	10	11	13	14
8						0	1	2	4	5	6	7	9	10	11	13	14	15	17	18
9					0	1	2	4	5	7	8	10	11	13	15	16	18	20	21	23
10					0	2	3	5	7	8	10	12	14	16	18	20	22	24	26	28
11					1	2	4	6	8	10	12	15	17	19	21	24	26	28	31	33
12					1	3	5	7	10	12	15	17	20	22	25	27	30	33	35	38
13				0	2	4	6	9	11	14	17	20	23	25	28	31	34	37	40	43
14				0	2	5	7	10	13	16	19	22	25	29	32	35	39	42	45	49
15				0	3	5	8	11	15	18	21	25	28	32	36	39	43	46	50	54
16				1	3	6	9	13	16	20	24	27	31	35	39	43	47	51	55	59
17				1	4	7	10	14	18	22	26	30	34	39	43	47	51	56	60	65
18				1	4	8	11	15	20	24	28	33	37	42	46	51	56	61	65	70
19				2	5	8	13	17	21	26	31	35	40	45	50	55	60	65	70	76
20				2	5	9	14	18	23	28	33	38	43	49	54	59	65	70	76	81

14.10 WILCOXON SIGNED-RANK TEST

Assumptions: Let $X_1, X_2, \ldots, X_n$ be a random sample from a continuous symmetric distribution.

Hypothesis test:

H_0: $\tilde{\mu} = \tilde{\mu}_0$

H_a: $\tilde{\mu} > \tilde{\mu}_0, \quad \tilde{\mu} < \tilde{\mu}_0, \quad \tilde{\mu} \neq \tilde{\mu}_0$

Rank the absolute differences $|X_1 - \tilde{\mu}_0|, |X_2 - \tilde{\mu}_0|, \ldots, |X_n - \tilde{\mu}_0|$. Equal absolute differences are assigned the mean rank for their positions.

TS: $T_+ =$ the sum of the ranks corresponding to the positive differences $(X_i - \tilde{\mu}_0)$.

RR: $T_+ \geq c_1, \quad T_+ \leq c_2, \quad T_+ \geq c$ or $T_+ \leq n(n+1) - c$

where c_1, c_2, and c are critical values for the Wilcoxon signed-rank statistic (see table on page 372) such that $\text{Prob}\,[T_+ \geq c_1] \approx \alpha$, $\text{Prob}\,[T_+ \leq c_2] \approx \alpha$, and $\text{Prob}\,[T_+ \geq c] \approx \alpha/2$.

Any observed difference $(x_i - \tilde{\mu}_0) = 0$ is excluded from the test and the sample size is reduced accordingly.

The normal approximation: When $n \geq 20$, T_+ has approximately a normal distribution with

$$\mu_{T_+} = \frac{n(n+1)}{4} \quad \text{and} \quad \sigma^2_{T_+} = \frac{n(n+1)(2n+1)}{24}. \tag{14.29}$$

The random variable

$$Z = \frac{T_+ - \mu_{T_+}}{\sigma_{T_+}} \tag{14.30}$$

has approximately a standard normal distribution when H_0 is true.

See D. J. Sheskin, *Handbook of Parametric and Nonparametric Statistical Procedures*, CRC Press LLC, Boca Raton, FL, 1997, pages 83–94.

CHAPTER 15

Quality Control and Risk Analysis

Contents

15.1 QUALITY ASSURANCE

15.1.1 Control charts

Expression	Meaning
$\overline{x}$ chart	control chart for means
R chart	control chart for sample ranges
σ chart	control chart for standard deviations
p chart	fraction defective chart
c chart	number of defects chart
LCL	lower control limit
UCL	upper control limit

Suppose the population mean μ and the population standard deviation σ are unknown. For k samples of size n let $\overline{x}_i$ and R_i be the sample mean and sample range for the ith sample, respectively. The *average* mean $\overline{x}$ and the *average* range $\overline{R}$ are defined by

$$\overline{x} = \frac{1}{k}\sum_{i=1}^{k}\overline{x}_i \qquad \overline{R} = \frac{1}{k}\sum_{i=1}^{k}R_i \qquad (15.1)$$

Suppose the population proportion p or the number of defects is unknown. For k samples assume sample i has n_i items, c_i defects, and e_i defective items.

383

Note that a single defective item may have many defects. Define $\bar{p}$ and $\bar{c}$ by

$$\bar{p} = \frac{e_1 + e_2 + \cdots + e_k}{n_1 + n_2 + \cdots + n_k} \qquad \bar{c} = \frac{1}{k}\sum_{i=1}^{k} c_i \qquad (15.2)$$

Type of chart central line	lower control limit (LCL)	upper control limit (UCL)
$\bar{x}$ chart (μ, σ known)		
μ	$\mu - A\sigma$	$\mu + A\sigma$
$\bar{x}$ chart (μ, σ unknown)		
$\bar{x}$	$\bar{x} - A_2\bar{R}$	$\bar{x} + A_2\bar{R}$
R chart (σ known)		
$d_2\sigma$	$D_1\sigma$	$D_2\sigma$
R chart (σ unknown)		
$\bar{R}$	$D_3\bar{R}$	$D_4\bar{R}$
p chart (based on past data)		
$\bar{p}$	$\min\left(0, \bar{p} - 3\sqrt{\frac{\bar{p}(1-\bar{p})}{n}}\right)$	$\bar{p} + 3\sqrt{\frac{\bar{p}(1-\bar{p})}{n}}$
number-of-defectives chart		
$n\bar{p}$	$n\bar{p} - 3\sqrt{n\bar{p}(1-\bar{p})}$	$n\bar{p} + 3\sqrt{n\bar{p}(1-\bar{p})}$
c chart		
$\bar{c}$	$\bar{c} - 3\sqrt{\bar{c}}$	$\bar{c} + 3\sqrt{\bar{c}}$

where n is the sample size, $A = 3/\sqrt{n}$, $D_1 = d_2 - 3d_3$, $D_2 = d_2 + 3d_3$, $D_3 = D_1/d_2$, and $D_4 = D_2/d_2$. Values of $\{A, A_2, d_2, d_3, D_1, D_2, D_3, D_4\}$ are given in Table 15.1[1].

---

[1]This table reproduced, by permission, from *ASTM Manual on Quality Control of Materials*, American Society for Testing and Materials, Philadelphia, PA, 1951.

n	Chart for averages			Chart for standard deviations					Chart for ranges				
	A	A_1	A_2	c_2	B_1	B_2	B_3	B_4	d_2	D_1	D_2	D_3	D_4
2	2.121	3.760	1.880	0.5642	0	1.843	0	3.267	1.128	0	3.686	0	3.267
3	1.732	2.394	1.023	0.7236	0	1.858	0	2.568	1.693	0	4.358	0	2.575
4	1.500	1.880	0.729	0.7979	0	1.808	0	2.266	2.059	0	4.698	0	2.282
5	1.342	1.596	0.577	0.8407	0	1.756	0	2.089	2.326	0	4.918	0	2.115
6	1.225	1.410	0.483	0.8686	0.026	1.711	0.030	1.970	2.534	0	5.078	0	2.004
7	1.134	1.277	0.419	0.8882	0.105	1.672	0.118	1.882	2.704	0.205	5.203	0.076	1.924
8	1.061	1.175	0.373	0.9027	0.167	1.638	0.185	1.815	2.847	0.387	5.307	0.136	1.864
9	1.000	1.094	0.337	0.9139	0.219	1.609	0.239	1.761	2.970	0.546	5.394	0.184	1.816
10	0.949	1.028	0.308	0.9227	0.262	1.584	0.284	1.716	3.078	0.687	5.469	0.223	1.777
11	0.905	0.973	0.285	0.9300	0.299	1.561	0.321	1.679	3.173	0.812	5.534	0.256	1.744
12	0.866	0.925	0.266	0.9359	0.331	1.541	0.354	1.646	3.258	0.924	5.592	0.284	1.716
13	0.832	0.884	0.249	0.9410	0.359	1.523	0.382	1.618	3.336	1.026	5.646	0.308	1.692
14	0.802	0.848	0.235	0.9453	0.384	1.507	0.406	1.594	3.407	1.121	5.693	0.329	1.671
15	0.775	0.816	0.223	0.9490	0.406	1.492	0.428	1.572	3.472	1.207	5.737	0.348	1.652

Table 15.1: Parameter values for control charts (reproduced by permission)

15.1.2   Abnormal distributions of points in control charts

Abnormality	Description
Sequence	Seven or more consecutive points on one side of the center line. Denotes the average value has shifted.
Bias	Fewer than seven consecutive points on one side of the center line, but most of the points are on that side. • 10 of 11 consecutive points • 12 or more of 14 consecutive points • 14 or more of 17 consecutive points • 16 or more of 20 consecutive points
Trend	Seven or more consecutive rising or falling points.
Approaching the limit	Two out of three or three or more out of seven consecutive points are more than two thirds the distance between the center line and a control limit.
Periodicity	The data points vary in a regular periodic pattern.

15.2   ACCEPTANCE SAMPLING

Expression	Meaning
AQL	acceptable quality level
AOQ	average outgoing quality
AOQL	average outgoing quality limit (maximum value of AOQ for varying incoming quality)
LTPD	lot tolerance percent defective
producer's risk	Type I error (percentage of "good" lots rejected)
consumer's risk	Type II error (percentage of "bad" lots accepted)

Military standard 105 D is a widely used sampling plan. There are three general levels of inspection corresponding to different consumer's risks. (Inspection level II is usually chosen; level I uses smaller sample sizes and level II uses larger sample sizes.) There are also three types of inspections: normal, tightened, and reduced. Tables are available for single, double, and multiple sampling.

To use MIL-STD-105 D for single sampling, determine the sample size code letter from Table 15.2. Using this sample size code letter find the sample size and the acceptance and rejection numbers from Table 15.3.

Lot or batch size			general inspection levels		
			I	II	III
2	to	8	A	A	B
9	to	15	A	B	C
16	to	25	B	C	D
26	to	50	C	D	E
51	to	90	C	E	F
91	to	150	D	F	G
151	to	280	E	G	H
281	to	500	F	H	J
501	to	1,200	G	J	K
1,201	to	3,200	H	K	L
3,201	to	10,000	J	L	M
10,001	to	35,000	K	M	N
35,001	to	150,000	L	N	P
150,001	to	500,000	M	P	Q
500,001	and	over	N	Q	R

Table 15.2: Sample size code letters for MIL-STD-105 D

Example 15.75: Suppose that MIL-STD-105D is to be used with incoming lots of 1,000 items, inspection level II is to be used in conjunction with normal inspection, and an AQL of 2.5 percent is desired. How should the inspections be carried out?

Solution:

(S1) From table 15.2 the sample size code letter is J.

(S2) From the table on page 388, for column J, the lot size should be 80. Using the row labeled 2.5 the acceptance number is 5 and the rejection number is 6.

(S3) Thus, if a single sample of size 80 (selected randomly from each lot of 1,000 items) contains 5 or fewer defectives then the lot is to be accepted. If it contains 6 or more defectives, then the lot is to be rejected.

15.2.1   Sequential sampling

We need to inspect a lot. Suppose the two sequences of numbers $\{a_n\}$ (for *accept*) and $\{r_n\}$ (for *reject*) are given. Elements from the lot are sequentially taken ($n = 1, 2, \ldots$); after each element is selected the total number of defective elements is determined. The lot is accepted after n elements if the number of defectives is less than or equal to a_n. The lot is rejected after n elements if the number of defectives is greater than or equal to r_n. Sampling continues as long as the number of defectives in the sample of size n falls between a_n and r_n.

Sample size code letter and sample size

AQL	R 2000	Q 1250	P 800	N 500	M 315	L 200	K 125	J 80	H 50	G 32	F 20	E 13	D 8	C 5	B 3	A 2
0.010	↓	0|1	↑	↑	↑	↑	↑	↑	↑	↑	↑	↑	↑	↑	↑	↑
0.015	↑	↓	0|1	↑	↑	↑	↑	↑	↑	↑	↑	↑	↑	↑	↑	↑
0.025	1|2	↑	↓	0|1	↑	↑	↑	↑	↑	↑	↑	↑	↑	↑	↑	↑
0.040	2|3	1|2	↑	↓	0|1	↑	↑	↑	↑	↑	↑	↑	↑	↑	↑	↑
0.065	3|4	2|3	1|2	↑	↓	0|1	↑	↑	↑	↑	↑	↑	↑	↑	↑	↑
0.10	5|6	3|4	2|3	1|2	↑	↓	0|1	↑	↑	↑	↑	↑	↑	↑	↑	↑
0.15	7|8	5|6	3|4	2|3	1|2	↑	↓	0|1	↑	↑	↑	↑	↑	↑	↑	↑
0.25	10|11	7|8	5|6	3|4	2|3	1|2	↑	↓	0|1	↑	↑	↑	↑	↑	↑	↑
0.40	14|15	10|11	7|8	5|6	3|4	2|3	1|2	↑	↓	0|1	↑	↑	↑	↑	↑	↑
0.65	21|22	14|15	10|11	7|8	5|6	3|4	2|3	1|2	↑	↓	0|1	↑	↑	↑	↑	↑
1.0	↓	21|22	14|15	10|11	7|8	5|6	3|4	2|3	1|2	↑	↓	0|1	↑	↑	↑	↑
1.5	↓	↓	21|22	14|15	10|11	7|8	5|6	3|4	2|3	1|2	↑	↓	0|1	↑	↑	↑
2.5	↓	↓	↓	21|22	14|15	10|11	7|8	5|6	3|4	2|3	1|2	↑	↓	0|1	↑	↑
4.0	↓	↓	↓	↓	21|22	14|15	10|11	7|8	5|6	3|4	2|3	1|2	↑	↓	0|1	↑
6.5	↓	↓	↓	↓	↓	21|22	14|15	10|11	7|8	5|6	3|4	2|3	1|2	↑	↓	0|1
10	↓	↓	↓	↓	↓	↓	21|22	14|15	10|11	7|8	5|6	3|4	2|3	1|2	↑	↑
15	↓	↓	↓	↓	↓	↓	↓	21|22	14|15	10|11	7|8	5|6	3|4	2|3	1|2	↑
25	↓	↓	↓	↓	↓	↓	↓	↓	21|22	14|15	10|11	7|8	5|6	3|4	2|3	1|2
40	↓	↓	↓	↓	↓	↓	↓	↓	↓	21|22	14|15	10|11	7|8	5|6	3|4	2|3
65	↓	↓	↓	↓	↓	↓	↓	↓	↓	↓	21|22	14|15	10|11	7|8	5|6	3|4
100	↓	↓	↓	↓	↓	↓	↓	↓	↓	↓	↓	21|22	14|15	10|11	7|8	5|6
150	↓	↓	↓	↓	↓	↓	↓	↓	↓	↓	↓	30|31	21|22	14|15	10|11	7|8
250	↓	↓	↓	↓	↓	↓	↓	↓	↓	↓	↓	44|45	30|31	21|22	14|15	10|11
400	↓	↓	↓	↓	↓	↓	↓	↓	↓	↓	↓	↓	44|45	30|31	21|22	14|15
650	↓	↓	↓	↓	↓	↓	↓	↓	↓	↓	↓	↓	↓	44|45	30|31	21|22
1000	↓	↓	↓	↓	↓	↓	↓	↓	↓	↓	↓	↓	↓	↓	44|45	30|31

AQL = Acceptable quality level (normal inspection).

Ac|Re = Accept if Ac or fewer are found, reject if Re or more are found.

← = Use first sampling procedure to left.

→ = Use first sampling procedure to right. If sample size equals, or exceeds, lot or batch size, do 100 percent inspection.

Table 15.3: Master table for single sampling inspection (normal inspection) MIL-STD-105 D.

A sequential sampling plan with AQL= p_0, LTPD= p_1, producer's risk α, and consumer's risk β is given by

$$a_n = \left| \frac{\log\left(\frac{\beta}{1-\alpha}\right) + n\log\left(\frac{1-p_0}{1-p_1}\right)}{\log\left(\frac{p_1}{p_0}\right) - \log\left(\frac{1-p_1}{1-p_0}\right)} \right|$$

$$r_n = \left| \frac{\log\left(\frac{1-\beta}{\alpha}\right) + n\log\left(\frac{1-p_0}{1-p_1}\right)}{\log\left(\frac{p_1}{p_0}\right) - \log\left(\frac{1-p_1}{1-p_0}\right)} \right|$$

(15.3)

15.2.1.1 Sequential probability ratio tests

A sequential probability ratio test is a sequential sampling strategy in which the ratio of probabilities (based on the hypotheses) is used. Given two simple hypotheses (H_0 and H_1) and m observations, compute

(a) P_{0m} = Prob (observations | H_0)

(b) P_{1m} = Prob (observations | H_1)

(c) $v_m = P_{1m}/P_{0m}$

and then follow the decision rule given by:

(a) If $v_m \geq \dfrac{1-\beta}{\alpha}$ then reject H_0.

(b) If $v_m \leq \dfrac{\beta}{1-\alpha}$ then reject H_1.

(c) If $\dfrac{\beta}{1-\alpha} < v_m < \dfrac{1-\beta}{\alpha}$ then make another observation.

Note that the number of samples taken is not fixed a priori, but is determined as sampling occurs.

Example 15.76: Lot acceptance

Let θ denote the fraction of defective items. Two simple hypotheses are H_0: $\theta = \theta_0 = 0.05$ and H_1: $\theta = \theta_1 = 0.15$. Choose $\alpha = .05$ and $\beta = .10$ (i.e., reject lot with $\theta = \theta_0$ about 5% of the time, accept lot with $\theta = \theta_1$ about 10% of the time). If, after m observations, there are d defective items, then

$$P_{im} = \binom{m}{d}\theta_i^d(1-\theta_i)^{m-d} \quad \text{and} \quad v_m = \left(\frac{\theta_1}{\theta_0}\right)^d\left(\frac{1-\theta_1}{1-\theta_0}\right)^{m-d}$$

or $v_m = 3^d(0.895)^{m-d}$ using the above numbers. The critical values are $\frac{\beta}{1-\alpha} = 0.105$ and $\frac{1-\beta}{\alpha} = 18$. The decision to perform another observation depends on whether or not

$$0.105 \leq 3^d(0.895)^{m-d} \leq 18$$

Taking logarithms, a $(m-d, d)$ control chart can be drawn with the the following lines: $d = 0.101(m-d) - 2.049$ and $d = 0.101(m-d) + 2.63$. On the figure below, a sample path leading to rejection of H_0 has been indicated:

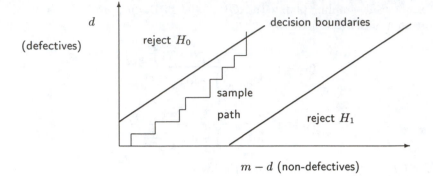

$$m - d \text{ (non-defectives)}$$

15.2.1.2   Two-sided mean test

Let X be normally distributed with unknown mean μ and known standard deviation σ. Consider the two simple hypotheses H_0: $\mu = \mu_0$ and H_1: $\mu = \mu_1$. If Y is the sum of the first m observations of X, then a (Y, m) control chart is constructed with the two lines:

$$Y = \frac{\mu_0 + \mu_1}{2} m + \frac{\sigma^2}{\mu_1 - \mu_0} \log \frac{\beta}{1 - \alpha}$$
$$Y = \frac{\mu_0 + \mu_1}{2} m + \frac{\sigma^2}{\mu_1 - \mu_0} \log \frac{1 - \beta}{\alpha}$$

$$(15.4)$$

15.2.1.3   One-sided variance test

Consider testing the null hypothesis that $\sigma = \sigma_1$ against the alternative that $\sigma = \sigma_2 > \sigma_1$. If α is the assigned risk of rejecting the null hypothesis when σ equals σ_1 and β is the assigned risk of accepting the null hypothesis when σ equal σ_2, then the sequential plan is as follows:

1. Define the quantities

$$g = 0.43429 \left(\frac{1}{\sigma_1^2} - \frac{1}{\sigma_2^2} \right)$$

$$a = \log_{10} \frac{1 - \beta}{\alpha} \qquad h_2 = \frac{2a}{g}$$

$$b = \log_{10} \frac{1 - \alpha}{\beta} \qquad h_1 = \frac{2b}{g}$$

$$(15.5)$$

$$s = \frac{\log_{10}(\sigma_2^2 / \sigma_1^2)}{g}$$

2. Define the acceptance and rejection lines as

$$Z_1^{(n)} = -h_1 + s(n - 1) \qquad \text{(lower)}$$
$$Z_2^{(n)} = h_2 + s(n - 1) \qquad \text{(upper)}$$

$$(15.6)$$

The sequential test is carried out as follows:

1. Let n stand for the number of sample items inspected
2. Let Z stand for the sum of squared deviations from the sample mean:

$$Z = \sum_{i=1}^{n} (x_i - \bar{x})^2 = \frac{n \sum_{i=1}^{n} x_i^2 - \left(\sum_{i=1}^{n} x_i \right)^2}{n} \tag{15.7}$$

3. Test against limits

   (a) If $Z < Z_1^{(n)}$ then accept the null hypothesis.
   (b) If $Z > Z_2^{(n)}$ then reject the null hypothesis.
   (c) If neither inequality is true, then take another sample.

15.3 RELIABILITY

1. The *reliability* of a product is the probability that the product will function within specified limits for at least a specified period of time.
2. A *series system* is one in which the entire system will fail if any of its components fail.
3. A *parallel system* is one in which the entire system will fail only if all of its components fail.
4. Let R_i denote the reliability of the ith component.
5. Let R_s denote the reliability of a series system.
6. Let R_p denote the reliability of a parallel system.

The *product law of reliabilities* states

$$R_s = \prod_{i=1}^{n} R_i \tag{15.8}$$

The *product law of unreliabilities* states

$$R_p = 1 - \prod_{i=1}^{n} (1 - R_i) \tag{15.9}$$

15.3.1 Failure time distributions

1. Let the probability of a component failing between times t and $t + \Delta t$ be $f(t)\Delta t$.
2. The probability that a component will fail on the interval from 0 to t is

$$F(t) = \int_0^t f(x)\, dx \tag{15.10}$$

3. The *reliability function* is the probability that a component survives to time t

$$R(t) = 1 - F(t) \tag{15.11}$$

4. The *instantaneous failure rate*, $Z(t)$, is the average rate of failure in the interval from t to $t + \Delta t$, given that the component survived to time t

$$Z(t) = \frac{f(t)}{R(t)} = \frac{f(t)}{1 - F(t)} \tag{15.12}$$

Note the relationships:

$$R(t) = e^{-\int_0^t Z(x)\,dx} \qquad\qquad f(t) = Z(t)e^{-\int_0^t Z(x)\,dx} \tag{15.13}$$

Example 15.77:   If $f(t) = \alpha\beta t^{\beta-1}e^{\alpha t^\beta}$ with $\alpha > 0$ and $\beta > 0$, the probability distribution function for a Weibull random variable, then the failure rate is $Z(t) = \alpha\beta t^{\beta-1}$ and $R(t) = e^{-\alpha t^\beta}$. Note that failure rate decreases with time if $\beta < 1$ and increases with time if $\beta > 1$.

15.3.1.1   *Use of the exponential distribution*

If the failure rate is a constant $Z(t) = \alpha$ (with $\alpha > 0$) then $f(t) = \alpha e^{-\alpha t}$ (for $t > 0$) which is the probability density function for an exponential random variable. If a failed component is replaced with another having the same constant failure rate α, then the occurrence of failures is a Poisson process. The constant $1/\alpha$ is called the *mean time between failures* (MTBF). The reliability function is $R(t) = e^{-\alpha t}$.

If a series system has n components, each with constant failure rate $\{\alpha_i\}$, then

$$R_s(t) = \exp\left(-\sum_{i=1}^{n} \alpha_i\right) \tag{15.14}$$

The MTBF for the series system is μ_s

$$\mu_s = \frac{1}{\frac{1}{\mu_1} + \frac{1}{\mu_2} + \cdots + \frac{1}{\mu_n}} \tag{15.15}$$

If a parallel system has n components, each with identical constant failure rate α, then the MTBF for the parallel system is μ_p

$$\mu_p = \frac{1}{\alpha}\left(1 + \frac{1}{2} + \cdots + \frac{1}{n}\right) \tag{15.16}$$

15.4   RISK ANALYSIS AND DECISION RULES

Suppose knowledge of a specific state of a system is desired, and those states can be delineated as $\{\theta_1, \theta_2, \ldots\}$. For example, in a weather application, the states might be θ_1 for *rain* and θ_2 for *no rain*. Decision rules are actions that may be taken based on the state of a system. For example, in deciding whether to go on a trip, there are the decision rules: *stay home, go with an umbrella*, and *go without an umbrella*.

Possible actions		System state	
		θ_1 (rain)	θ_2 (no rain)
Stay home	a_1	4	4
Go without an umbrella	a_2	5	0
Go with an umbrella	a_3	2	5

Table 15.4: An example loss function $\ell(\theta, a)$

A *loss function* is an arbitrary function that depends on a specific state and a decision rule. For example, consider the loss function $\ell(\theta, a)$ given in Table 15.4. It is possible to determine the "best" decision, under different models, even without obtaining any data.

- **Minimax principle**

  With this principle one should expect and prepare for the worst. That is, for each action it is possible to determine the minimum possible loss that may be incurred. This loss is assigned to each action; the action with the smallest (or minimum) maximum loss is the action chosen.

  For the data in Table 15.4 the maximum loss is 4 for action a_1 and 5 for either of the actions a_2 or a_3. Under a minimax principle, the chosen action would be a_1 and the minimax loss would be 4.

- **Minimax principle for mixed actions**

  It is possible to minimize the maximum loss when the action taken is a statistical distribution, $\mathbf{p}$, of actions. Assume that action a_i is taken with probability p_i (with $p_1 + p_2 + p_3 = 1$). Then the expected loss $L(\theta_i)$ is given by $L(\theta_i) = \mathrm{E}_a[\ell(\theta_i, a)] = p_1 \ell(\theta_i, a_1) + p_2 \ell(\theta_i, a_2) + p_3 \ell(\theta_i, a_3)$. The data in Table 15.4 result in the following expected losses:

  $$\begin{bmatrix} L(\theta_1) \\ L(\theta_2) \end{bmatrix} = p_1 \begin{bmatrix} 4 \\ 4 \end{bmatrix} + p_2 \begin{bmatrix} 5 \\ 0 \end{bmatrix} + p_3 \begin{bmatrix} 2 \\ 5 \end{bmatrix} \tag{15.17}$$

  It can be shown that the minimax point of this mixed action case has to satisfy $L(\theta_1) = L(\theta_2)$. Solving equation (15.17) with this constraint leads to $5p_2 = 3p_3$. Using this and $p_1 + p_2 + p_3 = 1$ in equation (15.17) results in $L(\theta_1) = L(\theta_2) = 4 - 7p_3/5$. This indicates that p_3 should be as large as possible. Hence, the maximum value is obtained by the mixed distribution $\mathbf{p} = (\frac{0}{8}, \frac{3}{8}, \frac{5}{8})$.

  Hence, if action a_2 is chosen 3/8's of the time, and action a_3 is chosen 5/8's of the time, then the minimax loss is equal to $L = 25/8$. This is a smaller loss than using a pure strategy of only choosing a single action.

- **Bayes actions**

  If the probability distribution of the states $\{\theta_1, \theta_2, \dots\}$ is given by the density function $g(\theta_i)$, then the loss has a known distribution with an expectation of $B(a) = E_i[\ell(\theta_i, a)] = \sum_i g(\theta_i)\ell(\theta_i, a)$. This quantity is known as the *Bayes loss* for action a. A Bayes action is an action that minimizes the Bayes loss.

  For example, assuming that the prior distribution is given by $g(\theta_1) = 0.4$ and $g(\theta_2) = 0.6$, then $B(a_1) = 4$, $B(a_2) = 2$, and $B(a_3) = 3.8$. This leads to the choice of action a_2.

A course of action can also be based on data about the states of interest. For example, a weather report Z will give data for the predictions of *rain* and *no rain*. Continuing the example, assume that the correctness of these predictions is given as follows:

		θ_1 (rain)	θ_2 (no rain)
Predict rain	z_1	0.8	0.1
Predict no rain	z_2	0.2	0.9

That is, when it will rain, then the prediction is correct 80% of the time.

A *decision function* is an assignment of data to actions. Since there are finitely many possible actions and finitely many possible values of Z, the number of decision functions is finite. In the example there are $3^2 = 9$ possible decision functions, $\{d_1, d_2, \dots, d_9\}$; they are defined to be:

	Decision functions								
	d_1	d_2	d_3	d_4	d_5	d_6	d_7	d_8	d_9
Predict z_1, take action	a_1	a_2	a_3	a_1	a_2	a_1	a_3	a_2	a_3
Predict z_2, take action	a_1	a_2	a_3	a_2	a_1	a_3	a_1	a_3	a_2

The *risk function* $R(\theta, d_i)$ is the expected value of the loss when a specific decision function is being used: $R(\theta, d_i) = E_Z[\ell(\theta, d_i(Z))]$. It is straightforward to compute the risk function for all values of $\{d_i\}$ and $\{a_j\}$. This results in the following values:

Risk function evaluation		
Decision Function	θ_1 (rain)	θ_2 (no rain)
d_1	4	4
d_2	5	0
d_3	2	5
d_4	4.2	0.4
d_5	4.8	3.6
d_6	3.6	4.9
d_7	2.4	4.1
d_8	4.4	4.5
d_9	2.6	0.5

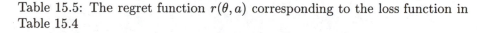

	θ_1 (rain)	θ_2 (no rain)
a_1	2	4
a_2	3	0
a_3	0	5

Table 15.5: The regret function $r(\theta, a)$ corresponding to the loss function in Table 15.4

This array can now be treated as though it gave the loss function in a no–data problem. The minimax principle for mixed action results in the "best" solution being rule d_3 for $\frac{7}{17}$'s of the time and rule d_9 for $\frac{10}{17}$'s of the time. This leads to a minimax loss of $\frac{40}{17}$. Before the data Z is received, the minimax loss was $\frac{25}{8}$. Hence, the data Z is "worth" $\frac{25}{8} - \frac{40}{17} = \frac{105}{136}$ in using the minimax approach.

The *regret function* (also called the *opportunity loss function*) $r(\theta, a)$ is the loss, $\ell(\theta, a)$, minus the minimum loss for the given θ: $r(\theta, a) = \ell(\theta, a) - \min_b \ell(\theta, b)$. For each state, the least loss is determined if that state were known to be true. This is the contribution to loss that even a good decision cannot avoid. The quantity $r(\theta, a)$ represents the loss that could have been avoided had the state been known—hence the term regret.

For the loss function example in Table 15.4, the minimum loss for $\theta = \theta_1$ is 2, and the minimum loss for $\theta = \theta_2$ is 0. Hence, the regret function is as given in Table 15.5.

Most of the computations performed for a loss function could also be performed with the risk function. If the minimax principle is used to determine the "best" action, then, in this example, the "best" action is a_2.

CHAPTER 16

General Linear Models

Contents

16.1   NOTATION

In this chapter, matrices are denoted by bold–face capital letters; for example, if the matrix $\mathbf{A}$ has m rows and n columns, then $\mathbf{A} = \mathbf{A}_{mn}$ and

$$\mathbf{A} = \begin{bmatrix} a_{11} & a_{12} & \cdots & a_{1n} \\ a_{21} & a_{22} & \cdots & a_{2n} \\ \vdots & \vdots & \ddots & \vdots \\ a_{m1} & a_{m2} & \cdots & a_{mn} \end{bmatrix}.$$

In general, column vectors will be denoted by lower–case bold–face letters. For example,

$$\mathbf{x}^{\mathrm{T}} = [x_1 \; x_2 \; \cdots \; x_n], \quad \boldsymbol{\beta}^{\mathrm{T}} = [\beta_1 \; \beta_2 \; \cdots \; \beta_k]. \tag{16.1}$$

If necessary, the number of rows in a column vector is indicated with a subscript, for example, $\mathbf{x}_n$ has n rows.

(1) Some special column vectors are $\mathbf{T}$ (vector of treatment totals), $\mathbf{B}$ (vector of block totals), $\mathbf{1}$ (vector of all ones), and $\mathbf{0}$ (vector of all zeros).

(2) $\mathbf{1}^{\mathrm{T}}\mathbf{A}$ is a row vector whose entries are the column sums of $\mathbf{A}$, and $\mathbf{A}\mathbf{1}$ denotes a column vector whose entries are the row sums of $\mathbf{A}$. $\mathbf{1}^{\mathrm{T}}\mathbf{A}\mathbf{1}$ denotes the sum of all the elements in the matrix $\mathbf{A}$.

(3) $\mathbf{A}^{\mathrm{T}}$ denotes the transpose of $\mathbf{A}$.

(4) $(\mathbf{A})_{ij} = a_{ij}$ denotes the element in the ith row and the jth column of $\mathbf{A}$.

(5) The identity matrix is denoted by $\mathbf{I}$. The order of the identity matrix may be indicated by a subscript, for example, $\mathbf{I}_n$ denotes an $n \times n$ identity matrix.

(6) $\mathbf{D}_x$ denotes a diagonal matrix with entries $x_1, x_2, \ldots, x_n$ (the subscript indicates the terms in the diagonal).

(7) A tilde placed above a matrix indicates the matrix is triangular. The matrix $\widetilde{\mathbf{T}}$ is a *lower* triangular matrix and $\widetilde{\mathbf{T}}^{\mathrm{T}}$ is an *upper* triangular matrix:

$$\widetilde{\mathbf{T}} = \begin{bmatrix} t_{11} & 0 & 0 & \cdots & 0 \\ t_{21} & t_{22} & 0 & \cdots & 0 \\ t_{31} & t_{32} & t_{33} & \cdots & 0 \\ \vdots & \vdots & \vdots & \ddots & \vdots \\ t_{n1} & t_{n2} & t_{n3} & \cdots & t_{nn} \end{bmatrix}, \qquad \widetilde{\mathbf{T}}^{\mathrm{T}} = \begin{bmatrix} t_{11} & t_{21} & t_{31} & \cdots & t_{n1} \\ 0 & t_{22} & t_{32} & \cdots & t_{n2} \\ 0 & 0 & t_{33} & \cdots & t_{n3} \\ \vdots & \vdots & \vdots & \ddots & \vdots \\ 0 & 0 & 0 & \cdots & t_{nn} \end{bmatrix}$$

16.2   THE GENERAL LINEAR MODEL

16.2.1   The simple linear regression model

Let $(x_1, y_1), (x_2, y_2), \ldots, (x_n, y_n)$ be n pairs of observations such that y_i is an observed value of the random variable Y_i. Assume there exist constants β_0

and β_1 such that

$$Y_i = \beta_0 + \beta_1 x_i + \epsilon_i \qquad (16.2)$$

where $\epsilon_1, \epsilon_2, \ldots, \epsilon_n$ are independent, normal random variables having mean 0 and variance σ^2.

Assumptions	
In terms of ϵ_i's	In terms of Y_i's
ϵ_i's are normally distributed	Y_i's are normally distributed
$E[\epsilon_i] = 0$	$E[Y_i] = \beta_0 + \beta_1 x_i$
$\text{Var}[\epsilon_i] = \sigma^2$	$\text{Var}[Y_i] = \sigma^2$
$\text{Cov}[\epsilon_i, \epsilon_j] = 0,\ i \neq j$	$\text{Cov}[Y_i, Y_j] = 0,\ i \neq j$

Using equation (16.2):

$$Y_1 = \beta_0 + \beta_1 x_1 + \epsilon_1$$
$$Y_2 = \beta_0 + \beta_1 x_2 + \epsilon_2$$
$$\vdots \qquad\qquad (16.3)$$
$$Y_n = \beta_0 + \beta_1 x_n + \epsilon_n$$

This set of equations may be written in matrix form:

$$
\begin{bmatrix} Y_1 \\ Y_2 \\ Y_3 \\ \vdots \\ Y_n \end{bmatrix}
=
\begin{bmatrix} 1 & x_1 \\ 1 & x_2 \\ 1 & x_3 \\ \vdots & \vdots \\ 1 & x_n \end{bmatrix}
\begin{bmatrix} \beta_0 \\ \beta_1 \end{bmatrix}
+
\begin{bmatrix} \epsilon_1 \\ \epsilon_2 \\ \epsilon_3 \\ \vdots \\ \epsilon_n \end{bmatrix}
$$
$$\mathbf{Y} \quad = \quad \mathbf{X} \qquad \boldsymbol{\beta} \quad + \quad \boldsymbol{\epsilon}$$

The matrix $\mathbf{X}$ is the design matrix and may also be written as $\mathbf{X} = [\mathbf{1}, \mathbf{x}]$, where $\mathbf{1}$ is a column vector containing all 1's and $\mathbf{x}$ is the column vector containing the x_i's.

The simple linear regression model, equation (16.2), is often written in the form

$$Y_i = \mu + \beta_1 (x_i - \bar{x}) + \epsilon_i \qquad (16.4)$$

where $\mu = \beta_0 + \beta_1 \overline{x}$. This model may also be written in matrix form:

$$
\begin{bmatrix} Y_1 \\ Y_2 \\ Y_3 \\ \vdots \\ Y_n \end{bmatrix}
=
\begin{bmatrix} 1 & (x_1 - \overline{x}) \\ 1 & (x_2 - \overline{x}) \\ 1 & (x_3 - \overline{x}) \\ \vdots & \vdots \\ 1 & (x_n - \overline{x}) \end{bmatrix}
\begin{bmatrix} \mu \\ \beta_1 \end{bmatrix}
+
\begin{bmatrix} \epsilon_1 \\ \epsilon_2 \\ \epsilon_3 \\ \vdots \\ \epsilon_n \end{bmatrix}
$$

$$
\mathbf{Y} \qquad = \qquad \mathbf{X} \qquad\qquad \beta \quad + \quad \epsilon
$$

where $\mathbf{X} = [\mathbf{1}, (\mathbf{x} - \overline{x}\mathbf{1})]$.

16.2.2   Multiple linear regression

Let there be n observations of the form $(x_{1i}, x_{2i}, \ldots, x_{ki}, y_i)$ such that y_i is an observed value of the random variable Y_i. Assume there exist constants $\beta_0, \beta_1, \ldots, \beta_k$ such that

$$
Y_i = \beta_0 + \beta_1 x_{1i} + \cdots + \beta_k x_{ki} + \epsilon_i \tag{16.5}
$$

where $\epsilon_1, \epsilon_2, \ldots, \epsilon_n$ are independent, normal random variables having mean 0 and variance σ^2.

Assumptions	
In terms of ϵ_i's	In terms of Y_i's
ϵ_i's are normally distributed	Y_i's are normally distributed
$\mathrm{E}[\epsilon_i] = 0$	$\mathrm{E}[Y_i] = \beta_0 + \beta_1 x_{1i} + \cdots + \beta_k x_{ki}$
$\mathrm{Var}[\epsilon_i] = \sigma^2$	$\mathrm{Var}[Y_i] = \sigma^2$
$\mathrm{Cov}[\epsilon_i, \epsilon_j] = 0,\ i \neq j$	$\mathrm{Cov}[Y_i, Y_j] = 0,\ i \neq j$

Using equation (16.5):

$$
\begin{aligned}
Y_1 &= \beta_0 + \beta_1 x_{11} + \beta_2 x_{21} + \cdots + \beta_k x_{k1} + \epsilon_1 \\
Y_2 &= \beta_0 + \beta_1 x_{12} + \beta_2 x_{22} + \cdots + \beta_k x_{k2} + \epsilon_2 \\
&\ \ \vdots \\
Y_n &= \beta_0 + \beta_1 x_{1n} + \beta_2 x_{2n} + \cdots + \beta_k x_{kn} + \epsilon_n
\end{aligned} \tag{16.6}
$$

This set of equations may be written in matrix form:

$$
\begin{bmatrix} Y_1 \\ Y_2 \\ \vdots \\ Y_n \end{bmatrix}
=
\begin{bmatrix}
1 & x_{11} & x_{21} & x_{31} & \cdots & x_{k1} \\
1 & x_{21} & x_{22} & x_{32} & \cdots & x_{k2} \\
\vdots & \vdots & \vdots & \vdots & \ddots & \vdots \\
1 & x_{1n} & x_{2n} & x_{3n} & \cdots & x_{kn}
\end{bmatrix}
\begin{bmatrix} \beta_0 \\ \beta_1 \\ \beta_2 \\ \vdots \\ \beta_k \end{bmatrix}
+
\begin{bmatrix} \epsilon_1 \\ \epsilon_2 \\ \epsilon_3 \\ \vdots \\ \epsilon_n \end{bmatrix}
$$

$$
\mathbf{Y} \qquad = \qquad\qquad\qquad \mathbf{X} \qquad\qquad\qquad \beta \quad + \quad \epsilon
$$

where the design matrix $\mathbf{X} = [\mathbf{1}, \mathbf{X}_{nk}]$ and $\mathbf{X}_{nk}$ is the matrix of observations on the independent variables.

16.2.3 One-way analysis of variance

Let there be k treatments (or populations), independent random samples of size n_i (where $i = 1, 2, \ldots, k$), from each population, and let $N = n_1 + n_2 + \cdots + n_k$. Let Y_{ij} be the jth random observation in the ith treatment group.

Assume a fixed effects experiment model:

$$Y_{ij} = \mu + \alpha_i + \epsilon_{ij} \tag{16.7}$$

where μ is the grand mean, α_i is the ith treatment effect, and ϵ_{ij} is the random error term. The ϵ_{ij}'s are assumed to be independent, normally distributed, with mean 0 and variance σ^2.

Using equation (16.7):

$$
\begin{aligned}
Y_{11} &= \mu + \alpha_1 & &+ \epsilon_{11} \\
Y_{12} &= \mu + \alpha_1 & &+ \epsilon_{12} \\
&\;\vdots & &\;\vdots \\
Y_{1n_1} &= \mu + \alpha_1 & &+ \epsilon_{1n_1} \\[4pt]
Y_{21} &= \mu & + \alpha_2 &+ \epsilon_{21} \\
Y_{22} &= \mu & + \alpha_2 &+ \epsilon_{22} \\
&\;\vdots & \;\vdots &\;\vdots \\
Y_{2n_2} &= \mu & + \alpha_2 &+ \epsilon_{2n_2} \\[6pt]
&\;\vdots & \;\vdots &\;\vdots \\[4pt]
Y_{k1} &= \mu & + \alpha_k &+ \epsilon_{k1} \\
Y_{k2} &= \mu & + \alpha_k &+ \epsilon_{k2} \\
&\;\vdots & \;\vdots &\;\vdots \\
Y_{kn_k} &= \mu & + \alpha_k &+ \epsilon_{kn_k}
\end{aligned}
$$

This set of equations may be written in matrix form:

$$
\begin{bmatrix}
Y_{11} \\
Y_{12} \\
\vdots \\
Y_{1n_1} \\
Y_{21} \\
Y_{22} \\
\vdots \\
Y_{2n_2} \\
\vdots \\
Y_{k1} \\
Y_{k2} \\
\vdots \\
Y_{kn_k}
\end{bmatrix}
=
\begin{bmatrix}
1 & 1 & 0 & 0 & \cdots & 0 & 0 \\
1 & 1 & 0 & 0 & \cdots & 0 & 0 \\
\vdots & \vdots & \vdots & \vdots & \ddots & \vdots & \vdots \\
1 & 1 & 0 & 0 & \cdots & 0 & 0 \\
1 & 0 & 1 & 0 & \cdots & 0 & 0 \\
1 & 0 & 1 & 0 & \cdots & 0 & 0 \\
\vdots & \vdots & \vdots & \vdots & \ddots & \vdots & \vdots \\
1 & 0 & 1 & 0 & \cdots & 0 & 0 \\
\vdots & \vdots & \vdots & \vdots & : & \vdots & \vdots \\
1 & 0 & 0 & 0 & \cdots & 0 & 1 \\
1 & 0 & 0 & 0 & \cdots & 0 & 1 \\
\vdots & \vdots & \vdots & \vdots & \ddots & \vdots & \vdots \\
1 & 0 & 0 & 0 & \cdots & 0 & 1
\end{bmatrix}
\begin{bmatrix}
\mu \\
\alpha_1 \\
\alpha_2 \\
\vdots \\
\alpha_k
\end{bmatrix}
+
\begin{bmatrix}
\epsilon_{11} \\
\epsilon_{12} \\
\vdots \\
\epsilon_{1n_1} \\
\epsilon_{21} \\
\epsilon_{22} \\
\vdots \\
\epsilon_{2n_2} \\
\vdots \\
\epsilon_{k1} \\
\epsilon_{k2} \\
\vdots \\
\epsilon_{kn_k}
\end{bmatrix}
$$

$$\mathbf{Y} \quad = \quad \mathbf{X} \qquad\qquad \boldsymbol{\beta} \quad + \quad \boldsymbol{\epsilon}$$

The design matrix $\mathbf{X}$ may be written as $\mathbf{X} = [\mathbf{1}_N, \mathbf{X}_{Nk}]$ where

$$
\mathbf{X}_{Nk} =
\begin{bmatrix}
\mathbf{1}_{n_1} & \mathbf{0}_{n_1} & \mathbf{0}_{n_1} & \cdots & \mathbf{0}_{n_1} \\
\mathbf{0}_{n_2} & \mathbf{1}_{n_2} & \mathbf{0}_{n_2} & \cdots & \mathbf{0}_{n_2} \\
\vdots & \vdots & \vdots & & \vdots \\
\mathbf{0}_{n_k} & \mathbf{0}_{n_k} & \mathbf{0}_{n_k} & \cdots & \mathbf{1}_{n_k}
\end{bmatrix}
\tag{16.8}
$$

and the parameter vector may be written as $\boldsymbol{\beta} = [\mu \ \boldsymbol{\alpha}]^{\mathrm{T}}$.

16.2.4   Two-way analysis of variance

Let Y_{ijk} be the kth random observation for the ith level of factor A and the jth level of factor B. Assume there are n_{ij} observations for the ij factor combination: $i = 1, 2, \ldots, a$, $j = 1, 2, \ldots, b$, $k = 1, 2, \ldots, n_{ij}$.

For simplicity, consider a fixed effects experiment model:

$$Y_{ijk} = \mu + \alpha_i + \beta_j + (\alpha\beta)_{ij} + \epsilon_{ijk} \tag{16.9}$$

where μ is the grand mean, α_i is the level i factor A effect, β_j is the level j factor B effect, $(\alpha\beta)_{ij}$ is the level ij interaction effect, and ϵ_{ijk} is the random error term. The ϵ_{ijk}'s are assumed to be independent, normally distributed, with mean 0 and variance σ^2.

Suppose $a = 2$ and $b = 3$. The model may be written in matrix form:

$$
\begin{bmatrix} y_{11} \\ y_{12} \\ y_{13} \\ y_{21} \\ y_{22} \\ y_{23} \end{bmatrix}
=
\begin{bmatrix}
1_{n_{11}} & 1 & 0 & 1 & 0 & 0 & 1 & 0 & 0 & 0 & 0 & 0 \\
1_{n_{12}} & 1 & 0 & 0 & 1 & 0 & 0 & 1 & 0 & 0 & 0 & 0 \\
1_{n_{13}} & 1 & 0 & 0 & 0 & 1 & 0 & 0 & 1 & 0 & 0 & 0 \\
1_{n_{21}} & 0 & 1 & 1 & 0 & 0 & 0 & 0 & 0 & 1 & 0 & 0 \\
1_{n_{22}} & 0 & 1 & 0 & 1 & 0 & 0 & 0 & 0 & 0 & 1 & 0 \\
1_{n_{23}} & 0 & 1 & 0 & 0 & 1 & 0 & 0 & 0 & 0 & 0 & 1
\end{bmatrix}
\begin{bmatrix}
\mu \\ \alpha_1 \\ \alpha_2 \\ \beta_1 \\ \beta_2 \\ \beta_3 \\ (\alpha\beta)_{11} \\ (\alpha\beta)_{12} \\ (\alpha\beta)_{13} \\ (\alpha\beta)_{21} \\ (\alpha\beta)_{22} \\ (\alpha\beta)_{23}
\end{bmatrix}
+
\begin{bmatrix}
\epsilon_{111} \\ \vdots \\ \epsilon_{11n_{11}} \\ \hline \epsilon_{121} \\ \vdots \\ \epsilon_{12n_{12}} \\ \hline \vdots \\ \epsilon_{231} \\ \vdots \\ \epsilon_{23n_{23}}
\end{bmatrix}
$$

$$\mathbf{Y} \quad = \quad \mathbf{X} \quad\quad\quad \boldsymbol{\beta} \quad + \quad \boldsymbol{\epsilon}$$

16.2.5 Analysis of covariance

The analysis of covariance procedure is a combination of analysis of variance and regression analysis. For example, consider a one-way classification with one independent variable. Let Y_{ij} be the jth random observation in the ith treatment group: $i = 1, 2, \ldots, a$, $j = 1, 2, \ldots, n_i$. The model is

$$Y_{ij} = \mu + \alpha_i + \beta x_{ij} + \epsilon_{ij} \tag{16.10}$$

where μ is the grand mean, α_i is the ith treatment effect, β is the regression coefficient of Y on X, and ϵ_{ij} is the random error term. The ϵ_{ij}'s are assumed to be independent, normally distributed, with mean 0 and variance σ^2.

Suppose $a = 3$, using equation (16.10):

$$
\begin{aligned}
Y_{11} &= \mu + \alpha_1 && + \beta x_{11} + \epsilon_{11} \\
Y_{12} &= \mu + \alpha_1 && + \beta x_{12} + \epsilon_{12} \\
&\ \ \vdots \quad\ \ \vdots \quad\ \ \vdots && \quad\ \ \vdots \quad\ \ \vdots \\
Y_{1n_1} &= \mu + \alpha_1 && + \beta x_{1n_1} + \epsilon_{1n_1} \\
\hline
Y_{21} &= \mu \quad\ + \alpha_2 && + \beta x_{21} + \epsilon_{21} \\
&\ \ \vdots \quad\ \ \vdots \quad\ \ \vdots && \quad\ \ \vdots \quad\ \ \vdots \\
Y_{2n_2} &= \mu \quad\ + \alpha_2 && + \beta x_{2n_2} + \epsilon_{2n_2} \\
\hline
Y_{31} &= \mu \quad\quad\quad\ + \alpha_3 && + \beta x_{31} + \epsilon_{31} \\
&\ \ \vdots \quad\ \ \vdots && \quad\ \ \vdots \quad\ \ \vdots \quad\ \ \vdots \\
Y_{3n_3} &= \mu \quad\quad\quad\ + \alpha_3 && + \beta x_{3n_3} + \epsilon_{3n_3}
\end{aligned}
$$

If $\mathbf{x}_1$, $\mathbf{x}_2$, and $\mathbf{x}_3$ denote the observations on the independent variable in each treatment group, this set of equations may be written in matrix form:

$$
\begin{bmatrix}
Y_{11} \\
Y_{12} \\
\vdots \\
Y_{1n_1} \\
\hline
Y_{21} \\
\vdots \\
Y_{2n_2} \\
\hline
Y_{31} \\
\vdots \\
Y_{3n_3}
\end{bmatrix}
=
\begin{bmatrix}
\mathbf{1}_{n_1} & 1 & 0 & 0 & \mathbf{x}_1 \\
\mathbf{1}_{n_2} & 0 & 1 & 0 & \mathbf{x}_2 \\
\mathbf{1}_{n_3} & 0 & 0 & 1 & \mathbf{x}_3
\end{bmatrix}
\begin{bmatrix}
\mu \\
\alpha_1 \\
\alpha_2 \\
\alpha_3 \\
\beta
\end{bmatrix}
+
\begin{bmatrix}
\epsilon_{11} \\
\epsilon_{12} \\
\vdots \\
\epsilon_{1n_1} \\
\hline
\epsilon_{21} \\
\vdots \\
\epsilon_{2n_2} \\
\hline
\epsilon_{31} \\
\vdots \\
\epsilon_{3n_3}
\end{bmatrix}
$$

$$\mathbf{Y} \quad = \quad \mathbf{X} \quad\quad \boldsymbol{\beta} \quad + \quad \boldsymbol{\epsilon}$$

16.3   SUMMARY OF RULES FOR MATRIX OPERATIONS

16.3.1   Linear combinations

Suppose $\mathbf{X}$ is a random vector: a vector whose elements are random variables. Let $\boldsymbol{\mu}$ be the vector of means and let $\boldsymbol{\Sigma}$ be the variance–covariance matrix, denoted

$$\mathrm{E}\left[\mathbf{X}\right] = \boldsymbol{\mu}, \qquad \mathrm{Cov}\left[\mathbf{X}\right] = \mathrm{E}\left[\mathbf{X}\mathbf{X}^{\mathrm{T}}\right] = \boldsymbol{\Sigma}. \tag{16.11}$$

For any conforming matrix $\mathbf{C}$, the linear combinations $\mathbf{Y} = \mathbf{C}\mathbf{X}$ have

$$\mathrm{E}\left[\mathbf{Y}\right] = \mathrm{E}\left[\mathbf{C}\mathbf{X}\right] = \mathbf{C}\boldsymbol{\mu}, \qquad \mathrm{Cov}\left[\mathbf{Y}\right] = \mathrm{Cov}\left[\mathbf{C}\mathbf{X}\right] = \mathbf{C}\boldsymbol{\Sigma}\mathbf{C}^{\mathrm{T}}. \tag{16.12}$$

The linear combinations $\mathbf{Z} = \mathbf{X}^{\mathrm{T}}\mathbf{C}$ have

$$\mathrm{E}\left[\mathbf{Z}\right] = \mathrm{E}\left[\mathbf{X}^{\mathrm{T}}\mathbf{C}\right] = \boldsymbol{\mu}^{\mathrm{T}}\mathbf{C}, \qquad \mathrm{Cov}\left[\mathbf{Z}\right] = \mathrm{Cov}\left[\mathbf{X}^{\mathrm{T}}\mathbf{C}\right] = \mathbf{C}^{\mathrm{T}}\boldsymbol{\Sigma}\mathbf{C}. \tag{16.13}$$

16.3.2   Determinants and partitioning of determinants

The *determinant* of a square matrix $\mathbf{X}$, denoted by $|\mathbf{X}|$ or $\det\left(\mathbf{X}\right)$, is a scalar function of $\mathbf{X}$ defined as

$$\det(\mathbf{X}) = \sum_{\sigma} \mathrm{sgn}(\sigma) x_{1,\sigma(1)} \, x_{2,\sigma(2)} \cdots x_{n,\sigma(n)} \tag{16.14}$$

where the sum is taken over all permutations σ of $\{1, 2, \ldots, n\}$. The signum function $\mathrm{sgn}(\sigma)$ is the number of successive transpositions required to change the permutation σ to the identity permutation. Note the properties of determinants: $|\mathbf{A}|\,|\mathbf{B}| = |\mathbf{A}\mathbf{B}|$ and $|\mathbf{A}| = |\mathbf{A}^{\mathrm{T}}|$.

Omitting the signum function in equation (16.14) yields the definition of the *permanent* of $\mathbf{X}$, given by $\mathrm{per}\,\mathbf{X} = \sum_{\sigma} x_{1,\sigma(1)} \cdots x_{n,\sigma(n)}$.

Suppose the matrix $\mathbf{X}$ can be *partitioned*, written as

$$\mathbf{X} = \begin{bmatrix} \mathbf{X}_{11} & \mathbf{X}_{12} \\ \mathbf{X}_{21} & \mathbf{X}_{22} \end{bmatrix}. \tag{16.15}$$

The determinant of $\mathbf{X}$ may be computed by

$$\begin{vmatrix} \mathbf{X}_{11} & \mathbf{X}_{12} \\ \mathbf{X}_{21} & \mathbf{X}_{22} \end{vmatrix} = |\mathbf{X}_{11}| \; |\mathbf{X}_{22} - \mathbf{X}_{21}\mathbf{X}_{11}^{-1}\mathbf{X}_{12}| \quad \text{if } \mathbf{X}_{11}^{-1} \text{ exists}$$
$$= |\mathbf{X}_{22}| \; |\mathbf{X}_{11} - \mathbf{X}_{12}\mathbf{X}_{22}^{-1}\mathbf{X}_{21}| \quad \text{if } \mathbf{X}_{22}^{-1} \text{ exists}. \tag{16.16}$$

16.3.3 Inverse of a partitioned matrix

Suppose the matrix $\mathbf{X}$ can be partitioned as in equation (16.15). The inverse of the matrix $\mathbf{X}$ may be written as

$$\begin{bmatrix} \mathbf{X}_{11} & \mathbf{X}_{12} \\ \mathbf{X}_{21} & \mathbf{X}_{22} \end{bmatrix}^{-1} = \begin{bmatrix} \mathbf{A} & \mathbf{B} \\ \mathbf{C} & \mathbf{D} \end{bmatrix} \quad \text{where}$$

$$\mathbf{A} = [\mathbf{X}_{11} - \mathbf{X}_{12}\mathbf{X}_{22}^{-1}\mathbf{X}_{21}]^{-1} \qquad \mathbf{D} = [\mathbf{X}_{22} - \mathbf{X}_{21}\mathbf{X}_{11}^{-1}\mathbf{X}_{12}]^{-1}$$
$$\mathbf{B} = -\mathbf{X}_{11}^{-1}\mathbf{X}_{12}\mathbf{D} \qquad \mathbf{C} = -\mathbf{X}_{22}^{-1}\mathbf{X}_{21}\mathbf{A}$$

16.3.3.1 *Symmetric case*

$$\begin{bmatrix} \mathbf{X}_{11} & \mathbf{X}_{12} \\ \mathbf{X}_{12}^{\mathrm{T}} & \mathbf{X}_{22} \end{bmatrix}^{-1} = \begin{bmatrix} \mathbf{A} & \mathbf{B} \\ \mathbf{B}^{\mathrm{T}} & \mathbf{D} \end{bmatrix} \quad \text{where}$$

$$\mathbf{A} = [\mathbf{X}_{11} - \mathbf{X}_{12}\mathbf{X}_{22}^{-1}\mathbf{X}_{12}^{\mathrm{T}}]^{-1} \qquad \mathbf{D} = [\mathbf{X}_{22} - \mathbf{X}_{12}^{\mathrm{T}}\mathbf{X}_{11}^{-1}\mathbf{X}_{12}]^{-1}$$
$$\mathbf{B} = -\mathbf{A}\mathbf{X}_{12}\mathbf{X}_{22}^{-1} = -\mathbf{X}_{11}^{-1}\mathbf{X}_{12}\mathbf{D}$$

16.3.4 Eigenvalues

If $\mathbf{A}$ is a $k \times k$ square matrix and $\mathbf{I}$ is the $k \times k$ identity matrix, then the scalers $\lambda_1, \lambda_2, \ldots, \lambda_k$ that satisfy the polynomial equation $|\mathbf{A} - \lambda\mathbf{I}|$ are the **eigenvalues** (or **characteristic roots**) of the matrix $\mathbf{A}$. The equation $|\mathbf{A} - \lambda\mathbf{I}|$ is a function of λ and is the **characteristic equation**.

Let $\mathrm{ch}(\mathbf{A})$ denote the characteristic roots of the matrix $\mathbf{A}$ and $\mathrm{tr}(\mathbf{A})$ denote the trace of $\mathbf{A}$.

(1) $\mathrm{ch}(\mathbf{AB}) = \mathrm{ch}(\mathbf{BA})$ except possibly for zero roots.

(2) $\mathrm{tr}(\mathbf{AB}) = \mathrm{tr}(\mathbf{BA})$

(3) If $\mathrm{ch}(\mathbf{A}) = \{\lambda\}_{i=1}^{n}$, then $\mathrm{ch}(\mathbf{A}^{-1}) = 1/\lambda_i$ and $\mathrm{ch}(\mathbf{I} \pm \mathbf{A}) = 1 \pm \lambda_i$ for $i = 1, 2, \ldots, n$.

16.3.5   Differentiation involving vectors/matrices

16.3.5.1   Definitions

(1) Let f be a real–valued function of $x_1, x_2, \ldots, x_n$. The symbol $\partial f/\partial \mathbf{x}$ denotes a column vector whose i^{th} element is $\partial f/\partial x_i$.

(2) Let f be a real–valued function of $x_{11}, x_{12}, \ldots, x_{1n}, x_{21}, \ldots, x_{2n}, \ldots, x_{m1}, x_{m2}, \ldots, x_{mn}$. The symbol $\partial f/\partial \mathbf{X}$ denotes a matrix whose (i, j) entry is $\partial f/\partial x_{ij}$.

   Note: If there are functional relationships between the elements of $\mathbf{X}$ (for example, in a symmetric matrix) these relationships are disregarded in the definition above. If $x_{ij} = x_{ji} = y_{ij}$ (y_{ij} is the symbol for the distinct variable that occurs in two places in $\mathbf{X}$) then $\partial f/\partial y_{ij} = (\partial f/\partial \mathbf{X})_{ij} + (\partial f/\partial \mathbf{X})_{ij}$.

(3) If $y_1, y_2, \ldots, y_n$ are functions of x, then $\partial \mathbf{y}/\partial x$ denotes the column vector whose i^{th} entry is $\partial y_i/\partial x$.

(4) If $y_{11}, y_{12}, \ldots, y_{1n}, y_{21}, \ldots, y_{2n}, \ldots, y_{m1}, \ldots, y_{mn}$ are functions of x, then $\partial \mathbf{Y}/\partial x$ denotes the matrix whose (i, j) entry is $\partial y_{ij}/\partial x$.

(5) If each of the quantities $y_1, y_2, \ldots, y_n$ is a function of the variables $x_1, x_2, \ldots, x_m$, then $\partial \mathbf{y}^{\text{T}}/\partial \mathbf{x}$ denotes an $m \times n$ matrix whose (i, j) entry is $\partial y_j/\partial x_i$.

16.3.5.2   Properties

Suppose $\mathbf{a}$, $\mathbf{b}$, $\mathbf{e}$, $\mathbf{x}$, $\mathbf{y}$, and $\mathbf{z}$ are column vectors, and $\mathbf{A}$, $\mathbf{Q}$, $\mathbf{X}$, and $\mathbf{Y}$ are matrices.

(1) $\dfrac{\partial(\mathbf{x}^{\text{T}}\mathbf{x})}{\partial \mathbf{x}} = 2\mathbf{x}$

(2) $\dfrac{\partial(\mathbf{x}^{\text{T}}\mathbf{Q}\mathbf{x})}{\partial \mathbf{x}} = \mathbf{Q}\mathbf{x} + \mathbf{Q}^{\text{T}}\mathbf{x}$

(3) $\partial(\mathbf{x}^{\text{T}}\mathbf{Q}\mathbf{x})/\partial \mathbf{x} = 2\mathbf{Q}\mathbf{x}$ if $\mathbf{Q}$ is symmetric.

(4) $\partial(\mathbf{a}^{\text{T}}\mathbf{x})/\partial \mathbf{x} = \mathbf{a}$

(5) $\partial(\mathbf{a}^{\text{T}}\mathbf{Q}\mathbf{x})/\partial \mathbf{x} = \mathbf{Q}^{\text{T}}\mathbf{a}$

(6) $\partial \operatorname{tr}(\mathbf{A}\mathbf{X})/\partial \mathbf{X} = \mathbf{A}^{\text{T}}$

(7) $\partial \operatorname{tr}(\mathbf{X}\mathbf{A})/\partial \mathbf{X} = \mathbf{A}^{\text{T}}$

(8) $\partial \ln|\mathbf{X}|/\partial \mathbf{X} = (\mathbf{X}^{\text{T}})^{-1}$ if $\mathbf{X}$ is square and nonsingular.

(9) $\dfrac{\partial \mathbf{y}}{\partial \mathbf{x}} = \dfrac{\partial \mathbf{z}^{\text{T}}}{\partial \mathbf{x}} \cdot \dfrac{\partial \mathbf{y}}{\partial \mathbf{z}}$ (Chain rule 1)

(10) $\dfrac{\partial(\mathbf{x}^{\text{T}}\mathbf{A})}{\partial \mathbf{x}} = \mathbf{A}$

(11) If $\mathbf{e} = \mathbf{B} - \mathbf{A}^T\mathbf{x}$, then

$$\frac{\partial(\mathbf{e}^T\mathbf{e})}{\partial\mathbf{x}} = \frac{\partial\mathbf{e}^T}{\partial\mathbf{x}} \cdot \frac{\partial(\mathbf{e}^T\mathbf{e})}{\partial\mathbf{e}} \qquad \text{(using property 9)}$$

$$= -2\mathbf{A}^T\mathbf{e} \qquad \text{(using properties 1 and 10)}$$

(12) If the scalar z is related to a scalar x through the variables y_{ij}, $i = 1, 2, \ldots, m; j = 1, 2, \ldots, n$, then

$$\frac{\partial z}{\partial x} = \text{tr}\left[\frac{\partial z}{\partial\mathbf{Y}} \cdot \frac{\partial\mathbf{Y}^T}{\partial x}\right] = \text{tr}\left[\frac{\partial z}{\partial\mathbf{Y}^T} \cdot \frac{\partial\mathbf{Y}}{\partial x}\right] \qquad (16.17)$$

This second chain rule is correct regardless of any functional relationships that may exist between the elements of $\mathbf{Y}$.

16.3.6   Additional definitions and properties

(1) If $\tilde{\mathbf{T}}$ is lower triangular, then $\tilde{\mathbf{T}}^{-1}$ is also lower triangular.

(2) A matrix $\mathbf{A}$ is

(a) *positive definite* if $\mathbf{x}^T\mathbf{A}\mathbf{x} > 0$ for all $\mathbf{x} \neq \mathbf{0}$.

(b) *positive semi-definite* if $\mathbf{x}^T\mathbf{A}\mathbf{x} \geq 0$ for all $\mathbf{x}$.

(3) For any matrix $\mathbf{Q}$, the dimension of the row (column) space of $\mathbf{Q}$ is the **row (column) rank** of $\mathbf{Q}$. (The row (column) rank of a matrix is also the number of linearly independent rows (columns) of that matrix.) The row rank and the column rank of any matrix $\mathbf{Q}$ are equal, and are the **rank** of the matrix $\mathbf{Q}$.

(4) If $\mathbf{Q}$ is a symmetric, positive-definite matrix, then there exists a unique real matrix $\tilde{\mathbf{T}}$ with positive diagonal entries such that $\mathbf{Q} = \tilde{\mathbf{T}}\tilde{\mathbf{T}}^T$. The matrix $\tilde{\mathbf{T}}$ may be obtained by using the forward Doolittle procedure: In each cycle, divide every element of the next–to–last row (the row which is immediately above the one beginning with 1) by the square–root of the *leading* (first) element. This technique produces $\tilde{\mathbf{T}}^T$ on the left and $\tilde{\mathbf{T}}^{-1}$ on the right–hand side.

(5) If $\mathbf{Q}$ is an $n \times n$ symmetric, positive semi-definite matrix of rank r, then the matrix $\tilde{\mathbf{T}}$ obtained via the forward Doolittle procedure will have zeros to the right of the rth column. $\mathbf{Q}$ may be written as

$$\mathbf{Q} = \begin{bmatrix} \tilde{\mathbf{T}}_1 \\ \mathbf{T}_2 \end{bmatrix} \begin{bmatrix} \tilde{\mathbf{T}}_1^T & \mathbf{T}_2^T \end{bmatrix} \qquad (16.18)$$

where only $\tilde{\mathbf{T}}_1$ is triangular. This computational procedure is important when determining characteristic roots. Suppose $\mathbf{A}$ and $\mathbf{B}$ are symmetric, and $\mathbf{A}$ is of rank $r < n$. To find the largest characteristic root of $\mathbf{A}\mathbf{B}$ first obtain the representation

$$\mathbf{A} = \begin{bmatrix} \tilde{\mathbf{T}}_1 \\ \mathbf{T}_2 \end{bmatrix} \begin{bmatrix} \tilde{\mathbf{T}}_1^T & \mathbf{T}_2^T \end{bmatrix} \qquad (16.19)$$

using the forward Doolittle procedure. The characteristic roots of $\mathbf{AB}$ may be found using

$$\text{ch}(\mathbf{AB}) = \text{ch}\left(\begin{bmatrix} \tilde{\mathbf{T}}_1^T & \mathbf{T}_2^T \end{bmatrix} \mathbf{B} \begin{bmatrix} \tilde{\mathbf{T}}_1 \\ \mathbf{T}_2 \end{bmatrix}\right) \tag{16.20}$$

where the right–hand matrix is of small order and symmetric.

16.4 PRINCIPLE OF MINIMIZING QUADRATIC FORMS AND GAUSS MARKOV THEOREM

16.4.1 Multivariate distributions

Suppose X is a random variable with mean μ and variance σ^2:

$$\text{E}[X] = \mu, \qquad \text{Var}[X] = \sigma^2. \tag{16.21}$$

(1) The *standardized* random variable $Y = (X - \mu)/\sigma$ has mean 0 and variance 1:

$$\text{E}[Y] = 0, \qquad \text{Var}[Y] = 1. \tag{16.22}$$

(2) If X is a normal random variable, then the random variable $Z^2 = (X - \mu)^2/\sigma^2$ has a chi–square distribution with one degree of freedom.

Suppose $\mathbf{X}$ is a random vector consisting of the p random variables $\{X_1, X_2, \ldots, X_p\}$: that is $\mathbf{X}^T = [X_1, X_2, \ldots, X_p]$. Let $\boldsymbol{\mu}$ be the vector of means and $\boldsymbol{\Sigma}$ be the variance–covariance matrix:

$$\text{E}[\mathbf{X}] = \boldsymbol{\mu}, \qquad \text{Cov}[\mathbf{X}] = \boldsymbol{\Sigma}. \tag{16.23}$$

Let $\boldsymbol{\Sigma} = \tilde{\mathbf{A}}\tilde{\mathbf{A}}^T$ where $\tilde{\mathbf{A}}$ is the lower triangular matrix obtained using the forward Doolittle analysis. Let $\mathbf{Y} = \tilde{\mathbf{A}}^{-1}(\mathbf{X} - \boldsymbol{\mu})$ and note $\boldsymbol{\Sigma}^{-1} = (\tilde{\mathbf{A}}^T)^{-1}\tilde{\mathbf{A}}^{-1} = (\tilde{\mathbf{A}}^{-1})^T\tilde{\mathbf{A}}^{-1}$.

(1) $\text{E}[\mathbf{Y}] = \tilde{\mathbf{A}}^{-1}\text{E}[\mathbf{X} - \boldsymbol{\mu}] = \mathbf{0}$.

(2) $\text{Cov}[\mathbf{Y}] = \tilde{\mathbf{A}}^{-1}\text{Cov}[\mathbf{X} - \boldsymbol{\mu}](\tilde{\mathbf{A}}^{-1})^T = \tilde{\mathbf{A}}^{-1}\text{Cov}[\mathbf{X} - \boldsymbol{\mu}](\tilde{\mathbf{A}}^T)^{-1}$

$\qquad\qquad = \tilde{\mathbf{A}}^{-1}\text{Cov}[\mathbf{X}](\tilde{\mathbf{A}}^T)^{-1}$

$\qquad\qquad = \tilde{\mathbf{A}}^{-1}\tilde{\mathbf{A}}\tilde{\mathbf{A}}^T(\tilde{\mathbf{A}}^T)^{-1} = \mathbf{I}$

(3) The expression

$$\mathbf{Y}^T\mathbf{Y} = (\mathbf{X} - \boldsymbol{\mu})^T(\widetilde{bfA}^{-1})^T\tilde{\mathbf{A}}^{-1}(\mathbf{X} - \boldsymbol{\mu})$$
$$= (\mathbf{X} - \boldsymbol{\mu})^T\boldsymbol{\Sigma}^{-1}(\mathbf{X} - \boldsymbol{\mu}) \tag{16.24}$$

is the **standard quadratic form**. If $\mathbf{X}$ has a multivariate normal distribution, then $\mathbf{Y}^T\mathbf{Y}$ has a chi–square distribution with p degrees of freedom.

16.4.2 The principle of least squares

Let $\mathbf{Y}$ be the random vector of responses, $\mathbf{y}$ be the vector of observed responses, β be the vector of regression coefficients, ϵ be the vector of random errors, and let $\mathbf{X}$ be the design matrix:

$$\mathbf{Y} = \begin{bmatrix} Y_1 \\ Y_2 \\ \vdots \\ Y_n \end{bmatrix} \quad \mathbf{y} = \begin{bmatrix} y_1 \\ y_2 \\ \vdots \\ y_n \end{bmatrix} \quad \beta = \begin{bmatrix} \beta_0 \\ \beta_1 \\ \vdots \\ \beta_k \end{bmatrix} \quad \epsilon = \begin{bmatrix} \epsilon_1 \\ \epsilon_2 \\ \vdots \\ \epsilon_n \end{bmatrix} \quad \mathbf{X} = \begin{bmatrix} 1 & x_{11} & x_{21} & \cdots & x_{k1} \\ 1 & x_{12} & x_{22} & \cdots & x_{k2} \\ \vdots & \vdots & \vdots & & \vdots \\ 1 & x_{1n} & x_{2n} & \cdots & x_{kn} \end{bmatrix}$$

The model may now be written as $\mathbf{Y} = \mathbf{X}\beta + \epsilon$ where $\epsilon \sim N_n(\mathbf{0}, \sigma^2\mathbf{I}_n)$ or equivalently $\mathbf{Y} \sim N_n(\mathbf{X}\beta, \sigma^2\mathbf{I}_n)$.

The sum of squared deviations about the true regression line is

$$\begin{aligned} S(\beta) &= \sum_{i=1}^{n}[y_i - (\beta_0 + \beta_1 x_{1i} + \cdots + \beta_k x_{ki})]^2 \\ &= (\mathbf{y}^T - \beta^T\mathbf{X}^T)(\mathbf{y} - \beta\mathbf{X}) = \mathbf{e}^T\mathbf{e} \end{aligned} \quad (16.25)$$

where $\mathbf{e}$ is the vector of observed errors. To minimize equation (16.25):

$$\frac{\partial(\mathbf{e}^T\mathbf{e})}{\partial\beta} = -2\mathbf{X}^T(\mathbf{y} - \mathbf{X}\beta) \quad (16.26)$$

The vector $\widehat{\beta}^T = (\widehat{\beta}_0, \widehat{\beta}_1, \ldots, \widehat{\beta}_k)$ that minimizes $S(\beta)$ is the vector of least squares estimates. Setting equation (16.26) equal to zero:

$$\begin{aligned} \mathbf{X}^T\left(\mathbf{y} - \mathbf{X}\widehat{\beta}\right) &= 0 \\ (\mathbf{X}^T\mathbf{X})\widehat{\beta} &= \mathbf{X}^T\mathbf{y}. \end{aligned} \quad (16.27)$$

The normal equations are given by equation (16.27). If the matrix $\mathbf{X}^T\mathbf{X}$ is non–singular, then

$$\widehat{\beta} = (\mathbf{X}^T\mathbf{X})^{-1}\mathbf{X}^T\mathbf{y}. \quad (16.28)$$

16.4.3 Minimum variance unbiased estimates

The minimum variance, unbiased, linear estimate of β is obtained by using a general form of the Gauss Markov theorem. Suppose

$$\mathbf{Y} = \mathbf{X}\beta + \epsilon, \quad E[\epsilon] = 0, \quad \text{Cov}[\mathbf{Y}] = \text{Cov}[\epsilon] = \sigma^2\mathbf{V} \quad (16.29)$$

where $\mathbf{V}$ is a square, symmetric, non–singular, $n \times n$ matrix with known entries. Therefore, the variance of Y_i (for $i = 1, 2, \ldots, n$) is known and $\text{Cov}[Y_i, Y_j]$, for all $i \neq j$, is known except for an arbitrary scalar multiple.

The best linear estimate of an arbitrary linear function $\mathbf{c}^{\mathrm{T}}\beta$ is $\mathbf{c}^{\mathrm{T}}\widehat{\beta}$ where $\widehat{\beta}$ minimizes the quadratic form $\epsilon^{\mathrm{T}}\mathbf{V}^{-1}\epsilon$. The standard quadratic form for ϵ is

$$(\epsilon - \mathrm{E}\,[\epsilon])^{\mathrm{T}}(\mathrm{Cov}\,[\epsilon])^{-1}(\epsilon - \mathrm{E}\,[\epsilon]) = (\epsilon - \mathbf{0})^{\mathrm{T}}(\sigma^2\mathbf{V})^{-1}(\epsilon - \mathbf{0})$$

$$= \frac{1}{\sigma^2}\epsilon^{\mathrm{T}}\mathbf{V}^{-1}\epsilon. \tag{16.30}$$

Minimizing the standard quadratic form is equivalent to minimizing $\epsilon^{\mathrm{T}}\mathbf{V}^{-1}\epsilon$, as stated in the Gauss Markov Theorem. In this case the normal equations are $\mathbf{X}^{\mathrm{T}}\mathbf{V}^{-1}\mathbf{X}\widehat{\beta} = \mathbf{X}^{\mathrm{T}}\mathbf{V}^{-1}\mathbf{y}$.

16.5   GENERAL LINEAR HYPOTHESIS OF FULL RANK

This section is concerned with the problem of testing hypotheses about certain parameters and the associated probability distributions.

16.5.1   Notation

A general null hypothesis is stated as $\mathbf{C}\beta = \mathbf{k}$ where

(1) $\mathbf{C} = \mathbf{C}_{n_h m}$, $(n_h \le m)$, is the hypothesis matrix and is of rank n_h.

(2) β is an $n \times 1$ column vector of parameters as defined in the general linear model.

(3) $\mathbf{k}$ is a vector of n_h known elements, usually equal to $\mathbf{0}$.

(4) n_h is the number of **degrees of freedom due to hypothesis**; the number of rows in the hypothesis matrix $\mathbf{C}$; the number of nonredundant statements in the null hypothesis.

(5) n_e is the number of **degrees of freedom due to error** and is equal to the number of observations minus the effective number of parameters.

Note: A composite hypothesis should *not* contain:

(1) contradictory statements like

$H_0\colon \beta_1 = \beta_2$ and $\beta_1 = 2\beta_2$ simultaneously,

(2) redundant statements like

$H_0\colon \beta_1 = \beta_2$ and $3\beta_1 = 3\beta_2$.

16.5.2   Simple linear regression

Model: Let $(x_1, y_1), (x_2, y_2), \ldots, (x_n, y_n)$ be n pairs of observations such that y_i is an observed value of the random variable Y_i. Assume there exist constants β_0 and β_1 such that

$$Y_i = \beta_0 + \beta_1 x_i + \epsilon_i, \quad \text{parameter vector } \beta = [\beta_0, \beta_1]^{\mathrm{T}} \tag{16.31}$$

where $\epsilon_1, \epsilon_2, \ldots, \epsilon_n$ are independent, normal random variables having mean 0 and variance σ^2.

Examples:

(1) H_0: $\beta_0 = 0$

H_a: $\beta_0 \neq 0$

General linear hypothesis: $n_h = 1$

$$[1, \ 0] \begin{bmatrix} \beta_0 \\ \beta_1 \end{bmatrix} = 0$$

$$\mathbf{C} \qquad \boldsymbol{\beta} \ = 0$$

(2) H_0: $\beta_1 = 0$

H_a: $\beta_1 \neq 0$

General linear hypothesis: $n_h = 1$

$$[0, \ 1] \begin{bmatrix} \beta_0 \\ \beta_1 \end{bmatrix} = 0$$

$$\mathbf{C} \qquad \boldsymbol{\beta} \ = 0$$

(3) H_0: $\beta_0 = \beta_1 = 0$ simultaneously

H_a: $\beta_i \neq 0$ for some i

General linear hypothesis: $n_h = 2$

$$\begin{bmatrix} 1 & 0 \\ 0 & 1 \end{bmatrix} \begin{bmatrix} \beta_0 \\ \beta_1 \end{bmatrix} = \begin{bmatrix} 0 \\ 0 \end{bmatrix}$$

$$\mathbf{C} \qquad \boldsymbol{\beta} \ = \ \mathbf{0}$$

(4) H_0: $\beta_0 = \beta_1$

H_a: $\beta_0 \neq \beta_1$

General linear hypothesis: $n_h = 1$

$$[1, \ -1] \begin{bmatrix} \beta_0 \\ \beta_1 \end{bmatrix} = 0$$

$$\mathbf{C} \qquad \boldsymbol{\beta} \ = 0$$

16.5.3 Analysis of variance, one-way anova

Model: Let there be k treatments, or populations, independent random samples of size n_i, $i = 1, 2, \ldots, k$, from each population, and let $N = n_1 + n_2 + \cdots + n_k$. Let Y_{ij} be the jth random observation in the ith treatment group. Assume a fixed effects experiment:

$$Y_{ij} = \mu + \alpha_i + \epsilon_{ij}, \quad i = 1, 2, \ldots, k, \quad j = 1, 2, \ldots, n_i$$
$$\text{parameter vector } \boldsymbol{\beta} = [\mu, \alpha_1, \alpha_2, \ldots, \alpha_k]^{\text{T}}$$

Examples:

(1) H_0: $\alpha_1 = \alpha_2 = \cdots = \alpha_k$

H_a: $\alpha_i \neq \alpha_j$ for some $i \neq j$

General linear hypothesis: $n_h = k - 1$

$$
\begin{bmatrix}
0 & 1 & -1 & 0 & 0 & \cdots & 0 \\
0 & 1 & 0 & -1 & 0 & \cdots & 0 \\
0 & 1 & 0 & 0 & -1 & \cdots & 0 \\
\vdots & \vdots & \vdots & \vdots & \vdots & \ddots & \vdots \\
0 & 1 & 0 & 0 & 0 & \cdots & -1
\end{bmatrix}
\begin{bmatrix}
\mu \\ \alpha_1 \\ \alpha_2 \\ \alpha_3 \\ \vdots \\ \alpha_k
\end{bmatrix}
=
\begin{bmatrix}
0 \\ 0 \\ 0 \\ 0 \\ \vdots \\ 0
\end{bmatrix}
$$

$$\mathbf{C}_{(k-1)(k+1)} \qquad\qquad \boldsymbol{\beta} \;=\; \mathbf{0}_{k-1}$$

(2) H_0: $\alpha_1 = \alpha_2 = \cdots = \alpha_k = 0$

H_a: $\alpha_i \neq 0$ for at least one i

General linear hypothesis: $n_h = k$

$$
\begin{bmatrix}
0 & 1 & 0 & 0 & 0 & \cdots & 0 \\
0 & 0 & 1 & 0 & 0 & \cdots & 0 \\
0 & 0 & 0 & 1 & 0 & \cdots & 0 \\
0 & 0 & 0 & 0 & 1 & \cdots & 0 \\
\vdots & \vdots & \vdots & \vdots & \vdots & \ddots & \vdots \\
0 & 0 & 0 & 0 & 0 & \cdots & 1
\end{bmatrix}
\begin{bmatrix}
\mu \\ \alpha_1 \\ \alpha_2 \\ \alpha_3 \\ \vdots \\ \alpha_k
\end{bmatrix}
=
\begin{bmatrix}
0 \\ 0 \\ 0 \\ 0 \\ \vdots \\ 0
\end{bmatrix}
$$

$$\mathbf{C}_{(k)(k+1)} \qquad\qquad \boldsymbol{\beta} \;=\; \mathbf{0}_{k}$$

(3) Suppose $i = 1, 2, 3, 4$.

H_0: $-\alpha_1 + 2\alpha_2 - \alpha_3 = 0$ (Quadratic contrast of three effects)

H_a: $-\alpha_1 + 2\alpha_2 - \alpha_3 \neq 0$

General linear hypothesis: $n_h = 1$

$$
\begin{bmatrix} 0, & -1, & +2, & -1, & 0 \end{bmatrix}
\begin{bmatrix}
\mu \\ \alpha_1 \\ \alpha_2 \\ \alpha_3 \\ \alpha_4
\end{bmatrix}
= 0
$$

$$\qquad\quad \mathbf{C} \qquad\qquad\quad \boldsymbol{\beta} \;= 0$$

16.5.4   Multiple linear regression

Model: Let there be n observations of the form $(x_{1i}, x_{2i}, \ldots, x_{ki}, y_i)$ such that y_i is an observed value of the random variable Y_i. Assume there exist constants $\beta_0, \beta_1, \ldots, \beta_k$ such that

$$Y_i = \beta_0 + \beta_1 x_{1i} + \cdots + \beta_k x_{ki} + \epsilon_i$$

$$\text{parameter vector } \boldsymbol{\beta} = [\beta_0, \ \beta_1, \ \beta_2, \ \ldots, \ \beta_k]^{\mathrm{T}}$$

where $\epsilon_1, \epsilon_2, \ldots, \epsilon_n$ are independent, normal random variables having mean 0 and variance σ^2.

Examples:

(1) H_0: $\beta_1 = 0$

 H_a: $\beta_1 \neq 0$

General linear hypothesis: $n_h = 1$

$$[0, \ 1, \ 0, \ 0, \cdots, \ 0] \begin{bmatrix} \beta_0 \\ \beta_1 \\ \beta_2 \\ \vdots \\ \beta_k \end{bmatrix} = 0$$

$$\mathbf{C} \qquad\qquad \boldsymbol{\beta} \quad = 0$$

(2) H_0: $\beta_1 = \beta_2 = \beta_3 = \cdots = \beta_k = 0$

 H_a: $\beta_i \neq 0$ for some i

General linear hypothesis: $n_h = k$

$$\begin{bmatrix} 0 & 1 & 0 & 0 & 0 & \cdots & 0 \\ 0 & 0 & 1 & 0 & 0 & \cdots & 0 \\ 0 & 0 & 0 & 1 & 0 & \cdots & 0 \\ 0 & 0 & 0 & 0 & 1 & \cdots & 0 \\ \vdots & \vdots & \vdots & \vdots & \vdots & \ddots & \vdots \\ 0 & 0 & 0 & 0 & 0 & \cdots & 1 \end{bmatrix} \begin{bmatrix} \beta_0 \\ \beta_1 \\ \beta_2 \\ \beta_3 \\ \vdots \\ \beta_k \end{bmatrix} = \begin{bmatrix} 0 \\ 0 \\ 0 \\ 0 \\ \vdots \\ 0 \end{bmatrix}$$

$$\mathbf{C}_{(k)(k+1)} \qquad\qquad \boldsymbol{\beta} \quad = \mathbf{0}_k$$

(3) H_0: $\beta_1 = \beta_2 = 0$

 H_a: $\beta_i \neq 0$ for some i

General linear hypothesis: $n_h = 2$

$$\begin{bmatrix} 0 & 1 & 0 & 0 & 0 & \cdots & 0 \\ 0 & 0 & 1 & 0 & 0 & \cdots & 0 \end{bmatrix} \begin{bmatrix} \beta_0 \\ \beta_1 \\ \beta_2 \\ \vdots \\ \beta_k \end{bmatrix} = \begin{bmatrix} 0 \\ 0 \end{bmatrix}$$

$$\mathbf{C} \qquad\qquad \boldsymbol{\beta} \quad = \mathbf{0}$$

16.5.5 Randomized blocks (one observation per cell)

Model: Let Y_{ij} be the random observation in the ith row and the jth column, $i = 1, 2, 3$ and $j = 1, 2, 3, 4$. Assume a fixed effects model:

$$Y_{ij} = \mu + \alpha_i + \beta_j + \epsilon_{ij}$$

parameter vector $\boldsymbol{\beta} = [\mu, \ \alpha_1, \ \alpha_2, \ \alpha_3, \ \beta_1, \ \beta_2, \ \beta_3, \ \beta_4]^{\text{T}}$

Examples:

(1) H_0: $\alpha_1 = \alpha_2 = \alpha_3$

   H_a: $\alpha_i \neq \alpha_j$ for some $i \neq j$

   General linear hypothesis: $n_h = 2$

$$
\begin{array}{cc}
\underbrace{\begin{bmatrix} 0 & 1 & -1 & 0 & 0 & 0 & 0 & 0 \\ 0 & 1 & 0 & -1 & 0 & 0 & 0 & 0 \end{bmatrix}}_{\mathbf{C}}
\underbrace{\begin{bmatrix} \mu \\ \alpha_1 \\ \alpha_2 \\ \alpha_3 \\ \beta_1 \\ \beta_2 \\ \beta_3 \\ \beta_4 \end{bmatrix}}_{\boldsymbol{\beta}}
= \begin{bmatrix} 0 \\ 0 \end{bmatrix} \\[2pt]
\phantom{\mathbf{C}} \qquad = 0
\end{array}
$$

(2) H_0: $-\alpha_1 + 2\alpha_2 - \alpha_3 = 0$ (quadratic contrast)

   H_a: $-\alpha_1 + 2\alpha_2 - \alpha_3 \neq 0$

   General linear hypothesis: $n_h = 1$

$$
\underbrace{[0,\ -1,\ 2,\ -1,\ 0,\ 0,\ 0,\ 0]}_{\mathbf{C}}
\underbrace{\begin{bmatrix} \mu \\ \alpha_1 \\ \alpha_2 \\ \alpha_3 \\ \beta_1 \\ \beta_2 \\ \beta_3 \\ \beta_4 \end{bmatrix}}_{\boldsymbol{\beta}} = 0
$$

$$
\mathbf{C} \qquad \boldsymbol{\beta} \ = 0
$$

16.5.6 Quadratic form due to hypothesis

For the general linear model

$$\mathbf{Y} = \mathbf{X}\boldsymbol{\beta} + \boldsymbol{\epsilon}, \quad \mathrm{E}\,[\mathbf{Y}] = \mathbf{X}\boldsymbol{\beta}, \quad \mathrm{Var}\,[\mathbf{Y}] = \sigma^2 \mathbf{I} \tag{16.32}$$

the normal equations are given by

$$(\mathbf{X}^\mathsf{T}\mathbf{X})\widehat{\boldsymbol{\beta}} = \mathbf{X}^\mathsf{T}\mathbf{Y}. \tag{16.33}$$

If the model is of full rank (($\mathbf{X}^\mathsf{T}\mathbf{X}$) has an inverse) then the estimate of β is

$$\widehat{\boldsymbol{\beta}} = (\mathbf{X}^\mathsf{T}\mathbf{X})^{-1}\mathbf{X}^\mathsf{T}\mathbf{Y} \tag{16.34}$$

and the variance–covariance matrix of $\widehat{\beta}$ is given by

$$
\begin{aligned}
\text{Var}\left[\widehat{\beta}\right] &= (\mathbf{X}^T\mathbf{X})^{-1}\,\text{Var}[\mathbf{X}^T\mathbf{Y}](\mathbf{X}^T\mathbf{X})^{-1} \\
&= (\mathbf{X}^T\mathbf{X})^{-1}\mathbf{X}^T\,\text{Var}\,[\mathbf{Y}]\mathbf{X}(\mathbf{X}^T\mathbf{X})^{-1} \\
&= \sigma^2(\mathbf{X}^T\mathbf{X})^{-1}\mathbf{X}^T\mathbf{X}(\mathbf{X}^T\mathbf{X})^{-1} \\
&= \sigma^2(\mathbf{X}^T\mathbf{X})^{-1}
\end{aligned}
\tag{16.35}
$$

If the null hypothesis is H_0: $\mathbf{C}\beta = \mathbf{0}$ then $\mathbf{C}\widehat{\beta}$ is an unbiased estimate of $\mathbf{C}\beta$ and

$$
\text{E}\left[\mathbf{C}\widehat{\beta}\right] = \mathbf{C}\beta = \mathbf{0} \tag{16.36}
$$

$$
\begin{aligned}
\text{Var}\left[\mathbf{C}\widehat{\beta}\right] &= \mathbf{C}\,\text{Var}[\widehat{\beta}]\mathbf{C}^T \\
&= \sigma^2\mathbf{C}(\mathbf{X}^T\mathbf{X})^{-1}\mathbf{C}^T,
\end{aligned}
\tag{16.37}
$$

$$
\left[\text{Var}\left[\mathbf{C}\widehat{\beta}\right]\right]^{-1} = \frac{1}{\sigma^2}[\mathbf{C}(\mathbf{X}^T\mathbf{X})^{-1}\mathbf{C}^T]^{-1}. \tag{16.38}
$$

Under the null hypothesis, the standard quadratic form is

$$
\text{SSH} = \frac{1}{\sigma^2}\widehat{\beta}^T\mathbf{C}^T[\mathbf{C}(\mathbf{X}^T\mathbf{X})^{-1}\mathbf{C}^T]^{-1}\mathbf{C}\widehat{\beta}. \tag{16.39}
$$

The expression in equation (16.39) is the **sum of squares due to hypothesis** (denoted SSH). If $\mathbf{Y}$ has a multivariate normal distribution, then SSH/σ^2 has a chi–square distribution with n_h degrees of freedom.

16.5.7 Sum of squares due to error

For the general linear model $\mathbf{Y} = \mathbf{X}\beta + \epsilon$ let $\mathbf{e} = \mathbf{Y} - \mathbf{X}\widehat{\beta}$, the error of estimation. The **sum of squares due to error** is given by

$$
\begin{aligned}
\text{SSE} &= \sum_{i=1}^{n} e_i^2 = \mathbf{e}^T\mathbf{e} \\
&= (\mathbf{Y} - \mathbf{X}\widehat{\beta})^T(\mathbf{Y} - \mathbf{X}\widehat{\beta}) \\
&= \mathbf{Y}^T\mathbf{Y} - 2\widehat{\beta}^T\mathbf{X}^T\mathbf{Y} + \widehat{\beta}^T\mathbf{X}^T\mathbf{X}\widehat{\beta} \\
&= \mathbf{Y}^T\mathbf{Y} - 2\widehat{\beta}^T\mathbf{X}^T\mathbf{Y} + \widehat{\beta}^T\mathbf{X}^T\mathbf{X}(\mathbf{X}^T\mathbf{X})^{-1}\mathbf{X}^T\mathbf{Y} \\
&= \mathbf{Y}^T\mathbf{Y} - 2\widehat{\beta}^T\mathbf{X}^T\mathbf{Y} + \widehat{\beta}^T\mathbf{I}\mathbf{X}^T\mathbf{Y} \\
&= \mathbf{Y}^T\mathbf{Y} - \widehat{\beta}^T\mathbf{X}^T\mathbf{Y}.
\end{aligned}
\tag{16.40}
$$

Thus, SSE is obtained by computing the sum of squares of all observations ($\mathbf{Y}^T\mathbf{Y}$) and subtracting the scalar product of the vector of estimates of $\widehat{\beta}$ and the vector on the right–hand side of the normal equations.

The sum of squares due to error may depend only on the model, and is determined once the model is stated. SSE is independent of any hypothesis which may be stated or tested.

If $\mathbf{Y}$ has a multivariate normal distribution, then SSE/σ^2 has a chi–square distribution with n_e degrees of freedom and is independent of any SSH.

16.5.8   Summary

For the general linear model

$$\mathbf{Y} = \mathbf{X}\boldsymbol{\beta} + \boldsymbol{\epsilon}, \qquad \text{E}[\mathbf{Y}] = \mathbf{X}\boldsymbol{\beta} \tag{16.41}$$

suppose the model is of full rank ($\mathbf{X}^\mathrm{T}\mathbf{X}$ is non–singular and thus has an inverse). If

$$\text{Var}[\mathbf{Y}] = \sigma^2 \mathbf{I} \quad \text{(homoscedasticity and independence)} \tag{16.42}$$

then the normal equations are given by

$$(\mathbf{X}^\mathrm{T}\mathbf{X})\widehat{\boldsymbol{\beta}}^\mathrm{T} = \mathbf{X}^\mathrm{T}\mathbf{Y} \tag{16.43}$$

and the estimate of β is

$$\widehat{\boldsymbol{\beta}} = (\mathbf{X}^\mathrm{T}\mathbf{X})^{-1}\mathbf{X}^\mathrm{T}\mathbf{Y}. \tag{16.44}$$

If the elements of $\mathbf{Y}$ are normally distributed, then the following hypothesis test may be conducted:

$$H_0: \mathbf{C}\boldsymbol{\beta} = \mathbf{0}$$
$$H_a: \mathbf{C}\boldsymbol{\beta} = \mathbf{d} \ \ (\neq \mathbf{0}) \tag{16.45}$$

This hypothesis matrix has n_h rows. If H_0 is consistent and contains no redundancies then n_h is the degrees of freedom due to hypothesis.

16.5.9   Computation procedure for hypothesis testing

A procedure for testing a hypothesis in a general linear model (equation (16.45)):

(1) Obtain the sum of squares due to hypothesis:

$$\text{SSH} = \widehat{\boldsymbol{\beta}}^\mathrm{T}\mathbf{C}^\mathrm{T}\left[\mathbf{C}(\mathbf{X}^\mathrm{T}\mathbf{X})^{-1}\mathbf{C}^\mathrm{T}\right]^{-1}\mathbf{C}\widehat{\boldsymbol{\beta}}. \tag{16.46}$$

(2) Obtain the sum of squares due to error:

$$\text{SSE} = \mathbf{Y}^\mathrm{T}\mathbf{Y} - \widehat{\boldsymbol{\beta}}^\mathrm{T}\mathbf{X}^\mathrm{T}\mathbf{Y}. \tag{16.47}$$

(3) Let $n_e = n - n_p =$ (sample size) $-$ (number of effective parameters in the model).

(4) If the null hypothesis, H_0, is true, then

$$\frac{\text{SSH}/n_k}{\text{SSE}/n_e} \tag{16.48}$$

has an F distribution with n_k and n_e degrees of freedom.

16.5.10 Regression significance test

For the general linear model

$$\mathbf{Y} = \mathbf{X}\beta + \epsilon, \quad \mathrm{E}\,[\mathbf{Y}] = \mathbf{X}\beta, \quad \mathrm{Var}[\mathbf{Y}] = \sigma^2\mathbf{I} \tag{16.49}$$

the normal equations, SSE, and an estimate of β are given by

$$\mathbf{X}^\mathrm{T}\mathbf{X}\widehat{\beta} = \mathbf{X}^\mathrm{T}\mathbf{Y}$$

$$\mathrm{SSE} = \mathbf{Y}^\mathrm{T}\mathbf{Y} - \widehat{\beta}^\mathrm{T}\mathbf{X}^\mathrm{T}\mathbf{Y} \tag{16.50}$$

$$\widehat{\beta} = (\mathbf{X}^\mathrm{T}\mathbf{X})^{-1}\mathbf{X}^\mathrm{T}\mathbf{Y}.$$

Suppose the reduced model is given by

$$\mathbf{Y} = \mathbf{X}\beta_r + \epsilon, \quad \mathbf{C}\beta_r = 0. \tag{16.51}$$

The sum of squares due to hypothesis is given by

$$\mathrm{SSH} = \mathrm{SSE(R)} - \mathrm{SSE}. \tag{16.52}$$

16.5.11 Alternate form of the distribution

The quotient

$$B = \frac{\mathrm{SSE}}{\mathrm{SSE} + \mathrm{SSH}} \tag{16.53}$$

has a beta distribution with parameters $n_e/2$ and $n_h/2$. Using this distribution, the hypothesis tests are lower-tailed; reject H_0 if the value of the test statistic B is smaller than the critical value. Thus the rejection region is given by

$$B = \frac{\mathrm{SSE}}{\mathrm{SSE(R)}} \le \beta\left(\frac{n_e}{2}, \frac{n_h}{2}\right) \quad \text{(usual notation) or,}$$

$$\le \beta^*(n_h, n_e) \quad \text{(beta percentage point), or}$$

$$\le I\left(\frac{n_e}{2}, \frac{n_h}{2}\right) \quad \text{(incomplete beta function).}$$

16.6 GENERAL LINEAR MODEL OF LESS THAN FULL RANK

A **singular** general linear model is not of full rank. For a general linear model

$$\mathbf{Y} = \mathbf{X}\beta + \epsilon \tag{16.54}$$

with normal equations

$$\mathbf{X}^\mathrm{T}\mathbf{X}\widehat{\beta} = \mathbf{X}^\mathrm{T}\mathbf{y} \tag{16.55}$$

suppose the rank of the design matrix $\mathbf{X}$ is r (with $r < m$). Then the matrix $(\mathbf{X}^\mathrm{T}\mathbf{X})$ is singular and has no inverse; there are no unbiased estimates for each β_i. However, there may exist unbiased estimates for certain *functions* of the β_i's.

16.6.1   Estimable function and estimability

Suppose $\mathbf{w}^T\boldsymbol{\beta}$ is a function of the β_i's where $\mathbf{w}^T$ is a given vector of *weights*. An estimator for $\mathbf{w}^T\boldsymbol{\beta}$ is a linear function of the Y's, $\mathbf{c}^T\mathbf{Y} = \mathbf{w}^T\widehat{\boldsymbol{\beta}}$, such that

$$\mathrm{E}\left[\mathbf{c}^T\mathbf{Y}\right] = \mathbf{w}^T\boldsymbol{\beta} \quad \text{for all } \boldsymbol{\beta} \tag{16.56}$$

and the variance is a minimum, i.e., $\mathrm{Var}\left[\mathbf{c}^T\mathbf{y}\right] = \text{minimum}$. The unbiasedness constraints are

$$\begin{aligned}
\mathrm{E}\left[\mathbf{c}^T\mathbf{Y}\right] &= \mathbf{w}^T\boldsymbol{\beta} \\
\mathbf{c}^T\mathrm{E}\left[\mathbf{Y}\right] &= \mathbf{w}^T\boldsymbol{\beta} \\
\mathbf{c}^T\mathbf{X}\boldsymbol{\beta} &= \mathbf{w}^T\boldsymbol{\beta} \quad \text{for all } \boldsymbol{\beta} \\
\mathbf{c}^T\mathbf{X} &= \mathbf{w}^T \\
\mathbf{c}^T\mathbf{X} - \mathbf{w}^T &= 0.
\end{aligned} \tag{16.57}$$

Therefore, the variance, $\mathrm{Var}\left[\mathbf{c}^T\mathbf{Y}\right] = \sigma^2\mathbf{c}^T\mathbf{c}$, must be a minimum subject to the constraints $\mathbf{c}^T\mathbf{X} = \mathbf{w}^T$.

The criterion function $\boldsymbol{\Phi}$ is

$$\boldsymbol{\Phi} = \frac{1}{2}\mathbf{c}^T\mathbf{c} - (\mathbf{c}^T\mathbf{X} - \mathbf{w}^T)\boldsymbol{\lambda} \quad \text{and} \tag{16.58}$$

$$\frac{\partial\boldsymbol{\Phi}}{\partial\mathbf{c}} = \mathbf{c} - \mathbf{X}\boldsymbol{\lambda}. \tag{16.59}$$

In equation (16.59), set the derivative equal to zero to obtain

$$\mathbf{X}\boldsymbol{\lambda} = \widehat{\mathbf{c}}. \tag{16.60}$$

Premultiply by $\mathbf{X}^T$:

$$\mathbf{X}^T\mathbf{X}\boldsymbol{\lambda} = \mathbf{X}^T\widehat{\mathbf{c}} \tag{16.61}$$

which is equal to $\mathbf{w}$ under the constraints. Therefore,

$$\mathbf{X}^T\mathbf{X}\boldsymbol{\lambda} = \mathbf{w}. \tag{16.62}$$

Equations (16.60) and (16.62) are the **conjugate normal equations**. If $\mathbf{X}$ has rank r (with $r < m$) there will always be r columns that form a basis with the remaining $m - r$ columns as an extension, linear combinations of the basis.

For the general linear model

$$\mathbf{Y} = \mathbf{X}\boldsymbol{\beta} + \boldsymbol{\epsilon} \tag{16.63}$$

partition the elements of $\boldsymbol{\beta}$ and the columns of $\mathbf{X}$ such that

$$\boldsymbol{\beta}^T = [\boldsymbol{\beta}_1^T, \boldsymbol{\beta}_2^T] \tag{16.64}$$

$$\mathbf{X} = [\mathbf{X}_1, \mathbf{X}_2] \tag{16.65}$$

with dimensions given by: $\boldsymbol{\beta}_1$ is $r \times 1$, $\boldsymbol{\beta}_2$ is $(m - r) \times 1$, $\mathbf{X}_1$ is $r \times r$, and $\mathbf{X}_2$ is $(m - r) \times r$, such that $\mathbf{X}_1$ is a basis for $\mathbf{X}$. The columns of $\mathbf{X}_2$ must be linear combinations of those in $\mathbf{X}_1$. Therefore, there exists a matrix $\mathbf{Q}_{r(m-r)}$ such that $\mathbf{X}_2 = \mathbf{X}_1 \mathbf{Q}$. If $\mathbf{X}_1$ and $\mathbf{X}_2$ are given, suppose

$$\mathbf{X}_2 = \mathbf{X}_1 \mathbf{Q} \quad \text{and} \tag{16.66}$$
$$\mathbf{X}_1^T \mathbf{X}_2 = \mathbf{X}_1^T \mathbf{X}_1 \mathbf{Q}$$
$$\mathbf{Q} = (\mathbf{X}_1^T \mathbf{X}_1)^{-1} \mathbf{X}_1^T \mathbf{X}_2. \tag{16.67}$$

Often, $\mathbf{Q}$ may be found by inspection.

Using equation (16.66) the matrix $\mathbf{X}$ may be written as

$$\mathbf{X} = \begin{bmatrix} \mathbf{X}_1 & \mathbf{X}_2 \end{bmatrix} = \begin{bmatrix} \mathbf{X}_1 & \mathbf{X}_1 \mathbf{Q} \end{bmatrix} = \mathbf{X}_1 \begin{bmatrix} \mathbf{I}_r & \mathbf{Q} \end{bmatrix} \tag{16.68}$$

and the conjugate normal equation (16.62) may be written as

$$\begin{bmatrix} \mathbf{X}_1^T \\ \mathbf{Q}^T \mathbf{X}_1^T \end{bmatrix} \begin{bmatrix} \mathbf{X}_1 & \mathbf{X}_1 \mathbf{Q} \end{bmatrix} \lambda = \begin{bmatrix} \mathbf{X}_1^T \mathbf{X}_1 & \mathbf{X}_1^T \mathbf{X}_1 \mathbf{Q} \\ \mathbf{Q}^T \mathbf{X}_1^T \mathbf{X}_1 & \mathbf{Q}^T \mathbf{X}_1^T \mathbf{X}_1 \mathbf{Q} \end{bmatrix} \lambda = \begin{bmatrix} \mathbf{w}_1 \\ \mathbf{w}_2 \end{bmatrix} \tag{16.69}$$

where $\mathbf{w}_1$ is $r \times 1$ and $\mathbf{w}_2$ is $(m - r) \times 1$. Equating components:

$$\begin{bmatrix} \mathbf{X}_1^T \mathbf{X}_1 & \mathbf{X}_1^T \mathbf{X}_1 \mathbf{Q} \end{bmatrix} \lambda = \mathbf{w}_1 \tag{16.70}$$
$$\begin{bmatrix} \mathbf{Q}^T \mathbf{X}_1^T \mathbf{X}_1 & \mathbf{Q}^T \mathbf{X}_1^T \mathbf{X}_1 \mathbf{Q} \end{bmatrix} \lambda = \mathbf{w}_2. \tag{16.71}$$

Premultiply equation (16.70) by $\mathbf{Q}^T$ to obtain

$$\begin{bmatrix} \mathbf{Q}^T \mathbf{X}_1^T \mathbf{X}_1 & \mathbf{Q}^T \mathbf{X}_1^T \mathbf{X}_1 \mathbf{Q} \end{bmatrix} \lambda = \mathbf{Q}^T \mathbf{w}_1. \tag{16.72}$$

In order for the system to be consistent, the condition

$$\mathbf{w}_2 = \mathbf{Q}^T \mathbf{w}_1 \tag{16.73}$$

must be true. Therefore, in the function $\mathbf{w}\beta$, the weight vector $\mathbf{w}$ must be of the form $\mathbf{w}^T = [\mathbf{w}_1^T, \mathbf{w}_2^T]$ where

$$\mathbf{w}_2^T = \mathbf{w}_1^T \mathbf{Q}. \tag{16.74}$$

If the vector of weights $\mathbf{w}$ is of this form then there is a linear unbiased (and mathematically consistent) estimate for the function $\mathbf{w}\beta$.

Equation (16.74) is the condition of **estimability** of a linear function. A function $\mathbf{w}^T \beta$ is **estimable** if $\mathbf{w}^T$ can be written as $\mathbf{w}^T = [\mathbf{w}_1^T, \mathbf{w}_2^T]$ where $\mathbf{w}_2^T$ is related to $\mathbf{w}_1^T$ in the same way as $\mathbf{X}_2$ is related to $\mathbf{X}_1$.

A parametric function is linearly **estimable** if there exists a linear combination of the observations whose expected value is equal to the function, i.e., if there exists an unbiased estimate.

If the function $\mathbf{w}^T \beta$ is estimable, equation (16.70) may be written as

$$(\mathbf{X}_1^T \mathbf{X}_1) \begin{bmatrix} \mathbf{I} & \mathbf{Q} \end{bmatrix} \lambda = \mathbf{w}_1. \tag{16.75}$$

The equation involving $\mathbf{w}_2$ may be disregarded since $\mathbf{w}_2$ is completely determined by the relation $\mathbf{w}_2 = \mathbf{Q}^T \mathbf{w}_1$. Therefore

$$\begin{bmatrix} \mathbf{I} & \mathbf{Q} \end{bmatrix} \lambda = (\mathbf{X}_1^T \mathbf{X}_1)^{-1} \mathbf{w}_1. \tag{16.76}$$

Rewriting the first conjugate normal equation (16.60) yields

$$\mathbf{X}\boldsymbol{\lambda} = \hat{\mathbf{c}}$$
$$[\mathbf{X}_1 \ \mathbf{X}_2]\,\boldsymbol{\lambda} = \hat{\mathbf{c}}$$
$$\mathbf{X}_1\,[\mathbf{I} \ \mathbf{Q}]\,\boldsymbol{\lambda} = \hat{\mathbf{c}}. \tag{16.77}$$

Using equation (16.76):

$$\mathbf{X}_1(\mathbf{X}_1^T\mathbf{X}_1)^{-1}\mathbf{w}_1 = \hat{\mathbf{c}} \quad \text{and} \tag{16.78}$$

$$\widehat{\mathbf{cy}} = \widehat{\mathbf{w}^T\boldsymbol{\beta}} = \mathbf{w}_1(\mathbf{X}_1^T\mathbf{X}_1)^{-1}\mathbf{X}_1^T\mathbf{y} \tag{16.79}$$

which is of the same form as in the non–singular case, except that $\mathbf{X}$ has been replaced by its basis $\mathbf{X}_1$ and in $\mathbf{w}$ only the first r elements are considered: $\mathbf{w}_1$.

The normal equations in the method of least squares,

$$(\mathbf{X}^T\mathbf{X})\widehat{\boldsymbol{\beta}}^T = \mathbf{X}^T\mathbf{y} \tag{16.80}$$

may be used formally in the reduced statement

$$(\mathbf{X}_1^T\mathbf{X}_1)\widehat{\boldsymbol{\beta}}_1^T = \mathbf{X}_1^T\mathbf{y}. \tag{16.81}$$

16.6.2   General linear hypothesis model of less than full rank

For a general linear model of less than full rank:

$$\mathbf{Y} = \mathbf{X}\boldsymbol{\beta} + \boldsymbol{\epsilon}$$
$$= [\mathbf{X}_1\mathbf{X}_2]\begin{bmatrix}\beta_1\\\beta_2\end{bmatrix} + \boldsymbol{\epsilon}$$
$$= [\mathbf{X}_1\mathbf{X}_1\mathbf{Q}]\begin{bmatrix}\beta_1\\\beta_2\end{bmatrix} + \boldsymbol{\epsilon} \tag{16.82}$$
$$= \mathbf{X}_1\beta_1 + \mathbf{X}_1\mathbf{Q}\beta_2 + \boldsymbol{\epsilon}$$
$$= \mathbf{X}_1(\beta_1 + \mathbf{Q}\beta_2) + \boldsymbol{\epsilon}.$$

Therefore, this general linear model may be written in the form $\mathbf{Y} = \mathbf{X}_1\boldsymbol{\beta}^* + \boldsymbol{\epsilon}$ where $\boldsymbol{\beta}^* = \beta_1 + \mathbf{Q}\beta_2$.

16.6.2.1   Sum of squares due to error

Since $\boldsymbol{\epsilon}$ has not changed in this model, the normal equations are

$$(\mathbf{X}_1^T\mathbf{X}_1)\widehat{\boldsymbol{\beta}}^* = \mathbf{X}_1^T\mathbf{y},$$

and the sum of squares due to error is

$$\text{SSE} = \mathbf{e}^T\mathbf{e} = \mathbf{y}^T\mathbf{y} - [\widehat{\boldsymbol{\beta}}^*]^T\mathbf{X}_1^T\mathbf{y}$$
$$= \mathbf{y}^T\mathbf{y} - (\mathbf{X}_1^T\mathbf{X}_1)^{-1}\mathbf{X}_1^T\mathbf{y}\mathbf{X}_1^T\mathbf{y}.$$

The expression SSE/σ^2 has a chi–square distribution with $n - r$ degrees of freedom where r is the rank of $\mathbf{X}$. The *effective* number of parameters in the singular model is only r, while the remaining $m-r$ parameters are determined in terms of the first r by the estimability condition.

16.6.2.2 Sum of squares due to hypothesis

Suppose the null hypothesis is given by

$$H_0: \mathbf{C}\beta = \mathbf{0} \quad \text{where} \quad \mathbf{C} = \begin{bmatrix} \mathbf{C}_1 & \mathbf{C}_2 \end{bmatrix} \tag{16.83}$$

and $\mathbf{C}_1$ has dimension $r \times r$ and $\mathbf{C}_2$ has dimension $r \times (m-r)$. Equation (16.83) implies

$$\begin{bmatrix} \mathbf{C}_1 & \mathbf{C}_2 \end{bmatrix} \begin{bmatrix} \beta_1 \\ \beta_2 \end{bmatrix} = \begin{bmatrix} 0 \\ 0 \end{bmatrix}. \tag{16.84}$$

The left–hand side of this equation must represent an estimable function, therefore

$$\mathbf{C}_2 = \mathbf{C}_1 \mathbf{Q}. \tag{16.85}$$

Equation (16.85) is the condition of **testability**: if

$$\mathbf{C} = \begin{bmatrix} \mathbf{c}_1^\mathrm{T} \\ \mathbf{c}_2^\mathrm{T} \\ \vdots \\ \mathbf{c}_{n_h}^\mathrm{T} \end{bmatrix}$$

where $(\mathbf{c}_i^\mathrm{T}\beta)$ is an estimable function $(i = 1, 2, \ldots, n_h)$, then the null hypothesis

$$H_0: \mathbf{C}_1\beta_1 + \mathbf{C}_2\beta_2 = \mathbf{0}$$

may be written as

$$H_0: \mathbf{C}_1\beta_1 + \mathbf{C}_1\mathbf{Q}\beta_2 = \mathbf{0} \quad \text{or simply}$$

$$H_0: \mathbf{C}_1\beta^* = \mathbf{0}, \quad \text{where} \quad \beta^* = (\beta_1 + \mathbf{Q}\beta_2).$$

Therefore, a null hypothesis $H_0: \mathbf{C}\beta = \mathbf{0}$ is *testable* if $\mathbf{C}\beta$ consists of n_h estimable functions, i.e., if $\mathbf{C}_2 = \mathbf{C}_1\mathbf{Q}$, where $\mathbf{C} = [\mathbf{C}_1, \mathbf{C}_2]$. The sum of squares due to hypothesis is given by

$$\text{SSH} = (\widehat{\beta}^*)^\mathrm{T}\mathbf{C}_1^\mathrm{T}[\mathbf{C}_1(\mathbf{X}_1^\mathrm{T}\mathbf{X}_1)^{-1}\mathbf{C}_1^\mathrm{T}]^{-1}\mathbf{C}_1\widehat{\beta}^*$$

where $\widehat{\beta}^* = (\mathbf{X}_1^\mathrm{T}\mathbf{X}_1)^{-1}\mathbf{X}_1^\mathrm{T}\mathbf{y}$. The expression SSH/σ^2 has a chi–square distribution with n_h degrees of freedom. If the null hypothesis is true, then

$$\frac{\text{SSH}/n_k}{\text{SSE}/n_e}$$

has an F distribution with n_k and n_e degrees of freedom.

16.6.3 Constraints and conditions

If the general linear model is singular of rank $r < m$, then $(m - r)$ constraints on the $\widehat{\boldsymbol{\beta}}_i$'s (the estimates) may be arbitrarily introduced, for example

$$\hat{\beta}_{r+1} = 0, \ldots, \hat{\beta}_m = 0 \quad \text{or} \tag{16.86}$$

$$\sum_{i=1}^m \hat{\beta}_i = 0, \quad \sum_{i=1}^m n_i \hat{\beta}_i = 0. \tag{16.87}$$

This procedure reparameterizes the model. The constraining functions are fairly arbitrary, but they must not be estimable functions, otherwise the resulting model will still be singular. To apply the constraints in equation (16.86), delete the last $(m - r)$ rows and columns of $\mathbf{X}^T\mathbf{X}$ and the last $(m - r)$ elements of $\mathbf{X}^T\mathbf{y}$. To apply the constraints in equation (16.87), add a constant to all elements of $\mathbf{X}^T\mathbf{X}$. This has no effect on the value of estimable functions or test statistics.

A different situation arises if conditions are placed on the parameters in a model, especially on interaction terms. In a two-factor, fixed effects experiment, the model is given by

$$Y_{ijk} = \mu + \alpha_i + \beta_j + (\alpha\beta)_{ij} + \epsilon_{ijk} \tag{16.88}$$

with assumptions on the interaction terms

$$\sum_{i=1}^a (\alpha\beta)_{ij} = \sum_{j=1}^b (\alpha\beta)_{ij} = 0. \tag{16.89}$$

The assumptions in equation (16.89) are often called **natural constraints** (even though the are neither natural nor constraints). These assumptions represent a set of conditions on the interactions, minimizing this effect (making SSH for interaction a minimum). Given these assumptions, the model is still singular, but can be made nonsingular by introducing the arbitrary constraints

$$\sum_{i=1}^a \hat{\alpha}_i = 0, \quad \sum_{j=1}^b \hat{\beta}_j = 0. \tag{16.90}$$

Using the different assumptions

$$\text{All } \alpha_i\text{'s} = 0, \quad \text{All } \beta_j\text{'s} = 0 \tag{16.91}$$

would simplify the model to a one-way anova.

CHAPTER 17

Miscellaneous Topics

Contents

17.1 GEOMETRIC PROBABILITY

1. Two points on a finite line:

   If A and B are uniformly chosen from the interval $[0, 1)$, and X is the distance between A and B (that is, $X = |A - B|$) then the probability density of X if $f_X(x) = 2(1 - x)$ for $0 \leq x \leq 1$.

2. Many points on a finite line:

   Suppose $n-1$ values are randomly selected from a uniform distribution on the interval $[0,1)$. These $n-1$ values determine n intervals.

   $P_k(x) =$ Probability (exactly k intervals have length larger than x)

   $$= \binom{n}{k}\left\{[1-kx]^{n-1} - \binom{n-1}{1}[1-(k+1)x]^{n-1} + \right.$$
   $$\left. \cdots + (-1)^s \binom{n-k}{s}[1-(k+s)x]^{n-1}\right\} \tag{17.1}$$

   where $s = \left\lfloor \dfrac{1}{x} - k\right\rfloor$. Using this, the probability that the largest interval length exceeds x is (for $0 \le x \le 1$):

   $$1 - P_0(x) = \binom{n}{1}(1-x)^{n-1} - \binom{n}{2}(1-2x)^{n-1} + \cdots \tag{17.2}$$

3. Points in the plane:

   Suppose the number of points in any region of area A of the plane is a Poisson random variable with mean λA (i.e., λ is the *density* of the points). Given a fixed point P define $R_1, R_2, \ldots$, to be the distance to the point nearest to P, second nearest to P, etc. The probability density function for R_s is (for $0 \le r \le \infty$):

   $$f_{R_s}(r) = \frac{2(\lambda\pi)^s}{(s-1)!} r^{2s-1} e^{-\lambda\pi r^2} \tag{17.3}$$

4. Points on a checkerboard:

   Consider the unit squares on a checkerboard and select one point uniformly in each square. The following results concern the distance between points, on average.

   (a) For adjacent squares (such as a black and white square) the mean distance between points is 1.088.

   (b) For diagonal squares (such as between two white squares) the mean between points is 1.473.

5. Points in three-dimensional space:

   Suppose the number of points in any volume V is a Poisson random variable with mean λV (i.e., λ is the *density* of the points). Given a fixed point P define $R_1, R_2, \ldots$, to be the distance to the point nearest to P, second nearest to P, etc. The probability density function for R_s is (for $0 \le r \le \infty$):

   $$f_{R_s}(r) = \frac{3\left(\frac{4}{3}\lambda\pi\right)^s}{(s-1)!} r^{3s-1} e^{-\frac{4}{3}\lambda\pi r^3} \tag{17.4}$$

6. Points in a cube:

   Choose two points uniformly in a unit cube. The distance between these points has mean 0.66171 and standard deviation 0.06214.

7. Points in n-dimensional cubes:

   Let two points be selected randomly from a unit n-dimensional cube. The expected distance between the points, $\Delta(n)$, is

   $$\Delta(1) = \frac{1}{3} \qquad\qquad\qquad \Delta(5) \approx 0.87852\ldots$$
   $$\Delta(2) \approx 0.54141\ldots \qquad\qquad \Delta(6) \approx 0.96895\ldots$$
   $$\Delta(3) \approx 0.66171\ldots \qquad\qquad \Delta(7) \approx 1.05159\ldots$$
   $$\Delta(4) \approx 0.77766\ldots \qquad\qquad \Delta(8) \approx 1.12817\ldots$$

8. Points on a circle:

   Select three points at random on a unit circle. These points determine a triangle with area A. The mean and variance of area are:

   $$\mu_A = \frac{3}{2\pi} \approx 0.4775$$
   $$\sigma_A^2 = \frac{3\left(\pi^2 - 6\right)}{8\pi^2} \approx 0.1470 \tag{17.5}$$

9. Particle in a box:

   A particle is bouncing randomly in a square box with unit side. On average, how far does it travel between bounces?

   Suppose the particle is initially at some random position in the box and is traveling in a straight line in a random direction and rebounds normally at the edges. Let θ be the angle of the point's initial vector. After traveling a distance r (where $r \gg 1$; think of many adjacent boxes and the particle exits each box and enters the next box), the point has moved $r\cos\theta$ horizontally and $r\sin\theta$ vertically, and thus has struck $r(\sin\theta + \cos\theta) + O(1)$ walls. Hence the average distance between walls is $1/(\sin\theta + \cos\theta)$. Averaging this over all angles results in

   $$\frac{2}{\pi} \int_0^{\pi/2} \frac{d\theta}{\sin\theta + \cos\theta} = \frac{2\sqrt{2}}{\pi} \ln(1 + \sqrt{2}) \approx 0.793515 \tag{17.6}$$

See J. G. Berryman, Random close packing of hard spheres and disks, *Physical Review A*, **27**, pages 1053–1061, 1983 and H. Solomon, *Geometric Probability*, SIAM, Philadelphia, PA, 1978.

17.2 INFORMATION AND COMMUNICATION THEORY

17.2.1 Discrete entropy

Suppose X is a discrete random variable that assumes n distinct values. Let $\mathbf{p}_X$ be the probability distribution for X and $\text{Prob}\,[X = x] = p_x$. The *entropy*

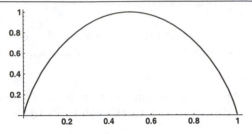

Figure 17.1: Binary entropy function

of the distribution is

$$H(\mathbf{p}_X) = -\sum_x p_x \log_2 p_x. \tag{17.7}$$

The units for entropy is *bits*. The maximum value of $H(\mathbf{p}_X)$ is $\log_2 n$ and is obtained when X is a discrete uniform random variable that assumes n values. Entropy measures how much information is gained from observing the value of X.

If X assumes only two values, $\mathbf{p}_X = (p, 1-p)$, and

$$H(\mathbf{p}_X) = H(p) = -p \log_2 p - (1-p) \log_2(1-p) \tag{17.8}$$

with a maximum at $p = 0.5$. A plot of $H(p)$ is in Figure 17.1.

Given two discrete random variables X and Y, $\mathbf{p}_{X \times Y}$ is the joint distribution of X and Y. The *mutual information* of X and Y is defined by

$$I(X,Y) = H(\mathbf{p}_X) + H(\mathbf{p}_Y) - H(\mathbf{p}_{X \times Y}) \tag{17.9}$$

Note that $I(X,Y) \geq 0$ and that $I(X,Y) = 0$ if and only if X and Y are independent. Mutual information gives the amount of information obtained about X after observing a value of Y (and vice versa).

Example 17.78:   *A coin weighing problem.*   There are 12 coins of which one is counterfeit, differing from the others by its weight. Using a balance but no weights, how many weighings are necessary to identify the counterfeit coin?

Solution:

(S1) Any of the 12 coins may turn out to be the counterfeit one, and it may be heavier or lighter than the genuine ones. Hence, there are 24 possible outcomes. For equal probabilities of these 24 outcomes, the entropy of the unknown result is then $\log_2 24 \approx 4.58$.

(S2) Each weighing process has three outcomes (equal weight, left side heavier, right side heavier). Using an assumption of equal probabilities gives an entropy of $\log_2 3 \approx 1.58$ per weighing. (Note that other assumptions will produce a smaller entropy.)

(S3) Therefore the minimal number of weighings cannot be less that $4.58/1.58 \approx 2.90$. Hence 3 weighings are needed. (In fact, 3 weighings are sufficient.)

17.2.2 Continuous entropy

For a d-dimensional continuous random variable $\mathbf{X}$, the entropy is

$$h(\mathbf{X}) = -\int_{\mathcal{R}^d} p(\mathbf{x}) \log p(\mathbf{x}) \, d\mathbf{x} \qquad (17.10)$$

Continuous entropy is not the the limiting case of the entropy of a discrete random variable. In fact, if X is the limit of the one-dimensional discrete random variable $\{X_n\}$, and the entropy of X is finite, then

$$\lim_{n \to \infty} (H(X_n) - n \log 2) = h(X) \qquad (17.11)$$

If $\mathbf{X}$ and $\mathbf{Y}$ are continuous d-dimensional random variables with density functions $p(\mathbf{x})$ and $q(\mathbf{y})$, then the *relative entropy* is

$$H(\mathbf{X}, \mathbf{Y}) = \int_{\mathcal{R}^d} p(\mathbf{x}) \log \frac{p(\mathbf{x})}{q(\mathbf{x})} \, d\mathbf{x} \qquad (17.12)$$

A d-dimensional Gaussian random variable $N(\mathbf{a}, \Gamma)$ has the density function

$$g(\mathbf{x}) = \frac{1.}{(2\pi)^{d/2} \sqrt{|\Gamma|}} \exp\left(-\frac{1}{2}(\mathbf{x} - \mathbf{a})^\mathrm{T} \Gamma^{-1}(\mathbf{x} - \mathbf{a})\right) \qquad (17.13)$$

where $\mathbf{a}$ is the vector of means and Γ is the positive definite covariance matrix.

1. If $\mathbf{X} = (X_1, X_2, \ldots, X_d)$ is a d-dimensional Gaussian random vector with distribution $N(\mathbf{a}, \Gamma)$ then

$$h(\mathbf{X}) = \frac{1}{2} \log \left((2\pi e)^d |\Gamma|\right) \qquad (17.14)$$

2. If $\mathbf{X}$ and $\mathbf{Y}$ are d-dimensional Gaussian random vectors with distributions $N(\mathbf{a}, \Gamma)$ and $N(\mathbf{b}, \Delta)$ then

$$\begin{aligned}
H(\mathbf{X}, \mathbf{Y}) = \frac{1}{2} \Big(&\log \frac{|\Delta|}{|\Gamma|} + \mathrm{tr}\left(\Gamma\left(\Delta^{-1} - \Gamma^{-1}\right)\right) \\
&+ (\mathbf{a} - \mathbf{b})^\mathrm{T} \Delta^{-1}(\mathbf{a} - \mathbf{b}) \Big)
\end{aligned} \qquad (17.15)$$

3. If $\mathbf{X}$ is a d-dimensional Gaussian random vector with distribution $N(\mathbf{a}, \Gamma)$, and if $\mathbf{Y}$ is a d-dimensional random vector with a continuous probability distribution having the same covariance matrix Γ, then

$$h(\mathbf{X}) \leq h(\mathbf{Y}) \qquad (17.16)$$

17.2.3 Channel capacity

The *transition probabilities* are defined by $t_{x,y} = \mathrm{Prob}\,[X = x \mid Y = y]$. The distribution $\mathbf{p}_X$ determines $\mathbf{p}_Y$ by $p_y = \sum_x t_{x,y} p_x$. The matrix $T = (t_{x,y})$ is the *transition matrix*. The matrix T defines a *channel* given by a transition diagram (input is X, output is Y). For example (here X and Y only assume two values):

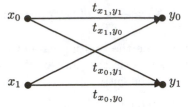

The *capacity* of the channel is defined as

$$C = \max_{\mathbf{p}_X} I(X, Y) \tag{17.17}$$

A channel is *symmetric* if each row is a permutation of the first row and the transition matrix is a symmetric matrix. The capacity of a symmetric channel is $C = \log_2 n - H(\mathbf{p})$, where $\mathbf{p}$ is the first row. The capacity of a symmetric channel is achieved with equally likely inputs. The channel below on the left is symmetric; both channels achieve capacity with equally likely inputs.

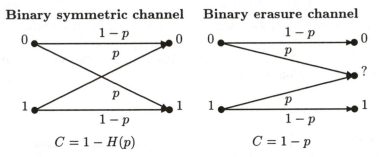

17.2.4 Shannon's theorem

Let both X and Y be discrete random variables with values in an alphabet A. A *code* is a set of n-tuples (*codewords*) with entries from A that is in one-to-one correspondence with M messages. The *rate* R of the code is defined as $\frac{1}{n} \log_2 M$. Assume that the codeword is sent via a channel with transition matrix T by sending each vector element independently. Define

$$e = \max_{\text{all codewords}} \text{Prob}\,[\text{codeword incorrectly decoded}]. \tag{17.18}$$

Then *Shannon's coding theorem* states:

(a) If $R < C$, then there is a sequence of codes with $n \to \infty$ such that $e \to 0$.

(b) If $R \geq C$, then e is always bounded away from 0.

17.3 KALMAN FILTERING

In the following model for $k \geq 0$:

$$\mathbf{x}_{k+1} = F_k \mathbf{x}_k + G_k \mathbf{w}_k$$
$$\mathbf{z}_k = H_k^T \mathbf{x}_k + \mathbf{v}_k \tag{17.19}$$

with the conditions:

1. The initial state $\mathbf{x}_0$ is a Gaussian random variable with mean $\mathbf{x}_0$ and covariance P_0, independent of $\{\mathbf{v}_k\}$ and $\{\mathbf{w}_k\}$.
2. The $\{\mathbf{v}_k\}$ and $\{\mathbf{w}_k\}$ are independent, zero mean, Gaussian white processes with

$$\mathrm{E}\left[\mathbf{v}_k \mathbf{v}_l^T\right] = R_k \delta_{kl} \quad \text{and} \quad \mathrm{E}\left[\mathbf{w}_k \mathbf{w}_l^T\right] = Q_k \delta_{kl} \tag{17.20}$$

an estimate of $\mathbf{x}_k$ from observation of the $\mathbf{z}_i$'s is desired.

1. Define Z_{k-1} to be the sequence of observed values $\{\mathbf{z}_0, \mathbf{z}_1, \ldots, \mathbf{z}_{k-1}\}$
2. Define the estimate of $\mathbf{x}_k$, conditioned on the $\mathbf{z}$ values (up to the $(k-1)^{\text{th}}$ value) to be $\widehat{\mathbf{x}}_{k/k-1} = \mathrm{E}\left[\mathbf{x}_k \mid Z_{k-1}\right]$. Similarly, define the estimate of $\mathbf{x}_k$, conditioned on the $\mathbf{z}$ values (up to the kth value) to be $\widehat{\mathbf{x}}_{k/k} = \mathrm{E}\left[\mathbf{x}_k \mid Z_k\right]$.
3. Define the *error covariance matrix* to be
   $\Sigma_{k/k-1} = \mathrm{E}\left[\left(\mathbf{x}_k - \widehat{\mathbf{x}}_{k/k-1}\right)\left(\mathbf{x}_k - \widehat{\mathbf{x}}_{k/k-1}\right)^T \mid Z_{k-1}\right]$. // Define $\Sigma_{k/k}$ in a similar way.
4. By convention, define $\widehat{\mathbf{x}}_{k/k-1}$ for $k = 0$ (i.e., $\widehat{\mathbf{x}}_{0/-1}$) to be $\overline{\mathbf{x}}_0 = \mathrm{E}\left[\mathbf{x}_0\right]$, i.e., the expected value of $\mathbf{x}_0$ given no measurements. Similarly, take $\Sigma_{0/-1}$ to be P_0.

The solution is given by (the intermediate matrix K_k is called the *gain matrix*)

$$\begin{aligned}
\widehat{\mathbf{x}}_{0/-1} &= \overline{\mathbf{x}}_0 \\
\Sigma_{0/-1} &= P_0 \\
\Omega_k &= H_k^T \Sigma_{k/k-1} H_k + R_k \\
K_k &= F_k \Sigma_{k/k-1} H_k \Omega_k^{-1} \\
\widehat{\mathbf{x}}_{k+1/k} &= \left(F_k - K_k H_k^T\right) \widehat{\mathbf{x}}_{k/k-1} + K_z \mathbf{z}_k \\
\widehat{\mathbf{x}}_{k/k} &= \widehat{\mathbf{x}}_{k/k-1} + \Sigma_{k/k-1} H_k \Omega_k^{-1} \left(\mathbf{z}_k - H_k^T \widehat{\mathbf{x}}_{k/k-1}\right) \\
\Sigma_{k/k} &= \Sigma_{k/k-1} - \Sigma_{k/k-1} H_k \Omega_k^{-1} H_k^T \Sigma_{k/k-1} \\
\Sigma_{k+1/k} &= F_k \Sigma_{k/k} F_k^T + G_k Q_k G_k^T
\end{aligned} \tag{17.21}$$

See B. D. O. Anderson and J. B. Moore, *Optimal Filtering*, Prentice–Hall, Inc., Englewood Cliffs, NJ, 1979.

17.3.1   Extended Kalman filtering

We have the following model for $k \geq 0$:

$$\begin{aligned} \mathbf{x}_{k+1} &= \mathbf{f}_k(\mathbf{x}_k) + \mathbf{g}_k(\mathbf{x}_k)\mathbf{w}_k \\ \mathbf{z}_k &= \mathbf{h}_k(\mathbf{x}_k) + \mathbf{v}_k \end{aligned} \tag{17.22}$$

with the usual assumptions. We presume the nonlinear functions $\{\mathbf{f}_k, \mathbf{g}_k, \mathbf{h}_k\}$ are sufficiently smooth and they can be expanded in Taylor series about the conditional means $\widehat{\mathbf{x}}_{k/k}$ and $\widehat{\mathbf{x}}_{k/k-1}$ as

$$\begin{aligned} \mathbf{f}_k(\mathbf{x}_k) &= \mathbf{f}_k(\widehat{\mathbf{x}}_{k/k}) + F_k(\mathbf{x}_k - \widehat{\mathbf{x}}_{k/k}) + \cdots \\ \mathbf{g}_k(\mathbf{x}_k) &= \mathbf{g}_k(\widehat{\mathbf{x}}_{k/k}) + \cdots = G_k + \cdots \\ \mathbf{h}_k(\mathbf{x}_k) &= \mathbf{h}_k(\widehat{\mathbf{x}}_{k/k-1}) + H_k^{\mathrm{T}}(\mathbf{x}_k - \widehat{\mathbf{x}}_{k/k-1}) + \cdots \end{aligned} \tag{17.23}$$

Neglecting higher order terms and assuming knowledge of $\widehat{\mathbf{x}}_{k/k}$ and $\widehat{\mathbf{x}}_{k/k-1}$ enables us to approximate the original system as

$$\begin{aligned} \mathbf{x}_{k+1} &= F_k\mathbf{x}_k + G_k\mathbf{w}_k + \mathbf{u}_k \\ \mathbf{z}_k &= H_k^{\mathrm{T}}\mathbf{x}_k + \mathbf{v}_k + \mathbf{y}_k \end{aligned} \tag{17.24}$$

where $\mathbf{u}_k$ and $\mathbf{y}_k$ are calculated from

$$\mathbf{u}_k = \mathbf{f}_k(\widehat{\mathbf{x}}_{k/k}) - F_k\widehat{\mathbf{x}}_{k/k} \quad \text{and} \quad \mathbf{y}_k = \mathbf{h}_k(\widehat{\mathbf{x}}_{k/k-1}) - H_k^{\mathrm{T}}\widehat{\mathbf{x}}_{k/k-1} \tag{17.25}$$

The Kalman filter for this approximate signal model is:

$$\begin{aligned} \widehat{\mathbf{x}}_{0/-1} &= \overline{\mathbf{x}}_0 \\ \Sigma_{0/-1} &= P_0 \\ \Omega_k &= H_k^{\mathrm{T}}\Sigma_{k/k-1}H_k + R_k \\ L_k &= \Sigma_{k/k-1}H_k\Omega_k^{-1} \\ \widehat{\mathbf{x}}_{k/k} &= \widehat{\mathbf{x}}_{k/k-1} + L_k\left(\mathbf{z}_k - \mathbf{h}_k(\widehat{\mathbf{x}}_{k/k-1})\right) \\ \widehat{\mathbf{x}}_{k+1/k} &= \mathbf{f}_k(\widehat{\mathbf{x}}_{k/k}) \\ \Sigma_{k/k} &= \Sigma_{k/k-1} - \Sigma_{k/k-1}H_k\Omega_k^{-1}H_k^{\mathrm{T}}\Sigma_{k/k-1} \\ \Sigma_{k+1/k} &= F_k\Sigma_{k/k}F_k^{\mathrm{T}} + G_kQ_kG_k^{\mathrm{T}} \end{aligned} \tag{17.26}$$

See B. D. O. Anderson and J. B. Moore, *Optimal Filtering*, Prentice–Hall, Inc., Englewood Cliffs, NJ, 1979.

17.4   LARGE DEVIATIONS (THEORY OF RARE EVENTS)

17.4.1   Theory

1. **Cramérs Theorem:** Let $\{X_i\}$ be a sequence of bounded, independent, identically distributed random variables with common mean m. Define

M_n to be the sample mean of the first n random variables:

$$M_n = \frac{1}{n}(X_1 + X_2 + \cdots + X_n) \tag{17.27}$$

The tails of the probability distribution for M_n decay exponentially, as $n \to \infty$, at a rate given by the convex *rate function* $I(x)$.

$$\text{Prob}[M_n > x] \sim e^{-nI(x)} \quad \text{for } x > m$$
$$\text{Prob}[M_n < x] \sim e^{-nI(x)} \quad \text{for } x < m \tag{17.28}$$

2. **Chernoff's Formula**: The rate-function $I(x)$ is related to the cumulant generating function $\lambda(\theta)$ (see page 38) via

$$I(x) = \max_{\theta}\{x\theta - \lambda(\theta)\}. \tag{17.29}$$

3. **Contraction Principle**: If $\{X_n\}$ satisfies a large deviation principle with rate function I and f is a continuous function, then $\{f(X_n)\}$ satisfies a large deviation principle with rate function J, where J is given by

$$J(y) = \min[I(x) \mid f(x) = y]. \tag{17.30}$$

17.4.2 Sample rate functions

1. Let $\{X_i\}$ be a sequence of Bernoulli random variables where p is the probability of obtaining a "1" and $(1-p)$ is the probability of obtaining a "0". Then $\lambda(\theta) = \ln[p \cdot e^{\theta} + (1-p) \cdot 1]$ and therefore

$$I(x) = x\ln\frac{x}{p} + (1-x)\ln\frac{1-x}{1-p} \tag{17.31}$$

(The maximum value of I occurs when θ is $\theta = \frac{\ln x}{p} - \frac{\ln(1-x)}{1-p}$.)

2. If the random variables in the sequence $\{X_i\}$ are all $N(\mu, \sigma^2)$ then

$$I(x) = \frac{1}{2}\left(\frac{x-\mu}{\sigma}\right)^2. \tag{17.32}$$

17.4.3 Example: Insurance company

Suppose an insurance company collects daily premiums as a constant rate p, and has daily claims total $Z \sim N(\mu, \sigma^2)$. The company would like to, naturally, avoid going bankrupt. The probability that the payments exceed income after T days is the probability that $\sum_{k=1}^{T} Z_k$ exceeds pT. For T large

$$\text{Prob}\left[\left(\frac{1}{T}\sum_{k=1}^{T} Z_k\right) > p\right] \sim e^{-TI(p)} \tag{17.33}$$

If an acceptable amount of risk is e^{-r}, then $e^{-TI(p)} = e^{-r}$, or $I(p) = r/T$. Using the rate function for a normal random variable, $r = \frac{T}{2}\left(\frac{p-\mu}{\sigma}\right)^2$, or

$$p = \mu + \sigma\sqrt{\frac{2r}{T}}. \tag{17.34}$$

The *safety loading* is defined by

$$\underbrace{\left(\frac{p-\mu}{\mu}\right)}_{\text{safety loading}} = \underbrace{\left(\frac{\sigma}{\mu}\right)}_{\text{size of fluctuations}} \quad \underbrace{\left(\sqrt{\frac{2r}{T}}\right)}_{\text{fixed by regulators}} \tag{17.35}$$

17.5   MARKOV CHAINS

A *discrete parameter stochastic process* is a collection of random variables $\{X(t),\ t = 0,\ 1,\ 2,\ \ldots\}$. The values of $X(t)$ are called the *states* of the process. The collection of states is called the *state space*. The values of t usually represent points in time. The number of states is either finite or countably infinite. A discrete parameter stochastic process is called a *Markov chain* if, for any set of n time points $t_1 < t_2 < \cdots < t_n$, the conditional distribution of $X(t_n)$ given values for $X(t_1)$, $X(t_2)$, $\ldots$, $X(t_{n-1})$ depends only on $X(t_{n-1})$. That is,

$$\text{Prob}\,[X(t_n) \le x_n \mid X(t_1) = x_1, \ldots, X(t_{n-1}) = x_{n-1}]$$
$$= \text{Prob}\,[X(t_n) \le x_n \mid X(t_{n-1}) = x_{n-1}]. \tag{17.36}$$

A Markov chain is said to be *stationary* if the value of the conditional probability $P\,[X(t_{n+1}) = x_{n+1} \mid X(t_n) = x_n]$ is independent of n. The following discussion will be restricted to stationary Markov chains.

17.5.1   Transition function

Let x and y be states and let $\{t_n\}$ be time points in $T = \{0, 1, 2, \ldots\}$. The *transition function*, $P(x,y)$, is defined by

$$P(x,y) = P_{n,n+1}(x,y) = \text{Prob}\,[X(t_{n+1}) = y \mid X(t_n) = x] \tag{17.37}$$

$P(x,y)$ is the probability that a Markov chain in state x at time n will be in state y at time $n + 1$. Some properties of the transition function are: $P(x,y) \ge 0$ and $\sum_y P(x,y) = 1$. The values of $P(x,y)$ are commonly called the *one-step transition probabilities*.

The function $\pi_0(x) = P(X(0) = x)$, with $\pi_0(x) \ge 0$ and $\sum_x \pi_0(x) = 1$, is called the *initial distribution* of the Markov chain. It is the probability distribution when the chain is started. Thus,

$$P\,[X(0) = x_0, X(1) = x_1, \ldots, X(n) = x_n]$$
$$= \pi_0(x_0)P_{0,1}(x_0, x_1)P_{1,2}(x_1, x_2)\cdots P_{n-1,n}(x_{n-1}, x_n). \tag{17.38}$$

17.5.2 Transition matrix

A convenient way to summarize the transition function of a Markov chain is by using the *one-step transition matrix*. It is defined as

$$P = \begin{bmatrix} P(0,0) & P(0,1) & \dots & P(0,n) & \dots \\ P(1,0) & P(1,1) & \dots & P(1,n) & \dots \\ \vdots & \vdots & \ddots & \vdots & \\ P(n,0) & P(n,1) & \dots & P(n,n) & \dots \\ \vdots & \vdots & & \vdots & \end{bmatrix}. \tag{17.39}$$

Define the *n–step transition matrix* by $P^{(n)}$ to be the matrix with entries

$$P^{(n)}(x,y) = \text{Prob}\left[X(t_{m+n}) = y \mid X(t_m) = x\right]. \tag{17.40}$$

This can be written using the one-step transition matrix as $P^{(n)} = P^n$.

Suppose the state space is finite. The one-step transition matrix is said to be *regular* if, for some positive power m, all of the elements of P^m are strictly positive.

Theorem 1 *(Chapman–Kolmogorov equation) Let $P(x,y)$ be the one-step transiton function of a Markov chain and define $P^0(x,y) = 1$, if $x = y$, and 0, otherwise. Then, for any pair of nonnegative integers s and t such that $s + t = n$,*

$$P^n(x,y) = \sum_z P^s(x,z)P^t(z,y). \tag{17.41}$$

17.5.3 Recurrence

Define the probability that a Markov chain starting in state x returns to state x for the first time after n steps by

$$f^n(x,x) = \text{Prob}\left[X(t_n) = x, X(t_{n-1}) \neq x, \dots, X(t_1) \neq x \mid X(t_0) = x\right]. \tag{17.42}$$

It follows that $P^n(x,x) = \sum_{k=0}^n f^k(x,x)P^{n-k}(x,x)$. A state x is said to be *recurrent* if $\sum_{n=0}^\infty f^n(x,x) = 1$. This means that a state x is recurrent if, after starting in x, the probability of returning to x after some finite length of time is one. A state which is not recurrent is said to be *transient*.

Theorem 2 *A state x of a Markov chain is recurrent if and only if $\sum_{n=1}^\infty P^n(x,x) = \infty$.*

Two states, x and y, are said to *communicate* if, for some $n \geq 0$, $P^n(x,y) > 0$. This theorem implies that if x is a recurrent state and x communicates with y, then y is also a recurrent state. A Markov chain is said to be *irreducible* if every state communicates with every other state and with itself.

Let x be a recurrent state and define T_x the (*return time*) to be the number of stages for a Markov chain to return to state x, having begun there. A recurrent state x is said to be *null recurrent* if $E[T_x] = \infty$. A recurrent state that is not null recurrent is said to be *positive recurrent*.

17.5.4   Stationary distributions

Let $\{X(t), t = 0, 1, 2, \ldots\}$ be a Markov chain having a one-step transition function of $P(x, y)$. A function $\pi(x)$ where each $\pi(x)$ is nonnegative, $\sum_x \pi(x)P(x, y) = \pi(y)$, and $\sum_y \pi(y) = 1$, is called a *stationary distribution*. If a Markov chain has a stationary distribution and $\lim_{n \to \infty} P^n(x, y) = \pi(y)$, then regardless of the initial distribution, $\pi_0(x)$, the distribution of $X(t_n)$ approaches $\pi(x)$ as n becomes infinite. When this happens $\pi(x)$ is often referred to as the *steady-state distribution*. The following categorizes those Markov chains that have stationary distributions.

Theorem 3 *Let X_P denote the set of positive recurrent states of a Markov chain.*

1. *If X_P is empty, the chain has no stationary distribution.*
2. *If X_P is a nonempty irreducible set, the chain has a unique stationary distribution.*
3. *If X_P is nonempty but not irreducible, the chain has an infinite number of distinct stationary distributions.*

The *period* of a state x is denoted by $d(x)$ and is defined to be the greatest common divisor of all integers $n \geq 1$ for which $P^n(x, x) > 0$. If $P^n(x, x) = 0$ for all $n \geq 1$ then define $d(x) = 0$. If each state of a Markov chain has $d(x) = 1$ the chain is said to be *aperiodic*. If each state has period $d > 1$ the chain is said to be *periodic* with period d. The vast majority of Markov chains encountered in practice are aperiodic. An irreducible, positive recurrent, aperiodic Markov chain always possesses a steady-state distribution.

An important special case occurs when the state space is finite. In this case, suppose that $X = \{1, 2, \ldots, K\}$. Let $\pi_0 = \{\pi_0(1), \pi_0(2), \ldots, \pi_0(K)\}$.

Theorem 4 *Let P be a regular one-step transition matrix and π_0 be an arbitrary vector of initial probabilities. Then $\lim_{n \to \infty} \pi_0(x)P^n = \mathbf{y}$, where $\mathbf{y}P = \mathbf{y}$, and $\sum_{i=1}^{K} \pi_0(t_i) = 1$.*

Example 17.79:

A Markov chain having three states, $\{0, 1, 2\}$, with a one-step transition matrix of

$P = \begin{bmatrix} 1/2 & 0 & 1/2 \\ 1/4 & 3/4 & 0 \\ 0 & 3/4 & 1/4 \end{bmatrix}$ is diagrammed as follows:

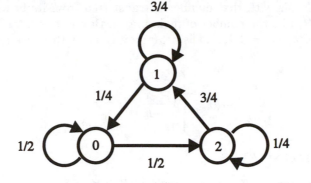

The one-step transition matrix gives a two–step transition matrix of

$$P^{(2)} = P^2 = \begin{bmatrix} 1/4 & 3/8 & 3/8 \\ 5/16 & 9/16 & 1/8 \\ 3/16 & 9/16 & 1/8 \end{bmatrix}$$

The one-step transition matrix is regular. This Markov chain is irreducible, and all three states are recurrent. In addition, all three states are positive recurrent. Since all states have period 1, the chain is aperiodic. The steady-state distribution is $\pi(0) = 3/11$, $\pi(1) = 6/11$, and $\pi(2) = 2/11$.

17.5.5   Random walks

Let $\{\eta(t_1), \eta(t_2), \ldots\}$ be independent random variables having a common density $f(x)$, and let t_1, t_2, ... be integers. Let $X(t_0)$ be an integer–valued random variable that is independent of $\eta(t_1)$, $\eta(t_2)$, ..., and $X(t_n) = X_0 + \sum_{i=1}^{n} \eta(t_i)$. The sequence $\{X(t_i), i = 0, 1, \ldots\}$ is called a *random walk*. An important special case is a *simple random walk*. It is defined by the following.

$$P(x, y) = \begin{cases} p & \text{if } y = x - 1 \\ r & \text{if } y = x \\ q & \text{if } y = x + 1 \end{cases}, \quad \text{where } p + q + r = 1, \quad P(0, 0) = p + r.$$

$$(17.43)$$

Here, an object begins at a certain point in a lattice and at each step either stays at that point or moves to a neighboring lattice point. In the case of a one– or two–dimensional lattice it turns out that if a random walk begins at a lattice point, x, it will return to that point with probability 1. In the case of a three–dimensional lattice the probability that it will return to its starting point is approximately 0.3405.

17.5.6   Ehrenfest chain

A simple model of gas exchange between two isolated bodies is as follows. Suppose that there are two boxes, Box I and Box II, where Box I contains K molecules numbered $1, 2, \ldots, K$ and Box II contains $N - K$ molecules numbered $K+1, K+2, \ldots, N$. A number is chosen at random from $\{1, 2, \ldots, N\}$, and the molecule with that number is transferred from its box to the other one. Let $X(t_n)$ be the number of molecules in Box I after n trials. Then the sequence $\{X(t_n), n = 0, 1, \ldots\}$ is a Markov chain with one–stage transition function of

$$P(x, y) = \begin{cases} \dfrac{x}{K} & y = x - 1, \\ 1 - \dfrac{x}{K} & y = x + 1, \\ 0 & \text{otherwise} \end{cases} \tag{17.44}$$

17.6   MARTINGALES

A stochastic process $\{Z_n \mid n \geq 1\}$ with $\mathrm{E}\left[|Z_n|\right] < \infty$ for all n is a

  (a) *martingale* process if $\mathrm{E}\left[Z_{n+1} \mid Z_1, Z_2, \ldots, Z_n\right] = Z_n$
  (b) *submartingale* process if $\mathrm{E}\left[Z_{n+1} \mid Z_1, Z_2, \ldots, Z_n\right] \geq Z_n$
  (c) *supermartingale* process if $\mathrm{E}\left[Z_{n+1} \mid Z_1, Z_2, \ldots, Z_n\right] \leq Z_n$

Azuma's inequality: Let $\{Z_n\}$ be a martingale process with mean $\mu = \mathrm{E}\left[Z_n\right]$. Let $Z_0 = \mu$ and suppose that $-\alpha_i \leq (Z_i - Z_{i-1}) \leq \beta_i$ for nonnegative constants $\{\alpha_i, \beta_i\}$ and $i \geq 1$. Then, for any $n \geq 0$ and $a > 0$:

  (a) $\mathrm{Prob}\left[X_n - \mu \geq a\right] \leq \exp\left(-2a^2 \middle/ \sum_{i=1}^{n} (\alpha_i + \beta_i)^2\right)$

  (b) $\mathrm{Prob}\left[X_n - \mu \leq -a\right] \leq \exp\left(-2a^2 \middle/ \sum_{i=1}^{n} (\alpha_i + \beta_i)^2\right)$

17.6.1   Examples of martingales

  (a) If $\{X_i\}$ are independent, mean zero random variables, and $Z_n = \sum_{i=1}^{n} X_i$, then $\{Z_n\}$ is a martingale.
  (b) If $\{X_i\}$ are independent random variables with $\mathrm{E}\left[X_i\right] = 1$, and $Z_n = \prod_{i=1}^{n} X_i$, then $\{Z_n\}$ is a martingale.
  (c) If $\{X, Y_i\}$ are arbitrary random variables with $\mathrm{E}\left[|X|\right] < \infty$, and $Z_n = \mathrm{E}\left[X \mid Y_1, Y_2, \ldots, Y_n\right]$, then $\{Z_n\}$ is a martingale.

17.7   MEASURE THEORETICAL PROBABILITY

  1. A *σ-field* of subsets of a set Ω is a collection $\mathcal{F}$ of subsets of Ω that contains ϕ (the empty set) as a member and is closed under complements and countable unions. If Ω is a topological space, the σ-field generated by the open subsets of Ω is called the Borel σ-field.

  2. A *probability measure* P on a σ-field $\mathcal{F}$ of subsets of a set Ω is a function from $\mathcal{F}$ to the unit interval $[0, 1]$ such that $P(\Omega) = 1$ and P of a countable union of disjoint sets $\{A_i\}$ equals the sum of $P(A_i)$.

3. A *probability space* is a triple $(\Omega, \mathcal{F}, P)$, where Ω is a set, $\mathcal{F}$ is a σ-field of subsets of Ω, and P is a probability measure on $\mathcal{F}$.

4. Given a probability space $(\Omega, \mathcal{F}, P)$ and a measurable space $(\Psi, \mathcal{G})$, a *random variable* from $(\Omega, \mathcal{F}, P)$ to $(\Psi, \mathcal{G})$ is a measurable function from $(\Omega, \mathcal{F}, P)$ to $(\Psi, \mathcal{G})$.

5. A random variable X from $(\Omega, \mathcal{F}, P)$ to $(\Psi, \mathcal{G})$ induces a probability measure on Ψ. The measure of a set A in $\mathcal{G}$ is simply $P(X^{-1}(A))$. This induced measure is called the distribution of X.

6. A real-valued function F defined on the set of real numbers $\mathcal{R}$ is called a distribution function for $\mathcal{R}$ if it is increasing and right-continuous and satisfies $\lim_{x \to -\infty} F(x) = 0$ and $\lim_{x \to \infty} F(x) = 1$.

Let Q be the distribution of X where X is a real valued random variable. Then the function $F\colon x \to Q((-\infty, x])$ is a distribution function. We call F the distribution function of X.

17.8 MONTE CARLO INTEGRATION TECHNIQUES

Random numbers may be used to approximate the value of a definite integral. Let g be an integrable function and define the integral I by

$$I = \int_B g(x)\,dx, \tag{17.45}$$

where B is a bounded region that may be enclosed in a rectangular parallelepiped R with volume $V(R)$. If $1_B(x)$ represents the indicator function of B,

$$1_B(x) = \begin{cases} 1 & \text{if } x \in B \\ 0 & \text{if } x \notin B \end{cases} \tag{17.46}$$

then the integral I may be written as

$$I = \int_R \Big(g(x)1_B(x)\Big)\,dx = \frac{1}{V(R)} \int_R \Big(g(x)1_B(x)V(R)\Big)\,dx \tag{17.47}$$

Equation (17.47) may be interpreted as an expected value of the function $h(X) = g(X)1_B(X)V(R)$ where the random variable X is uniformly distributed on the parallelepiped R (i.e., it has density function $1/V(R)$).

The expected value of $h(X)$ may be obtained by simulating random deviates from X, evaluating h at these points, and then computing the mean of the h values. If N trials are used, then the following estimate is obtained:

$$I \approx \hat{I} = \frac{1}{N} \sum_{i=1}^{N} h(x_i) = \frac{V(R)}{N} \sum_{i=1}^{N} g(x_i)1_B(x_i) \tag{17.48}$$

where each x_i is uniformly distributed in R.

See J. M. Hammersley and D. C. Handscomb, *Monte Carlo Methods*, John Wiley, 1965.

17.8.1   Importance sampling

Importance sampling is the term given to sampling from a non-uniform distribution so as to minimize the variance of the estimate for I in equation (17.45). Suppose a sample is selected from a distribution with density function $f(x)$. The integral I may be written as

$$I = \int_B \left(\frac{g(x)}{f(x)} \right) f(x)\, dx = \mathrm{E}_f \left[\frac{g(x)}{f(x)} \right] \tag{17.49}$$

where $\mathrm{E}_f[\cdot]$ denotes the expectation taken with respect to the density $f(x)$. That is, I is the mean of $g(x)/f(x)$ with respect to the distribution $f(x)$. Associated with this mean is the variance:

$$\sigma_f^2 = \mathrm{E}_f \left[\left\{ \frac{g(x)}{f(x)} - I \right\}^2 \right] = \mathrm{E}_f \left[\frac{g^2}{f^2} \right] - I^2 = \int_B \frac{g^2(x)}{f(x)}\, dx - I^2 \tag{17.50}$$

Approximations to I obtained by sampling from $f(x)$ will have errors that scale with σ_f.

A minimum variance estimator may be obtained by finding the density function $f(x)$ such that σ_f^2 is minimized. Using the calculus of variations the density function for the minimal estimator is

$$f_{\mathrm{opt}}(x) = C|g(x)| = \frac{|g(x)|}{\int_B |g(x)|\, dx} \tag{17.51}$$

where the constant C is chosen so that $f_{\mathrm{opt}}(x)$ is appropriately normalized. (Since $f_{\mathrm{opt}}(x)$ is a density function, it must integrate to unity.) While finding $f_{\mathrm{opt}}(x)$ is as difficult as determining the original integral I, equation (17.51) indicates that $f_{\mathrm{opt}}(x)$ should have the same general behavior as $|g(x)|$.

17.8.2   Hit-or-miss Monte Carlo method

The hit-or-miss Monte Carlo method is very inefficient but is easy to understand. Suppose that $0 \le f(x) \le 1$ when $0 \le x \le 1$. Defining

$$g(x,y) = \begin{cases} 0 & \text{if } f(x) < y, \\ 1 & \text{if } f(x) > y, \end{cases} \tag{17.52}$$

then $I = \int_0^1 f(x)\, dx = \int_0^1 \int_0^1 g(x,y)\, dy\, dx$. This integral may be estimated by

$$I \approx \widehat{I} = \frac{1}{n} \sum_{i=1}^n g(\xi_{2i-1}, \xi_{2i}) = \frac{n^*}{n} \tag{17.53}$$

where the $\{\xi_i\}$ are chosen independently and uniformly from the interval $[0,1]$. The summation in equation (17.53) reduces to the number of points in the unit square which are below the curve $y = f(x)$ (this defines n^*) divided by the total number of sample points (i.e., n).

17.9 QUEUING THEORY

The following diagram and notation are used to define a queue.

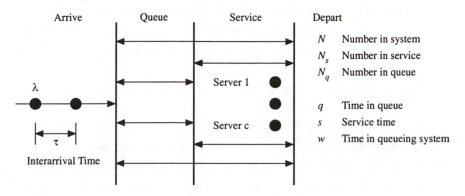

A queue is represented as $A/B/c/K/m/Z$ where

(a) A and B represent the interarrival times and service times:

D deterministic (constant) interarrival or service time distribution.

E_k Erlang–k interarrival or service time distribution (a gamma distribution with $\alpha = (k-1)$, $\beta = 1/\lambda k$ and density function $f(t) = \lambda k(\lambda kt)^{k-1}e^{-\lambda kt}/(k-1)!$ for $t > 0$.

GI general independent interarrival time.

G general service time distribution.

H_k k–stage hyperexponential interarrival or service time distribution (density function is $f(t) = \sum_{i=1}^{k} \alpha_i \mu_i e^{-\mu_i t}$ for $t \geq 0$).

M exponential interarrival or service time distribution.

(b) c is the number of identical servers.

(c) K is the system capacity.

(d) m is the number in the source.

(e) Z is the queue discipline:

FCFS first come, first served (also known as FIFO).
LIFO last in, first out.
RSS service in random order.
PRI priority service.

If all variables are not present, the last three above have the default values: $K = \infty$, $m = \infty$, and Z is RSS.

Note: The system includes both the queue and the service facility.

The variables of interest are:

(a) a_n: proportion of customers that find n customers already in the system when they arrive.

(b) c: number of servers in the service facility.

(c) d_n: proportion of customers leaving behind n customers in the system.

(d) K: maximum number of customers allowed in queueing system.

(e) L: mean number of customers in the steady-state system, $L = \mathrm{E}[N]$.

(f) L_q: mean number of customers in the steady-state queue, $L_q = \mathrm{E}[N_q]$.

(g) λ: mean arrival rate of customers to the system (number per unit time), $\lambda = 1/\mathrm{E}[\tau]$.

(h) μ: mean service rate per server (number per unit time), $\mu = 1/\mathrm{E}[s]$.

(i) N: random number of customers in system in steady state.

(j) N_a: random number of customers receiving service in steady state.

(k) N_q: random number of customers in queue in steady state.

(l) p_n: proportion of time the system contains n customers.

(m) $\pi_q(r)$: the queueing time that r percent of the customers do not exceed.

(n) $\pi_w(r)$: the system time that r percent of the customers do not exceed.

(o) q: random time a customer waits in the queue in order to begin service.

(p) q_n: probability that there are n customers in the system just before a customer enters.

(q) ρ: server utilization, the probability that any particular server is busy.

(r) s: random service time for one customer, $\mathrm{E}[s] = 1/\mu$.

(s) τ: random interarrival time, $\mathrm{E}[\tau] = 1/\lambda$.

(t) u: traffic intensity (units are *erlangs*) $u = \lambda/\mu$.

(u) W: mean time of customers in the system in steady state, $W = \mathrm{E}[w]$.

(v) w: total waiting time in the system, including queue and service times, $w = q + s$.

(w) W_q: mean time for customer in the queue in steady state, $W_q = \mathrm{E}[q]$.

Relationships between variables:

(a) Little's formula: $L = \lambda W$ and $L_q = \lambda W_q$.

(b) For Poisson arrivals: $p_n = a_n$.

(c) If customers arrive one at a time, and are served one at a time: $a_n = d_n$.

(d) $N = N_q + N_s$

(e) $W = W_q + W_s$

17.9.1 $M/M/1$ queue

Assume $\lambda < \mu$:

(a) $\rho = u/c = (\lambda/\mu)/c$

(b) $p_n = (1 - \rho)\rho^n$ for $n = 0, 1, 2, \ldots$

(c) $L = \rho/(1 - \rho)$

(d) $L_q = \rho^2/(1 - \rho)$

(e) $W = 1/\mu(1 - \rho)$

(f) $W_q = \rho/\mu(1 - \rho)$

(g) $\pi_q(r) = \max\left[0, W \log\left(\frac{100\rho}{100-r}\right)\right]$

(h) $\pi_w(r) = \max\left[0, W \log\left(\frac{100}{100-r}\right)\right]$

17.9.2 $M/M/1/K$ queue

Assume $K \geq 1$ and $N \leq K$:

(a) $\rho = (1 - p_K)u$

(b) $p_n = \begin{cases} \frac{(1-u)u^n}{1-u^{K+1}} & \text{if } \lambda \neq \mu \text{ and } n = 0, 1, \ldots, K \\ 1/(K+1) & \text{if } \lambda = \mu \text{ and } n = 0, 1, \ldots, K \end{cases}$

(c) $L = \begin{cases} \frac{u[1-(K+1)u^K+Ku^{K+1}]}{(1-u)(1-u^{K+1})} & \text{if } \lambda \neq \mu \\ K/2 & \text{if } \lambda = \mu \end{cases}$

(d) $L_q = L - (1 - p_0)$

(e) $\lambda_a = (1 - p_K)\lambda$ is the actual arrival rate at which customers enter the system.

(f) $W = L/\lambda_a$

(g) $W_q = L_q/\lambda_a$

(h) $q_n = p_n/(1 - p_K)$ for $n = 0, 1, \ldots, K - 1$

Note: p_K is the probability that an arriving customer is lost since there is no room in the queue.

17.9.3 $M/M/2$ queue

(a) $\rho = u/2$

(b) $p_0 = (1 - \rho)/(1 + \rho)$

(c) $p_n = 2(1 - \rho)\rho^n/(1 + \rho)$ for $n = 1, 2, 3, \ldots, c$

(d) $L = 2\rho/(1 - \rho^2)$

(e) $L_q = 2\rho^3/(1 - \rho^2)$

(f) $W = 1/\mu(1 - \rho^2)$

(g) $W_q = \rho^2/\mu(1 - \rho^2)$

17.9.4 $M/M/c$ queue

Erlang's C formula is the probability that all c servers are busy

$$C(c, u) = \left[1 + \frac{c!(1 - \rho)}{u^c} \sum_{n=0}^{c-1} \frac{u^n}{n!}\right]^{-1} \tag{17.54}$$

(a) $\rho = u/c$

(b) $u = \lambda/\mu$

(c) $p_0 = \dfrac{c!(1 - \rho)}{u^c} C(c, u) = \left[\dfrac{u^c}{c!(1 - \rho)} + \displaystyle\sum_{n=0}^{c-1} \dfrac{u^n}{n!}\right]^{-1}$

(d) $p_n = \begin{cases} \dfrac{u^n}{n!} p_0 & \text{for } n = 0, 1, \ldots, c \\ \dfrac{u^n}{c! c^{n-c}} p_0 & \text{for } n \geq c \end{cases}$

(e) $L = L_q + u$

(f) $L_q = \dfrac{uC(c,U)}{c(1-\rho)}$

(g) $W = W_q + 1/\mu$

(h) $W_q = \dfrac{C(c,u)}{c\mu(1-\rho)}$

(i) $\pi_q(90) = \max\left\{0, \dfrac{\ln[10\, C(c,U)]}{c\mu(1-\rho)}\right\}$

(j) $\pi_q(95) = \max\left\{0, \dfrac{\ln[20\, C(c,U)]}{c\mu(1-\rho)}\right\}$

17.9.5 $M/M/c/c$ queue

Erlang's B formula is the probability that all servers are busy

$$B(c, u) = \left[\frac{c!}{u^c} \sum_{n=0}^{c} \frac{u^n}{n!}\right]^{-1} \tag{17.55}$$

(a) $p_n = \left[\dfrac{n!}{u^n} \displaystyle\sum_{n=0}^{c} \dfrac{u^n}{n!}\right]^{-1}$ for $n = 0, 1, \ldots, c$

(b) $\lambda_a = \lambda(1 - B(c, u))$ is the average traffic rate experienced by the system.

(c) $\rho = \lambda_a/\mu c$

(d) $L = u[1 - B(c, u)]$

(e) $W = 1/\mu$

17.9.6 $M/M/c/K$ queue

(a) $p_0 = \left[\displaystyle\sum_{n=0}^{c} \dfrac{u^n}{n!} + \dfrac{u^c}{c!} \sum_{n=1}^{K-c} \left(\dfrac{u}{c}\right)^n\right]^{-1}$

(b) $p_n = \begin{cases} \frac{u^n}{n!} p_0 & \text{for } n = 0, 1, \ldots, c \\ \frac{u^c}{c!} \left(\frac{u}{c}\right)^{n-c} p_0 & \text{for } n = c+1, \ldots, K \end{cases}$

(c) $\lambda_a = \lambda(1 - p_K)$

(d) $\rho = (1 - p_K)u/c$

(e) $L = L_q + \sum_{n=0}^{c-1} n p_n + c\left(1 - \sum_{n=0}^{c-1} p_n\right)$

(f) $L_q = \frac{u^c p_0 u/c}{c!(1 - u/c)^2}\left[1 - \left(\frac{u}{c}\right)^{K-c+1} - (K - c + 1)\left(\frac{u}{c}\right)^{K-c}\left(1 - \frac{u}{c}\right)\right]$

(g) $W = L/\lambda_a$

(h) $W_q = L_q/\lambda_a$

17.9.7 $M/M/\infty$ queue

(a) $p_n = e^{-n} u^n/n!$ for $n = 0, 1, 2, \ldots$

(b) $L = u$

(c) $L_q = 0$

17.9.8 $M/E_k/1$ queue

(a) $p_n = (1 - \rho)\sum_{j=0}^{n}(-1)^{n-j} r^{n-j-1}\left(1 + \frac{\rho}{k}\right)^{kj}\left[\binom{kj}{n-j} r + \binom{kj}{n-j-1}\right]$ for
   $n = 0, 1, \ldots$, where $r = \rho/(k + \rho)$

(b) $L = L_q + \rho$

(c) $L_q = \lambda W_q$

(d) $W = W_q + 1/\mu$

(e) $W_q = \frac{\rho}{\mu(1-\rho)}\left(\frac{1+1/k}{2}\right)$

17.9.9 $M/D/1$ queue

(a) $p_0 = (1 - \rho)$

(b) $p_1 = (1 - \rho)(e^\rho - 1)$

(c) $p_n = (1 - \rho)\sum_{j=0}^{n}(-1)^{n-j}\left[\frac{(j\rho)^{n-j-1}(j\rho + n - j)e^{j\rho}}{(n - j)!}\right]$ for $n = 2, 3, \ldots$

(d) $L = \lambda W = L_q + \rho$

(e) $L_q = \lambda W_q = \frac{\rho^2}{2(1-\rho)}$

(f) $W = W_q + \frac{1}{\mu}$

(g) $W_q = \frac{\rho}{2\mu(1-\rho)}$

17.10 RANDOM MATRIX EIGENVALUES

1. Let A be a $n \times n$ matrix whose entries are independent standard normal deviates.

(a) The probability $p_{n,k}$ that A has k real eigenvalues has the form $r + s\sqrt{2}$, where r and s are rational. In particular, the probability that A has all real eigenvalues is

$$p_{n,n} = 2^{-n(n-1)/4} \qquad (17.56)$$

n	k	$p_{n,k}$	
1	1	1	1
2	2	$\frac{1}{2}\sqrt{2}$	≈ 0.707
	0	$1 - \frac{1}{2}\sqrt{2}$	≈ 0.293
3	3	$\frac{1}{4}\sqrt{2}$	≈ 0.354
	1	$1 - \frac{1}{4}\sqrt{2}$	≈ 0.646
4	4	$\frac{1}{8}$	0.125
	2	$-\frac{1}{4} + \frac{11}{16}\sqrt{2}$	≈ 0.722
	0	$\frac{9}{8} - \frac{11}{16}\sqrt{2}$	≈ 0.153

(b) The expected number of real eigenvalues of A is

$$E_n = \begin{cases} \sqrt{2} \displaystyle\sum_{k=0}^{n/2-1} \frac{(4k-1)!}{(4k)!!} & \text{when } n \text{ is even} \\[2mm] 1 + \sqrt{2} \displaystyle\sum_{k=1}^{(n-1)/2} \frac{(4k-3)!}{(4k-2)!!} & \text{when } n \text{ is odd} \end{cases} \qquad (17.57)$$

and $E_n \sim \sqrt{\frac{2n}{\pi}} \left(1 - \frac{3}{8n} - \frac{3}{128n^2} + \ldots \right)$ as $n \to \infty$.

(c) If λ_n denotes a real eigenvalue of A, then its marginal probability density $f_n(\lambda)$ is given by

$$f_n(\lambda) = \frac{1}{E_n} \left(\frac{1}{\sqrt{2\pi}} \left[\frac{\Gamma(n-1, \lambda^2)}{\Gamma(n-1)} \right] \right.$$
$$\left. + \frac{|\lambda^{n-1}| e^{-\lambda^2/2}}{\Gamma(n/2) 2^{n/2}} \left[\frac{\gamma((n-1)/2, \lambda^2/2)}{\Gamma((n-1)/2)} \right] \right) \qquad (17.58)$$

where $\gamma(a, x)$ and $\Gamma(a, x)$ are incomplete gamma functions (see page 519).

(d) If the elements of A have mean 0 and variance 1, and z is a scalar, then

$$\mathrm{E}\left[\det\left(A^2 + z^2 I\right)\right] = \mathrm{E}\left[\det\left(A + zI\right)^2\right] = \mathrm{per}\left(J + z^2 I\right)$$
$$= n!\, e_n\left(z^2\right) \tag{17.59}$$
$$= n!\, {}_1F_1\left(-1; \frac{1}{2}n; -\frac{1}{2}z^2 I\right)$$

where J is the matrix of all ones, "per" refers to the permanent of a matrix, $e_n(x) = \sum_{k=0}^{n} \frac{x^k}{k!}$ is the truncated Taylor series for e^x, and the hypergeometric function has a scalar multiple of the identity as its argument.

2. Let A be a random $n \times n$ matrix with randomly selected integer entries. Let $P(p, n)$ be the probability that $\det(A)$ is not congruent to 0 modulo p. Then

$$P(p, n) = \prod_{k=1}^{n} \left(1 - p^{-k}\right) \tag{17.60}$$

3. Let A be a random $n \times n$ complex matrix uniformly distributed on the sphere $1 = \|A\|_F = \sum_{i,j} |A_{ij}|^2$. Then,

$$\mathrm{E}\left[\det\left(A^H A\right) \mid \sigma_{\min}^2 = \lambda\right] = \sum_{r=0}^{n-1} \lambda^{n-r}(1 - n\lambda)^{n^2 + r - 1}$$
$$\times \frac{\Gamma(n^2)\Gamma(n+1)\Gamma(n+2)}{\Gamma(r+1)\Gamma(n-r)\Gamma(n^2+r-1)\Gamma(n+2-r)} \tag{17.61}$$

and if $E_n = \mathrm{E}\left[\det\left(A^H A\right)\right]$ then $E_n = \binom{n}{n^2+n-1}^{-1}$. The first few values of $1/E_n$ are $\{1, 10, 165, 3876, 118755, \ldots\}$.

4. Define the following matrices where $\tilde{N}(\mu, \sigma^2)$ refers to the distribution $X + iY$ where both X and Y are $N(\mu, \sigma^2)$:

(a) Gaussian matrix: $G(m, n)$, an $m \times n$ random matrix with iid elements which are $N(0, 1)$.

(b) Wishart matrix: $W(m, n)$, symmetric random matrix AA^T where A is $G(m, n)$.

(c) Gaussian orthogonal ensemble (GOE): an $m \times m$ random matrix $(A + A^T)/2$ where A is $G(m, m)$.

(d) Complex Gaussian matrix: $\tilde{G}(m, n)$, an $m \times n$ random matrix with iid elements which are $\tilde{N}(0, 1)$.

(e) Complex Wishart matrix: $\widetilde{W}(m, n)$, symmetric random matrix AA^H where A is $\tilde{G}(m, n)$.

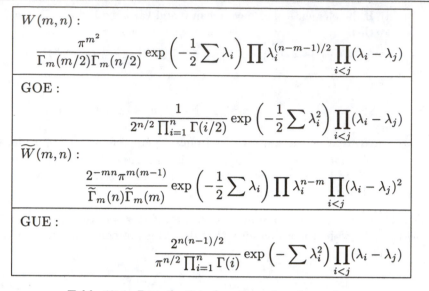

Table 17.1: Distribution functions for eigenvalues

(f) Gaussian unitary ensemble (GUE): an $m \times m$ random matrix $(A + A^{\mathrm{H}})/\sqrt{8}$ where A is $\widetilde{G}(m, m)$.

Then

(a) The joint densities of the eigenvalues $\lambda_1 \geq \cdots \geq \lambda_m$ for the above random matrices are in Table 17.1 where the unlabeled sums and products run from $i = 1$ to m and the complex multivariate gamma function is defined by

$$\widetilde{\Gamma}_m(a) = \pi^{m(m-1)/2} \prod_{i=1}^{m} \Gamma(a - i + 1) \qquad (17.62)$$

For the Wishart cases, this is the joint distribution for the non-negative eigenvalues. For the Gaussian cases, this is the joint distribution for the eigenvalues which may be anywhere on the real line.

(b) The probability density function of the smallest eigenvalue of a matrix from $W(m, m)$ is

$$f_{\lambda_{\min}}(\lambda) = \frac{m}{\sqrt{2\pi}} \Gamma\left(\frac{m+1}{2}\right) \lambda^{-1/2} e^{-\lambda m/2} U\left(\frac{m-1}{2}, -\frac{1}{2}, \frac{\lambda}{2}\right) \qquad (17.63)$$

Here, $U(a, b, z)$ is the Tricomi function (see section 18.9.3).

(c) If κ is the condition number of a matrix from $G(n,n)$, then the probability density function of κ/n converges to

$$f(x) = \frac{2x+4}{x^3} e^{-2/x-2/x^2} \tag{17.64}$$

and

$$E\left[\log \kappa\right] = \log n + c + o(1) \tag{17.65}$$

as $n \to \infty$ where $c \approx 1.537$.

See A. Edelman, How many eigenvalues of a random matrix are real?, *J. Amer. Math. Soc.*, **7**, 1994, pages 247–267, and A. Edelman, Eigenvalues and condition numbers of random matrices, *SIAM J. Matrix Anal. Appl.*, **9**, 1988, pages 543–560.

17.10.1 Random matrix products

Let $||\cdot||$ be a non-negative real-valued function of matrices that satisfies $||AB|| \le ||A|| \cdot ||B||$ when A and B are $d \times d$ matrices (i.e,, $||\cdot||$ could be a matrix norm). Let $\{A_i\}$ be iid random $d \times d$ matrices in which all d^2 elements of each A_i are iid normal random variables with mean 0 and variance 1. Then

$$\lim_{t \to \infty} \log \frac{||A_1 A_2 \cdots A_t||}{t} = \frac{1}{2}\left[1 + \psi\left(\frac{d}{2}\right)\right] \tag{17.66}$$

where ψ is the digamma function (see page 518).

See J. E. Cohen and C. M. Newman, The stability of large matrices and their products, *Ann. Probab.*, 1984, **12**, pages 283–310.

17.10.1.1 Vibonacci numbers

The Fibonacci numbers are defined by $f_n = f_{n-1} + f_{n-2}$ and $f_1 = f_2 = 1$. This formula can represented as the matrix product:

$$\begin{bmatrix} f_{n-1} \\ f_n \end{bmatrix} = \begin{bmatrix} 0 & 1 \\ 1 & 1 \end{bmatrix} \begin{bmatrix} f_{n-2} \\ f_{n-1} \end{bmatrix} \tag{17.67}$$

As $n \to \infty$, the ratio $\frac{f_{n+1}}{f_n} \sim \phi \approx 1.61803\ldots$ (the golden mean).

The vibonacci numbers are defined by $v_n = v_{n-1} \pm v_{n-1}$ where the $\pm$ sign is chosen randomly (50% of the time it is $-$ and 50% of the time it is $+$). This formula can represented as the matrix product:

$$\begin{bmatrix} v_{n-1} \\ v_n \end{bmatrix} = \begin{cases} \begin{bmatrix} 0 & 1 \\ 1 & 1 \end{bmatrix} \begin{bmatrix} v_{n-2} \\ v_{n-1} \end{bmatrix} & 50\% \text{ of the time} \\[2ex] \begin{bmatrix} 0 & 1 \\ 1 & -1 \end{bmatrix} \begin{bmatrix} v_{n-2} \\ v_{n-1} \end{bmatrix} & 50\% \text{ of the time} \end{cases} \tag{17.68}$$

As $n \to \infty$, the ratio $\frac{v_{n+1}}{v_n} \sim 1.1319882\ldots$ (Viswanath's constant).

See B. Hayes, The vibonacci numbers, *American Scientist*, **87**, July–August 1999, pages 296–301.

17.11   RANDOM NUMBER GENERATION

17.11.1   Methods of pseudorandom number generation

Depending on the application, either integers in some range or floating point numbers in $[0, 1)$ are the desired output from a pseudorandom number generator (PRNG). Since most PRNGs use integer recursions, a conversion into integers in a desired range or into a floating point number in $[0, 1)$ is required. If x_n is an integer produced by some PRNG in the range $0 \leq x_n \leq M - 1$, then an integer in the range $0 \leq x_n \leq N - 1$, with $N \leq M$, is given by $y_n = \lfloor \frac{N x_n}{M} \rfloor$. If $N \ll M$, then $y_n = x_n \pmod{N}$ may be used. Alternately, if a floating point value in $[0, 1)$ is desired, let $y_n = x_n/M$.

17.11.1.1   Linear congruential generators

Perhaps the oldest generator still in use is the linear congruential generator (LCG). The underlying integer recursion for LCGs is:

$$x_n = a x_{n-1} + b \pmod{M}. \tag{17.69}$$

Equation (17.69) defines a periodic sequence of integers modulo M starting with x_0, the initial seed. The constants of the recursion are referred to as the *modulus*, M, *multiplier*, a, and *additive constant*, b. If $M = 2^m$, a very efficient implementation is possible. Alternately, there are theoretical reasons why choosing M prime is optimal. Hence, the only moduli that are used in practical implementations are $M = 2^m$ or the prime $M = 2^p - 1$ (i.e., M is a Mersenne prime). With a Mersenne prime, modular multiplication can be implemented at about twice the computational cost of multiplication modulo 2^p.

Equation (17.69) yields a sequence $\{x_n\}$ whose period, denoted $\text{Per}(x_n)$, depends on M, a, and b. The values of the maximal period for the three most common cases used and the conditions required to obtain them are:

a	b	M	$\text{Per}(x_n)$
primitive root of M	anything	prime	$M - 1$
3 or 5 $\pmod 8$	0	2^m	2^{m-2}
1 $\pmod 4$	1 $\pmod 2$	2^m	2^m

A major shortcoming of LCGs modulo a power-of-two compared with prime modulus LCGs derives from the following theorem for LCGs:

Theorem 5 *Define the following LCG sequence:* $x_n = a x_{n-1} + b \pmod{M_1}$. *If M_2 divides M_1 and if $y_n = x_n \pmod{M_2}$, then y_n satisfies $y_n = a y_{n-1} + b$ $\pmod{M_2}$.*

Theorem 5 implies that the k least-significant bits of any power-of-two modulus LCG with $\mathrm{Per}(x_n) = 2^m = M$ has $\mathrm{Per}(y_n) = 2^k$, $0 < k \leq m$. Since a long period is crucial in PRNGs, when these types of LCGs are employed in a manner that makes use of only a few least-significant bits, their quality may be compromised. When M is prime, no such problem arises.

Since LCGs are in such common usage, the table below contains a list of parameter values mentioned in the literature. The Park–Miller LCG is widely considered to be a minimally acceptable PRNG.

a	b	M	Source
7^5	0	$2^{31} - 1$	Park–Miller
131	0	2^{35}	Neave
16333	25887	2^{15}	Oakenfull
3432	6789	9973	Oakenfull
171	0	30269	Wichman–Hill

17.11.1.2 Shift register generators

Another popular method of generating pseudorandom numbers is to use binary shift register sequences to produce pseudorandom bits. A binary shift register sequence (SRS) is defined by a binary recursion of the type:

$$x_n = x_{n-j_1} \oplus x_{n-j_2} \oplus \cdots \oplus x_{n-j_k}, \qquad j_1 < j_2 < \cdots < j_k = \ell. \qquad (17.70)$$

where $\oplus$ is the exclusive "or" operation. Note that $x \oplus y \equiv x + y \pmod 2$. Thus the new bit, x_n, is produced by adding k previously computed bits together modulo 2. The implementation of this recurrence requires keeping the last ℓ bits from the sequence in a shift register, hence the name. The longest possible period is equal to the number of nonzero ℓ-dimensional binary vectors, namely, $2^\ell - 1$.

A sufficient condition for achieving $\mathrm{Per}(x_n) = 2^\ell - 1$ is that the characteristic polynomial corresponding to equation (17.70) be primitive modulo 2. Since primitive trinomials of nearly all degrees of interest have been found, SRSs are usually implemented using two-term recursions of the form:

$$x_n = x_{n-k} \oplus x_{n-\ell}, \qquad 0 < k < \ell. \qquad (17.71)$$

In these two-term recursions, k is the lag and ℓ is the register length. Proper choice of the pair (ℓ, k) leads to SRSs with $\mathrm{Per}(x_n) = 2^\ell - 1$. Here is a list with suitable (ℓ, k) pairs:

Primitive trinomial exponents					
(5,2)	(7,1)	(7,3)	(17,3)	(17,5)	(17,6)
(31,3)	(31,6)	(31,7)	(31,13)	(127,1)	(521,32)

17.11.1.3 Lagged-Fibonacci generators

Another way of producing pseudorandom numbers is to use the lagged-Fibonacci methods. The term "lagged-Fibonacci" refers to two term recurrences of the form:

$$x_n = x_{n-k} \diamond x_{n-\ell}, \qquad 0 < k < \ell, \qquad (17.72)$$

where $\diamond$ refers to three common methods of combination: (1) addition modulo 2^m, (2) multiplication modulo 2^m, or (3) bit-wise exclusive 'OR'ing of m-long bit vectors. Combination method (3) can be thought of as a special implementation of a two-term SRS.

Using combination method (1) leads to additive lagged-Fibonacci sequences (ALFSs). If x_n satisfies:

$$x_n = x_{n-k} + x_{n-\ell} \pmod{2^m}, \qquad 0 < k < \ell, \qquad (17.73)$$

then the maximal period is $\text{Per}(x_n) = (2^\ell - 1)2^{m-1}$.

ALFS are especially suitable for producing floating point deviates using the real valued recursion $y_n = y_{n-k} + y_{n-\ell} \pmod 1$. This circumvents the need to convert from integers to floating point values and allows floating point hardware to be used. One caution with ALFS is that theorem 5 holds, and so the low-order bits have periods that are shorter than the maximal period. However, this is not nearly the problem as in the LCG case. With ALFSs the j least-significant bits will have period $(2^\ell - 1)2^{j-1}$, so if ℓ is large there really is no problem. Note that one can use the table of primitive trinomial exponents to find (ℓ, k) pairs that give maximal period ALF sequences.

17.11.2   Generating nonuniform random variables

In order to select random observations from an arbitrary distribution, suppose X has probability density function $f(x)$ and distribution function $F(x) = \int_{-\infty}^{x} f(u)\, du$. In the following we write "Y is $U[0,1)$" to mean Y is uniformly distributed on the interval $[0,1)$.

Two general techniques for converting uniform random variables into those from other distributions are:

1. The inverse transform method.

   If Y is $U[0,1)$, then the random variable $X = F^{-1}(Y)$ has probability density function $f(x)$.

2. The acceptance–rejection method.

   Suppose the density can be written as $f(x) = Ch(x)g(x)$ where $h(x)$ is the density of a computable random variable, $g(x)$ satisfies the inequality $0 < g(x) \le 1$, and $C^{-1} = \int_{-\infty}^{\infty} h(u)g(u)\, du$ is a normalization constant. If X is $U[0,1)$, Y has density $h(x)$, and if $X < g(Y)$, then X has density $f(x)$. Therefore, generate $\{X, Y\}$ pairs, reject both if $X \ge g(Y)$ and return X if $X < g(Y)$.

Examples of the inverse transform method:

1. (Exponential distribution) The exponential distribution with rate λ has $f(x) = \lambda e^{-\lambda x}$ (for $x \geq 0$) and $F(x) = 1 - e^{-\lambda x}$. Thus $u = F(x)$ can be solved to give $x = F^{-1}(u) = -\lambda^{-1} \ln(1 - u)$. If U is $U[0, 1)$ then so is $1 - U$. Hence $X = -\lambda^{-1} \ln U$ is exponentially distributed with rate λ.

2. (Normal distribution) Let Z_i be normally distributed with $f(z) = \frac{1}{\sqrt{2\pi}} e^{-z^2/2}$. The polar transformation produces random variables $R = \sqrt{Z_1^2 + Z_2^2}$ (exponentially distributed with $\lambda = 2$) and $\Theta = \arctan(Z_2/Z_1)$ (uniformly distributed on $[0, 2\pi)$). Inverting: $Z_1 = \sqrt{-2 \ln X_1} \cos 2\pi X_2$ and $Z_2 = \sqrt{-2 \ln X_1} \sin 2\pi X_2$ are normally distributed when X_1 and X_2 are $U[0, 1)$. (This is the Box–Muller technique.)

Examples of the rejection method:

1. (Exponential distribution with $\lambda = 1$)
   (a) Generate random numbers $\{U_i\}_{i=1}^N$ uniform on $[0, 1]$), stopping at $N = \min\{n \mid U_1 \geq U_2 \geq U_{n-1} < U_n\}$.
   (b) If N is even accept that run, and go to step (c). If N is odd reject the run, and return to step (a).
   (c) Set X equal to the number of failed runs plus the first random number in the successful run.

2. (Normal distribution)
   (a) Select two random variables (V_1, V_2) from $U[0, 1)$. Form $R = V_1^2 + V_2^2$.
   (b) If $R > 1$ then reject the (V_1, V_2) pair and select another pair.
   (c) If $R < 1$ then $X = V_1 \sqrt{-2 \dfrac{\ln R}{R}}$ has a $N(0, 1)$ distribution.

3. (Normal distribution)
   (a) Select two exponentially distributed random variables with rate 1: (V_1, V_2).
   (b) If $V_2 \geq (V_1 - 1)^2/2$, then reject the (V_1, V_2) pair and select another pair.
   (c) Otherwise, V_1 has a a $N(0, 1)$ distribution.

4. (Cauchy distribution) To generate values of X from $f(x) = \frac{1}{\pi(1+x^2)}$ on $-\infty < x < \infty$:
   (a) Generate random numbers U_1, U_2 (uniform on $[0, 1)$) and set $Y_1 = U_1 - \frac{1}{2}$, $Y_2 = U_2 - \frac{1}{2}$.
   (b) If $Y_1^2 + Y_2^2 \leq \frac{1}{4}$ then return $X = Y_1/Y_2$. Otherwise return to step 1.
   To generate values of X from a Cauchy distribution with parameters β and θ; $f(x) = \dfrac{\beta}{\pi[\beta^2 + (x - \theta)^2]}$ for $-\infty < x < \infty$; construct X as above and then use $\beta X + \theta$.

17.11.2.1   Discrete random variables

In general, the density function of a discrete random variable can be represented as a vector $\mathbf{p} = (p_0, p_1, \ldots, p_{n-1}, p_n)$ by defining the probabilities $\text{Prob}\,[x = j] = p_j$ (for $j = 0, \ldots, n$). The distribution function can be defined by the vector $\mathbf{c} = (c_0, c_1, \ldots, c_{n-1}, 1)$ where $c_j = \sum_{i=0}^{j} p_i$. Given this representation of $F(x)$ we can apply the inverse transform by computing X to be $U[0, 1)$, and then finding the index j so that $c_j \leq X < c_{j+1}$. In this case event j will have occurred. For example:

1. (Binomial distribution) The binomial distribution with n trials, with each trial having probability of success p, has $p_j = \binom{n}{j} p^j (1 - p)^{n-j}$ for $j = 0, \ldots, n$.

   (a) As an example, consider the result of flipping a fair coin. In 2 flips, the probability of obtaining $(0, 1, 2)$ heads are $\mathbf{p} = (\frac{1}{4}, \frac{1}{2}, \frac{1}{4})$. Hence $\mathbf{c} = (\frac{1}{4}, \frac{3}{4}, 1)$. If x (chosen from $U[0, 1)$) turns out to be say, 0.4, then "1 head" is returned (since $\frac{1}{4} \leq 0.4 < \frac{1}{2}$).

   (b) Note that when n is large it is costly to compute the density and distribution vectors. When n is large and relatively few binomially distributed pseudorandom numbers are desired, an alternative is to use the normal approximation to the binomial.

   (c) Alternately, one can form the sum $\sum_{i=1}^{n} \lfloor U_i + p \rfloor$, where each U_i is $U[0, 1)$.

2. (Geometric distribution) To simulate a value from $\text{Prob}\,[X = i] = p(1 - p)^{i-1}$ for $i \geq 1$; use $X = 1 + \left\lceil \dfrac{\log U}{\log(1 - p)} \right\rceil$.

3. (Poisson distribution) The Poisson distribution with mean λ has $p_j = \lambda^j e^{\lambda} / j!$ for $j \geq 0$. The Poisson distribution counts the number of events in a unit time interval if the times are exponentially distributed with rate λ. Thus if the times $\{T_i\}$ are exponentially distributed with rate λ, then J will be Poisson distributed with mean λ when $\sum_{i=0}^{J} T_i \leq 1 \leq \sum_{i=0}^{J+1} T_i$. Since $T_i = -\lambda^{-1} \ln U_i$, where U_i is $U[0, 1)$, the previous equation may be written as $\prod_{i=0}^{J} U_i \geq e^{-\lambda} \geq \prod_{i=0}^{J+1} U_i$. Hence, we can compute Poisson random variables by iteratively computing $P_J = \prod_{i=0}^{J} U_i$ until $P_J < e^{-\lambda}$. The first such J that makes this inequality true will have a Poisson distribution.

Random variables can be simulated using the following table (each U and U_i are uniform on the interval $[0, 1)$):

Using uniform random deviates to create random deviates from different distributions

Distribution	Density	Formula for deviate
Binomial	$p_j = \binom{n}{j} p^j (1-p)^{n-j}$	$\sum_{i=1}^{n} \lfloor U_i + p \rfloor$
Cauchy	$f(x) = \dfrac{\sigma}{\pi(x^2 + \sigma^2)}$	$\sigma \tan(\pi U)$
Exponential	$f(x) = \lambda e^{-\lambda x}$	$-\lambda^{-1} \ln U$
Pareto	$f(x) = \dfrac{ab^a}{x^{a+1}}$	$\dfrac{b}{U^{1/a}}$
Rayleigh	$\dfrac{x}{\sigma} e^{-x^2/2\sigma^2}$	$\sigma\sqrt{-\ln U}$

17.11.2.2   Testing pseudorandom numbers

The prudent way to check a complicated computation that makes use of pseudorandom numbers is to run it several times with different types of pseudorandom number generators and see if the results appear consistent across the generators. The fact that this is not always possible or practical has led researchers to develop statistical tests of randomness that should be passed by general purpose pseudorandom number generators. Some common tests are the spectral test, the equidistribution test, the serial test, the runs test, the coupon collector test (section 17.16.10), and the birthday spacing test (section 17.16.6).

17.11.3   References

1. L. Devroye, *Non-Uniform Random Variate Generation*, Springer–Verlag, New York, 1986.
2. S. K. Park and K. W. Miller, Random number generators: good ones are hard to find, *Communications of the ACM*, October 1988, Volume 31, Number 10, pages 1192–1201.

17.12   RESAMPLING METHODS

Assume the set $S = \{x_1, x_2, \ldots, x_n\}$ contains n sample values of the random deviate X. Resampling techniques use the values in S repeatedly to obtain an estimate of a statistic, and also the variance of that estimate. For example, when computing the sample mean of the values in S, the value of the sample standard deviation indicates the possible range of values for the true mean. However, when computing the sample median, there is no natural way (other than resampling) to find the variance of the estimate.

Suppose an estimate of the statistic $\theta(X)$ is desired. Let $\widehat{\theta}$ be a procedure (estimator) used for estimating θ.

1. Bootstrap

   To apply the bootstrap technique, m random sets of the same size are drawn from the set S with replacement, and $\widehat{\theta}$ is calculated for each sample. The bootstrapped estimate for $\theta(X)$ is the mean of the values of $\widehat{\theta}$ for each random set.

   Example 17.80:   Consider a sample of four data points: $\{1, 3, 5, 9\}$. The estimated median from this sample is 4.

   To estimate the variability of the estimate of the median, repeatedly sample with replacement from the four data points. For example:

   (a) $\{5, 9, 9, 9\}$, median is 9   (f) $\{3, 9, 9, 5\}$, median is 7
   (b) $\{1, 5, 1, 3\}$, median is 2   (g) $\{5, 3, 9, 1\}$, median is 4
   (c) $\{9, 3, 3, 9\}$, median is 6   (h) $\{3, 3, 9, 3\}$, median is 3
   (d) $\{3, 1, 5, 1\}$, median is 2   (i) $\{9, 5, 3, 3\}$, median is 4
   (e) $\{5, 1, 9, 5\}$, median is 5   (j) $\{3, 9, 9, 1\}$, median is 6

   These 10 estimates of the median have a mean of 4.8 and a standard deviation 2.3. This indicates the approximate variability of the estimate of the median.

2. Jackknife

   To apply the jackknife technique, $\widehat{\theta}$ is computed on the set S and also on the $n-1$ sets obtained from S by sequentially deleting each element. Hence, $\widehat{\theta}$ is computed for 1 sample of size n and n samples of size $n-1$. The jackknife estimate of $\theta(X)$ is the mean of the n values of $\widehat{\theta}$ that have been obtained.

See B. Efron, "Bootstrap methods, another look at the jackknife," *Annals of Statistics*, **7**, 1979, pages 1–26 and B. Efron, *The Jackknife, the Bootstrap and Other Resampling Plans*, Society for Industrial and Applied Mathematics, Philadelphia, 1982.

17.13   SELF-SIMILAR PROCESSES

17.13.1   Definitions

(a) A function $L(x)$ is *slowly varying* if, for all $x > x_0$,

$$\lim_{t \to \infty} \frac{L(tx)}{L(t)} = 1. \tag{17.74}$$

Slowly varying functions include $L(x) = c + o(1)$ for $x > 0$, $L(x) = \log x$ for $x > 1$, and $L(x) = 1/\log x$ for $x > 1$.

(b) A random variable X has a *heavy tailed distribution* if $\text{Prob}[X > x] = x^{-\alpha}L(x)$ for $\alpha > 0$ and $x > x_0$ where $L(x)$ is a slowly varying function.

17.13.2 Self-similar processes

A process $\{X_t\}_{t=0,1,2,\ldots}$ is *asymptotically self-similar* if the autocorrelation function, $r(k)$, has the form

$$r(k) \sim k^{-(2-2H)}L(k) \qquad \text{as } k \to \infty \qquad (17.75)$$

where $L(x)$ is a slowly varying function and the *Hurst parameter* H satisfies $1/2 < H < 1$. The process is *exactly self-similar* if

$$r(k) = \frac{1}{2}\left[(k+1)^{2H} - 2k^{2H} + (k-1)^{2H}\right]. \qquad (17.76)$$

Note: White noise has $r(k) = 0$, which corresponds to $H = 1/2$.

For any process $\{X_t\}_{t=0,1,2,\ldots}$, the *aggregated version* $\{X_t^{(m)}\}_{t=0,1,2,\ldots}$ is constructed by partitioning $\{X_t\}$ into non–overlapping blocks of m sequential elements and constructing a single element $X_t^{(m)}$ from the mean:

$$X_t^{(m)} = \frac{1}{m}\sum_{i=tm-m+1}^{tm} X_i \qquad (17.77)$$

Note: $\{X_t^{(m)}\}$ represents viewing $\{X_t\}$ on a time scale that is a factor of m coarser.

For a typical process, as m increases the autocorrelation of $\{X_t^{(m)}\}$ decreases until, in the limit, the elements of $\{X_t^{(m)}\}$ are uncorrelated. For a self-similar process, the processes $\{X_t\}$ and $\{X_t^{(m)}\}$ have the same autocorrelation function.

17.14 SIGNAL PROCESSING

17.14.1 Estimation

Let $\{e_t\}$ be a white noise process (so that $\text{E}[e_t] = \mu$, $\text{Var}[e_t] = \sigma^2$, and $\text{Cov}[e_t, e_s] = 0$ for $s \neq t$). Suppose that $\{X_t\}$ is a time series. A non-anticipating linear model presumes that $\sum_{u=0}^{\infty} h_u X_{t-u} = e_t$, where the $\{h_u\}$ are constants. This can be written $H(z)X_t = e_t$ where $H(z) = \sum_{u=0}^{\infty} h_u z^u$ and $z^n X_t = X_{t-n}$. Alternately, $X_t = H^{-1}(z)e_t$. In practice, several types of models are used:

1. AR(k): autoregressive model of order k. This assumes that $H(z) = 1 + a_1 z + \cdots + a_k z_k$ and so

$$X_t + a_1 X_{t-1} + \cdots + a_k X_{t-k} = e_t \qquad (17.78)$$

2. MA(l): moving average of order l. This assumes that

$H^{-1}(z) = 1 + b_1 z + \cdots + b_k z_k$ and so

$$X_t = e_t + b_1 e_{t-1} + \cdots + b_l e_{t-l} \qquad (17.79)$$

3. ARMA(k, l): mixed autoregressive/moving average of order (k, l). This assumes that $H^{-1}(z) = \frac{1 + b_1 z + \cdots + b_k z_k}{1 + a_1 z + \cdots + a_k z_k}$ and so

$$X_t + a_1 X_{t-1} + \cdots + a_k X_{t-k} = e_t + b_1 e_{t-1} + \cdots + b_l e_{t-l} \qquad (17.80)$$

17.14.2 Matched filtering (Wiener filter)

Let $S(t)$ represent a signal to be recovered, let $N(t)$ represent noise, and let $Y(t) = S(t) + N(t)$ represent the observable. A prediction of the signal is $S_p(t) = \int_0^\infty K(z)Y(t-z)\,dz$, where $K(z)$ is a filter. The mean square error is $E\left[(S(t) - S_p(t))^2\right]$; this is minimized by the optimal (Wiener) filter $K_{\text{opt}}(z)$. When X and Y are stationary, define their autocorrelation functions to be $R_{XX}(t - s) = E[X(t)X(s)]$ and $R_{YY}(t - s) = E[Y(t)Y(s)]$. If $\mathcal{F}$ represents the Fourier transform, then the optimal filter is given by

$$\mathcal{F}[K_{\text{opt}}(t)] = \frac{1}{2\pi} \frac{\mathcal{F}[R_{XX}(t)]}{\mathcal{F}[R_{YY}(t)]} \qquad (17.81)$$

For example, if X and N are uncorrelated, then

$$\mathcal{F}[K_{\text{opt}}(t)] = \frac{1}{2\pi} \frac{\mathcal{F}[R_{XX}(t)]}{\mathcal{F}[R_{XX}(t)] + \mathcal{F}[R_{NN}(t)]} \qquad (17.82)$$

In the case of no noise: $\mathcal{F}[K_{\text{opt}}(t)] = \frac{1}{2\pi}$, $K_{\text{opt}}(t) = \delta(t)$, and $S_p(t) = Y(t)$.

17.14.3 Median filter

A *median filter* replaces a value in a data set with the median of the entries surrounding that value. In the one-dimensional case it consists of sliding a window of an odd number of elements along the signal, replacing the center sample by the median of the samples in the window.

The median is a stronger "central indicator" than the average. The median is hardly affected by one or two discrepant values among the data values in the region. Consequently, median filtering is very effective at removing various kinds of noise. In two-dimensional data sets (such as images) median filtering is a nonlinear signal enhancement technique for the smoothing of signals, the suppression of impulse noise, and preserving of edges.

17.14.4 Mean filter

A *mean filter* or *averaging filter* replaces the values in a data set with the average of the entries surrounding that value. Thought of as a convolution filter, it is represented by a kernel which represents the shape and size of the neighborhood to be sampled when calculating the mean. In two-dimensional data sets (such as images) a 3×3 square kernel is often used:

$1/9$	$1/9$	$1/9$
$1/9$	$1/9$	$1/9$
$1/9$	$1/9$	$1/9$

A problem with the mean filter is that it blurs edges and other sharp details. An alternative is to use a median filter.

17.14.5   Spectral decomposition of stationary random functions

Any stationary function $X(t)$ can be written as

$$X(t) - \mu_X = \int_{-\infty}^{\infty} e^{i\omega t}\, d\Phi(\omega) \tag{17.83}$$

If the correlation function satisfies the equation $\int_{-\infty}^{\infty} |K_X(t)|\, dt < \infty$ then the increments $d\Phi(\omega)$ satisfy

$$E\left[d\Phi(\omega)\right] = 0$$
$$E\left[d\Phi^*(\omega_1)\, d\Phi(\omega_2)\right] = S_X(\omega)\delta(\omega_1 - \omega_2)\, d\omega_1\, d\omega_2 \tag{17.84}$$

where $*$ denotes the complex conjugate. Here, $S_X(\omega)$ is the *spectral density* of $X(t)$ and $\delta(x)$ denotes the δ-function.

The correlation function and the spectral density are related by mutually inverse Fourier transforms:

$$K_X(t) = \int_{-\infty}^{\infty} e^{i\omega t} S_X(\omega)\, d\omega, \quad \text{and} \quad S_X(\omega) = \frac{1}{2\pi}\int_{-\infty}^{\infty} e^{-i\omega t} K_X(t)\, dt \tag{17.85}$$

Note that:

(a) If $X(t)$ is a real function, then $S_X(\omega) = S_X(-\omega)$.

(b) The spectral density of $X'(t)$ is related to $S_X(\omega)$ by:

$$S_{X'}(\omega) = \omega^2 S_X(\omega) \tag{17.86}$$

17.15   STOCHASTIC CALCULUS

17.15.1   Brownian motion (Wiener processes)

Brownian motion $W(t)$ is a Gaussian random process that has a mean given by its starting point, $E\left[W(t)\right] = W_0 = W(t_0)$, a variance of $E\left[(W(t) - W_0)^2\right] = t - t_0$, and a covariance of $E\left[W(t)W(s)\right] = \min(t, s)$. The sample paths of $W(t)$ are continuous but not differentiable. Brownian motion is also called a Wiener process. Formally: Let $(\Omega, \mathcal{B}, \mathrm{Pr})$ be a Lebesgue probability space, and let $(\mathcal{R}, \mathcal{F}, m)$ represent the real numbers with Lebesgue measure m. Then a Brownian motion is a function $X(t, \omega) : \mathcal{R}^+ \times \Omega \to \mathcal{R}$ satisfying three conditions:

1. For any $0 < s < t$, $(X(t,\omega) - X(s,\omega))$ has a Gaussian distribution with mean zero and variance $m([s,t))$;
2. If $t_0 < t_1 < \cdots < t_k$, then $(X(t_j,\omega) - X(t_{j-1},\omega))_{j=1,2,\ldots,k}$ is an independent system;
3. $X(0,\omega) = 0$ for all $\omega \in \Omega$.

17.15.2   Brownian motion expectations

Define the following types of Brownian motions:

1. Brownian motion W_s
2. Brownian motion with drift $W_s^{(\mu)} = \mu s + W_s$
3. Reflecting Brownian motion $|W_s|$

For each, let the Brownian motion start at the location x. Define the following types of stopping times at which a process X will stop:

1. An *exponential stopping time* is the time τ given by $\mathrm{Prob}\,[\tau > t] = e^{-\lambda t}$ for $\lambda > 0$. (Here τ is independent of the process X).
2. A *first hitting time* is the first time that a process reaches some value; for example, $H_z = \min\{s \mid W_s = z\}$ is the first time when a Brownian motion reaches the value z.
3. A *first exit time* is the first time that a process leaves a region; for example, $H_{a,b} = \min\{s \mid W_s \notin (a,b)\}$ is the first time that a Brownian motion exits the interval (a,b).

We have the following expectations involving Brownian motion:

1. Brownian motion W_s:

   (a) Unconstrained

   i. $\mathrm{E}\left[e^{i\beta W_t}\right] = \exp\left(i\beta x - \dfrac{\beta^2 t}{2}\right)$

   ii. $\mathrm{E}\left[\exp\left(-\gamma \sup_{0<s<t} W_s\right)\right] = \exp\left(-\gamma x + \dfrac{\gamma^2 t}{2}\right) \mathrm{erfc}\left(\gamma\sqrt{\dfrac{t}{2}}\right)$

   iii. $\mathrm{E}\left[\exp\left(\gamma \inf_{0<s<t} W_s\right)\right] = \exp\left(\gamma x + \dfrac{\gamma^2 t}{2}\right) \mathrm{erfc}\left(\gamma\sqrt{\dfrac{t}{2}}\right)$

   (b) Exponential stopping times:

   i. $\mathrm{E}\left[e^{i\beta W_\tau}\right] = \dfrac{2\lambda}{2\lambda + \beta^2} e^{i\beta x}$

   ii. $\mathrm{E}\left[\exp\left(-\gamma \sup_{0<s<\tau} W_s\right)\right] = \dfrac{\sqrt{2\lambda}}{\gamma + \sqrt{2\lambda}} e^{-\gamma x}$

   iii. $\mathrm{E}\left[\exp\left(\gamma \inf_{0<s<\tau} W_s\right)\right] = \dfrac{\sqrt{2\lambda}}{\gamma + \sqrt{2\lambda}} e^{\gamma x}$

(c) First hitting time:

   i. $E\left[e^{-\alpha H_z}\right] = e^{-|x-z|\sqrt{2\alpha}}$

   ii. $E\left[\exp\left(-\gamma \sup\limits_{0<s<H_z} W_s\right)\right] = e^{-\gamma x}\left[1 - \gamma(x-z)\int\limits_{x-z}^{\infty} \frac{e^{-\gamma v}}{v}\,dv\right]$

   for $z \le x$

   iii. $E\left[\exp\left(\gamma \inf\limits_{0<s<H_z} W_s\right)\right] = e^{\gamma x}\left[1 - \gamma(z-x)\int\limits_{z-x}^{\infty} \frac{e^{-\gamma v}}{v}\,dv\right]$ for

   $x \le z$

   iv. $E\left[\exp\left(-\gamma \int\limits_0^{H_z} e^{2\beta W_s}\,ds\right)\right] = \begin{cases} \dfrac{I_0\left(\frac{\sqrt{2\gamma}}{|\beta|}e^{\beta x}\right)}{I_0\left(\frac{\sqrt{2\gamma}}{|\beta|}e^{\beta z}\right)}, & (z-x)\beta \ge 0 \\[4mm] \dfrac{K_0\left(\frac{\sqrt{2\gamma}}{|\beta|}e^{\beta x}\right)}{K_0\left(\frac{\sqrt{2\gamma}}{|\beta|}e^{\beta z}\right)}, & (z-x)\beta \le 0 \end{cases}$

(d) First exit time

   i. $E\left[e^{-\alpha H_{a,b}}\right] = \dfrac{\cosh\left((b-2x+a)\sqrt{\alpha/2}\right)}{\cosh\left((b-a)\sqrt{\alpha/2}\right)}$

   ii. $E\left[e^{\beta W_{H_{a,b}}}\right] = \dfrac{b-x}{b-a}e^{i\beta a} + \dfrac{x-a}{b-a}e^{i\beta b}$

2. Brownian motion with drift $W_s^{(\mu)} = \mu s + W_s$

(a) Unconstrained

   i. $E\left[e^{i\beta W_t^{(\mu)}}\right] = \exp\left(i\beta(x + \mu t) - \dfrac{\beta^2 t}{2}\right)$

(b) Exponential stopping times

   i. $E\left[e^{i\beta W_\tau^{(\mu)}}\right] = \dfrac{2\lambda}{2\lambda - 2i\beta\mu + \beta^2}e^{i\beta x}$

   ii. $E\left[\exp\left(-\gamma \sup\limits_{0<s<\tau} W_s^{(\mu)}\right)\right] = \dfrac{\sqrt{2\lambda + \mu^2} - \mu}{\gamma + \sqrt{2\lambda + \mu^2} - \mu}e^{-\gamma x}$

(c) First hitting time ($H_z = \min\{s \mid W_s^{(\mu)} = z\}$)

   i. $E\left[e^{-\alpha H_z}\right] = \exp\left(\mu(z-x) - |z-x|\sqrt{2\alpha + \mu^2}\right)$

(d) First exit time ($H_{a,b} = \min\{s \mid W_s^{(\mu)} \notin (a,b)\}$)

   i. $E\left[e^{-\alpha H_z}\right] = \dfrac{e^{\mu(a-x)}\sinh((b-x)\eta) + e^{\mu(b-x)}\sinh((x-a)\eta)}{\sinh((b-a)\eta)}$

   where $\eta = \sqrt{2\alpha + \mu^2}$

   ii. $E\left[e^{iW_{H_{a,b}}^{(\mu)}}\right] = \dfrac{\sinh\left((b-x)|\mu|\right)}{\sinh\left((b-a)|\mu|\right)}e^{\mu(a-x)+i\beta a}$

   $+ \dfrac{\sinh\left((x-a)|\mu|\right)}{\sinh\left((b-a)|\mu|\right)}e^{\mu(b-x)+i\beta b}$

3. Reflecting Brownian motion $|W_s|$

(a) Exponential stopping times

   i. $E\left[e^{-\beta|W_\tau|}\right] = \dfrac{\sqrt{2\lambda}}{2\lambda - \beta^2}\left(\sqrt{2\lambda}e^{-\beta x} - \beta e^{-\sqrt{2\lambda}x}\right)$ for $0 \le x$

See A. N. Borodin and P. Salminen, *Handbook of Brownian Motion — Facts and Formulae*, Birkhäuser Verlag, Boston, 1996.

17.15.3   Itô lemma

Consider the process $f(W) = \{F(W_t) \mid t \geq 0\}$ where f is a given smooth function and W is a Brownian motion. When f has two derivatives, the Itô formula is given by:

$$f(W_t) - f(W_0) = \int_0^t f'(W_s)\, dW_s + \frac{1}{2} \int_0^t f''(W_s)\, ds \qquad (17.87)$$

17.15.4   Stochastic integration

If $W(t)$ is a Wiener process and $G(t, W(t))$ is an arbitrary function, then the stochastic integral $I = \int_{t_0}^t G(s, W(s))\, dW(s)$ is defined as a limiting sum. Divide the interval $[t_0, t]$ into n sub-intervals: $t_0 \leq t_1 \leq \cdots \leq t_{n-1} \leq t_n = t$, and choose points $\{\tau_i\}$ that lie in each sub-interval: $t_{i-1} \leq \tau_i \leq t_i$. The stochastic integral I is defined as the limit of partial sums, $I = \lim_{n \to \infty} S_n$, with $S_n = \sum_{i=1}^n G(\tau_i, W(\tau_i))[W(t_i) - W(t_{i-1})]$.

Consider, for example, the special case of $G(t) = W(t)$. Then the expectation of S_n is

$$\begin{aligned}
\mathrm{E}\,[S_n] &= \mathrm{E}\left[\sum_{i=1}^n W(\tau_i)[W(t_i) - W(t_{i-1})]\right] \\
&= \sum_{i=1}^n [\min(\tau_i, t_i) - \min(\tau_i, t_{i-1})] \qquad (17.88) \\
&= \sum_{i=1}^n (\tau_i - t_{i-1}).
\end{aligned}$$

If $\tau_i = \alpha t_i + (1 - \alpha) t_{i-1}$ (where $0 < \alpha < 1$), then $\mathrm{E}\,[S_n] = \sum_{i=1}^n (t_i - t_{i-1})\alpha = (t - t_0)\alpha$. Hence, the value of S_n depends on α. For consistency, some specific choice must be made for the points $\{\tau_i\}$.

1. For the Itô stochastic integral (indicated by $_I\!\int$), we choose $\tau_i = t_{i-1}$ (i.e., $\alpha = 0$ in the above). That is:

$$\begin{aligned}
&_I\!\int_{t_0}^t G(s, W(s))\, dW(s) \\
&= \underset{n \to \infty}{\text{ms-lim}} \left\{ \sum_{i=1}^n G(t_{i-1}, W(t_{i-1}))[W(t_i) - W(t_{i-1})] \right\},
\end{aligned} \qquad (17.89)$$

where "ms-lim" refers to the mean square limit.

2. For the Stratonovich stochastic integral (indicated by $_S\!\int$), we choose $\tau_i = (t_i + t_{i-1})/2$ (i.e., $\alpha = \frac{1}{2}$ in the above). That is:

$$_S\!\int_{t_0}^{t} G(W(s), x)\, dW(s) \tag{17.90}$$

$$= \underset{n\to\infty}{\text{ms-lim}} \left\{ \sum_{i=1}^{n} G\left(t_{i-1}, w\left(\frac{t_i + t_{i-1}}{2}\right)\right) [W(t_i) - W(t_{i-1})] \right\}.$$

Example 17.81: Consider the integral $\int_{t_0}^{t} W(s)\, dW(s)$. The Itô and Stratonovich evaluations are

1. $_I\!\int_{t_0}^{t} W(s)\, dW(s) = \left[W^2(t) - W^2(t_0) - (t - t_0)\right]/2$.

2. $_S\!\int_{t_0}^{t} W(s)\, dW(s) = \left[W^2(t) - W^2(t_0)\right]/2$.

17.15.5 Stochastic differential equations

If a differential equation contains random terms, then the solution can only be described statistically. A linear ordinary differential equation for, say, $x(t)$ with linearly appearing white Gaussian noise terms can be converted to a parabolic partial differential equation whose solution is the probability density of $x(t)$. This equation is called a Fokker–Planck equation or a forward Kolmogorov equation.

Consider the linear differential system for the m component vector $\mathbf{X}(t)$

$$\frac{d}{dt}\mathbf{X}(t) = \mathbf{b}(t, \mathbf{X}) + \sigma(t, \mathbf{X})\,\mathbf{N}(t),$$
$$\mathbf{X}(t_0) = \mathbf{y}, \tag{17.91}$$

where $\sigma(t, \mathbf{X})$ is a real $m \times n$ matrix and $\mathbf{N}(t)$ is a vector of n independent white noise terms. That is,

$$\mathrm{E}\left[N_i(t)\right] = 0,$$
$$\mathrm{E}\left[N_i(t)N_j(t + \tau)\right] = \delta_{ij}\delta(\tau), \tag{17.92}$$

where δ_{ij} is the Kronecker delta, and $\delta(\tau)$ is the delta function. The Fokker–Planck equation corresponding to equation (17.91) is

$$\frac{\partial P}{\partial t} = -\sum_{i=1}^{m} \frac{\partial}{\partial x_i}(b_i P) + \frac{1}{2}\sum_{i,j=1}^{m} \frac{\partial^2}{\partial x_i \partial x_j}(a_{ij} P), \tag{17.93}$$

where $P = P(t, \mathbf{x})$ is a probability density and the matrix $A = (a_{ij})$ is defined by $A(t, \mathbf{x}) = \sigma(t, \mathbf{x})\sigma^{\mathrm{T}}(t, \mathbf{x})$. Any statistical information about $\mathbf{X}(t)$ that could be ascertained from equation (17.91) may be derived from $P(t, \mathbf{x})$.

In one dimension, the stochastic differential equation

$$\frac{dX}{dt} = f(X) + g(X)N(t), \tag{17.94}$$

with $X(0) = z$, corresponds to the Fokker–Planck equation

$$\frac{\partial P}{\partial t} = -\frac{\partial}{\partial x}(f(x)P) + \frac{1}{2}\frac{\partial^2}{\partial x^2}(g^2(x)P), \tag{17.95}$$

for $P(t, x)$ with $P(0, x) = \delta(x - z)$.

Example 17.82: The *Langevin equation*

$$X'' + \beta X' = N(t), \tag{17.96}$$

with the initial conditions $X(0) = 0$ and $X'(0) = u_0$ can be written as the vector system

$$\frac{d}{dt}\begin{bmatrix} X \\ U \end{bmatrix} = \begin{bmatrix} U \\ -\beta U \end{bmatrix} + \begin{bmatrix} 0 & 0 \\ 0 & 1 \end{bmatrix}\begin{bmatrix} N_1(t) \\ N_2(t) \end{bmatrix},$$
$$\begin{bmatrix} X \\ U \end{bmatrix}_{t=0} = \begin{bmatrix} 0 \\ u_0 \end{bmatrix}. \tag{17.97}$$

The Fokker–Planck equation for $P(t, x, u)$, the joint probability density of X and U at time t, is

$$\frac{\partial P}{\partial t} = -\frac{\partial}{\partial x}(uP) + \frac{\partial}{\partial u}(\beta u P) + \frac{1}{2}\frac{\partial^2 P}{\partial u^2},$$
$$P(0, x, u) = \delta(x)\delta(u - u_0). \tag{17.98}$$

This equation has the solution

$$P(t, x, u) = \frac{1}{\det D}\exp\left(-\begin{bmatrix} x - \mu_x \\ u - \mu_u \end{bmatrix} D \begin{bmatrix} x - \mu_x \\ u - \mu_u \end{bmatrix}^{\mathsf{T}}\right), \tag{17.99}$$

where $D = \begin{bmatrix} \sigma_{xx} & \sigma_{xu} \\ \sigma_{xu} & \sigma_{uu} \end{bmatrix}$, and the parameters $\{\mu_x, \mu_u, \sigma_{xx}, \sigma_{xu}, \sigma_{uu}\}$ are given by

$$\mu_x = \frac{u_0}{\beta}\left(1 - e^{-\beta t}\right),$$
$$\mu_u = u_0 e^{-\beta t},$$
$$\sigma_{xx}^2 = \frac{t}{\beta^2} - \frac{2}{\beta^3}\left(1 - e^{-\beta t}\right) + \frac{1}{2\beta^3}\left(1 - e^{-2\beta t}\right), \tag{17.100}$$
$$\sigma_{xu}^2 = \frac{1}{\beta^2}\left(1 - e^{-\beta t}\right) - \frac{1}{2\beta^2}\left(1 - e^{-2\beta t}\right),$$
$$\sigma_{uu}^2 = \frac{1}{2\beta}\left(1 - e^{-2\beta t}\right).$$

17.15.6   Motion in a domain

Consider a particle starting at $\mathbf{y}$ and randomly moving in a domain Ω. If the probability density of the location evolves according to

$$\frac{\partial P}{\partial t} = L[P] = -\sum_{i=1}^{m} b_i(\mathbf{y})\frac{\partial P}{\partial y_i} + \frac{1}{2}\sum_{i,j=1}^{m} a_{ij}(\mathbf{y})\frac{\partial^2 P}{\partial y_i \partial y_j}, \tag{17.101}$$

then

(a) The expectation of the exit time $w(\mathbf{y})$ is the solution of $L[w] = -1$ in Ω, with $w = 0$ on $\partial\Omega$.

(b) The probability $u(\mathbf{y})$ that the exit occurs on the boundary segment Γ is the solution of $L[u] = 0$ in Ω with

$$u(\mathbf{y}) = \begin{cases} 1 & \text{for } \mathbf{y} \in \Gamma \\ 0 & \text{for } \mathbf{y} \in \Omega/\Gamma \end{cases}.$$

17.15.7 Option Pricing

Let S represent the price of a share of stock, and presume that S follows a geometric Brownian motion $dS = \mu S\, dt + \sigma S\, dw$, where t is time, μ is a constant, and σ is a constant called the volatility. Let $V(S,t)$ be the value of a derivative security whose payoff is solely a function of S and t. Construct a portfolio consisting of V and Δ shares of stock. The value P of this portfolio is $P = V + S\Delta$. The random component of the portfolio increment (dP) can be removed by choosing $\Delta = -\partial V/\partial S$. The concept of arbitrage says that $dP = rP\, dt$, where r is the (constant) risk-free bank interest rate. Together this results in the Black–Scholes equation for option pricing (note that no transaction costs are included):

$$\frac{\partial V}{\partial t} + rS\frac{\partial V}{\partial S} + \frac{1}{2}\sigma^2 S^2 \frac{\partial^2 V}{\partial S^2} - rV = 0 \qquad (17.102)$$

If the asset pays a continuous dividend of $DS\, dt$ (i.e., this is proportional to the asset value S during the time period dt), then the equation is modified to become

$$\frac{\partial V}{\partial t} + (r - D)S\frac{\partial V}{\partial S} + \frac{1}{2}\sigma^2 S^2 \frac{\partial^2 V}{\partial S^2} - rV = 0 \qquad (17.103)$$

If E is the exercise price of the option, and T is the expiry date (the only date on which the option can be exercised), then the solution of the modified Black–Scholes equation is

$$V(S,t) = e^{-D(T-t)}SN(d_1) - Ee^{-r(T-t)}\Phi(d_2) \qquad (17.104)$$

where

$$d_1 = \frac{\log(S/E) + (r - D + \sigma^2/2)(T - t)}{\sigma\sqrt{T-t}}, \qquad d_2 = d_1 - \sigma\sqrt{T-t}, \quad (17.105)$$

and Φ is the cumulative probability distribution for the normal distribution.

17.16   CLASSIC AND INTERESTING PROBLEMS

17.16.1   Approximating a distribution

Given the moments of a distribution (μ, σ, and $\{\gamma_i\}$), the asymptotic probability density function is given by

$$
\begin{aligned}
f(t) = \phi(x) &- \frac{\gamma_1}{6}\phi^{(3)}(x) + \left[\frac{\gamma_2}{24}\phi^{(4)}(x) + \frac{\gamma_1^2}{72}\phi^{(6)}(x)\right] \\
&- \left[\frac{\gamma_3}{120}\phi^{(5)}(x) + \frac{\gamma_1\gamma_2}{144}\phi^{(7)}(x) + \frac{\gamma_1^3}{1296}\phi^{(9)}(x)\right] + \cdots
\end{aligned}
\tag{17.106}
$$

where $\phi(x) = \dfrac{1}{\sigma\sqrt{2\pi}}e^{-(x-\mu)^2/2\sigma^2}$ is the normal density function.

17.16.2   Averages over vectors

Let $\overline{f(\mathbf{n})}$ denote the average of the function f as the unit vector $\mathbf{n}$ varies uniformly in all directions in three dimensions. If $\mathbf{a}$, $\mathbf{b}$, $\mathbf{c}$, and $\mathbf{d}$ are constant three-dimensional vectors, then

$$
\begin{aligned}
\overline{|\mathbf{a}\cdot\mathbf{n}|^2} &= |\mathbf{a}|^2/3 \\
\overline{(\mathbf{a}\cdot\mathbf{n})(\mathbf{b}\cdot\mathbf{n})} &= (\mathbf{a}\cdot\mathbf{b})/3 \\
\overline{(\mathbf{a}\cdot\mathbf{n})\mathbf{n}} &= \mathbf{a}/3 \\
\overline{|\mathbf{a}\times\mathbf{n}|^2} &= 2\,|\mathbf{a}|^2/3 \\
\overline{(\mathbf{a}\times\mathbf{n})\cdot(\mathbf{b}\times\mathbf{n})} &= 2\mathbf{a}\cdot\mathbf{b}/3 \\
\overline{(\mathbf{a}\cdot\mathbf{n})(\mathbf{b}\cdot\mathbf{n})(\mathbf{c}\cdot\mathbf{n})(\mathbf{d}\cdot\mathbf{n})} &= [(\mathbf{a}\cdot\mathbf{b})(\mathbf{c}\cdot\mathbf{d}) + (\mathbf{a}\cdot\mathbf{c})(\mathbf{b}\cdot\mathbf{d}) \\
&\qquad + (\mathbf{a}\cdot\mathbf{d})(\mathbf{b}\cdot\mathbf{c})]/15
\end{aligned}
\tag{17.107}
$$

Now let $\overline{f(\mathbf{n})}$ denote the average of the function f as the unit vector $\mathbf{n}$ varies uniformly in all directions in two dimensions. If $\mathbf{a}$ and $\mathbf{b}$ are constant two-dimensional vectors, then

$$
\begin{aligned}
\overline{|\mathbf{a}\cdot\mathbf{n}|^2} &= |\mathbf{a}|^2/2 \\
\overline{(\mathbf{a}\cdot\mathbf{n})(\mathbf{b}\cdot\mathbf{n})} &= (\mathbf{a}\cdot\mathbf{b})/2 \\
\overline{(\mathbf{a}\cdot\mathbf{n})\mathbf{n}} &= \mathbf{a}/2
\end{aligned}
\tag{17.108}
$$

17.16.3   Bertrand's box "paradox"

Suppose there are three small boxes, each with two drawers, each containing a coin. One box has two gold coins, another has two silver coins, and the third has one gold and one silver coin. Suppose a box and a drawer are selected at random, and a gold coin is discovered. What is the probability that the other drawer also contains a gold coin?

If a gold coin is found, even though the box selected is not the one with two silver coins, it is not equally likely to be one of the remaining two boxes. Given a gold coin has been found, there are two ways this could happen if selecting a drawer from the G/G box, and only one way if selecting from the the the G/S box. Therefore, the probability the box is G/G is $2/3$ and the probability the other coin in the selected box is gold is $2/3$.

17.16.4 Bertrand's circle "paradox"

Consider the following problem:

> Given a circle of unit radius, find the probability that a *randomly* chosen chord will be longer than the side of an inscribed equilateral triangle.

There are at least three *solutions* to this problem:

(a) Randomly choose two points on a circle and measure the distance between them. The result depends only on the position of the second point relative to the first one (i.e., the first point can be fixed). Consider chords that emanate from a fixed first point; $1/3$ of the resulting chords will be longer that the side of an equilateral triangle. Therefore, the probability is $1/3$.

(b) A chord is completely specified by its midpoint. If the length of a chord exceeds the side of an equilateral triangle, then the midpoint must be inside a smaller circle of radius $1/2$. The area of the smaller circle is $1/4$ of the area of the original (unit) circle. Therefore, the probability is $1/4$.

(c) A chord is completely specified by its midpoint. If the length of a chord exceeds the side of an equilateral triangle, then the midpoint is within $1/2$ of the center of the circle. If the midpoints are distributed uniformly over the radius (instead of over the area, as assumed in ((b)), the probability is $1/2$.

It all depends on how *randomly* is defined. See M. Kac and S. M. Ulam, *Mathematics and Logic*, Dover Publications, NY, 1968.

17.16.5 Bingo cards: nontransitive

Consider the 4 bingo cards shown below (labeled A, B, C, and D), which were created by D. E. Knuth. Two players each select a bingo card. Numbers from 1 to 6 are randomly drawn without replacement, as they are in standard bingo. If a selected number is on a card, it is marked with a bean. The first player to complete a horizontal row wins. It can be shown that, statistically, card A beats card B, card B beats card C, card C beats card D, and card D beats card A.

A			B			C			D	
1	2		2	4		1	3		1	5
3	4		5	6		4	5		2	6

17.16.6   Birthday problem

The probability that n people have different birthdays is

$$q_n = \left(\frac{364}{365}\right) \cdot \left(\frac{363}{365}\right) \cdots \left(\frac{366-n}{365}\right) \tag{17.109}$$

Let $p_n = 1 - q_n$. For 23 people the probability of at least two people having the same birthday is more than half ($p_{23} = 1 - q_{23} > 1/2$).

n	10	20	23	30	40	50
p_n	0.117	0.411	0.507	0.706	0.891	0.970

The number of people needed to have a 50% chance of two people having the same birthday is 23. The number of people needed to have a 50% chance of three people having the same birthday is 88. For four, five, and six people having the same birthday the number of people necessary is 187, 313, and 460.

The number of people needed so that there is a 50% chance that two people have a birthday within one day of each other is 14. In general, in a n-day year the probability that p people all have birthdays at least k days apart (so $k = 1$ is the original birthday problem) is

$$\text{probability} = \binom{n - p(k-1) - 1}{p-1} \frac{(p-1)!}{n^{p-1}} \tag{17.110}$$

The non-equiprobable case was solved in M. Klamkin and D. Newman, Extensions of the birthday surprise, *J. Comb. Theory*, **3** (1967), pages 279–282.

17.16.7   Buffon's needle problem

A needle of length L is placed at random on a plane on which are drawn parallel lines a distance D apart. Assume that $L < D$ so that only one intersection is possible. The probability P that the needle intersects a line is

$$P = \frac{2L}{\pi D} \tag{17.111}$$

If a convex object of perimeter s, with a maximum diameter less than D, is randomly placed on the ruled plane, then the probability P that the object intersects a line is $P = s/\pi D$.

If a needle of length L is dropped on a rectilinear ruled grid, with line separations of a and b (with $L < a$ and $L < b$), then the probability that the needle will land on a line is $\dfrac{2L(a+b) - L^2}{ab\pi}$.

17.16.8   Card problems

17.16.8.1   Shuffling cards

A *riffle shuffle* takes a deck of n cards, splits it into two stacks, and then recombines them. We assume that the probability that the split occurs after the kth card has a binomial distribution. The recombination occurs by taking cards, one by one, off of the bottom of the two stacks, and placing them onto one stack; if there are k_1 and k_2 cards in the two stacks, then the probability of taking the next card from the first (resp. second) stack is $\frac{k_1}{k_1+k_2}$ (resp. $\frac{k_2}{k_1+k_2}$).

Let $\Omega_n = \{\pi_1, \pi_2, \ldots, \pi_{n!}\}$ denote the possible permutations of n cards. Let $f_X(\pi_i)$ be the probability that operator X (a riffle shuffle) produces the ordering π_i. The *variation distance* between the process X and a uniform distribution on Ω_n is

$$\|f_X - u\| = \frac{1}{2} \sum_{\pi \in \Omega_n} \left| f_X(\pi) - \frac{1}{n!} \right| \tag{17.112}$$

where the $\frac{1}{2}$ is used so that value falls in the interval $[0, 1]$. For the riffle shuffle equation (17.112) becomes

$$\|f_X - u\| = \frac{1}{2} \sum_{r=1}^{n} A_{n,r} \left| \binom{2^k + n - r}{n} \frac{1}{2^{nk}} - \frac{1}{n!} \right| \tag{17.113}$$

where the Eulerian numbers $\{A_{n,r}\}$ are defined by $A_{n,1} = 1$ and

$$A_{n,a} = a^n - \sum_{r=1}^{a-1} \binom{n+a-r}{n} A_{n,r} \tag{17.114}$$

Table 17.2 contains numerical values for a $n = 52$ card deck. Until 5 shuffles have occurred, the output of X is very far from random. After 5 shuffles, the distance from the random process is essentially halved each time a shuffle occurs.

See D. Bayer and P. Dianconis, Trailing the dovetail shuffle to its lair, *Annals of Applied Probability*, **2(2)**, 1992, pages 294–313.

17.16.8.2   Card games

(a) Poker hands

The number of distinct 13-card poker hands is $\binom{52}{5} = 2,598,960$.

Hand	Probability	Odds
royal flush	1.54×10^{-6}	649,739:1
straight flush	1.39×10^{-5}	72,192:1
four of a kind	2.40×10^{-4}	4,164:1
full house	1.44×10^{-3}	693:1
flush	1.97×10^{-3}	508:1
straight	3.92×10^{-3}	254:1
three of a kind	0.0211	46:1
two pair	0.0475	20:1
one pair	0.423	1.37:1

Number of riffle shuffles	Variation distance
1	1
2	1
3	1
4	0.99999953
5	0.924
6	0.614
7	0.334
8	0.167
9	0.0854
10	0.0429
11	0.0215
12	0.0108
13	0.00538
14	0.00269

Table 17.2: Number of riffle shuffles versus variation distance

(b) Bridge hands

The number of distinct 13-card bridge hands is $\binom{52}{13} = 635{,}013{,}559{,}600$. In bridge, the *honors* are the ten, jack, queen, king, or ace. Obtaining the three top cards (ace, king, and queen) of three suits and the ace, king, queen, and jack of the remaining suit is called *13 top honors*. Obtaining all cards of the same suit is called a *13-card suit*. Obtaining 12 cards of the same suit with ace high and the 13th card not an ace is called a *12-card suit, ace high*. Obtaining no honors is called a *Yarborough*.

Hand	Probability	Odds
13 top honors	6.30×10^{-12}	158,753,389,899:1
13-card suit	6.30×10^{-12}	158,753,389,899:1
12-card suit, ace high	2.72×10^{-9}	367,484,698:1
Yarborough	5.47×10^{-4}	1,827:1
four aces	2.64×10^{-3}	378:1
nine honors	9.51×10^{-3}	104:1

17.16.9   Coin problems

17.16.9.1   *Even odds from a biased coin*

If you have a coin biased toward heads, it is possible to get the equivalent of a fair coin with several tosses of the unfair coin.

Toss the biased coin twice. If both tosses give the same result, repeat this process (throw out the two tosses and start again). Otherwise, take the first of the two results as the toss of an unbiased coin.

17.16.9.2   Two heads in a row

What is the probability in n flips of a fair coin that there will be two heads in a row?

Consider strings of n H's and T's. We count the number of strings that do not contain HH and subtract it from the total number of such strings (which is 2^n). There must be no more than $n/2$ H's; otherwise two heads would be adjacent. If a string contains i H's, with no two of them in a row, then these H's must be placed between (or around) the $(n-i)$ T's present: there are $\binom{n-i+1}{i}$ ways to do this. Hence, the total number of strings that do not contain HH is $\displaystyle\sum_{i=0}^{n/2}\binom{n-i+1}{i} = F_{n+2}$ (where F_m is the mth Fibonacci number). Hence, the probability that n coin tosses will contain a HH is:
$$\frac{2^n - F_{n+2}}{2^n}.$$

17.16.10   Coupon collectors problem

There are n coupons that can be collected. At each time a random coupon is selected, with replacement. How long must one wait until they have a specified collection of coupons?

Define $W_{n,j}$ to be the number of time steps required until j different coupons are seen; then

$$\begin{aligned}
\mathrm{E}\left[W_{n,j}\right] &= n\sum_{i=1}^{j}\frac{1}{n-i+1}\\[2mm]
\mathrm{Var}\left[W_{n,j}\right] &= n\sum_{i=1}^{j}\frac{i-1}{(n-i+1)^2}
\end{aligned} \qquad (17.115)$$

When $j = n$, then all coupons are being collected and $\mathrm{E}\left[W_{n,n}\right] = nH_n$ with $H_n = 1+1/2+1/3+\cdots+1/n$. As $n \to \infty$, $\mathrm{E}\left[W_{n,n}\right] \sim n\log n$. Typical values are shown below.

n	$\mathrm{E}\left[W_{n,n}\right]$	$\sqrt{\mathrm{Var}\left[W_{n,n}\right]}$
2	3	1.41
5	11.4	5.02
10	29.3	11.2
50	225	62
100	519	126
200	1,176	254

In the unequal probabilities case, with p_i being the probability of obtaining coupon i,

$$E[W_{n,n}] = \int_0^\infty \left[1 - \prod_{i=1}^n \left(1 - e^{-p_i t}\right)\right] dt \qquad (17.116)$$

17.16.11 Dice problems

17.16.11.1 Dice: nontransitive (Efron)

(a) Let A have the two six-sided dice with sides of $\{0, 0, 4, 4, 4, 4\}$ and $\{2, 3, 3, 9, 10, 11\}$

(b) Let B have the two six-sided dice with sides of $\{3, 3, 3, 3, 3, 3\}$ and $\{0, 1, 7, 8, 8, 8\}$

(c) Let C have the two six-sided dice with sides of $\{2, 2, 2, 2, 6, 6\}$ and $\{5, 5, 6, 6, 6, 6\}$

(d) Let D have the two six-sided dice with sides of $\{1, 1, 1, 5, 5, 5\}$ and $\{4, 4, 4, 4, 12, 12\}$

Then the odds of

(a) A winning against B (c) C winning against D

(b) B winning against C (d) D winning against A

are all 2:1.

17.16.11.2 Dice: distribution of sums

When rolling two dice, the probability distribution of the sum is

$$\text{Prob[sum of } s] = \frac{6 - |s - 7|}{36} \qquad \text{for } 2 \le s \le 12 \qquad (17.117)$$

When rolling three dice, the probability distribution of the sum is

$$\text{Prob[sum of } s] = \frac{1}{216} \begin{cases} \frac{1}{2}(s-1)(s-2) & \text{for } 3 \le s \le 8 \\ -s^2 + 21s - 83 & \text{for } 9 \le s \le 14 \\ \frac{1}{2}(19-s)(20-s) & \text{for } 15 \le s \le 18 \end{cases} \qquad (17.118)$$

For 2 dice, the most common roll is a 7 (probability $1/6$). For 3 dice, the most common rolls are 10 and 11 (probability $1/8$ each). For 4 dice, the most common roll is a 14 (probability $73/648$).

17.16.11.3 Dice: same distribution

Ordinary six-sided dice have the numbers $\{1, 2, 3, 4, 5, 6\}$ on the sides. Rolling two dice and summing the numbers face up creates a distribution of values from 2 to 12 (see equation (17.117)). That same distribution of sums can be obtained by rolling two cubes, with the following numbers on their sides:

- $\{0, 1, 1, 2, 2, 3\}$ and $\{2, 4, 5, 6, 7, 9\}$
- $\{0, 2, 3, 4, 5, 7\}$ and $\{2, 3, 3, 4, 4, 5\}$
- $\{1, 2, 2, 3, 3, 4\}$ and $\{1, 3, 4, 5, 6, 8\}$ (Sicherman dice)

17.16.12 Ehrenfest urn model

Suppose there are two urns and $2R$ numbered balls, and suppose that there are $R + a$ balls in urn I and $R - a$ balls in urn II. Each second a ball is chosen at random and moved from its urn into the other urn. The procedure is repeated.

Consider the case of $R = 10,000$. If $a = 0$, then the expected time to return to the initial configuration is 175 seconds. If $a = R$, then the expected time to return to the initial configuration is approximately 10^{6000} years.

17.16.13 Envelope problem "paradox"

The envelope exchange problem is

> "Someone has prepared two envelopes containing money. One contains twice as much money as the other. You have decided to pick one envelope, but then the following argument occurs to you:
>
> > 'Suppose my chosen envelope contains X, then the other envelope either contains $X/2$ or $2X$. Both cases are equally likely, so my expectation if I take the other envelope is $\frac{1}{2}\left(\frac{X}{2}\right) + \frac{1}{2}(2X) = 1.25X$, which is higher than my current X. Hence, I should change my mind and take the other envelope.'
>
> But then I can apply the argument all over again. There must be sometime wrong."

There is, in fact, no contradiction. Switching the envelope or not switching the envelope results in the same expected return. See R. Christensen and J. Utts, Bayesian resolution of the 'Exchange Paradox', *The American Statistician*, 46 (4), 1992, pages 274–276.

17.16.14 Gambler's ruin problem

A gambler starts with z dollars. For each gamble: with probability p she wins one dollar, with probability q she loses one dollar. Gambling stops when she has either zero dollars, or a dollars.

If q_z denotes the probability of eventually stopping at $z = a$ ("gambler's success") then

$$
q_z = \begin{cases}
\dfrac{(q/p)^a - (q/p)^z}{(q/p)^a - 1} & \text{if } p \neq q \\[2ex]
1 - \dfrac{z}{a} & \text{if } p = q = \frac{1}{2}
\end{cases}
\tag{17.119}
$$

For example:

	p	q	z	a	q_z
fair	0.5	0.5	9	10	.900
game	0.5	0.5	90	100	.900
	0.5	0.5	900	1000	.900
	0.5	0.5	9000	10000	.900
biased	0.4	0.6	90	100	.017
game	0.4	0.6	90	99	.667

17.16.15 Gender distributions

For these problems, there is a 50/50 chance of male or female on each birth.

1. Hospital deliveries: Every day a large hospital delivers 1000 babies and a small hospital delivers 100 babies. Which hospital has a better chance of having the same number of boys as girls?

   The small one. If $2n$ babies are born, then the probability of an even split is $\dfrac{\binom{2n}{n}}{2^{2n}}$. This is a decreasing function of n. As n goes to infinity the probability of an even split approaches zero (although it is still the most likely event).

2. Family planning

   Suppose that in large society of people, every family continues to have children until they have a girl, then they stop having children. After many generations of families, what is the ratio of males to females?

   The ratio will be 50-50; half of all conceptions will be male, half female.

3. If a person has two children and

   (a) The older one is a girl, then the probability that both children are girls is $1/2$.
   (b) At least one is a girl, then the probability that both children are girls is $1/3$.

17.16.16 Holtzmark distribution: stars in the galaxy

The force $\mathbf{f}$ on a unit mass due to a star of mass m at a location $\mathbf{r}$ is $\mathbf{f} = \frac{Gm}{|\mathbf{r}|^3}\mathbf{r}$. The probability density for the magnitude of the force, $f = |\mathbf{f}|$, due to a uniform distribution of stars in the galaxy, with random masses, with λ stars per unit volume, is given by

$$p(f) = \frac{1}{4\pi a^2 \beta^2} H(\beta)$$

$$a = \frac{4\lambda}{15}(2\pi G)^{3/2} \, \mathrm{E}\left[m^{3/2}\right] \tag{17.120}$$

$$\beta = f/a^{2/3}$$

$$H(\beta) = \frac{2}{\pi\beta}\int_0^\infty x\, e^{-(x/\beta)^{3/2}} \sin x \, dx$$

Note that

$$H(\beta) \sim \begin{cases} \frac{4\beta^2}{3\pi} + O(\beta^4) & \text{as } \beta \to 0 \\[2mm] \frac{15}{8}\sqrt{\frac{2}{\pi}}\beta^{-5/2} + O(\beta^{-4}) & \text{as } \beta \to \infty \end{cases} \tag{17.121}$$

See S. Chandrasekhar, Stochastic problems in physics and astronomy, *Reviews of Modern Physics*, **15**, number 1, January 1943, pages 1–89.

17.16.17 Large-scale testing

17.16.17.1 Infrequent success

Suppose that a disease occurs in one person out of every 1000. Suppose that a test for this disease has a type I and a type II error of .01 (that is, $\alpha = \beta = 0.01$). Imagine that 100,000 people are tested. Of the 100 people who have the disease, 99 will be diagnosed as having it. Of the 99,900 people who do not have the disease, 999 will be diagnosed as having it. Hence, only $\frac{99}{1098} \approx 9\%$ of the people who test positive for the disease actually have it.

17.16.17.2 Pooling of blood samples

A large population of persons is to be screened for the presence of some condition using a blood test which registers positive if a sample contains blood from a person having the condition and registers negative otherwise. To implement the screening, the blood specimens of groups of k persons are pooled, and the pooled blood samples are tested. If a pooled sample registers negative, the group is cleared; otherwise each person in the group is tested. If the probability of a random person having a positive test is p, what group size k^* minimizes the expected number of tests per person? If $f(k)$ is the expected number of tests per person, then

$$f(k) = \begin{cases} 1 & \text{if } k = 1 \\ 1 - (1-p)^k + 1/k & \text{if } k \geq 2 \end{cases} \tag{17.122}$$

If $\{\{x\}\} = x - \lfloor x \rfloor$ denotes that fractional part of x, then

(a) If $p > 1 - \left(\frac{1}{3}\right)^{1/3} \approx 0.3066$ then $k^* = 1$.

(b) If $p < 1 - \left(\frac{1}{3}\right)^{1/3}$ then either

- $k^* = 1 + \lfloor p^{-1/2} \rfloor$; or
- $k^* = 2 + \lfloor p^{-1/2} \rfloor$.

(c) If $\left\{\left\{p^{-1/2}\right\}\right\} < \dfrac{\lfloor p^{-1/2} \rfloor}{2 \lfloor p^{-1/2} \rfloor + \{\{p^{-1/2}\}\}}$, then $k^* = 1 + \lfloor p^{-1/2} \rfloor$.

(d) It is never the case that $k^* = 2$.

Some values of k^*:

p	.0001	.0005	.001	.005	.01	.05	.1	.2	.3	.4	.5
k^*	101	45	32	15	11	5	4	3	1	1	1

See S. M. Samuels, The exact solution to the two-stage group-testing problem, *Technometrics*, 1978, **20**, pages 497–500.

17.16.18 Leading digit distribution

17.16.18.1 Ratio of uniform numbers

If X and Y are chosen uniformly from the interval $(0, 1]$, what is the probability that the ratio Y/X starts with a 1? What is the probability that the ratio Y/X starts with a 9?

(a) If Y/X has a leading digit of 1, then Y/X must lie in one of the intervals $\{\ldots, [0.1, 0.2), [1, 2), [10, 20), \ldots\}$. This corresponds to Y/X lying in one of several triangles with height 1 and bases on either the right or top edges of the square. The triangles are defined by the lines ($Y = 0.1X$, $Y = 0.2X$, and $X = 1$), ($Y = X$, $Y = X$, and $Y = 1$), ($Y = 10X$, $Y = 20Xl$, and $Y = 1$), $\ldots$. The bases along the right edge have lengths $0.1 + 0.01 + \cdots = 1/9$. The bases along the top edge have lengths $0.5 + 0.05 + \cdots = 5/9$. For a total base length of $6/9$ and a height of 1, the total area is $1/3$. The total area of the square is 1, hence the probability that Y/X starts with a 1 is $1/3$.

(b) In this case, similar triangles can be drawn; on the right edge the total length is $1/9$. On the top edge the length is $1/90 + 1/900 + \cdots = 1/81$. Total base length is $10/81$ and the total area (and hence probability of a leading 9) is $5/81 \approx 0.061728$.

17.16.18.2 Benford's law

Let P_d be the probability that the leading digit of a set of numbers is d. Assuming that the probability distribution of leading digits is scale-invariant (common, for example, for data that are dimensional), then P_d is approximately given by $P_d \approx \log_{10}\left(\frac{d+1}{d}\right)$. Hence,

P_1	P_2	P_3	P_4	P_5	P_6	P_7	P_8	P_9
0.301	0.176	0.125	0.097	0.079	0.067	0.058	0.051	0.048

17.16.19 Lotteries

Consider a collection of n numbers. A *player* chooses a subset T, called a *ticket*, of these numbers of size k. The *house* chooses a subset S of these numbers of size p (these are the *winning numbers*). The *payoff* depends on the size of the intersection between S and T. If the intersection is i elements, then the payoff is w_i.

If n, p, and k are fixed, then

1. There are $\binom{n}{p}$ possible tickets.
2. The number of ways that i of the player's k choices are among the selected p values is $\binom{k}{i}\binom{n-k}{p-i}$.
3. The probability that i of the player's k choices are among the selected p values is $\dfrac{\binom{k}{i}\binom{n-k}{p-i}}{\binom{n}{p}}$.
4. The *expected return* is $\displaystyle\sum_{i=0}^{p} w_i \dfrac{\binom{k}{i}\binom{n-k}{p-i}}{\binom{n}{p}}$.
5. Each i-subset has the same probability $\dfrac{\binom{p}{i}}{\binom{n}{i}}$ of appearing in the p-subset chosen.

It is possible to obtain a large number of tickets and obtain few matches. If $n = 49$ and $p = 6$ then there are 13,983,816 possible tickets.

1. There are $\binom{6}{0}\binom{43}{6} = 6,096,454$ tickets that have no matches with the winning numbers.
2. There are $\binom{6}{0}\binom{43}{6} + \binom{6}{1}\binom{43}{5} = 11,872,042$ tickets that have, at most, only a 1-match with the winning numbers.
3. There are $\binom{6}{0}\binom{43}{6} + \binom{6}{1}\binom{43}{5} + \binom{6}{2}\binom{43}{4} = 13,723,192$ tickets that have, at most, only a 2-match with the winning numbers.

A *lottery wheel* is a system of buying multiple tickets to guarantee a minimum match. For example, if $n = 49$ and $p = 6$ then it is possible to obtain 174 tickets and guarantee at least a 3-match.

See C. J. Colbourn and J. H. Dinitz, *CRC Handbook of Combinatorial Designs*, CRC Press, Boca Raton, FL, 1996, pages 578–581.

17.16.20   Match box problem

A certain mathematician carries one match box in his right pocket and one in his left pocket. When he wants a match, he selects a pocket at random. Each box initially contains N matches. The probability u_r that a selected box is empty on the rth trial is

$$u_r = \binom{N-r}{N} 2^{-2N+r} \tag{17.123}$$

17.16.21   Maximum entropy distributions

Define the entropy of a probability density function to be
$J = -\int f(x) \log[f(x)] \, dx$.

(a) Among all probability density functions restricted to the interval $[-1,1]$, the one of minimum entropy has the form $f(x) = C$; that is, it is the

uniform distribution: $f(x) = \begin{cases} \frac{1}{2} & \text{for } -1 \le x \le 1 \\ 0 & \text{otherwise} \end{cases}$

(b) Among all probability density functions restricted to the interval $[-1,1]$ and having a given mean of μ, the one of minimum entropy has the

form $f(x) = Ce^{\lambda x}$; that it $f(x) = \begin{cases} \frac{\lambda}{2 \sinh \lambda} e^{\lambda x} & \text{for } -1 \le x \le 1 \\ 0 & \text{otherwise} \end{cases}$ where

$\mu = \frac{\lambda - \tanh \lambda}{\lambda \tanh \lambda}$.

(c) Among all probability density functions restricted to the interval $[-1,1]$ that have a mean of zero and a given variance, the one of minimum entropy has the form $f(x) = Ce^{\lambda_1 x + \lambda_2 x^2}$

See F. E. Udwadia, Some results on maximum entropy distributions for parameters known to lie in certain intervals, *SIAM Review*, **31**, Number 1, March 1989, SIAM, Philadelphia, pages 103–109.

17.16.22   Monte Hall problem

Consider a game in which there are three doors: one door has a prize, two doors have no prize. A player selects one of the three doors. Suppose someone opens one of the two unselected doors that contains no prize. The player is then asked if he would like to exchange the originally selected door for the remaining unopened door. To increase the chance of winning, the player should switch doors: without switching the probability of winning is $1/3$; by switching the probability of winning is $2/3$.

17.16.23   Multi-armed bandit problem

In the multi-armed bandit problem, a gambler must decide which arm of K non-identical slot machines to play in a sequence of trials so as to maximize his reward. This problem has received much attention because of the simple model it provides of the trade-off between exploration (trying out each arm to find the best one) and exploitation (playing the arm believed to give the best payoff).

17.16.24   Parking problem

Let $E(x)$ denote the expected number of cars of length 1 which can be parked on a block of length x if cars park randomly (and with a uniform distribution in the available space). From the integral equation

$$E(x) = 1 + \frac{2}{x-1} \int_0^{x-1} E(t)\, dt \tag{17.124}$$

consider the Laplace transform

$$\int_0^\infty e^{-sx} E(x)\, dx = \frac{e^{-s}}{s} \int_s^\infty \exp\left(-2 \int_s^t \frac{1-e^{-u}}{u}\, du\right) dt. \tag{17.125}$$

This implies $E(x) \sim cx$ as $x \to \infty$ (where $c \approx 0.7476$).

17.16.25   Passage problems

Given a random function $X(t)$, a passage (time) at a given level a is when a graph of $X(t)$ crosses the horizontal line $X = A$ from below. Define the derivative of X to be $V(t) = \frac{dX(t)}{dt}$, and let $f(x, v \mid t)$ be the probability density of x and v at time t (define the probability density for X to be $f(x)$).

The probability that a passage (time) lies in the time interval dt about the time t is $p(a \mid t)\, dt$ and

$$p(a \mid t) = \int_0^\infty f(a, v \mid t)\, v\, dv \tag{17.126}$$

(a) For normal functions

$$p(a \mid t) = \frac{1}{2\pi} \exp\left(-\frac{(a - \bar{x})^2}{2\sigma_x^2}\right) \sqrt{\frac{1}{K_x(t,t)} \left.\frac{\partial^2 K_x(t_1, t_2)}{\partial t_1\, \partial t_2}\right|_{t_1=t_2=t}} \tag{17.127}$$

where K_x is the correlation function.

(b) For normal stationary functions

$$p(a \mid t) = p(a) = \frac{1}{2\pi} \frac{\sigma_v}{\sigma_x} \exp\left(-\frac{(a - \bar{x})^2}{2\sigma_x^2}\right) \tag{17.128}$$

(c) The number of passages of a stationary function during a time interval T is $\overline{N}_a = Tp(a)$.

(d) The average duration τ_a of a passage of a stationary function is

$$\tau_a = \frac{\int_a^\infty f(x)\, dx}{p(a)} \tag{17.129}$$

For normal stationary functions:

$$\tau_a = \pi \frac{\sigma_x}{\sigma_v} \exp\left(-\frac{(a - \overline{x})^2}{2\sigma_x^2}\right) \left[1 - \Phi\left(\frac{a - \overline{x}}{\sigma_x}\right)\right] \tag{17.130}$$

(e) Finding the average number of minima of a random differentiable function can be reduced to passage problems since a maxima is achieved when the first derivative passes through zero from above. For a small average number of passages during a time interval T, the approximate probability Q for non-occurrence of any run during this interval is $Q = e^{-\overline{N}_a}$ (i.e., the number of passages in the given interval can be approximated as a Poisson distribution).

17.16.26 Proofreading mistakes

Suppose that proofreader A finds a mistakes, and proofreader B finds b mistakes. Assume that there are c overlaps, mistakes that both A and B found. The approximate number of total mistakes is $\dfrac{ab}{c}$ and the approximate number of mistakes missed by both A and B is $\dfrac{(a - c)(b - c)}{c}$.

17.16.27 Raisin cookie problem

A baker creates enough cookie dough for 1000 raisin cookies. The number of raisins to be added to the dough, R, is to be determined.

1. In order to be 99% certain that the *first* cookie will have at least one raisin, then $1 - \left(\frac{999}{1000}\right)^R \geq 0.99$, or $R \geq 4603$.

2. In order to be 99% certain that *every* cookie will have at least one raisin, then $P(C, R) \geq 0.99$, where C is the number of cookies and $P(C, R) = C^{-R} \sum_{i=0}^{C} \binom{C}{i}(-1)^i (C - i)^R$. Or, $R \geq 11508$.

17.16.28 Random sequences

17.16.28.1 Long runs

For a biased coin with a probability of heads p, let R_n denote that length of the longest run of heads in the first n tosses of the coin. It has been shown:

$$\text{Prob}\left[\lim_{n \to \infty} \frac{R_n}{\log_{1/p}(n)} = 1\right] = 1 \tag{17.131}$$

See R. Arratia and M. S. Waterman, An Erdős–Rényi law with shifts, *Advances in Mathematics*, **55**, 1985, pages 13–23.

17.16.28.2 Waiting times: two types of characters

In a Bernoulli process with a probability of success p and a probability of failure $q = 1 - p$, the probability that a run of α consecutive successes occurs before a run of β consecutive failures is

$$\text{probability} = \frac{p^{\alpha-1}\left(1 - q^{\beta}\right)}{p^{\alpha-1} - q^{\beta-1}} \tag{17.132}$$

The expected waiting time until either run occurs is

$$\text{expected waiting time} = \frac{\left(1 - p^{\alpha}\right)\left(1 - q^{\beta}\right)}{p^{\alpha}q + pq^{\beta} - p^{\alpha}q^{\beta}} \tag{17.133}$$

The waiting times for strings of two characters and for strings of many types of characters can be very different.

17.16.28.3 Waiting times: many types of characters

Assume there are N different characters, each equally likely to occur. Let $S = (a_1, a_2, \cdots, a_n)$ be a sequence of characters, some of which may be equal. If $S_m = (a_1, a_2, \ldots, a_m)$, for $1 \le m \le n$, appears both at the beginning and at the end of S, then S_m is an *overlapping* subsequence of S. The mean waiting time until S appears depends on the lengths of the overlapping subsequences. If the lengths of the overlapping subsequences are $\{k_1, k_2, \ldots, k_r\}$ then,

$$\text{expected waiting time} = \sum_{i=0}^{r} N^{k_r} \tag{17.134}$$

Example 17.83: Choosing letters randomly from the alphabet, it will take on the average,

(a) 26^3 letters to get APE

(b) $26^3 + 26$ letters to get DAD

(c) $26^{11} + 26^4 + 26$ letters to get ABRACADABRA.

Example 17.84: Flipping a fair coin, it will take on the average,

(a) 2^4 flips to get HHTT

(b) $2^4 + 2^1$ flips to get HTHH

(c) $2^4 + 2^2$ flips to get TH TH

(d) $2^4 + 2^3 + 2^2 + 2^1$ flips to get HHHH

See G. Blom, On the mean number of random digits until a given sequence occurs, *J. Appl. Prob.*, **19**, 1982, pages 136–143 and G. Blom and D. Thorurn, How many random digits are required until given sequences are obtained?, *J. Appl. Prob.*, **19**, 1982, pages 518–531.

$$[AA] = 1\ 0\ 1\ 0 = 10 \qquad [AB] = 0\ 1\ 0\ 1 = 5$$
$$A = \mathrm{T\,H\,T\,H} \qquad\qquad A = \mathrm{T\,H\,T\,H}$$
$$A = \mathrm{T\,H\,T\,H} \qquad\qquad B = \mathrm{H\,T\,H\,H}$$
$$[BB] = 1\ 0\ 0\ 1 = 9 \qquad [BA] = 0\ 0\ 0\ 0 = 0$$
$$B = \mathrm{H\,T\,H\,H} \qquad\qquad B = \mathrm{H\,T\,H\,H}$$
$$B = \mathrm{H\,T\,H\,H} \qquad\qquad A = \mathrm{T\,H\,T\,H}$$

Figure 17.2: Computation of leading numbers

17.16.28.4   First random sequence

Let a fair coin be flipped until a given sequence appears. Let two competitors, P and Q, choose sequences of heads and tails; the winner is the one whose sequence appears first. Conway's leading number algorithm indicates who is likely to win:

Given two sequences $A = (a_1, \ldots, a_m)$ and $B = (b_1, \ldots, b_m)$, define the *leading number* of A over B as a binary integer $\epsilon_n \epsilon_{n-1} \cdots \epsilon_1$ via

$$\epsilon_i = \begin{cases} 1 & \text{if } 1 \le i \le \min(m, n) \text{ and the two sequence} \\ & (a_{m-i+1}, \ldots, a_m) \text{ and } (b_1, \ldots, b_i) \text{ are identical} \\ 0 & \text{otherwise} \end{cases}$$

$$(17.135)$$

Then, let $[AB]$ denote the leading number of A over B. The odds for B to precede A in a symmetric Bernoulli process are

$$([AA] - [AB]) : ([BB] - [BA]) \qquad\qquad (17.136)$$

Example 17.85:   Consider the two different 4-tuples $A = $THTH and $B = $HTHH. Figure 17.2 contains the needed leading number computations so that $([AA] - [AB]) :$ $([BB] - [BA])$ (in equation 17.136) becomes $(10 - 5) : (9 - 0)$ or $5 : 9$. Hence the odds that A occurs before B is $9/14$ (see section 3.3.5).

Note that sequence $A = $THTH has a waiting time (see section 17.16.28.2) of 20 while the sequence $B = $HTHH has a waiting time of 18. Hence, in this case, an event that is less frequent in the long run is likely to happen before a more frequent event.

Table 17.3 has the probability of Q winning in a double game. Table 17.4 has the probability of Q winning in a triplet game. When playing a triplet game, note that whatever triplet the first player chooses, the second player can choose a better one.

$P \setminus Q$	HH	HT	TH	TT
HH	—	$1/2$	$1/4$	$1/2$
HT	$1/2$	—	$1/2$	$3/4$
TH	$3/4$	$1/2$	—	$1/2$
TT	$1/2$	$1/4$	$1/2$	—

Table 17.3: Probability of P winning in a double game

$P \setminus Q$	HHH	HHT	HTH	HTT	THH	THT	TTH	TTT
HHH	—	$1/2$	$2/5$	$2/5$	$1/8$	$5/12$	$3/10$	$1/2$
HHT	$1/2$	—	$2/3$	$2/3$	$1/4$	$5/8$	$1/2$	$7/10$
HTH	$3/5$	$1/3$	—	$1/2$	$1/2$	$1/2$	$3/8$	$7/12$
HTT	$3/5$	$1/3$	$1/2$	—	$1/2$	$1/2$	$3/4$	$7/8$
THH	$7/8$	$3/4$	$1/2$	$1/2$	—	$1/2$	$1/3$	$3/5$
THT	$7/12$	$3/8$	$1/2$	$1/2$	$1/2$	—	$1/3$	$3/5$
TTH	$7/10$	$1/2$	$5/8$	$1/4$	$2/3$	$2/3$	—	$1/2$
TTT	$1/2$	$3/10$	$5/12$	$1/8$	$2/5$	$2/5$	$1/2$	—

Table 17.4: Probability of P winning in a triplet game

17.16.29   Random walks

17.16.29.1   Random walk on a grid

Consider a random walk of uniform step lengths on a one-, two-, or three-dimensional grid. Each step goes in any of the 2, 4, or 8 directions randomly. In one and two dimensions, the random walk will, with probability 1, return to the starting location. In three dimensions the probability of a return to the origin is approximately 0.34053.

The probability of return to the origin in d dimensions (for $d = 4, 5, \dots$) is $p(d) = 1 - \frac{1}{u(d)}$ where $u(d) = \int_0^\infty \left[I_0 \left(\frac{t}{d} \right) \right]^d e^{-t}\, dt$. The approximate probabilities of returning to the origin are as follows:

dimension	4	5	6	7	8
probability	0.20	0.136	0.105	0.0858	0.0729

17.16.29.2   Random walk in two dimensions (Rayleigh problem)

Consider a two-dimensional random walk in which each step, $\mathbf{x}_i$, is of length a and is taken in a random direction. The position after n steps is $(X_1, X_2) = \sum_{i=1}^n \mathbf{X}_i$. The radial distance after n steps is $R_n = \sqrt{X_1^2 + X_2^2}$. If $p(\mathbf{x})$ is the probability density for a single step and $p_n(\mathbf{x})$ is the probability density after

n steps, then

$$p(\mathbf{x}) = \frac{1}{2\pi a}\delta(|\mathbf{x}| - a)$$

$$\phi(\boldsymbol{\xi}) = \int_{-\infty}^{\infty}\int_{-\infty}^{\infty} p(\mathbf{x})\,e^{i\boldsymbol{\xi}\cdot\mathbf{x}}\,d\mathbf{x} = J_0\left(a\,|\boldsymbol{\xi}|\right)$$

$$p_n(\mathbf{x}) = \frac{1}{2\pi}\int_0^{\infty}\rho\,J_0\left(\rho\,|\mathbf{x}|\right)\,[J_0(\rho a)]^n\,d\rho \qquad (17.137)$$

$$\mathrm{Prob}\,[R_N < r] = r\int_0^{\infty} J_1(ru)\,[J_0(au)]^n\,du$$

Note that

$$\mathrm{E}\,[X_1] = \mathrm{E}\,[X_2] = 0$$

$$\mathrm{E}\,[X_1 X_2] = 0$$

$$\mathrm{E}\left[X_1^2\right] = \mathrm{E}\left[X_2^2\right] = \frac{na^2}{2} \qquad (17.138)$$

$$\mathrm{E}\left[R_n^2\right] = na^2$$

If, instead, the ith step has length a_i, then

$$\mathrm{Prob}\,[R_N < r] = r\int_0^{\infty} J_1(ru)\left[J_0(a_1 u)\cdot J_0(a_2 u)\cdots J_0(a_n u)\right]\,du \qquad (17.139)$$

17.16.29.3   *Random walk in three dimensions*

Consider a three-dimensional random walk in which each step, $\mathbf{X}_i$, is of length a and is taken in a random direction. If $p(\mathbf{x})$ is the probability density for a single step and $p_n(\mathbf{x})$ is the probability density after n steps, then

$$p(\mathbf{x}) = \frac{1}{4\pi a^2}\delta(|\mathbf{x}| - a)$$

$$\phi(\boldsymbol{\xi}) = \int_{-\infty}^{\infty}\int_{-\infty}^{\infty}\int_{-\infty}^{\infty} p_{\mathbf{x}}(\mathbf{x})e^{i\boldsymbol{\xi}\cdot\mathbf{x}}\,d\mathbf{x} = \frac{1}{a\,|\boldsymbol{\xi}|}\sin(a\boldsymbol{\xi}) \qquad (17.140)$$

$$p_n(\mathbf{x}) = \frac{1}{2\pi^2}\int_0^{\infty}\rho^2\left[\frac{\sin(a\rho)}{a\rho}\right]^n\frac{\sin(|\mathbf{x}|\,\rho)}{|\mathbf{x}|\,\rho}\,d\rho$$

Note that

$$p_2(\mathbf{x}) = \begin{cases} \dfrac{1}{8\pi a^2\,|\mathbf{x}|} & \text{for } |\mathbf{x}| \leq 2a \\[2ex] 0 & \text{for } |\mathbf{x}| > 2a \end{cases} \qquad (17.141)$$

$$p_3(\mathbf{x}) = \begin{cases} \dfrac{1}{8\pi a^3} & \text{for } 0 \le |\mathbf{x}| \le a \\[2mm] \dfrac{3a - |\mathbf{x}|}{16\pi a^3 |\mathbf{x}|} & \text{for } a < |\mathbf{x}| \le 3a \\[2mm] 0 & \text{for } 3a < |\mathbf{x}| \end{cases} \tag{17.142}$$

17.16.29.4 Self-avoiding walks

Consider a random walk of n unit length steps on a d-dimensional grid. A walk in which no grid point is visited twice is called a *self-avoiding walk*; the number of such walks is denoted $c(n)$. Known values include: $c(0) = 1$, $c(1) = 2d$, and $c(2) = 2d(d - 1)$. For each d there is a non-zero constant μ_d (the *connective constant*) such that $c(n) \sim (\mu_d)^{1/n}$ as $n \to \infty$. Current bounds include:

$$2.62 < \mu_2 < 2.70$$
$$4.57 < \mu_3 < 4.75$$
$$6.74 < \mu_4 < 6.82 \tag{17.143}$$
$$8.82 < \mu_5 < 8.87$$
$$10.87 < \mu_6 < 10.89$$

See J. Noonan, New upper bounds for the connective constants of self-avoiding walks, *J. Statist. Physics*, **91**, 1998, pages 871–888.

17.16.30 Relatively prime integers

Given two integers chosen at random, the probability that they are relatively prime is $[\zeta(2)]^{-1} = 6/\pi^2 \approx 0.608$. Given three integers chosen at random, the probability they have no common factor other than 1 is $[\zeta(3)]^{-1} \approx 1.202^{-1} \approx 0.832$.

17.16.31 Roots of a random polynomial

Consider the polynomial $f(x; \mathbf{a}) = a_0 + a_1 x + \cdots + a_{n-1} x^{n-1}$ where the $\{a_i\}$ are chosen randomly. To determine the expected number of real roots, suppose:

(a) The $\{a_i\}$ are chosen uniformly distributed on the interval $(-1, 1)$; or

(b) The $\{a_i\}$ are chosen equal to $+1$ and -1 with equal probability; or

(c) The $\{a_i\}$ are chosen to be independent standard normal coefficients

then the expected number of real roots is

$$E\,[\text{number of real roots}] = \frac{4}{\pi} \int_0^1 \sqrt{\frac{1}{(1 - t^2)^2} - \frac{(n + 1)^2 t^{2n}}{(1 - t^{2n+2})^2}} \; dt$$

$$\sim \frac{2}{\pi} \log n + C + \frac{2}{n\pi} + O\left(\frac{1}{n^2}\right) \qquad \text{as } n \to \infty \tag{17.144}$$

where $C = 0.6257358072\ldots$.

Consider a random polynomial of degree n with coefficients that are independent and identically distributed normal random variables. Define $m \neq 0$ to be the mean divided by the standard deviation ($m = \mu/\sigma$). Then the expected number of real roots as $n \to \infty$ is

$$\frac{1}{\pi} \log n + \frac{C+1}{2} - \frac{2+\gamma}{\pi} - \frac{2}{\pi} \log|m| + O\left(\frac{1}{n}\right) \tag{17.145}$$

where γ is Euler's constant.

See M. Kac, On the average number of real roots of a random algebraic equation, *Bulletin of the AMS*, **49**, pages 314–320, 1943, and A. T. Bharucha-Reid and M. Sambandham, *Random Polynomials*, Academic Press, New York, Chapter 4, pages 49–102, 1986.

17.16.32   Roots of a random quadratic

If A, B, and C are independent random variables uniformly distributed on $(0, 1)$ then the probability that $Ax^2 + Bx + C = 0$ has real roots is $(5 + 3\ln 4)/36$.

17.16.33   Simpson paradox

Consider the comparison of a new treatment with an old treatment, and suppose that the following results are obtained:

All patients	improved	not improved	percent improved
New treatment	20	20	50%
Old treatment	24	16	60%

On the basis of these data, one would be inclined to say that the new treatment is not better than the old treatment. Now consider a *disaggregation* of these data into the following two sub-groups:

Young patients	improved	not improved	percent improved
New treatment	12	18	40%
Old treatment	3	7	30%

Old patients	improved	not improved	percent improved
New treatment	8	2	80%
Old treatment	21	9	70%

From these data, one would be inclined to say that the new treatment is better than the old treatment for both young and old patients. This is an example of Simpson's paradox, and it is not a small sample effect.

Simpson's paradox reflects the fact that it is possible for all three of the following inequalities to hold simultaneously:

$$\text{Prob}[I \mid A, B] > \text{Prob}[I \mid A, B']$$
$$\text{Prob}[I \mid A', B] > \text{Prob}[I \mid A', B'] \qquad (17.146)$$
$$\text{Prob}[I \mid B] < \text{Prob}[I \mid B']$$

However, if $\text{Prob}[A \mid B] = \text{Prob}[A \mid B']$ then it is not possible for the 3 equations in (17.146) to hold. Practically, this means that the relative proportions (new treatment versus old treatment) should be the same for both new and old patients.

17.16.34 Secretary call problem

Suppose a secretary must be hired and n applicants are in line. Each applicant will be interviewed one at a time. Following each interview an immediate decision is made to hire or reject the candidate. Once a candidate has left, he cannot be called back. The following strategy may be used to find the *best* secretary.

> For sufficiently large n, the interviewer should consider n/e candidates, and then select the next candidate that is better than any seen previously. Using this strategy, the probability of selecting the best candidate is approximately $1/e$.

See J. S. Rose, The secretary problem with a call option, *Operations Research Letters*, Vol. 3, No. 5, December 1984.

17.16.35 Waiting for a bus

Suppose buses pass a certain corner with an average time between them of 20 minutes. What is the average time that one would expect to wait for a bus?

If the buses are exactly 20 minutes apart, then the average waiting time is 10 minutes. If the buses' arrival pattern is a Poisson distribution, then the average waiting time is 20 minutes. If the buses arrival pattern is a hyperexponential distribution, then the average waiting time may exceed 20 minutes.

17.17 ELECTRONIC RESOURCES

Thanks are extended to John C. Pezzullo for supplying the source for much of this section.

17.17.1 Statlib

Statlib, located at `http://lib.stat.cmu.edu` is a system for distributing statistical software, datasets, and information by electronic mail, FTP and WWW. The main web page includes pointers to:

(a) The *apstat* collection; a nearly complete set of algorithms published in *Applied Statistics*.

(b) The *CMLIB* collection of non-proprietary Fortran subprograms solving a variety of mathematical and statistical problems (originally produced by NIST).

(c) The *DASL* library of datafiles and stories that illustrate the use of basic statistics methods.

(d) A collection of interesting datasets, from classics like the Stanford heart transplant data, to the complete data from several textbooks. For example, over 6Mb of data and descriptions are available from *Case Studies in Biometry*, by N. Lange, L. Ryan, L. Billard, and D. Brillinger, John Wiley & Sons, New York, 1994.

(e) A collection of designs, programs, and algorithms for creating designs for statistical experiments.

(f) Algorithms from *Applied Statistics Algorithms* by P. Griffiths and I. D. Hill, Ellis Horwood, Chichester, 1985.

(g) A collection of software related to articles published in the *Journal of the American Statistical Association*.

(h) The *jcgs* archive of contributed datasets and software and abstracts from the *Journal of Computational Graphics and Statistics*.

(i) The *jqt* collection of algorithms from articles published in the *Journal of Quality Technology*.

(j) *P-Stat* functions and related software.

(k) Software and macros for the *Genstat* language.

(l) Macros, software, and algorithms for the *GLIM* statistical package.

(m) Software and extensions for the S (*Splus*) language. Over 130 separate packages including many novel statistical ideas.

(n) The *Sapaclisp* collection of Lisp functions that can be used for computations described in *Spectral Analysis for Physical Applications: Multitaper and Conventional Univariate Techniques*, by D. B. Percival and A. T. Walden, Cambridge University Press, Cambridge, England, 1993.

(o) The *MacAnova* statistical system (for Mac, Windows, and Unix).

(p) Macros for *Minitab*.

(q) Materials related to the *Stata* statistical package.

(r) *Lisp-Stat*, which is an extensible environment for statistical computing and dynamic graphics based on Lisp.

17.17.2   Uniform resource locators

The following is a list of interesting Uniform Resource Locators (URL):

(a) `http://www.stat.berkeley.edu/~stark/SticiGui/Text/gloss.htm`   A very useful glossary.

(b) http://www.yahoo.com/Science/Mathematics/
A very large list of useful sites relating to mathematics. It is perhaps
the best place to start researching an arbitrary mathematical question
not covered elsewhere in this list.

(c) http://daisy.uwaterloo.ca/~alopez-o/math-faq/
math-faq.html The *FAQ* (frequently asked questions) listing from the
newsgroup *sci.math*.

(d) http://e-math.ams.org/
The American Mathematical Society home page, with information about
AMS-TEX, the Combined Membership List of the AMS, Math Reviews
subject classifications, preprints, etc.

(e) http://www.siam.org/
The Society for Industrial and Applied Mathematics.

(f) http://gams.cam.nist.gov/
The Guide to Available Mathematical Software (GAMS).

(g) http://www.york.ac.uk/depts/maths/histstat/welcome.htm
The early history of statistics: a collection of seminal papers by Bayes,
Pascal, Laplace, Legendre, and others.

(h) http://www-sci.lib.uci.edu/HSG/RefCalculators.html#STAT
Martindale's reference desk—calculators on–line—statistics: the largest
compendia of calculating web pages.

(i) http://www.fedstats.gov
FedStats: master page maintained by the Federal Interagency Council
on Statistical Policy to provide easy access to the full range of statistics
and information produced by more than 70 agencies in the United States
Government.

(j) http://members.aol.com/johnp71/javastat.html
A collection of web pages with links to many statistical sites on the web,
maintained by John C. Pezzullo.

17.17.3 Interactive demonstrations and tutorials

(a) **Statistics tutorials**
http://204.215.60.174/tutindex.html
These briefly explain the use and interpretation of standard statistical
analysis techniques, using the WINKS program from TexaSoft. Tutorials include:

- descriptive statistics program
- Friedman's test
- independent group t-test
- Kruskal–Wallis test

- Mann–Whitney test
- one-way ANOVA
- paired t-test
- Pearson's correlation coefficient

- simple linear regression   - single sample t-test

(b) **Simulation and demonstrations**
    `http://www.ruf.rice.edu/~lane/stat_sim/index.html`
    From the creator of the *HyperStat Online* statistics book. More than a
    dozen applets; including:

- 2×2 contingency tables
- a *small* effect size can make a
  large difference
- bin widths
- chi–square test of deviations
  from expected frequencies
- components of r
- confidence interval for a
  proportion
- confidence intervals
- cross validation

- histograms
- mean and median
- normal approximation to
  binomial distribution
- regression by eye
- reliability and regression
  analysis
- repeated measures
- restriction of range
- sampling distribution
  simulation

(c) **Virtual fly lab**
    `http://vflylab.calstatela.edu/edesktop/VirtApps/VflyLab/`
    `IntroVflyLab.html`
    Simulate classic drosophila genetics experiments: Specify characteristics
    of a hypothetical male and female fly; simulate the offspring of a mating;
    select a pair of offspring to breed; simulate the offspring of their mat-
    ing; and compare the observed frequencies with those predicted from
    Mendelian genetics.
    `http://www.users.on.net/zhcchz/java/quincunx/quincunx.1.html`
    Illustration of the central limit theorem in Java.

(d) `http://www.math.csusb.edu/faculty/stanton/m262/`
    `probstat.html`   Miscellaneous Java demos.

(e) `http://www.ms.uky.edu/~mai/java/AppletIndex.html`
    Five demonstrations of how samples of increasing size approach a theo-
    retical distribution:

- Empirical
- Kaplan–Meier
- Nelson–Aalen

- interval censored
- doubly-censored data

(f) `http://www.ms.uky.edu/~mai/java/AppletIndex.html`
    Three demonstrations of famous random processes:

- 1-dimensional and 2-dimensional Brownian motion
- Buffon needle experiment
- Galton's ball–drop *quincunx*

(g) `http://www.ms.uky.edu/~lancastr/java/javapage.html`
    Three bootstrap demonstrations all using the Exp(1) distribution:

- central limit theorem
- parametric bootstrap of sample mean
- nonparametric bootstrap of sample mean

(h) http://www-stat.stanford.edu/~naras/jsm/NormalDensity/
NormalDensity.html
Experiments with the normal distribution.

(i) http://www.stat.uiuc.edu/~stat100/java/guess/
PPApplet.html   Linear regression and correlation demo in Java—
click a bunch of points onto the screen; as each is entered the computer
immediately computes and displays an adjusted regression line (with
equation and correlation coefficient).

(j) http://stat-www.berkeley.edu/users/stark/Java/
Correlation.html   Similar to above, but lets you simulate points
from distributions with known correlation coefficient.

(k) http://www.ruf.rice.edu/~lane/stat_sim/sampling_dist/
sampling_demo.html
Simulation of various sampling distributions in Java.

(l) http://www.stat.uiuc.edu/~stat100/java/box/BoxApplet.html
Simulation in Java of drawing objects (with replacement) from a "box"
whose contents you can define.

(m) http://www.stat.sc.edu/~west/javahtml/epid.html
Simulate the course of an epidemic in Java.

(n) http://huizen.dds.nl/~berrie/
A collection of QuickTime movies illustrating various statistical concepts.

(o) http://members.aol.com/Trane64/java/CardPowersim.html
Monte–Carlo p versus sample size simulation for survey questionnaire
results, with graphical output (in Java).

17.17.4   Online textbooks, reference manuals, and journals

(a) **Glossary of statistical terms used in evidence-based medicine**
http://www.musc.edu/muscid/glossary.html
Can be printed out into a handy 12-page mini-handbook of statistical
concepts and terminology.

(b) **HyperStat statistics textbook**
http://ruf.rice.edu/~lane/hyperstat/contents.html
A well-designed and well-constructed "hyper-text-book."

(c) **Statistics at Square One**
http://www.bmj.com/collections/statsbk/index.shtml
An excellent online textbook.

(d) **Electronic Statistics Textbook**
http://www.statsoft.com/textbook/stathome.html
(by StatSoft) Very extensive and well-organized (can also be downloaded for quicker access from your hard drive).

(e) **Instat Guide**
http://www.graphpad.com/instatman
Instat Guide to Choosing and Interpreting Statistical Tests, from Graphpad.

(f) **Glossary of over 30 statistical terms**
http://www.animatedsoftware.com/statglos/statglos.htm
Explanations vary from short definitions to extensive explanations, with graphics and hyperlinks.

(g) **Introduction to Probability**
http://www.geom.umn.edu/docs/snell/chance/teaching_aids/
probability_book/pdf.html
A complete book by Grinstead and Snell in Adobe Acrobat and Postscript format. It can be downloaded all at once, or chapter by chapter.

(h) **Graphpad web site of statistical resources**
http://www.graphpad.com/www/
Short articles, book chapters, bibliographies, and (commercial) software. Well-written, down-to-earth, and helpful.

(i) **InterStat (Statistics on the internet)**
http://interstat.stat.vt.edu/intersta.htm/
An online journal where one can publish or read about any aspect of statistical research or innovative method. Abstracts of articles can be viewed; from there you can read or download the article (or any comments written about it) in *pdf* and *ps* formats.

(j) **Journal of Statistics Education**
http://www.stat.ncsu.edu/info/jse/
A refereed electronic journal on postsecondary teaching of statistics.

(k) http://www.psychstat.smsu.edu/sbk00.htm
Introductory Statistics: Concepts, Models, and Applications by David W. Stockburger. In-depth coverage, with extensive use of web technology (animated graphics, interactive calculating pages).

(l) http://www.psychstat.smsu.edu/multibook/mlt00.htm
Multivariate Statistics: Concepts, Models, and Applications, by David W. Stockburger In-depth coverage, with extensive use of web technology (animated graphics, interactive calculating pages).

(m) http://trochim.human.cornell.edu/kb/kbhome.htm
The Knowledge Base— An Online Research Methods Textbook, by William M. K. Trochim. An online textbook for an introductory course in research methods.

(n) `http://www.cern.ch/Physics/DataAnalysis/BriefBook/`
Data Analysis BriefBook: A condensed handbook, or an extended glossary, written in encyclopedic format, covering subjects in statistics, computing, analysis, and related fields; meant to be both introduction and reference for data analysts, scientists and engineers (from CERN).

(o) `http://www.cne.gmu.edu/modules/dau/stat/`
Statistics refresher tutorial.

(p) `http://www.wrightslaw.com/advoc/articles/`
`tests_measurements.html`
Statistics of performance measurement, designed for parents of children with special educational needs, but also worthwhile for a good introductory presentation of basic statistical concepts.

(q) `http://nimitz.mcs.kent.edu/~blewis/stat/scon.html`
Guide to Basic Laboratory Statistics (in Java).

(r) `http://duke.usask.ca/~rbaker/stats.html`
Basic Principles of Statistical Analysis.

17.17.5   Free statistical software packages

A selection of free software packages.

(a) **ViSta**
`http://forrest.psych.unc.edu/research/ViSta.html`
A VIsual STAtistics program for Windows, Mac, and Unix. Features a structured desktop.

(b) **WebStat**
`http://www.stat.sc.edu/~west/webstat/`
Java-based statistical computing environment for the web. Requires a browser, but can be run offline.

(c) **MacANOVA**
`http://www.stat.umn.edu/~gary/macanova/macanova.home.html`
Available for Windows and Mac. This is a complete programming language.

(d) **Scilab**
`http://www-rocq.inria.fr/scilab/`
Available for Windows, Mac, and Unix. This is a programming language with MatLab-like syntax, hundreds of built-in functions and libraries, 3D graphics, and symbolic capabilities through a Maple interface.

(e) **SISA**
`http://home.clara.net/sisa/pasprog.htm`
Simple Interactive Statistical Analysis for DOS from Daan Uitenbroek. Collection of modules for many types of hypothesis testing, sample size calculations, and survey design. Includes many analyses not usually found elsewhere.

(f) **Anderson statistical archives**

http://odin.mdacc.tmc.edu/anonftp/page_2.html

A large collection of statistical programs for DOS and Mac with Fortran and C source, from the Biomathematics Department of the M. D. Anderson Cancer Center.

(g) **STPLAN**

http://odin.mdacc.tmc.edu/anonftp/page_2.html

Performs power, sample size, and related calculations needed to plan studies. Covers a wide variety of situations, including studies whose outcomes involve the binomial, poisson, normal, and log-normal distributions, or are survival times or correlation coefficients. For DOS and Mac with Fortran and C source.

(h) **EpiInfo**

http://www.cdc.gov/epo/epi/epiinfo.htm

A set of programs for word processing, data management, and epidemiologic analysis, designed for public health professionals. Consists of Epi Info (forms design, data entry, data management), Epi Map (generated geographical, map-based output), SSS1 (Box–Jenkins time series analysis, MMWR graphs, trend analysis, and 2-source comparisons).

(i) **Rweb**

http://www.math.montana.edu/Rweb/

A Web based interface to the "R" statistical programming language (similar to S or S-plus).

(j) **ARIMA**

ftp://ftp.census.gov/pub/ts/x12a/

A seasonal adjustment program for PC and Unix, developed by the Census Bureau.

(k) **G*Power**

http://www.psychologie.uni-trier.de:8000/projects/gpower.html

A general Power Analysis program for DOS and Macintosh. Performs high-precision analysis for t-tests, F-tests, chi–square tests. Computes power, sample sizes, alpha, beta, and alpha/beta ratios. Contains web-based tutorial and reference manual.

(l) **JDB**

http://www.isi.edu/~johnh/SOFTWARE/JDB/index.html

Relational database and elementary statistics for unix. Useful for manipulating experimental data (joining files, cleaning data, reformatting for input into other programs). Computes basic statistics (mean, std. dev., confidence intervals, quartiles, n-tiles, percentiles, histograms, correlations, z-scores, t-scores).

(m) **EasyMA**

http://www.spc.univ-lyon1.fr/~mcu/easyma/

DOS program for the meta-analysis of clinical trials results. Developed to help physicians and medical researchers to synthesize evidence in clinical or therapeutic research.

(n) **Meta-analysis 5.3**
http://www.yorku.ca/faculty/academic/schwarze/meta_e.htm
DOS software for meta-analysis, perhaps the most frequently used meta-analysis software in the world. Can select the analysis of exact p values or effect sizes (d or r, with a cluster size option). Plots a stem-and-leaf display of correlation coefficients. A utility menu allows various transformations and preliminary computations that are typically required before meta-analysis can be performed.

(o) **First Bayes**
http://www.maths.nott.ac.uk/personal/aoh/1b.html
A windows application for elementary Bayesian Statistics. Performs most standard, elementary Bayesian analyses, including: plotting and summarizing distributions, defining and examining arbitrary mixtures of distributions, analysis of two kinds of linear model (one or more normal samples with common but unknown variance, and simple linear regression), examination of marginal distributions for arbitrary linear combinations of the location parameters, and the generation of predictive distributions.

(p) **MANET**
http://www1.math.uni-augsburg.de/Manet/
Missings Are Now Equally Treated: Mac software for interactive graphics tools for data sets with missing values. Generates missing values chart, histograms and barcharts, boxplots and dotplots, scatterplots, mosaic plots, polygon plots, highlighted boxplots, interactive trellis displays, traces, context-sensitive interrogation, cues, redframing, selection sequences.

(q) **WinSAAM**
http://www.winsaam.com/#WinSAAM
Windows version of SAAM (System Analysis and Modeling Software). Lets you create mathematical models, design and simulate experiments, and analyze data. Models can contain differential equations. Graphic and tabular output is provided.

(r) **Boomer**
http://www.cpb.uokhsc.edu/common/anonymous/MFB/toc.html
Non-linear regression program for analysis of pharmacokinetic and pharmacodynamic data. Includes normal fitting, Bayesian estimation, or simulation-only, with integrated or differential equation models. Allows selection of weighting schemes and methods for numerical integration. Available for Macintosh and PC. Online manual, tutorial, and sample data sets.

(s) **Serpik Graph** `http://www.chat.ru/~mserpik/index.htm`
A very sophisticated function graphing program.

17.17.6 Demonstration statistical software packages

The following are a selection of "demonstration versions" or "student versions" of commercial packages. They can be freely downloaded, but are usually restricted or limited in some way.

(a) **AssiStat**
`http://users.aol.com/micrometr/assistat.htm`
A Windows package, designed to complement a typical statistical package. It covers topics not normally in a primary analysis package such as correction of correlations for restriction in range or less-than-perfect reliability.

(b) **CART**
`http://www.salford-systems.com/`
Salford Systems flagship decision-tree software, combines an easy-to-use GUI with advanced features for data mining, data pre-processing, and predictive modeling.

(c) **InStat**
`http://www.graphpad.com/instat3/instatdemo.htm`
INstant STATistics: A package from GraphPad Software for Windows and Mac. Demo version disables printing, saving, and exporting capabilities.

(d) **MiniTab**
`http://www.minitab.com/products/minitab/rel12/index.htm`
aA program for Windows. Good coverage of industrial quality control analyses.

(e) **Prism**
`http://www.graphpad.com/prism/pdemo.htm`
A package from from GraphPad Software for Windows and Mac. Performs basic biostatistics, fits curves and creates publication quality scientific graphs in one complete package.

(f) **QuickStat**
`http://www.camcode.com/arcus.html`
A program for Windows. Does not assume statistical knowledge. Gives advice on experimental design, analysis, and interpretation from the integrated statistical help system. Results presented in plain language. For the more experienced user, Arcus is an extremely useful statistical toolbox which covers biomedical statistical methods.

(g) **Rasch**
`http://mesa.spc.uchicago.edu/`
Measurement software deals with the various nuances of constructing optimal rating scales from a number of (usually) dichotomous measure-

ments, such as responses to questions in a survey or test.

(h) **STATISTICA demo**
http://www.statsoft.com/download.html#demo
Has many features of the "Basic Statistics" module of STATISTICA.

(i) **Statlets**
http://www.statlets.com/
A Java statistics program.

(j) **SPSS demos**
http://www.spss.com/cool/index.htm
Numerous demo packages from products acquired by SPSS (all for the PC except one):

- allCLEAR 3.5 and 4.5
- GOLDMineR
- DeltaGraph (Mac)
- LogXact 2.1
- PeakFit 4.06
- QI Analyst 3.5
- Remark Office OMR 3.0
- SamplePower 1.2
- SigmaGel 1.0
- SPSS Diamond

- SigmaPlot 4.0
- SigmaScan Pro 4.0
- SigmaStat 2.0
- SmartViewer
- StatXact 3.1
- SYSTAT 7.0
- TableCurve 2D 4.07
- TableCurve 3D 3.01
- WesVar Complex Samples

17.18 TABLES

17.18.1 Random deviates

Consider the two data sequences $S_{\text{continuous unif}}$ and $S_{\text{std normal}}$ defined by:

(a) Continuous uniform random numbers on the interval $(0,1)$ are generated by a linear congruential generator with parameters: $x_0 = 1$, $a = 99$, $b = 0$, and $M = 2^{20}$; see section 17.11.1.1. The sequence begins $\{1, 99, 9801, 970299, 639185, 364755, \ldots\}$. Dividing this sequence by M yields numbers on the interval $(0,1)$. Hence, $S_{\text{continuous unif}} = \{.00000095, .000094, .0093, .9253, .6095, .3478, \ldots\}$

(b) $S_{\text{std normal}}$ is then obtained by using the Box–Muller method (see section 17.11.2). Specifically, if the values in $S_{\text{continuous uniform}}$ are written as $\{u_1, v_1, u_2, v_2, \ldots\}$ then standard normal variables are given by $\sqrt{-2\ln u_i} \cos 2\pi v_i$. Hence, $S_{\text{std normal}} = \{5.266, 2.727, -.574, -.763, -.140, \ldots\}$.

The following random units are obtained from the third digit of the first 2,500 values in $S_{\text{continuous unif}}$. The standard normal values are the first 500 values in $S_{\text{std normal}}$.

2,500 Random units

Line	Col = (1)	(2)	(3)	(4)	(5)	(6)	(7)	(8)	(9)	(10)
1	00959	77100	36484	06101	34734	18924	28001	87674	16684	07027
2	08149	50395	26971	29957	25235	85468	85629	57477	68311	33626
3	49324	30579	12667	28910	42885	78563	10340	77508	96233	84084
4	03207	57615	90513	02808	68168	12164	81531	90806	01306	94806
5	43855	79617	46906	90009	48557	21458	31805	99078	15841	12303
6	56825	52535	70234	98035	96147	93443	08513	81271	04047	29078
7	91822	62169	10152	29526	48393	07763	12402	43920	06443	44055
8	96171	27730	38041	12364	85443	55515	26148	85117	49046	28379
9	81089	47387	46900	96147	68169	37767	82207	74778	21357	54056
10	00214	98330	43526	10401	39687	16238	43098	63975	59317	90397
11	77471	67351	83417	65376	79343	15268	37451	36632	52965	29032
12	69857	73393	24342	43527	49655	46460	12109	36747	80829	80614
13	91127	97088	07052	71213	52345	20899	50674	29653	04919	99244
14	17793	53476	95434	42814	82423	29613	77770	99328	90929	23462
15	44797	23004	91138	43018	14720	10925	53861	22495	14266	06873
16	80021	02262	28002	01277	17361	81989	28191	10545	83231	47679
17	23856	96770	76740	42257	70503	07524	93800	19562	32113	34636
18	56168	45893	93645	99333	25071	39446	51115	57579	68558	23032
19	72348	46166	92883	40513	01003	86613	28231	06276	40051	70357
20	17468	91307	08718	73452	48267	56730	11798	23490	06958	14224
21	44671	49100	23878	73037	99519	35718	60484	31475	64635	55380
22	86905	30471	97545	47616	05116	03575	52029	32216	41781	57265
23	20038	92416	54157	50748	30886	01292	12454	75935	00909	19524
24	10567	25587	13156	25831	92058	37872	42508	41351	08577	58216
25	15270	23258	16824	62411	69868	29323	31010	81249	49061	48966
26	76791	13967	66020	83633	02426	45880	19989	10388	97013	10831
27	55235	76804	47756	62552	49756	71619	83809	06429	62513	28251
28	57021	01659	32822	72473	47212	10174	98107	61187	58584	86515
29	03229	35982	13383	95360	59148	56016	53365	46634	65098	90531
30	75332	71330	68576	04241	75405	24229	22471	98754	66221	90688
31	38634	66207	19246	60714	07599	82754	80271	84538	74552	84207
32	97519	96693	14402	82045	52534	92484	09281	89645	15765	63189
33	68223	30041	74131	58640	27944	02945	05826	02617	61112	86956
34	61103	01559	62880	84410	72932	44660	38281	62393	36801	73695
35	17932	68475	38125	31372	72691	00476	93257	16191	06430	83422
36	89736	05257	17573	50608	65491	50562	43426	61796	57616	95798
37	57969	44305	67666	37122	42042	64781	47188	44139	16099	13088
38	83902	88463	16427	12809	71391	83774	07016	99953	55540	55620
39	60705	47476	55277	62116	24849	32422	78322	78196	26035	00704
40	01078	04490	01831	63010	66789	79496	89329	23308	23914	60209
41	51923	25930	42663	19843	63039	37580	26048	50867	89446	58697
42	00304	09740	80005	82576	33389	36698	97231	84403	83824	01422
43	56659	01069	57000	39995	11739	63783	76384	46473	73722	43182
44	02302	91182	53872	17030	03905	55025	76676	22339	83255	42673
45	17892	88588	76728	89583	58411	98924	22062	35871	27748	06631
46	00996	60460	33299	81480	14693	89001	78743	23702	98548	29161
47	24141	34012	93370	57766	01847	60396	75707	17819	46328	88234
48	51133	50180	02363	88535	44723	75027	82942	64448	14215	49756
49	05261	99025	29539	35100	86954	23969	84766	55919	91772	54978
50	23374	37461	32685	96100	85808	20952	17928	94719	50170	04314

500 Random normal numbers with $\mu = 0$, $\sigma = 1$

Line	Col = (1)	(2)	(3)	(4)	(5)	(6)	(7)	(8)	(9)	(10)
1	5.266	2.727	−0.574	−0.763	−0.140	−0.645	−2.263	1.252	−2.044	−1.055
2	−0.099	0.811	1.207	−0.287	1.809	0.839	3.572	0.078	0.773	0.795
3	−0.436	0.022	−1.077	0.766	0.556	−1.075	0.706	−0.572	0.980	−1.021
4	1.261	−0.685	0.364	−1.045	1.151	−0.747	−0.050	−2.057	0.564	−1.714
5	−1.395	−2.157	−0.494	−1.670	0.032	−0.128	0.561	0.181	0.612	−0.099
6	−0.424	0.894	1.277	−0.047	0.709	1.871	−2.185	−1.179	−0.121	−0.882
7	0.664	−1.257	0.612	0.025	−0.210	2.123	−0.214	2.638	0.754	−1.035
8	−1.042	1.385	0.677	0.307	−1.218	0.135	1.399	0.124	0.722	−0.719
9	0.432	−2.265	−0.226	1.120	0.216	0.620	0.008	−1.319	−0.816	0.897
10	−0.976	0.057	−1.004	−0.585	1.426	−0.330	−0.506	0.934	−0.169	−1.126
11	−1.350	1.221	0.131	0.317	−0.977	0.360	−1.105	0.914	0.240	0.335
12	−0.400	−0.595	−1.737	−0.814	−0.234	1.423	0.693	−0.289	0.107	0.564
13	0.127	−1.512	0.066	−0.005	2.384	1.103	−0.333	−0.288	−0.098	1.524
14	0.388	−0.366	−0.157	−0.200	−0.565	0.372	−0.981	0.175	−0.193	0.612
15	1.729	−0.169	0.907	0.183	−0.270	−0.947	0.215	−1.625	0.176	0.242
16	0.036	−1.458	−0.821	0.945	−0.046	1.573	1.968	−0.464	−0.882	1.741
17	0.548	−0.325	−0.599	−0.795	0.586	1.610	1.032	−0.189	−0.252	−0.699
18	−1.009	−0.102	−0.064	−0.661	0.366	0.212	−0.880	1.825	−1.637	1.538
19	−1.053	−1.050	−0.036	−0.482	−0.175	0.579	1.041	0.316	0.298	−0.838
20	1.076	−0.520	0.626	−0.187	−0.107	−0.267	−0.922	0.009	−0.510	0.352
21	−0.839	−1.043	−0.938	0.528	−1.290	0.692	−1.089	−0.968	0.435	−0.102
22	−1.528	0.303	0.526	−1.717	−0.545	−1.138	−0.428	0.558	0.335	0.254
23	−0.043	−0.391	−1.238	−0.449	1.081	−0.375	−0.829	0.635	−0.098	0.126
24	0.384	−1.481	0.037	0.038	−0.669	−0.698	−0.070	0.061	0.041	0.751
25	−2.603	0.156	0.611	−0.138	−1.676	−0.073	1.491	−1.177	0.620	−0.914
26	−1.072	1.243	−0.714	−1.498	0.232	−1.446	−0.415	0.711	1.109	−1.220
27	−1.650	1.330	0.299	0.408	−2.151	1.257	0.267	−0.678	−0.010	−0.541
28	0.052	−1.202	−1.635	0.495	−0.384	0.417	−0.316	0.493	−0.528	0.479
29	1.458	0.704	−0.689	−0.580	0.034	1.022	−0.161	−0.122	0.532	0.763
30	0.094	1.669	0.377	−1.093	1.349	0.284	0.530	−0.967	1.496	−0.688
31	2.339	1.534	−1.522	−0.078	0.633	0.225	−0.959	0.934	−0.889	−0.602
32	−0.153	−0.535	−0.026	−1.107	−1.187	0.917	0.045	−0.301	0.356	1.042
33	1.131	−0.313	0.444	0.451	−0.051	1.719	0.744	−0.137	0.989	−1.067
34	−0.641	−0.402	0.184	−0.815	0.678	0.045	−1.626	−0.413	0.383	−0.505
35	0.700	0.826	−0.609	−0.971	0.017	−0.359	0.165	2.421	1.404	0.462
36	−1.902	−1.028	−1.636	−0.384	−0.900	0.466	1.761	−0.412	−0.471	0.077
37	2.053	−0.333	0.200	0.138	−0.205	−1.281	0.292	−0.614	−0.412	2.186
38	−0.938	−0.357	−0.764	0.742	0.251	−0.938	−0.803	−0.301	−0.789	0.723
39	0.951	0.229	−0.107	−0.707	0.469	0.069	0.859	−0.674	−0.406	−0.835
40	1.644	−0.247	−1.016	−0.761	−0.881	−1.342	−0.737	−0.534	−1.414	0.008
41	1.601	−1.484	−0.114	−0.064	0.076	0.694	0.788	−1.181	−0.114	−0.098
42	−1.666	0.998	−0.323	0.360	1.075	−0.912	0.227	2.022	0.693	0.584
43	1.307	0.553	1.461	1.297	−0.085	−0.201	0.155	−0.895	−2.334	−1.281
44	0.548	−0.623	−1.971	−0.993	−0.350	−0.195	0.923	0.118	1.602	0.858
45	0.476	1.318	0.146	−1.564	−0.432	−1.733	0.609	−1.462	−0.473	1.087
46	−1.303	−0.567	0.325	0.725	−0.026	0.322	−1.127	0.942	−0.339	−0.336
47	0.150	2.147	0.652	0.910	−0.970	−0.560	0.885	−0.997	0.826	−1.420
48	−2.351	−0.813	1.751	−0.980	−0.291	1.930	−0.320	0.337	0.121	1.108
49	0.309	0.720	−0.304	−0.190	−3.205	−0.408	1.503	0.718	0.194	0.526
50	1.672	−1.693	−1.064	−0.623	0.201	2.392	−0.255	−1.015	1.252	1.796

17.18.2   Permutations

This table contains the number of permutations of n distinct things taken m at a time, given by (see section 3.2.3):

$$P(n,m) = \frac{n!}{(n-m)!} = n(n-1)\cdots(n-m+1) \qquad (17.147)$$

Permutations $P(n,m)$

n	$m=0$	1	2	3	4	5	6	7	8
0	1								
1	1	1							
2	1	2	2						
3	1	3	6	6					
4	1	4	12	24	24				
5	1	5	20	60	120	120			
6	1	6	30	120	360	720	720		
7	1	7	42	210	840	2520	5040	5040	
8	1	8	56	336	1680	6720	20160	40320	40320
9	1	9	72	504	3024	15120	60480	181440	362880
10	1	10	90	720	5040	30240	151200	604800	1814400
11	1	11	110	990	7920	55440	332640	1663200	6652800
12	1	12	132	1320	11880	95040	665280	3991680	19958400
13	1	13	156	1716	17160	154440	1235520	8648640	51891840
14	1	14	182	2184	24024	240240	2162160	17297280	121080960
15	1	15	210	2730	32760	360360	3603600	32432400	259459200

Permutations $P(n,m)$

n	$m=9$	10	11	12	13
9	362880				
10	3628800	3628800			
11	19958400	39916800	39916800		
12	79833600	239500800	479001600	479001600	
13	259459200	1037836800	3113510400	6227020800	6227020800
14	726485760	3632428800	14529715200	43589145600	87178291200
15	1816214400	10897286400	54486432000	217945728000	653837184000

17.18.3   Combinations

This table contains the number of combinations of n distinct things taken m at a time, given by (see section 3.2.5):

$$C(n,m) = \binom{n}{m} = \frac{n!}{m!(n-m)!} \qquad (17.148)$$

Combinations $C(n, m)$

n	$m = 0$	1	2	3	4	5	6	7
1	1	1						
2	1	2	1					
3	1	3	3	1				
4	1	4	6	4	1			
5	1	5	10	10	5	1		
6	1	6	15	20	15	6	1	
7	1	7	21	35	35	21	7	1
8	1	8	28	56	70	56	28	8
9	1	9	36	84	126	126	84	36
10	1	10	45	120	210	252	210	120
11	1	11	55	165	330	462	462	330
12	1	12	66	220	495	792	924	792
13	1	13	78	286	715	1287	1716	1716
14	1	14	91	364	1001	2002	3003	3432
15	1	15	105	455	1365	3003	5005	6435
16	1	16	120	560	1820	4368	8008	11440
17	1	17	136	680	2380	6188	12376	19448
18	1	18	153	816	3060	8568	18564	31824
19	1	19	171	969	3876	11628	27132	50388
20	1	20	190	1140	4845	15504	38760	77520
21	1	21	210	1330	5985	20349	54264	116280
22	1	22	231	1540	7315	26334	74613	170544
23	1	23	253	1771	8855	33649	100947	245157
24	1	24	276	2024	10626	42504	134596	346104
25	1	25	300	2300	12650	53130	177100	480700
26	1	26	325	2600	14950	65780	230230	657800
27	1	27	351	2925	17550	80730	296010	888030
28	1	28	378	3276	20475	98280	376740	1184040
29	1	29	406	3654	23751	118755	475020	1560780
30	1	30	435	4060	27405	142506	593775	2035800
31	1	31	465	4495	31465	169911	736281	2629575
32	1	32	496	4960	35960	201376	906192	3365856
33	1	33	528	5456	40920	237336	1107568	4272048
34	1	34	561	5984	46376	278256	1344904	5379616
35	1	35	595	6545	52360	324632	1623160	6724520
36	1	36	630	7140	58905	376992	1947792	8347680
37	1	37	666	7770	66045	435897	2324784	10295472
38	1	38	703	8436	73815	501942	2760681	12620256
39	1	39	741	9139	82251	575757	3262623	15380937
40	1	40	780	9880	91390	658008	3838380	18643560
41	1	41	820	10660	101270	749398	4496388	22481940
42	1	42	861	11480	111930	850668	5245786	26978328
43	1	43	903	12341	123410	962598	6096454	32224114
44	1	44	946	13244	135751	1086008	7059052	38320568
45	1	45	990	14190	148995	1221759	8145060	45379620
46	1	46	1035	15180	163185	1370754	9366819	53524680
47	1	47	1081	16215	178365	1533939	10737573	62891499
48	1	48	1128	17296	194580	1712304	12271512	73629072
49	1	49	1176	18424	211876	1906884	13983816	85900584
50	1	50	1225	19600	230300	2118760	15890700	99884400

Combinations $C(n, m)$

n	$m = 8$	9	10	11	12
8	1				
9	9	1			
10	45	10	1		
11	165	55	11	1	
12	495	220	66	12	1
13	1287	715	286	78	13
14	3003	2002	1001	364	91
15	6435	5005	3003	1365	455
16	12870	11440	8008	4368	1820
17	24310	24310	19448	12376	6188
18	43758	48620	43758	31824	18564
19	75582	92378	92378	75582	50388
20	125970	167960	184756	167960	125970
21	203490	293930	352716	352716	293930
22	319770	497420	646646	705432	646646
23	490314	817190	1144066	1352078	1352078
24	735471	1307504	1961256	2496144	2704156
25	1081575	2042975	3268760	4457400	5200300
26	1562275	3124550	5311735	7726160	9657700
27	2220075	4686825	8436285	13037895	17383860
28	3108105	6906900	13123110	21474180	30421755
29	4292145	10015005	20030010	34597290	51895935
30	5852925	14307150	30045015	54627300	86493225
31	7888725	20160075	44352165	84672315	141120525
32	10518300	28048800	64512240	129024480	225792840
33	13884156	38567100	92561040	193536720	354817320
34	18156204	52451256	131128140	286097760	548354040
35	23535820	70607460	183579396	417225900	834451800
36	30260340	94143280	254186856	600805296	1251677700
37	38608020	124403620	348330136	854992152	1852482996
38	48903492	163011640	472733756	1203322288	2707475148
39	61523748	211915132	635745396	1676056044	3910797436
40	76904685	273438880	847660528	2311801440	5586853480
41	95548245	350343565	1121099408	3159461968	7898654920
42	118030185	445891810	1471442973	4280561376	11058116888
43	145008513	563921995	1917334783	5752004349	15338678264
44	177232627	708930508	2481256778	7669339132	21090682613
45	215553195	886163135	3190187286	10150595910	28760021745
46	260932815	1101716330	4076350421	13340783196	38910617655
47	314457495	1362649145	5178066751	17417133617	52251400851
48	377348994	1677106640	6540715896	22595200368	69668534468
49	450978066	2054455634	8217822536	29135916264	92263734836
50	536878650	2505433700	10272278170	37353738800	121399651100

Combinations $C(n, m)$

n	$m = 13$	14	15	16	17
13	1				
14	14	1			
15	105	15	1		
16	560	120	16	1	
17	2380	680	136	17	1
18	8568	3060	816	153	18
19	27132	11628	3876	969	171
20	77520	38760	15504	4845	1140
21	203490	116280	54264	20349	5985
22	497420	319770	170544	74613	26334
23	1144066	817190	490314	245157	100947
24	2496144	1961256	1307504	735471	346104
25	5200300	4457400	3268760	2042975	1081575
26	10400600	9657700	7726160	5311735	3124550
27	20058300	20058300	17383860	13037895	8436285
28	37442160	40116600	37442160	30421755	21474180
29	67863915	77558760	77558760	67863915	51895935
30	119759850	145422675	155117520	145422675	119759850
31	206253075	265182525	300540195	300540195	265182525
32	347373600	471435600	565722720	601080390	565722720
33	573166440	818809200	1037158320	1166803110	1166803110
34	927983760	1391975640	1855967520	2203961430	2333606220
35	1476337800	2319959400	3247943160	4059928950	4537567650
36	2310789600	3796297200	5567902560	7307872110	8597496600
37	3562467300	6107086800	9364199760	12875774670	15905368710
38	5414950296	9669554100	15471286560	22239974430	28781143380
39	8122425444	15084504396	25140840660	37711260990	51021117810
40	12033222880	23206929840	40225345056	62852101650	88732378800
41	17620076360	35240152720	63432274896	103077446706	151584480450
42	25518731280	52860229080	98672427616	166509721602	254661927156
43	36576848168	78378960360	151532656696	265182149218	421171648758
44	51915526432	114955808528	229911617056	416714805914	686353797976
45	73006209045	166871334960	344867425584	646626422970	1103068603890
46	101766230790	239877544005	511738760544	991493848554	1749695026860
47	140676848445	341643774795	751616304549	1503232609098	2741188875414
48	192928249296	482320623240	1093260079344	2254848913647	4244421484512
49	262596783764	675248872536	1575580702584	3348108992991	6499270398159
50	354860518600	937845656300	2250829575120	4923689695575	9847379391150

CHAPTER 18

Special Functions

Contents

The following notation is used throughout this chapter:

(a) $\mathcal{N}$ denotes the *natural numbers*.

(b) $\mathcal{R}$ denotes the *real numbers*.

(c) Re(z) denotes the real part of the complex quantity z.

18.1 BESSEL FUNCTIONS

18.1.1 Differential equation

The Bessel differential equation is

$$z^2 y'' + z y' + (z^2 - \nu^2)y = 0 \qquad (18.1)$$

It has a regular singularity at $z = 0$ and an irregular singularity at $z = \infty$. The solutions are denoted by $J_\nu(z)$ and $Y_\nu(z)$; these are the ordinary Bessel functions. Other solutions are $J_{-\nu}(z)$ and $Y_{-\nu}(z)$. If ν is an integer

$$J_{-n}(z) = (-1)^n J_n(z), \quad n = 0, 1, 2, \ldots \qquad (18.2)$$

Complete solutions to Bessel's equation may be written as

$$\begin{aligned} c_1 J_\nu(z) + c_2 J_{-\nu}(z), & \quad \text{if } \nu \text{ is not an integer,} \\ c_1 J_\nu(z) + c_2 Y_\nu(z), & \quad \text{for any value of } \nu. \end{aligned} \qquad (18.3)$$

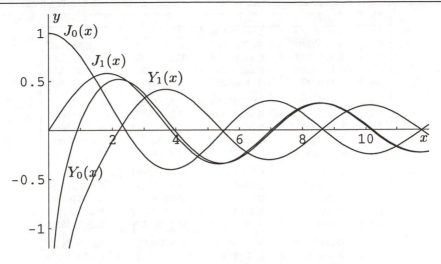

Figure 18.1: Bessel functions $J_0(x)$, $J_1(x)$, $Y_0(x)$, $Y_1(x)$, for $0 \leq x \leq 12$.

18.1.2 Series expansions

For any complex number z

$$J_\nu(z) = \left(\frac{1}{2}z\right)^\nu \sum_{n=0}^\infty \frac{(-1)^n \left(\frac{1}{2}z\right)^{2n}}{\Gamma(n+\nu+1)\,n!}$$

$$J_0(z) = 1 - \left(\frac{1}{2}z\right)^2 + \frac{1}{2!\,2!}\left(\frac{1}{2}z\right)^4 - \frac{1}{3!\,3!}\left(\frac{1}{2}z\right)^6 + \dots \tag{18.4}$$

$$J_1(z) = \frac{1}{2}z\left[1 - \frac{1}{1!\,2!}\left(\frac{1}{2}z\right)^2 + \frac{1}{2!\,3!}\left(\frac{1}{2}z\right)^4 - \frac{1}{3!\,4!}\left(\frac{1}{2}z\right)^6 + \dots\right]$$

18.1.3 Recurrence relations

The term *cylinder function* refers to any of the functions $J_\nu(z)$, $Y_\nu(z)$, $H_\nu^{(1)}(z)$, or $H_\nu^{(2)}(z)$. If $C_\nu(z)$ denotes any of the cylinder functions, then

$$\begin{aligned}
C_{\nu-1}(z) + C_{\nu+1}(z) &= \frac{2\nu}{z}C_\nu(z), \\
C_{\nu-1}(z) - C_{\nu+1}(z) &= 2C_\nu'(z), \\
C_\nu'(z) &= C_{\nu-1}(z) - \frac{\nu}{z}C_\nu(z), \\
C_\nu'(z) &= -C_{\nu+1}(z) + \frac{\nu}{z}C_\nu(z).
\end{aligned} \tag{18.5}$$

x	$J_0(x)$	$J_1(x)$	$Y_0(x)$	$Y_1(x)$
0.0	1.00000000	0.00000000	$-\infty$	$-\infty$
0.2	0.99002497	0.09950083	-1.08110532	-3.32382499
0.4	0.96039823	0.19602658	-0.60602457	-1.78087204
0.6	0.91200486	0.28670099	-0.30850987	-1.26039135
0.8	0.84628735	0.36884205	-0.08680228	-0.97814418
1.0	0.76519769	0.44005059	0.08825696	-0.78121282
1.2	0.67113274	0.49828906	0.22808350	-0.62113638
1.4	0.56685512	0.54194771	0.33789513	-0.47914697
1.6	0.45540217	0.56989594	0.42042690	-0.34757801
1.8	0.33998641	0.58151695	0.47743171	-0.22366487
2.0	0.22389078	0.57672481	0.51037567	-0.10703243
2.2	0.11036227	0.55596305	0.52078429	0.00148779
2.4	0.00250768	0.52018527	0.51041475	0.10048894
2.6	-0.09680495	0.47081827	0.48133059	0.18836354
2.8	-0.18503603	0.40970925	0.43591599	0.26354539
3.0	-0.26005195	0.33905896	0.37685001	0.32467442
3.2	-0.32018817	0.26134325	0.30705325	0.37071134
3.4	-0.36429560	0.17922585	0.22961534	0.40101529
3.6	-0.39176898	0.09546555	0.14771001	0.41539176
3.8	-0.40255641	0.01282100	0.06450325	0.41411469
4.0	-0.39714981	-0.06604333	-0.01694074	0.39792571
4.2	-0.37655705	-0.13864694	-0.09375120	0.36801281
4.4	-0.34225679	-0.20277552	-0.16333646	0.32597067
4.6	-0.29613782	-0.25655284	-0.22345995	0.27374524
4.8	-0.24042533	-0.29849986	-0.27230379	0.21356517
5.0	-0.17759677	-0.32757914	-0.30851763	0.14786314

Table 18.1: Values of the Bessel functions J_0, J_1, Y_0, and Y_1.

18.1.4   Behavior as $z \to 0$

If $\mathrm{Re}(\nu) > 0$, then

$$
J_\nu(z) \sim \frac{\left(\frac{1}{2}z\right)^\nu}{\Gamma(\nu+1)}, \qquad Y_\nu(z) \sim -\frac{1}{\pi}\Gamma(\nu)\left(\frac{2}{z}\right)^\nu \tag{18.6}
$$

The same relations hold as $\mathrm{Re}(\nu) \to \infty$, with z fixed.

18.1.5   Integral representations

Let $\mathrm{Re}(z) > 0$ and ν be any complex number

$$
\begin{aligned}
J_\nu(z) &= \frac{1}{\pi}\int_0^\pi \cos(\nu\theta - z\sin\theta)\,d\theta - \frac{\sin\nu\pi}{\pi}\int_0^\infty e^{-\nu t - z\sinh t}\,dt, \\
Y_\nu(z) &= \frac{1}{\pi}\int_0^\pi \sin(z\sin\theta - \nu\theta)\,d\theta - \int_0^\infty \left(e^{\nu t} + e^{-\nu t}\cos\nu\pi\right)e^{-z\sinh t}\,dt.
\end{aligned} \tag{18.7}
$$

When $\nu = n$ (integer) the second integral in the first relation is zero.

18.1.6 Fourier expansion

For any complex z

$$e^{-iz \sin t} = \sum_{n=-\infty}^{\infty} e^{-int} J_n(z) \tag{18.8}$$

with Parseval relation

$$\sum_{n=-\infty}^{\infty} J_n^2(z) = 1. \tag{18.9}$$

18.1.7 Asymptotic expansions

For large positive values of x:

$$J_\nu(x) = \sqrt{\frac{2}{\pi x}} \left[\cos\left(x - \frac{1}{2}\nu\pi - \frac{1}{4}\pi\right) + \mathcal{O}(x^{-1}) \right],$$

$$Y_\nu(x) = \sqrt{\frac{2}{\pi x}} \left[\sin\left(x - \frac{1}{2}\nu\pi - \frac{1}{4}\pi\right) + \mathcal{O}(x^{-1}) \right]. \tag{18.10}$$

18.1.8 Half-order Bessel functions

For integer values of n let

$$j_n(z) = \sqrt{\pi/(2z)}\, J_{n+\frac{1}{2}}(z), \qquad y_n(z) = \sqrt{\pi/(2z)}\, Y_{n+\frac{1}{2}}(z). \tag{18.11}$$

Then

$$j_0(z) = y_{-1}(z) = \frac{\sin z}{z}, \qquad y_0(z) = -j_{-1}(z) = -\frac{\cos z}{z}, \tag{18.12}$$

and for $n = 0, 1, 2, \ldots$

$$j_n(z) = (-z)^n \left(\frac{1}{z}\frac{d}{dz}\right)^n \frac{\sin z}{z}, \qquad y_n(z) = -(-z)^n \left(\frac{1}{z}\frac{d}{dz}\right)^n \frac{\cos z}{z}. \tag{18.13}$$

The functions $j_n(z)$ and $y_n(z)$ both satisfy

$$z[f_{n-1}(z) + f_{n+1}(z)] = (2n+1)f_n(z)$$

$$nf_{n-1}(z) - (n+1)f_{n+1}(z) = (2n+1)f_n'(z) \tag{18.14}$$

18.1.9 Modified Bessel functions

The differential equation for the modified Bessel functions is

$$z^2 y'' + z y' - (z^2 + \nu^2)y = 0 \tag{18.15}$$

This differential equation has solutions $I_\nu(z)$ and $K_\nu(z)$ defined by

$$I_\nu(z) = \left(\frac{z}{2}\right)^\nu \sum_{n=0}^{\infty} \frac{(z/2)^{2n}}{\Gamma(n+\nu+1)\, n!}, \qquad K_\nu(z) = \frac{\pi}{2} \frac{I_{-\nu}(z) - I_\nu(z)}{\sin \nu\pi}, \tag{18.16}$$

x	$e^{-x}I_0(x)$	$e^{-x}I_1(x)$	$e^x K_0(x)$	$e^x K_1(x)$
0.0	1.00000000	0.00000000	∞	∞
0.2	0.82693855	0.08228312	2.14075732	5.83338603
0.4	0.69740217	0.13676322	1.66268209	3.25867388
0.6	0.59932720	0.17216442	1.41673762	2.37392004
0.8	0.52414894	0.19449869	1.25820312	1.91793030
1.0	0.46575961	0.20791042	1.14446308	1.63615349
1.2	0.41978208	0.21525686	1.05748453	1.44289755
1.4	0.38306252	0.21850759	0.98807000	1.30105374
1.6	0.35331500	0.21901949	0.93094598	1.19186757
1.8	0.32887195	0.21772628	0.88283353	1.10480537
2.0	0.30850832	0.21526929	0.84156822	1.03347685
2.2	0.29131733	0.21208773	0.80565398	0.97377017
2.4	0.27662232	0.20848109	0.77401814	0.92291367
2.6	0.26391400	0.20465225	0.74586824	0.87896728
2.8	0.25280553	0.20073741	0.72060413	0.84053006
3.0	0.24300035	0.19682671	0.69776160	0.80656348
3.2	0.23426883	0.19297862	0.67697511	0.77628028
3.4	0.22643140	0.18922985	0.65795227	0.74907206
3.6	0.21934622	0.18560225	0.64045596	0.72446066
3.8	0.21290013	0.18210758	0.62429158	0.70206469
4.0	0.20700192	0.17875084	0.60929767	0.68157595
4.2	0.20157738	0.17553253	0.59533899	0.66274241
4.4	0.19656556	0.17245023	0.58230127	0.64535587
4.6	0.19191592	0.16949973	0.57008720	0.62924264
4.8	0.18758620	0.16667571	0.55861332	0.61425660
5.0	0.18354081	0.16397227	0.54780756	0.60027386

Table 18.2: Values of the Bessel functions I_0, I_1, K_0, and K_1.

where the right-hand side should be determined by a limiting process when ν assumes integer values. When $n = 0, 1, 2, \ldots$

$$K_n(z) = (-1)^{n+1} I_n(z) \ln \frac{z}{2} + \frac{1}{2} \left(\frac{2}{z}\right)^n \sum_{k=0}^{n-1} \frac{(n-k-1)!}{k!} \left(-\frac{z^2}{4}\right)^k$$

$$+ \frac{(-1)^n}{2} \left(\frac{z}{2}\right)^n \sum_{k=0}^{\infty} [\psi(k+1) + \psi(n+k+1)] \frac{(z/2)^{2k}}{k! \, (n+k)!}. \tag{18.17}$$

18.1.9.1 Relation to ordinary Bessel functions

$$I_\nu(z) = \begin{cases} e^{-\frac{1}{2}\nu\pi i} J_\nu\left(z e^{\frac{1}{2}\pi i}\right), & \text{for } -\pi < \arg z \le \frac{1}{2}\pi \\ e^{\frac{3}{2}\nu\pi i} J_\nu\left(z e^{-\frac{3}{2}\pi i}\right), & \text{for } \frac{1}{2}\pi < \arg z \le \pi \end{cases} \tag{18.18}$$

For $n = 0, 1, 2, \ldots$

$$I_n(z) = i^{-n} J_n(iz), \qquad I_{-n}(z) = I_n(z) \tag{18.19}$$

For any ν

$$K_{-\nu}(z) = K_\nu(z) \tag{18.20}$$

18.1.9.2   Recursion relations

$$I_{\nu-1}(z) - I_{\nu+1}(z) = \frac{2\nu}{z} I_\nu(z), \quad K_{\nu+1}(z) - K_{\nu-1}(z) = \frac{2\nu}{z} K_\nu(z)$$

$$I_{\nu-1}(z) + I_{\nu+1}(z) = 2I'_\nu(z), \quad K_{\nu-1}(z) + K_{\nu+1}(z) = -2K'_\nu(z) \tag{18.21}$$

18.1.9.3   Integrals

$$I_\nu(z) = \frac{1}{\pi} \int_0^\pi e^{z\cos\theta} \cos(\nu\theta)\, d\theta - \frac{\sin\nu\pi}{\pi} \int_0^\infty e^{-\nu t - z\cosh t}\, dt$$

$$K_\nu(z) = \int_0^\infty e^{-z\cosh t} \cosh(\nu t)\, dt \tag{18.22}$$

When $\nu = n$ (integer) the second integral in the first relation disappears.

18.2   BETA FUNCTION

The beta function is defined by

$$B(p, q) = \int_0^1 t^{p-1}(1-t)^{q-1}\, dt, \quad \mathrm{Re}(p) > 0, \quad \mathrm{Re}(q) > 0 \tag{18.23}$$

The beta function is related to the gamma function as follows:

$$B(p, q) = \frac{\Gamma(p)\,\Gamma(q)}{\Gamma(p+q)} \tag{18.24}$$

Values of the beta function are in Table 18.3.

18.2.1   Other integrals

Other integral representations for the beta function include (in all cases $\mathrm{Re}(p) > 0, \mathrm{Re}(q) > 0$):

(a) $B(p, q) = 2 \int_0^{\pi/2} \sin^{2p-1}\theta \, \cos^{2q-1}\theta \, d\theta$

(b) $B(p, q) = \int_0^\infty \frac{t^{p-1}}{(t+1)^{p+q}}\, dt$

(c) $B(p, q) = \int_0^\infty e^{-pt}\,(1 - e^{-t})^{q-1}\, dt$

(d) $B(p, q) = r^q(r + 1)^p \int_0^1 \frac{t^{p-1}(1-t)^{q-1}}{(r+t)^{p+q}}\, dt, \quad r > 0$

18.2.2   Properties

The beta function has the properties:

(a) $B(p, q) = B(q, p)$

(b) $B(p, q + 1) = \frac{q}{p} B(p + 1, q) = \frac{q}{p+q} B(p, q)$

p	$q = 0.100$	0.200	0.300	0.400	0.500	0.600	0.700	0.800	0.900	1.000
0.1	19.715	14.599	12.831	11.906	11.323	10.914	10.607	10.365	10.166	10.000
0.2	14.599	9.502	7.748	6.838	6.269	5.872	5.576	5.345	5.157	5.000
0.3	12.831	7.748	6.010	5.112	4.554	4.169	3.883	3.661	3.482	3.333
0.4	11.906	6.838	5.112	4.226	3.679	3.303	3.027	2.813	2.641	2.500
0.5	11.323	6.269	4.554	3.679	3.142	2.775	2.506	2.299	2.135	2.000
0.6	10.914	5.872	4.169	3.303	2.775	2.415	2.154	1.954	1.796	1.667
0.7	10.607	5.576	3.883	3.027	2.506	2.154	1.899	1.705	1.552	1.429
0.8	10.365	5.345	3.661	2.813	2.299	1.954	1.705	1.517	1.369	1.250
0.9	10.166	5.157	3.482	2.641	2.135	1.796	1.552	1.369	1.226	1.111
1.0	10.000	5.000	3.333	2.500	2.000	1.667	1.429	1.250	1.111	1.000
1.2	9.733	4.751	3.099	2.279	1.791	1.468	1.239	1.069	0.938	0.833
1.4	9.525	4.559	2.921	2.113	1.635	1.321	1.101	0.938	0.813	0.714
1.6	9.355	4.404	2.779	1.982	1.513	1.208	0.994	0.837	0.718	0.625
1.8	9.213	4.276	2.663	1.875	1.415	1.117	0.909	0.758	0.644	0.556
2.0	9.091	4.167	2.564	1.786	1.333	1.042	0.840	0.694	0.585	0.500
2.2	8.984	4.072	2.480	1.710	1.264	0.979	0.783	0.641	0.536	0.455
2.4	8.890	3.989	2.406	1.644	1.205	0.925	0.734	0.597	0.495	0.417
2.6	8.805	3.915	2.340	1.586	1.153	0.878	0.692	0.558	0.460	0.385
2.8	8.728	3.848	2.282	1.534	1.107	0.837	0.655	0.525	0.430	0.357
3.0	8.658	3.788	2.230	1.488	1.067	0.801	0.622	0.496	0.403	0.333

Table 18.3: Values of the beta function $B(p,q)$.

(c) $B(p,q)\,B(p+q,r) = \frac{\Gamma(p)\,\Gamma(q)\,\Gamma(r)}{\Gamma(p+q+r)}$

18.3   CEILING AND FLOOR FUNCTIONS

The ceiling function of x, denoted $\lceil x \rceil$, is the least integer that is not smaller than x. For example, $\lceil \pi \rceil = 4$, $\lceil 5 \rceil = 5$, and $\lceil -1.5 \rceil = -1$.

The floor function of x, denoted $\lfloor x \rfloor$, is the largest integer that is not larger than x. For example, $\lfloor \pi \rfloor = 3$, $\lfloor 5 \rfloor = 5$, and $\lfloor -1.5 \rfloor = -2$.

18.4   DELTA FUNCTION

The delta function of x, denoted $\delta(x)$, is defined by $\delta(x) = 0$ when $x \neq 0$, and $\int_{-\infty}^{\infty} \delta(x)\,dx = 1$.

18.5   ERROR FUNCTIONS

The error function, erf(x), and the complementary error function, erfc(x), are defined by:

$$\text{erf}\,x = \frac{2}{\sqrt{\pi}} \int_0^x e^{-t^2}\,dt \qquad \text{erfc}\,x = \frac{2}{\sqrt{\pi}} \int_x^{\infty} e^{-t^2}\,dt \qquad (18.25)$$

These functions have the properties:

(a) erf x + erfc $x = 1$,

(b) $\operatorname{erf}(-x) = -\operatorname{erf} x$,

(c) $\operatorname{erfc}(-x) = 2 - \operatorname{erfc} x$

The error function is related to the normal probability distribution function as follows:

$$\Phi(x) = \frac{1}{2}\left[1 + \operatorname{erf}\left(\frac{x}{\sqrt{2}}\right)\right] \tag{18.26}$$

18.5.1 Expansions

The error function has the series expansion

$$\operatorname{erf}(x) = \frac{2}{\sqrt{\pi}} \sum_{n=0}^{\infty} \frac{(-1)^n\, x^{2n+1}}{(2n+1)\, n!} = \frac{2}{\sqrt{\pi}}\left(x - \frac{x^3}{3} + \frac{1}{2!}\frac{x^5}{5} - \frac{1}{3!}\frac{x^7}{7} + \cdots\right)$$

$$= \frac{2}{\sqrt{\pi}} \sum_{n=0}^{\infty} \frac{\Gamma(\frac{3}{2})\, e^{-x^2}}{\Gamma(n + \frac{3}{2})}\, x^{2n+1} = \frac{2}{\sqrt{\pi}} e^{-x^2}\left(x + \frac{2}{3}x^3 + \frac{4}{15}x^5 \cdots\right) \tag{18.27}$$

and the asymptotic expansion (for $z \to \infty$, with $|\arg z| < \frac{3}{4}\pi$):

$$\operatorname{erfc}(z) \sim \frac{2}{\sqrt{\pi}} \frac{e^{-z^2}}{2z} \sum_{n=0}^{\infty} \frac{(-1)^n\, (2n)!}{n!(2z)^{2n}}$$

$$\sim \frac{2}{\sqrt{\pi}} \frac{e^{-z^2}}{2z}\left(1 - \frac{1}{z^2} + \frac{6}{z^4} - \frac{15}{8\, z^6} + \cdots\right) \tag{18.28}$$

18.5.2 Special values

Special values of the error function include:

(a) $\operatorname{erf}(\pm\infty) = \pm 1$ (d) $\operatorname{erf}(0) = 0$

(b) $\operatorname{erfc}(-\infty) = 2$ (e) $\operatorname{erfc}(0) = 1$

(c) $\operatorname{erfc}(\infty) = 0$

Note that $\operatorname{erf}(x_0) = \operatorname{erfc}(x_0) = \frac{1}{2}$ for $x_0 = 0.476936\ldots$

18.6 EXPONENTIAL FUNCTION

18.6.1 Exponentiation

For a any real number and m a positive integer, the exponential a^m is defined as

$$a^m = \underbrace{a \cdot a \cdot a \cdots a}_{m\ \text{terms}} \tag{18.29}$$

The three laws of exponents are:

1. $a^n \cdot a^m = a^{m+n}$

2. $\dfrac{a^m}{a^n} = \begin{cases} a^{m-n} & \text{if } m > n, \\ 1 & \text{if } m = n, \\ \dfrac{1}{a^{n-m}} & \text{if } m < n. \end{cases}$

3. $(a^m)^n = a^{(mn)}$

The n-th root function is defined as the inverse of the n-th power function. That is

$$\text{if } b^n = a, \text{ then } b = \sqrt[n]{a} = a^{(1/n)}. \tag{18.30}$$

If n is odd, there will be a unique real number satisfying the above definition of $\sqrt[n]{a}$, for any real value of a. If n is even, for positive values of a there will be two real values for $\sqrt[n]{a}$, one positive and one negative. By convention, the symbol $\sqrt[n]{a}$ is understood to mean the positive value. If n is even and a is negative, then there are no real values for $\sqrt[n]{a}$.

To extend the definition to include a^t (for t not necessarily an integer), in such a way as to maintain the laws of exponents, the following definitions are required (where we now restrict a to be positive, p to be an odd number, and q to be an even number):

$$a^0 = 1 \qquad a^{p/q} = \sqrt[q]{a^p} \qquad a^{-t} = \frac{1}{a^t} \tag{18.31}$$

With these restrictions, the second law of exponents can be written as: $\dfrac{a^m}{a^n} = a^{m-n}$.

If $a > 1$ then the function a^x is monotone increasing while, if $0 < a < 1$ then the function a^x is monotone decreasing.

18.6.2   Definition of e^z

$$\exp(z) = e^z = \lim_{m \to \infty} \left(1 + \frac{z}{m}\right)^m = 1 + z + \frac{z^2}{2!} + \frac{z^3}{3!} + \frac{z^4}{4!} + \dots \tag{18.32}$$

18.6.3   Derivative and integral of e^z

The derivative of e^z is $\dfrac{de^x}{dx} = e^z$. The integral of e^z is $\displaystyle\int e^x \, dx = e^z + C$.

18.6.4   Circular functions and exponentials

$$\cos z = \frac{e^{iz} + e^{-iz}}{2}, \qquad e^{iz} = \cos z + i \sin z \tag{18.33}$$

$$\sin z = \frac{e^{iz} - e^{-iz}}{2i} \qquad e^{-iz} = \cos z - i \sin z \tag{18.34}$$

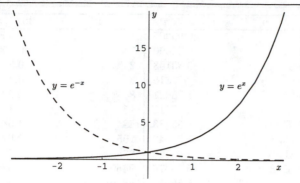

Figure 18.2: Graphs of e^x and e^{-x}

If $z = x + iy$, then

$$e^z = e^x e^{iy} = e^x(\cos y + i \sin y) \tag{18.35}$$

18.6.5 Hyperbolic functions

$$\cosh z = \frac{e^z + e^{-z}}{2}, \qquad e^z = \cosh z + \sinh z \tag{18.36}$$

$$\sinh z = \frac{e^z - e^{-z}}{2} \qquad e^{-z} = \cosh z - \sinh z \tag{18.37}$$

18.7 FACTORIALS AND POCHHAMMER'S SYMBOL

The factorial of n, denoted $n!$, is the product of all integers less than or equal to n: $n! = n \cdot (n-1) \cdot (n-2) \cdots 2 \cdot 1$. The double factorial of n, denoted $n!!$, is the product of every other integer: $n!! = n \cdot (n-2) \cdot (n-4) \cdots$, where the last element in the product is either 2 or 1, depending on whether n is even or odd. The generalization of the factorial function is the gamma function (see section 18.8). When n is an integer, $\Gamma(n) = (n-1)!$. A table of values is in Table 18.4.

The shifted factorial (also called the rising factorial or Pochhammer's symbol) is denoted by $(n)_k$ (sometimes $n^{\underline{k}}$), and is defined as

$$(n)_k = \underbrace{n \cdot (n+1) \cdot (n+2) \cdots}_{k \text{ terms}} = \frac{(n+k-1)!}{(n-1)!} = \frac{\Gamma(n+k)}{\Gamma(n)} \tag{18.38}$$

18.8 GAMMA FUNCTION

The gamma function is defined by

$$\Gamma(z) = \int_0^\infty t^{z-1} e^{-t} \, dt, \quad z = x + iy, \quad x > 0 \tag{18.39}$$

n	$n!$	$\log_{10} n!$	$n!!$	$\log_{10} n!!$
0	1	0.00000	1	0.00000
1	1	0.00000	1	0.00000
2	2	0.30103	2	0.30103
3	6	0.77815	3	0.47712
4	24	1.38021	8	0.90309
5	120	2.07918	15	1.17609
6	720	2.85733	48	1.68124
7	5040	3.70243	105	2.02119
8	40320	4.60552	384	2.58433
9	3.6288×10^5	5.55976	945	2.97543
10	3.6288×10^6	6.55976	3840	3.58433
11	3.9917×10^7	7.60116	10395	4.01682
12	4.7900×10^8	8.68034	46080	4.66351
13	6.2270×10^9	9.79428	1.3514×10^5	5.13077
14	8.7178×10^{10}	10.94041	6.4512×10^5	5.80964
15	1.3077×10^{12}	12.11650	2.0270×10^6	6.30686
16	2.0923×10^{13}	13.32062	1.0322×10^7	7.01376
17	3.5569×10^{14}	14.55107	3.4459×10^7	7.53731
18	6.4024×10^{15}	15.80634	1.8579×10^8	8.26903
19	1.2165×10^{17}	17.08509	6.5473×10^8	8.81606
20	2.4329×10^{18}	18.38612	3.7159×10^9	9.57006
25	1.5511×10^{25}	25.19065	7.9059×10^{12}	12.89795
50	3.0414×10^{64}	64.48307	5.2047×10^{32}	32.71640
100	9.3326×10^{157}	157.97000	3.4243×10^{79}	79.53457
150	5.7134×10^{262}	262.75689	9.3726×10^{131}	131.97186
500	1.2201×10^{1134}	1134.0864	5.8490×10^{567}	567.76709
1000	4.0239×10^{2567}	2567.6046	3.9940×10^{1284}	1284.6014

Table 18.4: Values of the factorial function ($n!$) and the double factorial function ($n!!$).

The gamma function can also be defined by products:

(a) $\displaystyle \Gamma(z) = \lim_{n\to\infty} \frac{n!\, n^z}{z(z+1)\cdots(z+n)}$

(b) $\displaystyle \frac{1}{\Gamma(z)} = z\, e^{\gamma z} \prod_{n=1}^{\infty} \left[\left(1 + \frac{z}{n}\right) e^{-z/n}\right]$, where γ is Euler's constant

18.8.1   Other integrals for the gamma function

Other integral representations for the gamma function include:

(a) $\displaystyle \Gamma(z) \cos\left(\frac{1}{2}\pi z\right) = \int_0^{\infty} t^{z-1}\,\cos t\, dt$ for $0 < \mathrm{Re}(z) < 1$

(b) $\displaystyle \Gamma(z) \sin\left(\frac{1}{2}\pi z\right) = \int_0^{\infty} t^{z-1}\,\sin t\, dt$ for $-1 < \mathrm{Re}(z) < 1$

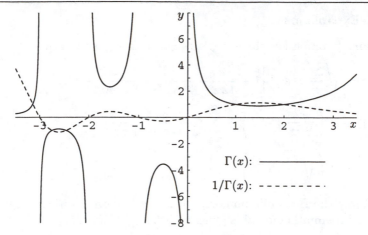

Figure 18.3: Graphs of $\Gamma(x)$ and $1/\Gamma(x)$, for x real.

18.8.2 Properties

$$\Gamma'(1) = \int_0^\infty \ln t\, e^{-t}\, dt = -\gamma \tag{18.40}$$

Multiplication formula:

$$\Gamma(2z) = \pi^{-\frac{1}{2}}\, 2^{2z-1}\, \Gamma(z)\, \Gamma\left(z + \frac{1}{2}\right) \tag{18.41}$$

Reflection formulas:

$$\Gamma(z)\, \Gamma(1-z) = \frac{\pi}{\sin \pi z},$$
$$\Gamma\left(\frac{1}{2} + z\right)\Gamma\left(\frac{1}{2} - z\right) = \frac{\pi}{\cos \pi z}, \tag{18.42}$$
$$\Gamma(z - n) = (-1)^n \Gamma(z)\frac{\Gamma(1-z)}{\Gamma(n+1-z)} = \frac{(-1)^n\, \pi}{\sin \pi z\, \Gamma(n+1-z)}$$

The gamma function has the recursion formula:

$$\Gamma(z + 1) = z\, \Gamma(z) \tag{18.43}$$

The relation $\Gamma(z) = \Gamma(z+1)/z$ can be used to define the gamma function in the left half plane, $z \neq 0, -1, -2, \ldots$

The gamma function has simple poles at $z = -n$, (for $n = 0, 1, 2, \ldots$), with the respective residues $(-1)^n/n!$. That is,

$$\lim_{z \to -n} (z + n)\Gamma(z) = \frac{(-1)^n}{n!} \tag{18.44}$$

18.8.3 Expansions

The gamma function has the series expansion, for $z \to \infty$, $|\arg z| < \pi$:

$$\Gamma(z) \sim \sqrt{\frac{2\pi}{z}}\, z^z\, e^{-z} \left[1 + \frac{1}{12\,z} + \frac{1}{288\,z^2} - \frac{139}{51\,840\,z^3} + \cdots \right]$$

$$\ln \Gamma(z) \sim \ln \left(\sqrt{\frac{2\pi}{z}} z^z e^{-z} \right) + \sum_{n=1}^{\infty} \frac{B_{2n}}{2n\,(2n-1)} \frac{1}{z^{2n-1}}, \tag{18.45}$$

$$\sim \ln \left(\sqrt{\frac{2\pi}{z}} z^z e^{-z} \right) + \frac{1}{12z} - \frac{1}{360z^3} + \frac{1}{1\,260z^5} - \frac{1}{1\,680z^7} + \cdots,$$

where B_n are the Bernoulli numbers. If $z = n$, a large positive integer, then a useful approximation for $n!$ is given by Stirling's formula:

$$n! \sim \sqrt{2\pi n}\, n^n\, e^{-n}, \quad n \to \infty \tag{18.46}$$

18.8.4 Special values

$$\Gamma(n+1) = n! \quad \text{if } n = 0, 1, 2, \ldots, \text{where } 0! = 1,$$

$$\Gamma(1) = 1, \quad \Gamma(2) = 1, \quad \Gamma(3) = 2, \quad \Gamma\left(\frac{1}{2}\right) = \sqrt{\pi},$$

$$\Gamma\left(m + \frac{1}{2}\right) = \frac{1 \cdot 3 \cdot 5 \cdots (2m-1)}{2^m} \sqrt{\pi}, \quad m = 1, 2, 3, \ldots,$$

$$\Gamma\left(-m + \frac{1}{2}\right) = \frac{(-1)^m 2^m}{1 \cdot 3 \cdot 5 \cdots (2m-1)} \sqrt{\pi}, \quad m = 1, 2, 3, \ldots$$

$$\begin{aligned}
\Gamma(\tfrac{1}{4}) &= 3.62560\,99082 & \Gamma(\tfrac{1}{3}) &= 2.67893\,85347 \\
\Gamma(\tfrac{1}{2}) &= \sqrt{\pi} = 1.77245\,38509 & \Gamma(\tfrac{2}{3}) &= 1.35411\,79394 \\
\Gamma(\tfrac{3}{4}) &= 1.22541\,67024 & \Gamma(\tfrac{3}{2}) &= \sqrt{\pi}/2 = 0.88622\,69254
\end{aligned}$$

See table 18.5.

18.8.5 Digamma function

The digamma function $\varphi(x)$ is defined by

$$\varphi(x) = \frac{d \ln \Gamma(z)}{dz} = \frac{\Gamma'(z)}{\Gamma(z)} \tag{18.47}$$

n	$\Gamma(n)$	n	$\Gamma(n)$	n	$\Gamma(n)$	n	$\Gamma(n)$
1.00	1.0000	1.25	.9064	1.50	.8862	1.75	.9191
1.01	.9943	1.26	.9044	1.51	.8866	1.76	.9214
1.02	.9888	1.27	.9025	1.52	.8870	1.77	.9238
1.03	.9835	1.28	.9007	1.53	.8876	1.78	.9262
1.04	.9784	1.29	.8990	1.54	.8882	1.79	.9288
1.05	.9735	1.30	.8975	1.55	.8889	1.80	.9314
1.06	.9687	1.31	.8960	1.56	.8896	1.81	.9341
1.07	.9642	1.32	.8946	1.57	.8905	1.82	.9368
1.08	.9597	1.33	.8934	1.58	.8914	1.83	.9397
1.09	.9555	1.34	.8922	1.59	.8924	1.84	.9426
1.10	.9514	1.35	.8912	1.60	.8935	1.85	.9456
1.11	.9474	1.36	.8902	1.61	.8947	1.86	.9487
1.12	.9436	1.37	.8893	1.62	.8959	1.87	.9518
1.13	.9399	1.38	.8885	1.63	.8972	1.88	.9551
1.14	.9364	1.39	.8879	1.64	.8986	1.89	.9584
1.15	.9330	1.40	.8873	1.65	.9001	1.90	.9618
1.16	.9298	1.41	.8868	1.66	.9017	1.91	.9652
1.17	.9267	1.42	.8864	1.67	.9033	1.92	.9688
1.18	.9237	1.43	.8860	1.68	.9050	1.93	.9724
1.19	.9209	1.44	.8858	1.69	.9068	1.94	.9761
1.20	.9182	1.45	.8857	1.70	.9086	1.95	.9799
1.21	.9156	1.46	.8856	1.71	.9106	1.96	.9837
1.22	.9131	1.47	.8856	1.72	.9126	1.97	.9877
1.23	.9108	1.48	.8857	1.73	.9147	1.98	.9917
1.24	.9085	1.49	.8859	1.74	.9168	1.99	.9958
1.25	.9064	1.50	.8862	1.75	.9191	2.00	1.0000

Table 18.5: Values of the gamma function.

18.8.6 Incomplete gamma functions

The incomplete gamma functions are defined by

$$
\gamma(a, x) = \int_0^x e^{-t} t^{a-1}\, dt
$$
$$
\Gamma(a, x) = \Gamma(a) - \gamma(a, x) = \int_x^\infty e^{-t} t^{a-1}\, dt
$$

(18.48)

18.9   HYPERGEOMETRIC FUNCTIONS

Recall the geometric series and the binomial expansion ($|z| < 1$):

$$(1-z)^{-1} = \sum_{n=0}^{\infty} z^n, \qquad (1-z)^{-a} = \sum_{n=0}^{\infty} \binom{-a}{n}(-z)^n = \sum_{n=0}^{\infty}(a)_n \frac{z^n}{n!} \quad (18.49)$$

where Pochhammer's symbol, $(a)_n$, is defined in section 18.7.

18.9.1   Generalized hypergeometric function

The generalized hypergeometric function is defined by:

$$
\begin{aligned}
{}_pF_q\left(a_1, a_2, \ldots, a_p;\ b_1, b_2, \ldots, b_q;\ x\right) &= {}_pF_q\left[\begin{array}{c} a_1, \ldots, a_p;\ x \\ b_1, \ldots, b_q \end{array}\right] \\
&= \sum_{k=0}^{\infty} \frac{\prod_{i=1}^{p}(a_i)_k}{\prod_{i=1}^{q}(b_i)_k} \frac{x^k}{k!} \qquad (18.50) \\
&= \sum_{k=0}^{\infty} \frac{(a_1)_k \cdots (a_p)_k}{(b_1)_k \cdots (b_q)_k} \frac{x^k}{k!}
\end{aligned}
$$

where $(n)_k$ represents Pochhammer's symbol. Usually, ${}_2F_1(a, b; c; x)$ is called "the" hypergeometric function; this is also called the Gauss hypergeometric function.

18.9.2   Gauss hypergeometric function

The Gauss hypergeometric function is

$$F(a, b; c; x) = {}_2F_1(a, b; c; x) = \sum_{n=0}^{\infty} \frac{(a)_n (b)_n}{(c)_n} \frac{x^n}{n!} \qquad (18.51)$$

18.9.2.1   *Special cases*

$$
\begin{aligned}
F(a, b; b; x) &= (1-x)^{-a}, \\
F(1, 1; 2; x) &= -\frac{\ln(1-x)}{x}, \\
F(\tfrac{1}{2}, 1; \tfrac{3}{2}; x^2) &= \frac{1}{2x} \ln\left(\frac{1+x}{1-x}\right), \\
F(\tfrac{1}{2}, 1; \tfrac{3}{2}; -x^2) &= \frac{\arctan x}{x}, \qquad\qquad (18.52) \\
F(\tfrac{1}{2}, \tfrac{1}{2}; \tfrac{3}{2}; x^2) &= \frac{\arcsin x}{x}, \\
F(\tfrac{1}{2}, \tfrac{1}{2}; \tfrac{3}{2}; -x^2) &= \frac{\ln(x + \sqrt{1+x^2})}{x}.
\end{aligned}
$$

18.9.2.2 *Functional relations*

$$F(a, b; c; x) = (1-x)^{-a} F\left(a, c-b; c; \frac{x}{x-1}\right)$$

$$= (1-x)^{-b} F\left(c-a, b; c; \frac{x}{x-1}\right) \qquad (18.53)$$

$$= (1-x)^{c-a-b} F(c-a, c-b; c; x).$$

18.9.3 Confluent hypergeometric functions

The confluent hypergeometric functions, M and U, are defined by

$$M(a, c, z) = \lim_{z \to \infty} F\left(a, b; c; \frac{z}{b}\right)$$

$$U(a, c, z) = \frac{\Gamma(1-c)}{\Gamma(a-c+1)} M(a, c, z) + \frac{\Gamma(c-1)}{\Gamma(a)} z^{1-c} M(a-c, 2-c, z) \qquad (18.54)$$

Sometimes the notation ψ is used for M.

Sometimes $U(a, b, z)$ is called the Tricomi function; it is the unique solution to $zw'' + (b-z)w' - aw = 0$ with $U(a, b, 0) = \Gamma(1-b)/\Gamma(1+a-b)$ and $U(a, b, \infty) = 0$.

18.10   LOGARITHMIC FUNCTIONS

18.10.1   Definition of the natural log

The natural logarithm (also known as the Naperian logarithm) of z is written as $\ln z$ or as $\log_e z$. It is sometimes written $\log z$ (this is also used to represent a "generic" logarithm, a logarithm to any base). One definition is

$$\ln z = \int_1^z \frac{dt}{t}, \qquad (18.55)$$

where the integration path from 1 to z does not cross the origin or the negative real axis.

For complex values of z the natural logarithm, as defined above, can be represented in terms of it's magnitude and phase. If $z = x + iy = re^{i\theta}$, then $\ln z = \ln r + i\theta$, where $r = \sqrt{x^2 + y^2}$, $x = r\cos\theta$, and $y = r\sin\theta$.

18.10.2   Special values

$$\lim_{\epsilon \to 0} (\ln \epsilon) = -\infty \qquad\qquad \ln 1 = 0 \qquad\qquad \ln e = 1$$

$$\ln(-1) = i\pi + 2\pi i k \qquad\qquad \ln(\pm i) = \pm\frac{i\pi}{2} + 2\pi i k$$

18.10.3 Logarithms to a base other than e

The logarithmic function to the base a, written $\log_a$, is defined as

$$\log_a z = \frac{\log_b z}{\log_b a} = \frac{\ln z}{\ln a} \tag{18.56}$$

Note the properties:

(a) $\log_a a^p = p$

(b) $\log_a b = \dfrac{1}{\log_b a}$

(c) $\log_{10} z = \dfrac{\ln z}{\ln 10} = (\log_{10} e) \ln z \approx (0.4342944819\ldots) \ln z$

(d) $\ln z = (\ln 10) \log_{10} z \approx (2.3025850929\ldots) \log_{10} z$

18.10.4 Relation of the logarithm to the exponential

For real values of z the *logarithm* is a monotonic function, as is the exponential. Any monotonic function has a single–valued inverse function; the natural logarithm is the inverse of the exponential. That is, if $x = e^y$, then $y = \ln x$ and $x = e^{\ln x}$. The same inverse relations exist for bases other than e. For example, if $u = a^w$, then $w = \log_a u$ and $u = a^{\log_a u}$.

18.10.5 Identities

$$\log_a z_1 z_2 = \log_a z_1 + \log_a z_2 \qquad \text{for } (-\pi < \arg z_1 + \arg z_2 < \pi)$$

$$\log_a \frac{z_1}{z_2} = \log_a z_1 - \log_a z_2 \qquad \text{for } (-\pi < \arg z_1 - \arg z_2 < \pi)$$

$$\log_a z^n = n \log_a z \qquad\qquad \text{for } (-\pi < n \arg z < \pi), \text{ when } n \text{ is an integer}$$

18.10.6 Series expansions for the natural logarithm

$$\ln (1 + z) = z - \frac{1}{2}z^2 + \frac{1}{3}z^3 - \cdots, \qquad\qquad\qquad \text{for } |z| < 1$$

$$\ln z = \left(\frac{z-1}{z}\right) + \frac{1}{2}\left(\frac{z-1}{z}\right)^2 + \frac{1}{3}\left(\frac{z-1}{z}\right)^3 + \cdots \quad \text{for } \mathrm{Re}(z) \geq \frac{1}{2}$$

18.10.7 Derivative and integration formulae

$$\frac{d \ln z}{dz} = \frac{1}{z} \qquad \int \frac{dz}{z} = \ln |z| + C \qquad \int \ln z \, dz = z \ln |z| - z + C \tag{18.57}$$

n	1	2	3	4	5	6	7	8	9	10
$p(n)$	1	2	3	5	7	11	15	22	30	42

n	11	12	13	14	15	16	17	18	19	20
$p(n)$	56	77	101	135	176	231	297	385	490	627

n	21	22	23	24	25	26	27	28	29	30
$p(n)$	792	1002	1255	1575	1958	2436	3010	3718	4565	5604

n	31	32	33	34	35	40	45	50
$p(n)$	6842	8349	10143	12310	14883	37338	89134	204226

Table 18.6: Values of the partition function.

18.11 PARTITIONS

A partition of a number n is a representation of n as the sum of any number of positive integral parts (for example: $5 = 4 + 1 = 3 + 2 = 3 + 1 + 1 = 2 + 2 + 1 = 2 + 1 + 1 + 1 = 1 + 1 + 1 + 1 + 1$). The number of partitions of n is $p(n)$ (for example, $p(5) = 7$). The number of partitions of n into at most m parts is equal to the number of partitions of n into parts which do not exceed m; this is denoted $p_m(n)$ (for example, $p_3(5) = 5$ and $p_2(5) = 3$).

The generating functions for $p(n)$ and $p_m(n)$ are

$$1 + \sum_{n=1}^{\infty} p(n)x^n = \frac{1}{(1-x)(1-x^2)(1-x^3)\cdots} \qquad (18.58)$$

$$1 + \sum_{n=1}^{\infty} \sum_{m=1}^{n} p_m(n)x^n t^m = \frac{1}{(1-tx)(1-tx^2)(1-tx^3)\cdots} \qquad (18.59)$$

18.12 SIGNUM FUNCTION

The signum function indicates whether the argument is greater than or less than zero:

$$\text{sgn}(x) = \begin{cases} 1 & x > 0 \\ -1 & x < 0 \end{cases} \qquad (18.60)$$

18.13 STIRLING NUMBERS

There are two types of Stirling numbers.

18.13.1 Stirling numbers

The number $(-1)^{n-m} \begin{bmatrix} n \\ m \end{bmatrix}$ is the number of permutations of n symbols which have exactly m cycles. The term $\begin{bmatrix} n \\ m \end{bmatrix}$ is called a Stirling number. It can be

n	$m = 1$	2	3	4	5	6	7
1	1						
2	-1	1					
3	2	-3	1				
4	-6	11	-6	1			
5	24	-50	35	-10	1		
6	-120	274	-225	85	-15	1	
7	720	-1764	1624	-735	175	-21	1
8	-5040	13068	-13132	6769	-1960	322	-28
9	40320	-109584	118124	-67284	22449	-4536	546
10	-362880	1026576	-1172700	723680	-269325	63273	-9450

Table 18.7: Table of Stirling numbers $\left[\begin{smallmatrix} n \\ m \end{smallmatrix}\right]$.

numerically evaluated as (see Table 18.7):

$$\begin{bmatrix} n \\ m \end{bmatrix} = \sum_{k=0}^{n-m} (-1)^k \binom{n-1+k}{n-m+k}\binom{2n-m}{n-m-k}\begin{Bmatrix} n-m-k \\ k \end{Bmatrix} \qquad (18.61)$$

where $\left\{\begin{smallmatrix} n-m-k \\ k \end{smallmatrix}\right\}$ is a Stirling cycle number (see Table 18.8).

(1) There is the recurrence relation: $\begin{bmatrix} n+1 \\ m \end{bmatrix} = \begin{bmatrix} n \\ m-1 \end{bmatrix} - n\begin{bmatrix} n \\ m \end{bmatrix}$.

(2) The factorial polynomial is defined as $x^{(n)} = x(1-x)\cdots(x-n+1)$ with $x^{(0)} = 1$ by definition. If $n > 0$, then

$$x^{(n)} = \begin{bmatrix} n \\ 1 \end{bmatrix} x + \begin{bmatrix} n \\ 2 \end{bmatrix} x^2 + \cdots + \begin{bmatrix} n \\ n \end{bmatrix} x^n \qquad (18.62)$$

For example: $x^{(3)} = x(x-1)(x-2) = 2x - 3x^2 + x^3 = \begin{bmatrix} 3 \\ 1 \end{bmatrix} x + \begin{bmatrix} 3 \\ 2 \end{bmatrix} x^2 + \begin{bmatrix} 3 \\ 2 \end{bmatrix} x^3$

(3) Stirling numbers satisfy $\sum_{n=m}^{\infty} \begin{bmatrix} n \\ m \end{bmatrix} \frac{x^n}{n!} = \frac{(\log(1+x))^m}{m!}$ for $|x| < 1$.

Example 18.86: For the 4 element set $\{a, b, c, d\}$ there are $\begin{bmatrix} 4 \\ 2 \end{bmatrix} = 11$ permutations containing exactly 2 cycles. They are:

$$\begin{pmatrix} 1\,2\,3\,4 \\ 2\,3\,1\,4 \end{pmatrix} = (123)(4), \qquad \begin{pmatrix} 1\,2\,3\,4 \\ 3\,1\,2\,4 \end{pmatrix} = (132)(4), \qquad \begin{pmatrix} 1\,2\,3\,4 \\ 3\,2\,4\,1 \end{pmatrix} = (134)(2),$$

$$\begin{pmatrix} 1\,2\,3\,4 \\ 4\,2\,1\,3 \end{pmatrix} = (143)(2), \qquad \begin{pmatrix} 1\,2\,3\,4 \\ 2\,4\,3\,1 \end{pmatrix} = (124)(3), \qquad \begin{pmatrix} 1\,2\,3\,4 \\ 4\,1\,3\,2 \end{pmatrix} = (142)(3),$$

$$\begin{pmatrix} 1\,2\,3\,4 \\ 1\,3\,4\,2 \end{pmatrix} = (234)(1), \qquad \begin{pmatrix} 1\,2\,3\,4 \\ 1\,4\,2\,3 \end{pmatrix} = (243)(1), \qquad \begin{pmatrix} 1\,2\,3\,4 \\ 2\,1\,4\,3 \end{pmatrix} = (12)(34),$$

$$\begin{pmatrix} 1\,2\,3\,4 \\ 3\,4\,1\,2 \end{pmatrix} = (13)(24), \qquad \begin{pmatrix} 1\,2\,3\,4 \\ 4\,3\,2\,1 \end{pmatrix} = (14)(23).$$

n	$m = 1$	2	3	4	5	6	7
1	1						
2	1	1					
3	1	3	1				
4	1	7	6	1			
5	1	15	25	10	1		
6	1	31	90	65	15	1	
7	1	63	301	350	140	21	1
8	1	127	966	1701	1050	266	28
9	1	255	3025	7770	6951	2646	462
10	1	511	9330	34105	42525	22827	5880
11	1	1023	28501	145750	246730	179487	63987
12	1	2047	86526	611501	1379400	1323652	627396
13	1	4095	261625	2532530	7508501	9321312	5715424
14	1	8191	788970	10391745	40075035	63436373	49329280
15	1	16383	2375101	42355950	210766920	420693273	408741333

Table 18.8: Table of Stirling cycle numbers $\left\{ {n \atop m} \right\}$.

18.13.2 Stirling cycle numbers

The Stirling cycle number, $\left\{ {n \atop m} \right\}$, is the number of ways to partition n into m blocks. (Equivalently, it is the number of ways that n distinguishable balls can be placed in m indistinguishable cells, with no cell empty.) The Stirling cycle numbers can be numerically evaluated as (see Table 18.8):

$$\left\{ {n \atop m} \right\} = \frac{1}{m!} \sum_{i=0}^{m} (-1)^{m-i} \binom{m}{i} i^n \tag{18.63}$$

Ordinary powers can be expanded in terms of factorial polynomials. If $n > 0$, then

$$x^n = \left\{ {n \atop 1} \right\} x^{(1)} + \left\{ {n \atop 2} \right\} x^{(2)} + \cdots + \left\{ {n \atop n} \right\} x^{(n)} \tag{18.64}$$

For example: $x^3 = \left\{ {3 \atop 1} \right\} x^{(1)} + \left\{ {3 \atop 2} \right\} x^{(2)} + \left\{ {3 \atop 3} \right\} x^{(3)}$

Example 18.87: Placing the 4 distinguishable balls $\{a, b, c, d\}$ into 2 distinguishable cells, so that no cell is empty, can be done in $\left\{ {4 \atop 2} \right\} = 7$ ways. These are (vertical bars delineate the cells):

| $ab \mid cd$ | $ad \mid bc$ | $ac \mid bd$ | $a \mid bcd$ |
| $b \mid acd$ | $c \mid abd$ | $d \mid abc$ |

18.14 SUMS OF POWERS OF INTEGERS

Define $s_k(n) = 1^k + 2^k + \cdots + n^k = \displaystyle\sum_{m=1}^{n} m^k$. Properties include:

(a) $s_k(n) = (k+1)^{-1}[B_{k+1}(n+1) - B_{k+1}(0)]$, where the B_k are Bernoulli polynomials

n	$\sum_{k=1}^{n} k$	$\sum_{k=1}^{n} k^2$	$\sum_{k=1}^{n} k^3$	$\sum_{k=1}^{n} k^4$	$\sum_{k=1}^{n} k^5$
1	1	1	1	1	1
2	3	5	9	17	33
3	6	14	36	98	276
4	10	30	100	354	1300
5	15	55	225	979	4425
6	21	91	441	2275	12201
7	28	140	784	4676	29008
8	36	204	1296	8772	61776
9	45	285	2025	15333	120825
10	55	385	3025	25333	220825
11	66	506	4356	39974	381876
12	78	650	6084	60710	630708
13	91	819	8281	89271	1002001
14	105	1015	11025	127687	1539825
15	120	1240	14400	178312	2299200
16	136	1496	18496	243848	3347776
17	153	1785	23409	327369	4767633
18	171	2109	29241	432345	6657201
19	190	2470	36100	562666	9133300
20	210	2870	44100	722666	12333300

Table 18.9: Sums of powers of integers

(b) Writing $s_k(n)$ as $\sum_{m=1}^{k+1} a_m n^{k-m+2}$ there is the recursion formula:

$$s_{k+1}(n) = \left(\frac{k+1}{k+2}\right) a_1 n^{k+2} + \cdots + \left(\frac{k+1}{k}\right) a_3 n^k$$

$$+ \cdots + \left(\frac{k+1}{2}\right) a_{k+1} n^2 + \left[1 - (k+1) \sum_{m=1}^{k+1} \frac{a_m}{k+3-m}\right] n \quad (18.65)$$

$$s_1(n) = 1 + 2 + 3 + \cdots + n = \frac{1}{2}n(n+1)$$

$$s_2(n) = 1^2 + 2^2 + 3^2 + \cdots + n^2 = \frac{1}{6}n(n+1)(2n+1)$$

$$s_3(n) = 1^3 + 2^3 + 3^3 + \cdots + n^3 = \frac{1}{4}(n^2(n+1)^2) = [s_1(n)]^2$$

$$s_4(n) = 1^4 + 2^4 + 3^4 + \cdots + n^4 = \frac{1}{5}(3n^2 + 3n - 1)s_2(n)$$

$$s_5(n) = 1^5 + 2^5 + 3^5 + \cdots + n^5 = \frac{1}{12}n^2(n+1)^2(2n^2 + 2n - 1)$$

$$s_6(n) = \frac{n}{42}(n+1)(2n+1)(3n^4 + 6n^3 - 3n + 1)$$

$$s_7(n) = \frac{n^2}{24}(n+1)^2(3n^4 + 6n^3 - n^2 - 4n + 2)$$

$$s_8(n) = \frac{n}{90}(n+1)(2n+1)(5n^6 + 15n^5 + 5n^4 - 15n^3 - n^2 + 9n - 3)$$

$$s_9(n) = \frac{n^2}{20}(n+1)^2(2n^6 + 6n^5 + n^4 - 8n^3 + n^2 + 6n - 3)$$

$$s_{10}(n) = \frac{n}{66}(n+1)(2n+1)(3n^8 + 12n^7 + 8n^6 - 18n^5$$
$$- 10n^4 + 24n^3 + 2n^2 - 15n + 5)$$

18.15 TABLES OF ORTHOGONAL POLYNOMIALS

In the following:

- H_n are Hermite polynomials
- L_n are Laguerre polynomials
- P_n are Legendre polynomials
- T_n are Chebyshev polynomials
- U_n are Chebyshev polynomials

$H_0 = 1$

$H_1 = 2x$

$H_2 = 4x^2 - 2$

$H_3 = 8x^3 - 12x$

$H_4 = 16x^4 - 48x^2 + 12$

$H_5 = 32x^5 - 160x^3 + 120x$

$H_6 = 64x^6 - 480x^4 + 720x^2 - 120$

$H_7 = 128x^7 - 1344x^5 + 3360x^3 - 1680x$

$H_8 = 256x^8 - 3584x^6 + 13440x^4 - 13440x^2 + 1680$

$H_9 = 512x^9 - 9216x^7 + 48384x^5 - 80640x^3 + 30240x$

$H_{10} = 1024x^{10} - 23040x^8 + 161280x^6 - 403200x^4 + 302400x^2 - 30240$

$x^{10} = (30240H_0 + 75600H_2 + 25200H_4 + 2520H_6 + 90H_8 + H_{10})/1024$

$x^9 = (15120H_1 + 10080H_3 + 1512H_5 + 72H_7 + H_9)/512$

$x^8 = (1680H_0 + 3360H_2 + 840H_4 + 56H_6 + H_8)/256$

$x^7 = (840H_1 + 420H_3 + 42H_5 + H_7)/128$

$x^6 = (120H_0 + 180H_2 + 30H_4 + H_6)/64$

$x^5 = (60H_1 + 20H_3 + H_5)/32$

$x^4 = (12H_0 + 12H_2 + H_4)/16$

$x^3 = (6H_1 + H_3)/8$

$x^2 = (2H_0 + H_2)/4$

$x = (H_1)/2$

$1 = H_0$

$L_0 = 1$

$L_1 = -x + 1$

$L_2 = (x^2 - 4x + 2)/2$

$L_3 = (-x^3 + 9x^2 - 18x + 6)/6$

$L_4 = (x^4 - 16x^3 + 72x^2 - 96x + 24)/24$

$L_5 = (-x^5 + 25x^4 - 200x^3 + 600x^2 - 600x + 120)/120$

$L_6 = (x^6 - 36x^5 + 450x^4 - 2400x^3 + 5400x^2 - 4320x + 720)/720$

$x^6 = 720L_0 - 4320L_1 + 10800L_2 - 14400L_3 + 10800L_4 - 4320L_5 + 720L_6$

$x^5 = 120L_0 - 600L_1 + 1200L_2 - 1200L_3 + 600L_4 - 120L_5$

$x^4 = 24L_0 - 96L_1 + 144L_2 - 96L_3 + 24L_4$

$x^3 = 6L_0 - 18L_1 + 18L_2 - 6L_3$

$x^2 = 2L_0 - 4L_1 + 2L_2$

$x = L_0 - L_1$

$1 = L_0$

$P_0 = 1$

$P_1 = x$

$x^{10} = (4199P_0 + 16150P_2 + 15504P_4 + 7904P_6 + 2176P_8 + 256P_{10})/46189$

$x^9 = (3315P_1 + 4760P_3 + 2992P_5 + 960P_7 + 128P_9)/12155$

$P_2 = (3x^2 - 1)/2$

$P_3 = (5x^3 - 3x)/2$

$P_4 = (35x^4 - 30x^2 + 3)/8$

$P_5 = (63x^5 - 70x^3 + 15x)/8$

$P_6 = (231x^6 - 315x^4 + 105x^2 - 5)/16$

$P_7 = (429x^7 - 693x^5 + 315x^3 - 35x)/16$

$P_8 = (6435x^8 - 12012x^6 + 6930x^4 - 1260x^2 + 35)/128$

$P_9 = (12155x^9 - 25740x^7 + 18018x^5 - 4620x^3 + 315x)/128$

$P_{10} = (46189x^{10} - 109395x^8 + 90090x^6 - 30030x^4 + 3465x^2 - 63)/256$

$x^8 = (715P_0 + 2600P_2 + 2160P_4 + 832P_6 + 128P_8)/6435$

$x^7 = (143P_1 + 182P_3 + 88P_5 + 16P_7)/429$

$x^6 = (33P_0 + 110P_2 + 72P_4 + 16P_6)/231$

$x^5 = (27P_1 + 28P_3 + 8P_5)/63$

$x^4 = (7P_0 + 20P_2 + 8P_4)/35$

$x^3 = (3P_1 + 2P_3)/5$

$x^2 = (P_0 + 2P_2)/3$

$x = P_1$

$1 = P_0$

$T_0 = 1$

$T_1 = x$

$T_2 = 2x^2 - 1$

$T_3 = 4x^3 - 3x$

$T_4 = 8x^4 - 8x^2 + 1$

$T_5 = 16x^5 - 20x^3 + 5x$

$T_6 = 32x^6 - 48x^4 + 18x^2 - 1$

$T_7 = 64x^7 - 112x^5 + 56x^3 - 7x$

$T_8 = 128x^8 - 256x^6 + 160x^4 - 32x^2 + 1$

$T_9 = 256x^9 - 576x^7 + 432x^5 - 120x^3 + 9x$

$T_{10} = 512x^{10} - 1280x^8 + 1120x^6 - 400x^4 + 50x^2 - 1$

$x^{10} = (126T_0 + 210T_2 + 120T_4 + 45T_6 + 10T_8 + T_{10})/512$

$x^9 = (126T_1 + 84T_3 + 36T_5 + 9T_7 + T_9)/256$

$x^8 = (35T_0 + 56T_2 + 28T_4 + 8T_6 + T_8)/128$

$x^7 = (35T_1 + 21T_3 + 7T_5 + T_7)/64$

$x^6 = (10T_0 + 15T_2 + 6T_4 + T_6)/32$

$x^5 = (10T_1 + 5T_3 + T_5)/16$

$x^4 = (3T_0 + 4T_2 + T_4)/8$

$x^3 = (3T_1 + T_3)/4$

$x^2 = (T_0 + T_2)/2$

$x = T_1$

$1 = T_0$

$U_0 = 1$

$U_1 = 2x$

$U_2 = 4x^2 - 1$

$U_3 = 8x^3 - 4x$

$U_4 = 16x^4 - 12x^2 + 1$

$U_5 = 32x^5 - 32x^3 + 6x$

$U_6 = 64x^6 - 80x^4 + 24x^2 - 1$

$U_7 = 128x^7 - 192x^5 + 80x^3 - 8x$

$U_8 = 256x^8 - 448x^6 + 240x^4 - 40x^2 + 1$

$U_9 = 512x^9 - 1024x^7 + 672x^5 - 160x^3 + 10x$

$U_{10} = 1024x^{10} - 2304x^8 + 1792x^6 - 560x^4 + 60x^2 - 1$

$x^{10} = (42U_0 + 90U_2 + 75U_4 + 35U_6 + 9U_8 + U_{10})/1024$

$x^9 = (42U_1 + 48U_3 + 27U_5 + 8U_7 + U_9)/512$

$x^8 = (14U_0 + 28U_2 + 20U_4 + 7U_6 + U_8)/256$

$x^7 = (14U_1 + 14U_3 + 6U_5 + U_7)/128$

$x^6 = (5U_0 + 9U_2 + 5U_4 + U_6)/64$

$x^5 = (5U_1 + 4U_3 + U_5)/32$

$x^4 = (2U_0 + 3U_2 + U_4)/16$

$x^3 = (2U_1 + U_3)/8$

$x^2 = (U_0 + U_2)/4$

$x = (U_1)/2$

$1 = U_0$

18.16   REFERENCES

1. M. Abramowitz and I. A. Stegun, *Handbook of mathematical functions*, NIST, Washington, DC, 1964,

2. A. Erdélyi (ed.), Bateman Manuscript Project, *Tables of integral transforms*, in 3 volumes, McGraw–Hill, New York, 1954.

3. N. M. Temme, *Special functions: An introduction to the classical functions of mathematical physics*, John Wiley & Sons, New York, 1996.

List of Notation

Symbols

D_n: derangement 28
d_n: proportion of customers 442
$\mathbf{D}_x$: diagonal matrix 398

E

e_{ij}: observed value 282
$E[\,]$: expectation 34
e_i: residual 263
E_k: Erlang-k service time 441
erf: error function 512
erfc: complementary error function
        512

F

$\mathcal{F}$: Fourier transform 458
FCFS: first come, first served ... 441
FIFO: first in, first out 441
f_k: frequency 9
$F(x)$: cumulative distribution
        function 33
$F(x_1, x_2, \ldots, x_n)$: cumulative
        distribution function 39, 40
$f(x_1, x_2, \ldots, x_n)$: probability density
        function 40
$f(x \mid y)$: conditional probability .. 47

G

G: general service time distribution
        441
g_1: coefficient of skewness 18
g_2: coefficient of skewness 19
GI: general interarrival time 441
GM: geometric mean 10

H

$H(\mathbf{p}_X)$: entropy 428
H_0: null hypothesis 227
H_a: alternative hypothesis 227
H_k: k-stage hyperexponential service
        time 441
HM: harmonic mean 10
$H_n(x)$: Hermite polynomial 527

I

I: identity matrix 398
$I(X, Y)$: mutual information 428
iid: independent and identically
        distributed 57
$I_\nu(z)$: Bessel function 509

IQR: interquartile range 15

J

J: determinant of the Jacobian .. 54
$j_n(z)$: half order Bessel function 509
$J_\nu(z)$: Bessel function 506

K

K: system capacity 441
k: parameter of a BIBD 325
K_k: Kalman gain matrix 431
$K_\nu(z)$: Bessel function 509
$K_X(t_1, t_2)$: correlation function .. 44

L

L: average number of customers. 442
$L(\theta)$: expected loss function 393
$L(\theta)$: likelihood function 231
$\ell(\theta, a)$: loss function 393
λ: average arrival rate 442
λ_{lower}: confidence interval 202
λ_{upper}: confidence interval 202
LCG: linear congruential generator
        450
LCL: lower control limit 383
LIFO: last in, first out 441
$L_n(x)$: Laguerre polynomial 527
ln: logarithm 521
log: logarithm 521
$\log_b$: logarithm to base b 522
L_q: average number of customers 442
LTPD: lot tolerance percent
        defective 386

M

M: exponential service time 441
M: hypergeometric function 521
m: number in the source 441
$MA(l)$: moving average 457
$M/D/1$: queue 445
MD: mean deviation 13
$M/E_k/1$: queue 445
mgf: moment generating function 36
MLE: maximum likelihood estimator
        189
$M/M/1$: queue 443
$M/M/1/K$: queue 443
$M/M/2$: queue 443
$M/M/c$: queue 444

Index